Noise in nonlinear dynamical systems

Volume 2

Theory of noise induced processes in special applications

Noise in nonlinear dynamical systems

Volume 2

Theory of noise induced processes in special applications

Edited by

Frank Moss, *Professor of Physics,*
University of Missouri at St. Louis

and

P. V. E. McClintock, *Reader in Physics,*
University of Lancaster

The right of the
University of Cambridge
to print and sell
all manner of books
was granted by
Henry VIII in 1534.
The University has printed
and published continuously
since 1584.

CAMBRIDGE UNIVERSITY PRESS

Cambridge New York New Rochelle
Melbourne Sydney

CAMBRIDGE UNIVERSITY PRESS
Cambridge, New York, Melbourne, Madrid, Cape Town, Singapore, São Paulo, Delhi

Cambridge University Press
The Edinburgh Building, Cambridge CB2 8RU, UK

Published in the United States of America by Cambridge University Press, New York

www.cambridge.org
Information on this title: www.cambridge.org/9780521118521

First published 1989
This digitally printed version 2009

A catalogue record for this publication is available from the British Library

Library of Congress Cataloguing in Publication data

Noise in nonlinear dynamical systems.
Includes indexes.
Contents: v. 1. Theory of continuous Fokker–Planck
systems – v. 2. Theory of noise induced processes in
special applications.
1. Fluctuations (Physics) – Collected works.
2. Nonlinear theories – Collected works. I. Moss, Frank,
1934– . II. McClintock, P.V.E.
QC6.4.F58N64 003 87–34856

ISBN 978-0-521-35229-1 hardback
ISBN 978-0-521-11852-1 paperback

Contents

Contributors

G. Broggi
Physik Institut der Universität
Schönberggasse 9
CH-8001 Zurich
Switzerland

M. Büttiker
IBM Thomas J. Watson Research Center
POB 218
Yorktown Heights
NY 10598
USA

Edward Celarier
Chemical Physics Theory Group
Department of Chemistry
University of Toronto
Toronto
Ontario M5S 1Al
Canada

M. I. Dykman
Institute of Semiconductors
Academy of Sciences of the UkrSSR
pr. Nauki, 115
Kiev, 252028
USSR

Ronald F. Fox
School of Physics
Georgia Institute of Technology
Atlanta
GA 30332
USA

Contributors

Peter Hänggi
Lehrstuhl für Theoretische Physik
Universität Augsburg
Memminger Strasse 6
D-8900 Ausburg
FRG

Werner Horsthemke
Center for Studies in Statistical Mechanics
Department of Physics
University of Texas at Austin
Austin
TX 78712
USA

Raymond Kapral
Chemical Physics Theory Group
Department of Chemistry
University of Toronto
Toronto
Ontario M5S 1Al
Canada

Edgar Knobloch
Department of Physics
University of California
Berkeley
CA 94720
USA

Dilip K. Kondepudi
Department of Chemistry
Wake Forest University
PO Box 7486
Winston-Salem
NC 27109
USA

M. A. Krivoglaz
Institute of Metal Physics
Academy of Sciences of the UkrSSR
pr. Nauki, 115
Kiev, 252028
USSR

Contributors

René Lefever
Service de Chimie Physique II
C.P. 231, Campus Plaine
Université Libre de Bruxelles
B-1050 Brussels
Belgium

M. Lücke
Institut für Theoretische Physik
Universität des Saarlandes
D-6600 Saarbrucken
FRG

L. A. Lugiato
Dipartimento de Fisica del Politecnico
Corso Duca degli Abruzzi 24
I-10129 Torino
Italy

M. Merri
Dipartimento di Fisica dell'Universita
Via Celoria 16
I-20133 Milano
Italy

Toyonori Munakata
Department of Applied Mathematics and Physics
Kyoto University
Kyoto 606
Japan

M. A. Pernigo
Max-Planck Institut für Quantenoptik
D-8046 Garching bei München
FRG

H. Risken
Abteilung für Theoretische Physik
Universität Ulm
D-7900 Ulm
FRG

W. Schleich
Max-Planck Institut für Quantenoptik
D-8046 Garching bei München
FRG

and

Center for Advanced Studies
Department of Physics and Astronomy
University of New Mexico
Albuquerque
NM 87131
USA

S. M. Soskin
Institute of Semiconductors
Academy of Sciences of the UkrSSR
pr. Nauki, 115
Kiev, 252028
USSR

Peter Talkner
Institut für Theoretische Physik
Universität Basel
Klingelbergstrasse 82
CH-4056 Basel
Switzerland

K. Vogel
Abteilung für Theoretische Physik
Universität Ulm
D-7900 Ulm
FRG

Jeffrey B. Weiss
Department of Physics
University of California
Berkeley
CA 94720
USA

Kurt Wiesenfeld
School of Physics
Georgia Institute of Technology
Atlanta
GA 30332
USA

Preface

All macroscopic physical systems are subject to fluctuations or noise. One of the most useful and interesting developments in modern statistical mechanics has been the realization that even complex nonequilibrium systems can often be reduced to equivalent ones of only a few degrees of freedom by the elimination of dynamically nonrelevant variables. Theoretical descriptions of such contracted systems necessarily begin with a set of either continuous or discrete dynamical equations which can then be used to describe noise driven systems with the inclusion of random terms. Studies of these stochastic dynamical equations have expanded rapidly in the past two decades, so that today an exuberant theoretical activity, a few experiments, and a remarkably large number of applications, some with challenging technological implications, are evident.

The purpose of these volumes is twofold. First we hope that their publication will help to stimulate new experimental activity by contrasting the smallness of the number of existing experiments with the many research opportunities raised by the chapters on applications. Secondly, it has been our aim to collect together in one place a complete set of authoritative reviews with contributions representative of all the major practitioners in the field. We recognize that as an inevitable consequence of the intended comprehensiveness, there will be few readers who will wish to digest these volumes in their entirety. We trust, instead, that readers will be stimulated to choose from the many possibilities for new research represented herein, and that they will find all the specialized tools, be they experimental or theoretical, that they are likely to require.

Although there is a strong underlying theme running through all three volumes – the influence of noise on dynamical systems – each chapter should be considered as a self-contained account of the authors' most important research in the field, and hence can be read either alone or in concert with the others. In view of this, the editors have chosen not to attempt to impose a uniform style; nor have they insisted on any standard set of mathematical symbols or notation. The discerning reader will detect points of detail on which full concordance appears to be lacking, especially in Volume 1. This should certainly be regarded as the signature of an active and challenging theoretical activity in a rapidly expanding field of study. In selecting the contributors, the editors have made special efforts to include younger authors with new ideas and perspectives along with the more active seasoned veterans, in the confident belief that the field will be invigorated by their contributions.

Finally, it is our pleasure to acknowledge the many valuable suggestions made by our colleagues and contributors, from which we have greatly benefited. Special thanks are due to R. Mannella for constructive criticism and helpful comments at various stages of this enterprise.

Completion of this work owes much to the generous support provided by the North Atlantic Treaty Organization (under grant RG85/0070) and the UK Science and Engineering Research Council (under GR/D/61925).

Frank Moss and P. V. E. McClintock
Lancaster, May 1987

Introduction to Volume 2

While in Volume 1 the treatments are confined exclusively to Fokker–Planck systems, a range of contemporary problems, indicative of the rich diversity of applications of noise driven dynamics, are reviewed in this volume. Though most problems currently treated are classical, recent work on dissipative quantum tunnelling has focused attention on quantum mechanical applications of stochastic dynamics. Chapter 1 builds on the theory of the Anderson–Kubo oscillator, generalized to account for colored noise effects, with applications to molecular spectroscopy. Helpful comparisons with the classical theory of Brownian motion are drawn. Chapter 2 reviews the important and highly topical problem of the crossover from classical Kramers-type thermal activation to the quantum tunnelling of particles out of a metastable state. Having set the stage with a discussion of classical Brownian motion in a periodic potential, the theory is applied to the diffusion of an impurity hopping among interstitial sites in a crystal lattice. An interesting non-Markovian application follows wherein the impurity is driven by a random force weighted around a characteristic frequency. Enhanced activation rates are observed when the noise correlation becomes comparable to the inverse characteristic frequency. This review concludes with a discussion of the formidable problem of quantum tunnelling in the low damping regime. Even classical treatments of this topic are fraught with difficulties, as the following review in Chapter 3 demonstrates. Building on Kramers' original treatment, the escape rate of particles from a metastable state is discussed in the limits of low and moderately high damping. An improved approximation, taking into account the nonuniform distribution of particle densities for energies larger than the barrier height, which successfully interpolates between these extremes, is reviewed.

The following two chapters consider an entirely new problem – that of discrete dynamics driven by high dimensional noise. In recent years, considerable advances in the understanding of complex dynamical systems have resulted from the realization that one can often replace such systems with a discrete map of one or a few dimensions. Yet real complex, macroscopic systems are always subject to noise; hence the study of noise driven discrete maps becomes important. In Chapter 4, the dynamics of one-dimensional maps with additive, Gaussian, white noise are considered, and a theory is developed to calculate the moments of the noise induced fluctuations around stable attractors. Weak noise results in random, 'out-of-order' jumps among the attractors and occasional escapes to neighboring attractors. The basins of stability around the attractors and the statistical density of the escape rates are determined numerically. In Chapter 5 an analytic theory of the mean first passage time for noise induced jumps from one attractor to another is developed, and the invariant density of the fluctuations within a basin of attraction is obtained. The theory is applied to several example maps with the mean first passage time calculations applied to weakly nonlinear circle maps. These two chapters convey entirely new results which demonstrate the remarkable sensitivity of discrete dynamical systems to external noise.

Chapters 6 and 7 provide current reviews of bifurcation processes in the presence of noise. The former begins with a methodical development of the bifurcation theory of periodically driven deterministic systems which is then generalized to include parametric noise modulation. Finally, a discrete dynamical system – the logistic map – is considered with both additive and multiplicative noise within the framework of a small noise intensity perturbation theory. The theory is tested against extensive numerical simulations. The next chapter is an entertaining account of the author's attempts to locate the precise point (in parameter space) of a period doubling bifurcation in the presence of noise. It turns out that noise causes the system to 'anticipate' the bifurcation, as signalled by the appearance of a 'bump' in the power spectrum at the period doubled frequency well before the bifurcation parameter has reached its (deterministic) critical value. A lucid development of the theory of these 'noisy precursors' for various types of bifurcations follows, wherein universal scaling behavior is demonstrated and explained. Finally the theory is applied (without noise) to a very practical idea: the use of bifurcating systems as small signal amplifiers. The principle is demonstrated with an electronic, driven Duffing oscillator.

Chapter 8 is a review and an update of the state of the theory of noise induced transitions: a class of phenomena whereby noise qualitatively changes the (deterministic) behavior of a system. This is manifest as a noise induced shift of an instability toward higher mean values of a bifurcation parameter (a postponement) or as the appearance in certain systems of new statistically favored states at some critical noise intensity. The existing body of theory is carefully reviewed and generalized to include colored Gaussian and Poisson white noise. The behavior of example systems with monochromatic, periodic modulations is contrasted to that with broadband noise. Current experimental evidence for noise induced transitions is briefly discussed. (In this context, see also Chapters 1–3 of Volume 3.) A review of noise induced transitions applied to chemical systems is offered in Chapter 9. Such systems, both in reality and as they are modelled, are characteristically of higher dimension than the typical system encountered so far in these volumes. Consequently a more complex dynamics is encountered as evidenced by noise induced transitions between fixed points and limit cycles, among multiple limit cycles, to chaos and among neighboring chaotic attractors. The basin boundaries of such attractors are also more complex than is the case for lower dimensional systems. The use of one-dimensional discrete maps as representations of the complex dynamical system is also discussed. The focus, however, is on the mechanisms for noise induced transitions, several of which are identified and elucidated with applications to specific transitions. Two archetypal models for chemical dynamics are frequently used to illustrate these processes: the Rossler system, and the Golubitsky–Keyfitz model.

Noise induced state selection is the topic of Chapter 10, which commences with the observation that macroscopic states of near perfect symmetry are certain to be very delicate or 'sensitive'. This means that, near bifurcations, even very small symmetry breaking fields can have a statistically profound influence as the system evolves toward some final state. The theory of bi- and multistable state selection in the presence of noise is developed. Applications to biochemical selection processes, chirality selection and such practical matters as noise averaging switching are discussed.

Applications in optics are covered in the next two chapters. Chapter 11 is devoted to a theoretical analysis of the colored noise dithered ring laser gyroscope using the matrix continued fraction method. In this system, noise is deliberately introduced in order to overcome the well known phase locking problem which limits the accuracy of such gyroscopes at small rotation velocities. The stationary two-dimensional statistical densities are obtained as well as the time and noise averaged beat frequency. Chapter 12 deals with the influence of white and colored noise on dispersive and absorptive optical bistability. One can immediately appreciate, from a careful reading of this chapter, that nonlinear optics offer an extremely fruitful ground for the exploration of a wide variety of exotic effects arising from multiplicative colored noise in nonlinear dynamical systems. A lucid theoretical development, which maintains contact with the physical applications leads to discussions of colored noise induced transitions, transient bimodality (also observed experimentally, as discussed in Chapter 6 of Volume 3), noise enhanced switching and finally a swept parameter laser model with noise.

In the last chapter of Volume 2, the bistable archetype is revisited, this time in the form of the white noise driven Duffing oscillator. Calculations of the transition probabilities in such systems have challenged theorists since the time of Kramers (see also Chapter 3 of this volume) but an exact solution has not been forthcoming. Chapter 13 presents accurate analytic approximations for the transition rates and the power spectra. Apart from the matrix continued fraction method this is the only treatment to date which has successfully predicted all the observed features of these quantities.

We conclude by remarking that only a few of the applications herein presented have actually been subject to experimental tests. The contents of this volume therefore present a variety of opportunities for experimentalists.

1 Stochastic processes in quantum mechanical settings

RONALD F. FOX

1.1 Introduction

The apparent generality of the title to this chapter is appropriately reduced by sufficiently precisely defining the terms used. We have in mind the natural extension to quantum mechanical settings of the ideas usually covered in a discussion of stochastic processes in classical mechanical settings. In classical mechanical settings, the fundamental process is Brownian motion, which describes the influence of external fluctuating forces on a small particle. The mathematical apparatus developed for the study of Brownian motion has been generalized and applied to irreversible thermodynamics (Onsager and Machlup, 1953), fluctuating hydrodynamics (Fox and Uhlenbeck, 1970), light scattering from fluids and mixtures (Berne and Pecora, 1976) and chemical reactions (McQuarrie and Keizer, 1981). In each case, the stochastic process involves fluctuations generated by thermal motions. We are not considering the stochastic trajectories (chaos) of classically deterministic motion in non-linear systems (Lichtenberg and Lieberman, 1983), which is an entirely separate context for stochastic processes. Our goal in this chapter is to study the effects of thermal fluctuations in quantum mechanical settings. Therefore, we exclude a discussion of interpreting quantum mechanics as a classical stochastic motion (Nelson, 1966), an idea which should have been put to rest by Grabert, Hänggi and Talkner (1979). We also exclude a discussion of quantum chaos, which is currently a very active area of research (Zaslavsky, 1981).

Just as Brownian motion provides the paradigm for stochastic processes in classical mechanical settings, the Anderson–Kubo oscillator (Anderson, 1954; Kubo, 1954; Kubo, Toda and Nashitsume, 1985) is the paradigm for stochastic processes in quantum mechanical settings. It requires a shift from additive stochastic processes which are used in classical settings to multiplicative stochastic processes for quantum settings (Fox, 1978). Thus, the shift in setting requires a significant adjustment of the mathematical formalism. The first four sections of this chapter are concerned with the mathematical side of this adjustment. The remaining sections are devoted to several concrete physical applications of the results.

1.2 The Anderson–Kubo oscillator

The prototype process for stochastic processes in classical mechanical settings is Langevin's equation for Brownian motion (Fox, 1978), an additive stochastic process. It describes the motion of a Brownian particle immersed in a thermal fluid. The equation of motion is

$$M\frac{\mathrm{d}}{\mathrm{d}t}\vec{v} = -\alpha\vec{v} + \tilde{\vec{F}}(t) \tag{1.2.1}$$

in which M is the mass, $\vec{v}$ is the velocity, α is the damping parameter, and $\tilde{\vec{F}}(t)$ is the stochastic force caused by collisions between the Brownian particle and the fluid molecules. This is an inhomogeneous equation for $\vec{v}$ with a stochastic inhomogeneity. This makes it an *additive* stochastic process. Here, and in the following, stochastic differential equations are to be interpreted in the sense of Stratonovich (see Chapter 2, Volume 1, and Arnold, 1974) which justifies writing them in the form of (1.2.1). The stochastic properties of $\vec{v}$ induced by $\tilde{\vec{F}}$ are determined from the properties given for $\tilde{\vec{F}}$. In this case $\tilde{\vec{F}}$ is taken to be Gaussian white noise with zero mean which translates into

$$\langle \tilde{F}_i(t) \rangle = 0, \quad i = 1, 2, 3 \tag{1.2.2}$$

$$\langle \tilde{F}_i(t)\tilde{F}_j(s) \rangle = 2k_B T\alpha\,\delta_{ij}\delta(t-s), \quad i, j = 1, 2, 3, \tag{1.2.3}$$

where k_B is Boltzmann's constant and T is the absolute temperature. These two equations express the totality of stochastic information contained in $\tilde{\vec{F}}$. Equation (1.2.3) is an example of a fluctuation–dissipation relation (of the second kind; Kubo, Toda and Nashitsume, 1985). The essential feature of additive stochastic processes, the fluctuation–dissipation relation (1.2.3), connects a phenomenologically introduced damping parameter, α, with the strength of the fluctuations.

The prototype process for stochastic processes in quantum mechanical settings is the Anderson–Kubo oscillator (Anderson, 1954; Kubo, 1954), a *multiplicative* stochastic process. It describes phase fluctuations in a complex amplitude factor. These fluctuations are caused by stochastic interactions with reservoir particles. The equation of motion is

$$\frac{\mathrm{d}}{\mathrm{d}t}a(t) = \mathrm{i}(\omega_0 + \tilde{\omega}(t))a(t) \tag{1.2.4}$$

in which ω_0 is a constant and $\tilde{\omega}(t)$ is a Gaussian, white noise frequency fluctuation with zero mean. The complex amplitude, $a(t)$, evolves according to (1.2.4) which does not contain any damping parameter. Moreover, (1.2.4) is homogeneous, which makes it a multiplicative stochastic process (Fox, 1978). The stochastic properties of a are induced by $\tilde{\omega}$ which possesses the properties

$$\langle \tilde{\omega}(t) \rangle = 0 \tag{1.2.5}$$

$$\langle \tilde{\omega}(t)\tilde{\omega}(s) \rangle = 2\lambda\delta(t-s), \tag{1.2.6}$$

in which λ measures the strength of the frequency fluctuations. It is crucial to understanding the physical significance of this process *not* to think of this oscillator in terms appropriate for the study of the Brownian motion of a classical harmonic oscillator.

The basic difference in character between (1.2.1) and (1.2.4) leads to distinctly different solutions and averaged behavior. The solution to (1.2.1) is

$$\vec{v}(t) = \exp\left[-\frac{\alpha}{M}t\right]\vec{v}(0) + \int_0^t ds \exp\left[-\frac{\alpha}{M}(t-s)\right]\frac{\tilde{\vec{F}}(s)}{M}, \quad (1.2.7)$$

whereas the solution to (1.2.4) is

$$a(t) = \exp\left[i\omega_0 + i\int_0^t ds\tilde{\omega}(s)\right]a(0). \quad (1.2.8)$$

The averaged value of (1.2.7) is simple because it is *linear* in $\tilde{F}$:

$$\langle \vec{v}(t) \rangle = \exp\left[-\frac{\alpha}{M}t\right]\vec{v}(0). \quad (1.2.9)$$

The average to (1.2.8) is also simple to express, although more difficult to obtain because it is *non-linear* in $\tilde{\omega}$:

$$\langle a(t) \rangle = \exp[i\omega_0 t - \lambda t]a(0). \quad (1.2.10)$$

We will justify this later. Note for now, however, that averaging a multiplicative process solution *creates damping*, i.e. the factor $\exp[-\lambda t]$. Since the strength of the damping derives from the strength of the fluctuations, we have an example of the multiplicative fluctuation–dissipation relation.

1.3 General multiplicative stochastic processes

The extended applicability of both additive and multiplicative stochastic processes in areas of physics, as well as in other sciences, is a result of increased numbers of dimensions. The general additive stochastic process can be rendered (Fox, 1978)

$$\frac{d}{dt}a_i = -G_{ij}a_j + \tilde{F}_i, \quad (1.3.1)$$

in which repeated indices are summed. $\tilde{F}_i$ is characterized as Gaussian white noise:

$$\langle \tilde{F}_i(t) \rangle = 0 \quad (1.3.2)$$

$$\langle \tilde{F}_i(t)\tilde{F}_j(s) \rangle = 2Q_{ij}\delta(t-s). \quad (1.3.3)$$

Moreover, the equilibrium distribution for the a_i's is given by the Einstein–

Boltzmann–Planck distribution

$$W(a_1 \ldots a_N) = \left(\frac{\| E \|}{(2\pi)^N} \right)^{1/2} \exp[-\tfrac{1}{2} a_i E_{ij} a_j] \tag{1.3.4}$$

in which $\| E \|$ is the determinant of the matrix E_{ij}, the Einstein matrix (also called the entropy matrix). The fluctuation–dissipation relation (of the second kind) now reads:

$$G_{ij}(\mathrm{E}^{-1})_{jk} + (E^{-1})_{ij} G_{kj} = 2Q_{ik}, \tag{1.3.5}$$

which is conveniently expressed

$$\mathbf{GE}^{-1} + \mathbf{E}^{-1}\mathbf{G}^{\dagger} = 2\mathbf{Q}. \tag{1.3.6}$$

The general solution and its averages are obtained from,

$$\vec{a}(t) = \exp[-t\mathbf{G}]\vec{a}(0) + \int_0^t \mathrm{d}s \exp[-(t-s)\mathbf{G}]\tilde{\vec{F}}(s) \tag{1.3.7}$$

the straightforward extension of (1.2.7).

The situation for the multiplicative case is fundamentally different. The general multiplicative process can be rendered

$$\frac{\mathrm{d}}{\mathrm{d}t} a_i = M_{ij} a_j + \tilde{M}_{ij}(t) a_j, \tag{1.3.8}$$

in which M_{ij} is a constant matrix and $\tilde{M}_{ij}(t)$ is a stochastic, Gaussian, white noise matrix. Again, repeated indices are summed. The properties of $\tilde{M}_{ij}$ are expressed by

$$\langle \tilde{M}_{ij}(t) \rangle = 0 \tag{1.3.9}$$

$$\langle \tilde{M}_{ij}(t) \tilde{M}_{kl}(s) \rangle = 2Q_{ijkl}\,\delta(t-s). \tag{1.3.10}$$

This is as far as the extension of (1.2.4) goes without difficulty. The problem that arises in this multiplicative case is the potential non-commutativity of $\tilde{\mathbf{M}}(t)$ with $\tilde{\mathbf{M}}(s)$ for $t \neq s$. There is no analog to this problem in (1.3.1), the additive case. The potential non-commutativity of $\tilde{\mathbf{M}}$ at different times prevents us from expressing the solution to (1.3.8) as an ordinary exponential such as in (1.2.8). Instead, the time ordered exponential (Fox, 1978) is required:

$$\vec{a}(t) = \underset{\leftarrow}{T} \exp\left[\int_0^t \mathrm{d}s (\mathbf{M} + \tilde{\mathbf{M}}(s)) \right] \vec{a}(0). \tag{1.3.11}$$

This expression is further complicated by the possibility that $\mathbf{M}$ and $\tilde{\mathbf{M}}(s)$ do not commute. As we shall see, such non-commutativity occurs naturally in quantum mechanical problems, although there is no intrinsic reason at this stage to think of (1.3.8) as restricted to quantum mechanics. In fact, we have intentionally omitted the factor of i, seen in (1.2.4), in order to separate the mathematical ideas of multiplicative processes and the particular connection with quantum mechanics. Not all multiplicative processes are quantum mechanical!

Non-commutativity not only complicates evaluation of the solutions and their averages for multiplicative processes, it also complicates the idea of a Gaussian process itself. Each of the complications is overcome by two powerful mathematical constructs: operator cumulants and characteristic functionals.

1.4 Operator cumulants and characteristic functionals

In order to clearly express the results we will need regarding cumulants for systems of many dimensions, it is useful to review these results for systems of one dimension. Even this much more modest objective is clarified by starting with a time-independent random process, $\tilde{x}$. Suppose that $\tilde{x}$ is said to be Gaussian. This means that it has a distribution function which is Gaussian:

$$W(x) = (2\pi\sigma^2)^{-1/2} \exp\left[-\frac{(x-\bar{x})^2}{2\sigma^2} \right]. \tag{1.4.1}$$

Therefore, its first two moments are

$$\langle \tilde{x} \rangle = \bar{x} \tag{1.4.2}$$

$$\langle \tilde{x}^2 \rangle = \bar{x}^2 + \sigma^2. \tag{1.4.3}$$

In fact, it is possible (Fox, 1978) to express any moment in terms of $\bar{x}$ and σ^2:

$$\langle (\tilde{x})^n \rangle = \sum_{m_1 + 2m_2 = n} \left[\frac{n!}{m_1! m_2! 2^{m_2}} (\bar{x})^{m_1} (\sigma^2)^{m_2} \right] \tag{1.4.4}$$

in which the sum is over all non-negative integers m_1 and m_2 such that $m_1 + 2m_2 = n$. These results can be generated from the *characteristic function*, $\Phi(k)$, defined by

$$\begin{aligned} \Phi(k) &\equiv \int_{-\infty}^{\infty} \mathrm{d}x\, W(x) \exp(ikx) \\ &= \exp[-\tfrac{1}{2}\sigma^2 k^2 + ik\bar{x}]. \end{aligned} \tag{1.4.5}$$

The moments are now given by

$$\langle (\tilde{x})^n \rangle = (-\mathrm{i})^n \frac{\mathrm{d}^n}{\mathrm{d}k^n} \Phi(k) \bigg|_{k=0}, \tag{1.4.6}$$

which means k is set equal to zero after the derivatives have been taken. Up to this point there is no advantage to using the distribution function or the characteristic function. Both are Gaussian *because* both are exponentials of quadratic forms.

Cumulants may be introduced whenever an average of an exponential is performed. For example, cumulants are defined in order to evaluate

$$\left.\begin{aligned}\langle e^{\tilde{x}}\rangle &\equiv \int_{-\infty}^{\infty} dx\, W(x) e^{x} \\ &= \sum_{n=0}^{\infty} \frac{1}{n!}\langle(\tilde{x})^n\rangle,\end{aligned}\right\} \tag{1.4.7}$$

which has been expressed as an infinite summation over the moments of $\tilde{x}$. The definition of the corresponding cumulant expansion (Fox, 1978) is

$$\langle e^{\tilde{x}}\rangle \equiv \exp\left[\sum_{n=0}^{\infty} \frac{1}{n!}\langle(\tilde{x})^n\rangle_c\right]. \tag{1.4.8}$$

In order that (1.4.7) and (1.4.8) agree, very *non-linear* relations must exist between the moments and the cumulants. These relations go both ways:

$$\langle(\tilde{x})^n\rangle = \sum_{\sum_{l=1}^{\infty} lm_l = n} n! \prod_{l=1}^{\infty} \frac{1}{(l!)^{m_l} m_l!}(\langle(\tilde{x})^l\rangle_c)^{m_l} \tag{1.4.9}$$

and

$$\langle(\tilde{x})^n\rangle_c = \sum_{\sum_{l=1}^{\infty} lm_l = n} n!(-1)^{p-1}(p-1)! \prod_{l=1}^{\infty} \frac{1}{(l!)^{m_l} m_l!}(\langle(\tilde{x})^l\rangle)^{m_l}, \tag{1.4.10}$$

in which $\sum_{l=1}^{\infty} lm_l = n$ is a partition of the integer n into smaller integers l with multiplicities m_l, and $p \equiv \sum_{l=1}^{\infty} m_l$ for any particular partition.

For a simple one-dimensional Gaussian stochastic process which is independent of time, the cumulants of order higher than the second vanish identically. The expression in (1.4.4) yields identically zero on the right-hand side of (1.4.10) when $n > 2$. This property is sometimes also taken to be synonymous with Gaussianness, along with the Gaussian forms of $W(x)$ and $\Phi(k)$. In fact, if we now use the cumulant expansion to compute $\Phi(k)$ in accord with (1.4.5), then we get

$$\left.\begin{aligned}\Phi(k) &= \langle \exp[ik\tilde{x}]\rangle \\ &= \exp\left[\sum_{n=1}^{\infty} \frac{(ik)^n}{n!}\langle(\tilde{x})^n\rangle_c\right] \\ &= \exp[ik\bar{x} - \tfrac{1}{2}\sigma^2 k^2],\end{aligned}\right\} \tag{1.4.11}$$

which agrees with (1.4.5) and reflects the results:

$$\left.\begin{aligned}&\langle\tilde{x}\rangle_c = \langle\tilde{x}\rangle = \bar{x} \\ &\langle(\tilde{x})^2\rangle_c = \langle(\tilde{x})^2\rangle - \langle\tilde{x}\rangle^2 = \sigma^2 \\ &\langle(\tilde{x})^n\rangle_c = 0 \quad \text{for} \quad n \geqslant 3.\end{aligned}\right\} \tag{1.4.12}$$

Let us now generalize these results to include time-dependence, but still restrict the dimension to one. Let $\tilde{\omega}(t)$ be a time-dependent, Gaussian white noise:

$$\langle\tilde{\omega}(t)\rangle = \omega_0 \tag{1.4.13}$$

$$\langle\tilde{\omega}(t)\tilde{\omega}(s)\rangle = 2\lambda\delta(t-s) + \omega_0^2. \tag{1.4.14}$$

A characteristic functional exists which is defined by

$$\Phi[K(t)] = \left\langle \exp\left(\mathrm{i}\int_0^\infty \mathrm{d}s\, k(s)\tilde{\omega}(s) \right) \right\rangle, \tag{1.4.15}$$

in which the average is to be computed relative to a distribution *functional.* Since we are dealing with Gaussians, all of this can be done with *functional integrals*, i.e. path integrals. Write the distribution functional

$$W[\omega(t)] = N\exp\left[-\frac{1}{2\lambda}\int_0^\infty \mathrm{d}s\, \omega^2(s) \right], \tag{1.4.16}$$

which is normalized by (Fox, 1986a)

$$N^{-1} = \iint \mathscr{D}\omega \exp\left[-\frac{1}{2\lambda}\int_0^\infty \mathrm{d}s\, \omega^2(s) \right]. \tag{1.4.17}$$

Thus, $\Phi[k(t)]$ is given by

$$\begin{aligned} \Phi[k(t)] &= \left\langle \exp\left(\mathrm{i}\int_0^\infty \mathrm{d}s\, k(s)\tilde{\omega}(s) \right) \right\rangle \\ &= \iint \mathscr{D}\omega\, W[\omega(t)] \exp\left(\mathrm{i}\int_0^\infty \mathrm{d}s\, k(s)\omega(s) \right). \end{aligned} \tag{1.4.18}$$

Moments are generated in accord with (δ denotes functional differentiation)

$$\left\langle \prod_{l=1}^{n} \tilde{\omega}(t_l) \right\rangle = (-\mathrm{i})^n \frac{\delta^n}{\prod_{l=1}^{n} \delta k(t_l)} \Phi[k(t)] \bigg|_{k(t)=0}, \tag{1.4.19}$$

which is the functional analog of (1.4.6). Even cumulants may be defined for this case by

$$\begin{aligned} \left\langle \exp\left[\int_0^t \mathrm{d}s\, \tilde{\omega}(s) \right] \right\rangle &= \sum_{n=0}^{\infty} \left[\frac{1}{n!} \left\langle \left(\int_0^t \mathrm{d}s\, \tilde{\omega}(s) \right)^n \right\rangle \right] \\ &\equiv \exp\left[\sum_{n=1}^{\infty} \frac{1}{n!} \left\langle \left(\int_0^t \mathrm{d}s\, \tilde{\omega}(s) \right)^n \right\rangle_{\mathrm{c}} \right]. \end{aligned} \tag{1.4.20}$$

This yields, for example,

$$\langle\tilde{\omega}(t)\rangle_{\mathrm{c}} = \langle\tilde{\omega}(t)\rangle = \omega_0 \tag{1.4.21}$$

and

$$\langle \tilde{\omega}(t)\tilde{\omega}(s)\rangle_c = 2\lambda\delta(t-s). \tag{1.4.22}$$

In fact, for Gaussian $\tilde{\omega}(t)$ it can again be shown that higher order cumulants vanish identically. Thus, in one dimension, the introduction of time-dependence introduces functionals for functions but otherwise does not materially alter anything. There are still three equivalent ways in which Gaussianness may be expressed: (*a*) Gaussian W; (*b*) Gaussian Φ; and (*c*) vanishing cumulants after the second order.

The generalization to higher dimension, however, introduces significant changes. Return to the stochastic matrices of (1.3.8), $\tilde{M}_{ij}(t)$. These may be thought of as matrix representations in a specific basis for a stochastic operator $\tilde{\mathbf{M}}(t)$. In one dimension, the stochastic object, e.g. $\tilde{\omega}(t)$, the characteristic functional, e.g. $\Phi[k(t)]$, and the cumulants, e.g. $\langle \tilde{\omega}(t_1)\tilde{\omega}(t_2)\tilde{\omega}(t_3)\rangle_c$, are scalar quantities. In higher dimension, the stochastic object, e.g. $\tilde{\mathbf{M}}(t)$, is an operator, and its cumulants are generally operator products. The characteristic functional, however, is still a scalar. To construct it, an operator analog for k in (1.4.5) and for $k(t)$ in (1.4.15) is required:

$$\Phi[\mathbf{K}(t)] \equiv \left\langle \exp\left[i \int_0^\infty ds\, K_{\alpha\beta}(s)\tilde{M}_{\alpha\beta}(s) \right]\right\rangle, \tag{1.4.23}$$

in which a double summation over α and β is intended. Now, even though $\tilde{\mathbf{M}}(t)$ is an operator, the combination $K_{\alpha\beta}\tilde{M}_{\alpha\beta}$ is scalar. Let the first two moments of $\tilde{\mathbf{M}}(t)$ be

$$\langle \tilde{\mathbf{M}}(t)\rangle = \mathbf{M}^0 \tag{1.4.24}$$

and

$$\langle \tilde{M}_{\alpha\beta}(t)\tilde{M}_{\theta\tau}(s)\rangle = 2Q_{\alpha\beta\theta\tau}\delta(t-s) + M^0_{\alpha\beta}M^0_{\theta\tau}. \tag{1.4.25}$$

These moments suffice for a Gaussian operator, because Gaussianness is defined by the requirement that $\Phi[\mathbf{K}(t)]$ be Gaussian. Since $K_{\alpha\beta}\tilde{M}_{\alpha\beta}$ is a Gaussian scalar, our previous results hold and

$$\Phi[\mathbf{K}(t)] = \exp\left[i \int_0^\infty ds\, K_{\alpha\beta}(s)M^0_{\alpha\beta} - \frac{1}{2}\int_0^\infty ds\, K_{\alpha\beta}(s)K_{\mu\nu}(s)Q_{\alpha\beta\mu\nu}\right], \tag{1.4.26}$$

which is clearly Gaussian. Moments now follow from

$$\left\langle \prod_{l=1}^{n} \tilde{M}_{\alpha_l\beta_l}(t_l)\right\rangle = (-i)^n \frac{\delta^n}{\prod_{l=1}^{n} \delta K_{\alpha_l\beta_l}(t_l)} \Phi[\mathbf{K}(t)]\Big|_{K(t)=0}. \tag{1.4.27}$$

If, however, we wish to average an expression such as the solution to (1.3.8) given by (1.3.11), then we first note that the argument of the exponential is an operator, not a scalar, and the exponential is necessarily time ordered. The *time ordered* cumulant expansion for operators is expressed

$$\left\langle \underset{\leftarrow}{T}\exp\left[\int_0^t \mathrm{d}s\,\tilde{\mathbf{M}}(s)\right]\right\rangle = 1 + \sum_{n=1}^{\infty}\int_0^t \mathrm{d}s_1 \int_0^{s_1} \mathrm{d}s_2 \cdots \times \int_0^{s_{n-1}} \mathrm{d}s_n \langle \tilde{\mathbf{M}}(s_1)\ldots\tilde{\mathbf{M}}(s_n)\rangle$$

$$\equiv \underset{\leftarrow}{T}\exp\left[\sum_{n=1}^{\infty}\int_0^t \mathrm{d}s\,\mathbf{C}^{(n)}(s)\right]. \tag{1.4.28}$$

The analogs to (1.4.9) and (1.4.10) become (Steiger and Fox, 1982)

$$\langle \tilde{\mathbf{M}}(t_1)\ldots\tilde{\mathbf{M}}(t_n)\rangle = \sum_{\text{partitions of } n} \langle \tilde{\mathbf{M}}(t_{i_{11}})\ldots\tilde{\mathbf{M}}(t_{i_{1n_1}})\rangle_c \cdots \times \langle \tilde{\mathbf{M}}(t_{i_{k1}})\ldots\tilde{\mathbf{M}}(t_{i_{kn_k}})\rangle_c, \tag{1.4.29}$$

in which the sum over partitions of n means a division of the first n integers into k subsets: $(i_{11}\ldots i_{1n})\ldots(i_{k1}\ldots i_{kn_k})$, which are *ordered* by $i_{ls} < l_{lr}$ for $s < r$ and $i_{ls} < i_{l'1}$ for $l < l'$. The inverse of this is

$$\langle \tilde{\mathbf{M}}(t_1)\ldots\tilde{\mathbf{M}}(t_n)\rangle_c = \sum_{\text{partitions of } n} (-1)^{k-1} \sum_p \prod_{s=1}^{k} \times \left\langle \prod_{l=1}^{n_{p(s)}} \tilde{\mathbf{M}}(t_{i_{p(s),l}})\right\rangle \tag{1.4.30}$$

in which the partitions of n are the same as above for (1.4.29), $i_{11} = 1$, and p is a permutation on the integers $2, 3, \ldots, k$ $(p(1) \equiv 1)$. These formulas are much more complex than are (1.4.9) and (1.4.10), although they reduce to (1.4.9) and (1.4.10) when commutativity is not an issue. Generally, non-commutativity of $\tilde{\mathbf{M}}$ with itself at different times makes it extremely important to respect the order of the terms in the *highly non-linear products* in (1.4.29) and (1.4.30). The first two operator cumulants corresponding with (1.4.24) and (1.4.25) are

$$\left.\begin{aligned} &\langle \tilde{\mathbf{M}}(t)\rangle_c = \mathbf{M}^0 \\ \text{and} \\ &\langle \tilde{\mathbf{M}}_{\alpha\beta}(t)\tilde{\mathbf{M}}_{\theta\tau}(s)\rangle_c = 2\theta_{\alpha\beta\theta\tau}\,\delta(t-s). \end{aligned}\right\} \tag{1.4.31}$$

Now, if $\tilde{\mathbf{M}}(t)$ is Gaussian, then *by definition* $\Phi[\mathbf{K}(t)]$ is Gaussian. This does not by itself imply that the operator cumulant expansion for $\langle \exp[\int_0^t \mathrm{d}s\,\tilde{\mathbf{M}}(s)]\rangle$ terminates at the second cumulant, as it does for scalar processes (Fox, 1978, 1979). Indeed, all higher order cumulants are non-vanishing *unless* we are dealing with white noise. Stochastic processes in quantum mechanics are often approximated by white noise, but in reality they are non-white or *colored*. This means that in place of the Dirac delta correlation in (1.4.25) or (1.4.31), there is a finite correlation with correlation time τ:

$$2D\delta(t-s) \rightarrow \frac{D}{\tau}\exp\left[-\frac{|t-s|}{\tau}\right]. \tag{1.4.32}$$

For Gaussian, *colored noise* operators of this type, the operator cumulant of order $2m$, i.e. $\mathbf{C}^{(2m)}$, is of order $m-1$ in τ, i.e. proportional to τ^{m-1}. For the second cumulant, $m=1$, and the τ- dependence is not manifest, but for higher order, $n>2$, the result depends explicitly on τ. In the white noise limit $\tau\to 0$, and these higher order operator cumulants *vanish anyway*! We will return to the physical implications of these remarks later.

1.5 The stochastic Schrödinger equation

As was indicated in the introduction the terminology stochastic Schrödinger equation has a special meaning in this chapter. It refers to a phenomenological treatment of the quantum mechanics of a subsystem which is in interaction with a reservoir of other particles which could in principle be treated as a complicated many body problem. The many body problem is so complicated, however, that no real progress can be made with the exact Hamiltonian formalism so that we instead replace all of the reservoir dynamics by an effective stochastic contribution to the Hamiltonian. This parallels the philosophy employed in justifying the Langevin theory for Brownian motion in which the dynamical influences of the fluid molecules on the Brownian particle are modeled by a stochastic force. In the quantum mechanical setting, several new features emerge in the implementation of this approach. The first new feature has been previewed at length in the preceding sections in which *multiplicative* stochastic processes have been described and characterized. A second new feature is that there is a demarcation between the microcanonical and the canonical equilibrium distributions. One type of stochastic Schrödinger equation creates subsystem relaxation which results in the microcanonical equilibrium distribution whereas a second type produces the canonical equilibrium.

We begin with the simplest stochastic Schrödinger equation (Faid, 1986; Faid and Fox, 1986, 1987; Fox, 1978)

$$i\hbar\frac{\partial}{\partial t}\psi = \mathbf{H}_0\psi + \tilde{\mathbf{H}}(t)\psi. \tag{1.5.1}$$

$\mathbf{H}_0$ is the subsystem Hamiltonian and it can be expressed in terms of its eigenkets

$$\mathbf{H}_0 = \sum_i \hbar\omega_i |i\rangle\langle i| \tag{1.5.2}$$

in which the energy eigenvalues have been expressed as $E_i = \hbar\omega_i$. The stochastic Hamiltonian, $\tilde{\mathbf{H}}(t)$, represents the perturbation caused by interaction with a many body reservoir. It is assumed to be Gaussian white noise with zero mean, which means

$$\langle\tilde{\mathbf{H}}(t)\rangle = 0 \tag{1.5.3}$$

and

$$\langle \tilde{H}_{ij}(t)\tilde{H}_{kl}(s)\rangle = 2Q_{ijkl}\delta(t-s). \tag{1.5.4}$$

It is also stipulated that $\tilde{\mathbf{H}}(t)$ is Hermitian, i.e. $\tilde{\mathbf{H}}^{\dagger}(t) = \tilde{\mathbf{H}}(t)$.

In quantum mechanics, we do not make measurements of ψ, but instead make measurements of operator expectation values for operator observables. For this reason, it is appropriate to express stochastic quantum mechanics in terms of the density operator $\rho = |\psi\rangle\langle\psi|$:

$$i\hbar\frac{\partial}{\partial t}\rho = [\mathbf{H}_0, \rho] + [\tilde{\mathbf{H}}(t), \rho]. \tag{1.5.5}$$

If there were no stochasticity then (1.5.1) and (1.5.5) would merely represent equivalent expressions for identical physics. However, because there is stochasticity and because we will be averaging over this stochasticity before we are through, it is essential to distinguish between the average of a product and the product of averages. Upon averaging, (1.5.1) and (1.5.5) yield very different consequences. Averaging (1.5.1), which is too soon, will create decay of total probability, whereas averaging (1.5.5) will be shown to preserve total probability, while at the same time it *creates* relaxation to equilibrium. Moreover, no phenomenological relaxation parameters appear in (1.5.5), as they do in other approaches such as the Wigner–Weisskopf theory for relaxation of unstable states (Sakurai, 1967).

The stochastic average of expectation values will be determined by $\langle\rho(t)\rangle$. Since (1.5.5) is a multiplicative stochastic process, the formal solution may be rendered in terms of the time ordered exponential of the commutator operators

$$\left.\begin{aligned} &[\mathbf{H}_0, \bullet] \equiv \mathbf{H}_0\bullet - \bullet\mathbf{H}_0 \\ \text{and}\qquad &[\tilde{\mathbf{H}}(t), \bullet] \equiv \tilde{\mathbf{H}}(t)\bullet - \bullet\tilde{\mathbf{H}}(t), \end{aligned}\right\} \tag{1.5.6}$$

in which the dots indicate the placement of operators upon which these commutator operators (superoperators) act. In fact, it proves technically convenient to transform (1.5.5) into the *interaction picture* (Fox, 1978, 1986c) with respect to $\mathbf{H}_0$ before averaging. The averaging is executed by using operator cumulants (Fox, 1978, 1986c), and because we assume white noise in (1.5.4) the operator cumulant expansion terminates *exactly* after the second cumulant (see preceding section for discussion). The result is

$$i\hbar\frac{\partial}{\partial t}\langle\rho(t)\rangle = [\mathbf{H}_0, \langle\rho(t)\rangle] + i\hbar\mathbf{R}\langle\rho(t)\rangle, \tag{1.5.7}$$

in which $\mathbf{R}$ is the relaxation superoperator generated by the second operator cumulant (see (1.4.28)):

$$\mathbf{R} \equiv \mathbf{C}^{(2)}(t) = \frac{-1}{\hbar^2}\int_0^t \mathrm{d}s\,\langle[\tilde{\mathbf{H}}(t), \bullet][\tilde{\mathbf{H}}(s), \bullet]\rangle$$

$$= -\frac{1}{\hbar^2}\int_0^t ds\{\langle \mathbf{H}(t)\tilde{\mathbf{H}}(s)\rangle \bullet + \bullet\langle \tilde{\mathbf{H}}(s)\tilde{\mathbf{H}}(t)\rangle$$

$$- \langle \tilde{\mathbf{H}}(t)\bullet\tilde{\mathbf{H}}(s)\rangle - \langle \tilde{\mathbf{H}}(s)\bullet\tilde{\mathbf{H}}(t)\rangle\}. \quad (1.5.8)$$

This result yields an **R** which is independent of t because of the white noise assumption (1.5.4). It has an explicit representation with eigenstate indices

$$i\hbar\frac{\partial}{\partial t}\langle \rho_{ij}(t)\rangle = \hbar(\omega_i - \omega_j)\langle \rho_{ij}(t)\rangle + i\hbar R_{ijkl}\langle \rho_{kl}(t)\rangle, \quad (1.5.9)$$

in which repeated k's and l's are summed and

$$R_{ijkl} \equiv = -\frac{1}{\hbar^2}\{Q_{immk}\delta_{lj} + \delta_{ik}Q_{lmmj} - 2Q_{iklj}\}, \quad (1.5.10)$$

in which repeated m's are summed.

A look at (1.5.4) confirms that $Q_{ijkl} = Q_{klij}$. This means that the Q_{iklj} terms in (1.5.10) can be rewritten as Q_{ljik}. With this change, it is easy to prove that

$$\sum_i R_{iikl} = 0 \quad (1.5.11)$$

which has the consequence:

$$\text{Trace } \langle \rho(t)\rangle = \sum_i \langle \rho_{ii}(t)\rangle = 1 \quad (1.5.12)$$

for all t; i.e. total probability is conserved.

Even though **R** does not dissipate probability, it does drive $\langle \rho(t)\rangle$ towards equilibrium. This is shown by considering an arbitrary operator **X**, and looking at Trace $\mathbf{X}^\dagger\mathbf{R}\mathbf{X}$. From (1.5.8) we see that

$$\text{Trace } \mathbf{X}^\dagger\mathbf{R}\mathbf{X} = -\frac{1}{\hbar^2}\int_0^t ds \,\text{Trace } \mathbf{X}^\dagger\langle[\tilde{\mathbf{H}}(t),[\tilde{\mathbf{H}}(s),\mathbf{X}]]\rangle$$

$$= -\frac{1}{\hbar^2}\int_0^t ds \,\text{Trace}\,\langle[\mathbf{X}^\dagger,\tilde{\mathbf{H}}(t)][\tilde{\mathbf{H}}(s),\mathbf{X}]\rangle$$

$$= -\frac{1}{\hbar^2}\int_0^t ds \,\text{Trace}\,\langle[\tilde{\mathbf{H}}(t),\mathbf{X}]^\dagger[\tilde{\mathbf{H}}(s),\mathbf{X}]\rangle \leqslant 0 \quad (1.5.13)$$

because Trace $\mathbf{A}[\mathbf{B},\mathbf{C}] = $ Trace $[\mathbf{A},\mathbf{B}]\mathbf{C}$ and $[\mathbf{A},\mathbf{B}]^\dagger = [\mathbf{B}^\dagger,\mathbf{A}^\dagger]$ are true for arbitrary operators **A**, **B** and **C**; $\tilde{\mathbf{H}}(t)$ is Hermitian; and the stochastic average is Dirac delta correlated, i.e. only $t = s$ contributes. This result implies that the superoperator **R** has eigenoperators with non-positive eigenvalues.

The equilibrium produced by this relaxation is the microcanonical density operator. Suppose that $\mathbf{H}_0$ has a Hilbert space of dimension N. We may be thinking about a highly degenerate energy level or a cluster of levels with energy spacing very small compared with $k_B T$, i.e. a high temperature limit. Then the microcanonical density matrix is

$$\langle \rho_{ij}(t) \rangle_{t \to \infty} \to \frac{1}{N} \delta_{ij}. \tag{1.5.14}$$

This is precisely the zero eigenvalue eigenmatrix of R_{ijkl}:

$$R_{ijkl} \langle \rho_{kl}(\infty) \rangle = \frac{1}{N} R_{ijkl} \delta_{kl} = 0 \tag{1.5.15}$$

by an argument closely paralleling the argument used to justify (1.5.11). Any other density matrix for the initial state will relax into the microcanonical density matrix according to (1.5.7).

Extension of these results to *temperature-dependence* requires a more elaborate stochastic Schrödinger equation than that given by (1.5.1) (Faid and Fox, 1987; Fox, 1978). As will be shown below, the canonical equilibrium density operator will be approached by relaxation only if we explicitly bring into consideration the Hilbert space for the reservoir. However, we do this in a phenomenological model way. Let the total Hamiltonian for the subsystem and the reservoir act in the direct product Hilbert space which has one factor for the subsystem states and the other factor for the reservoir's phenomenological states. These latter states are not the true many body states of the reservoir but are instead assumed to be adequately modeled by the states of independent bosonic quasi-particles. Therefore, we write

$$\mathbf{H}(t) = \mathbf{H}_0 \otimes \mathbf{1}_R + \mathbf{1}_S \otimes \sum_k (\tilde{\mathbf{H}}^k(t) b_k + \tilde{\mathbf{H}}^{k\dagger}(t) b_k^\dagger) \tag{1.5.16}$$

in which $\mathbf{1}_R$ is the identity on the reservoir Hilbert space and $\mathbf{1}_S$ is the identity on the subsystem Hilbert space; $\mathbf{H}_0$ is the subsystem alone Hamiltonian; and the last factor contains the subsystem–reservoir quasi-particle interaction. This last term contains terms with b_k and $b_k^\dagger$ factors which act in the reservoir Hilbert space and represent, respectively, quasi-particle annihilation and creation. The other factors, $\tilde{\mathbf{H}}^k(t)$ and $\tilde{\mathbf{H}}^{k\dagger}(t)$, are operators acting in the subsystem Hilbert space. They are not Hermitian individually, although their combination in (1.5.16) is Hermitian overall. The time evolution of the full density operator follows from the equation

$$i\hbar \frac{\partial}{\partial t} \rho(t) = [\mathbf{H}(t), \rho(t)] \tag{1.5.17}$$

with the special initial condition

$$\rho(0) = \rho_S(0) \otimes \prod_k \exp[-\beta\hbar\omega_k b_k^\dagger b_k](1 - \exp(-\beta\hbar\omega_k)), \tag{1.5.18}$$

in which $\rho_S(0)$ is an *arbitrary* subsystem density operator factor, whereas the rest is the canonical density operator for bosonic quasi-particles at temperature $T \equiv 1/k_B\beta$, and with single boson energies $\hbar\omega_k$.

In this enlarged setting, the averaged subsystem density operator is obtained by the *combined effects* of averaging over the stochastic $\tilde{H}^k(t)$'s and tracing over the reservoir quasi-particle states. This generates the averaged *reduced* density operator

$$\mathbf{D}(t) \equiv \text{Trace}_{\mathbf{R}}\langle \rho(t) \rangle \equiv \langle\langle \rho(t) \rangle\rangle. \tag{1.5.19}$$

This combination, denoted above by $\langle\langle \ldots \rangle\rangle$, can be viewed as a new kind of averaging with respect to which we can define a new kind of cumulant (Faid and Fox, 1987; Fox, 1978).

The stochastic properties of $\tilde{\mathbf{H}}^k(t)$ (and $\tilde{\mathbf{H}}^k(t)^\dagger$) are that they are Gaussian, white noise with zero mean. This means

$$\langle \tilde{\mathbf{H}}^k(t) \rangle = 0 \tag{1.5.20}$$

and

$$\langle \tilde{H}^{k*}_{ji}(t) \tilde{H}^k_{mn}(s) \rangle = 2Q^k_{ijmn}\delta(t-s). \tag{1.5.21}$$

Implementation of the ordered cumulant approach is complicated by extra structure created by the boson operators, but nevertheless tractable (Faid and Fox, 1987; Fox, 1978, 1986c), and yields the exact, averaged (and reduced) equation

$$i\hbar\frac{\partial}{\partial t}\mathbf{D}(t) = [\mathbf{H}_0, \mathbf{D}(t)] + i\hbar\mathbf{R}\mathbf{D}(t) \tag{1.5.22}$$

in which $\mathbf{R}$ is given by

$$\begin{aligned}\mathbf{R}_{ijmn} = -\frac{1}{\hbar^2}\sum_k \{ & \delta_{jn}(Q^k_{illm}n_k + Q^k_{lmil}(n_k+1)) \\ & + \delta_{im}(Q^k_{nllj}n_k + Q^k_{ljnl}(n_k+1)) \\ & - 2Q^k_{njim}n_k - 2Q^k_{imnj}(n_k+1)\},\end{aligned} \tag{1.5.23}$$

in which n_k is given by

$$n_k = \frac{1}{\exp(\beta\hbar\omega_k) - 1}. \tag{1.5.24}$$

Conservation of total probability is again guaranteed because $\sum_i R_{iimn} = 0$. Moreover, the canonical subsystem density operator is

$$\mathbf{D}_{\text{can}} = \frac{1}{Z}\exp[-\beta\mathbf{H}_0] \tag{1.5.25}$$

where

$$Z = \text{Trace}_{\mathbf{S}}\exp[-\beta\mathbf{H}_0]. \tag{1.5.26}$$

In the $\mathbf{H}_0$ eigenstate basis used to write the indexed expression for $\mathbf{R}$ in (1.5.23), $\mathbf{D}_{\text{can}}$ becomes

$$\mathbf{D}_{\text{can}} \rightarrow \frac{1}{Z}\exp[-\beta E_m]\delta_{mn} \tag{1.5.27}$$

where

$$\mathbf{H}_0|m\rangle = E_m|m\rangle. \tag{1.5.28}$$

In order to prove that this canonical density operator is a solution to (1.5.22), additional properties of $\tilde{\mathbf{H}}^k(t)b_k$ are required. The subsystem matrix elements of this operator, $\tilde{H}^k_{ij}(t)b_k$, represent the possibility that bosonic quasi-particle k is absorbed while the subsystem makes a transition from state $|j\rangle$ to state $|i\rangle$, i.e.

$$\tilde{H}^k_{ij}b_k = \langle i|\tilde{\mathbf{H}}^k(t)b_k|j\rangle. \tag{1.5.29}$$

Similarly

$$\tilde{H}^{k*}_{ij}b^\dagger_k = \langle i|\tilde{\mathbf{H}}^{k\dagger}(t)b^\dagger_k|j\rangle. \tag{1.5.30}$$

In this phenomenological model of the reservoir–subsystem coupling, it is necessary to assume energy conservation as well as the earlier stochastic properties. This means

$$\tilde{H}^k_{ij}b_k = 0 \quad \text{unless} \quad E_i - E_j = \hbar\omega_k \tag{1.5.31}$$

and

$$\tilde{H}^{k*}_{ji}b^\dagger_k = 0 \quad \text{unless} \quad E_i - E_j = -\hbar\omega_k. \tag{1.5.32}$$

Note especially that $\tilde{H}^{k\dagger}_{ij} \neq \tilde{H}^k_{ij}$ because $\tilde{\mathbf{H}}^k$ is *not* Hermitian. These conditions simplify Q^k_{ijmn} so that

$$Q^k_{ijmn} = 0 \quad \text{unless} \quad E_m - E_n = \hbar\omega_k \ \textit{and}\ E_i - E_j = -\hbar\omega_k. \tag{1.5.33}$$

If a pair of indices are identical, then (1.5.33) implies (for unsummed l)

$$Q^k_{ljil} = \delta_{ij}Q^k_{ljjl} \quad \text{and} \quad Q^k_{illj} = \delta_{ij}Q^k_{jllj}. \tag{1.5.34}$$

This reduces R_{ijmn} in (1.5.23) to

$$R_{ijmn} = -\frac{1}{\hbar^2}\sum_k \{\delta_{jn}\delta_{im}[(Q^k_{illi} + Q^k_{jllj})n_k + (Q^k_{liil} + Q^k_{ljjl})(n_k+1)] - 2Q^k_{njim}n_k - 2Q^k_{imnj}(n_k+1)\}, \tag{1.5.35}$$

with the further constraints implicit in (1.5.33). We are now in a position to claim that the canonical equilibrium, (1.5.27), satisfies $\mathbf{RD}_{\text{can}} = 0$, i.e.

$$\sum_m R_{ijmm}\frac{1}{Z}\exp[-\beta E_m] = 0. \tag{1.5.36}$$

Any other density matrix for the initial subsystem state will relax into the canonical density matrix. We will return to this point in the next section.

1.6 Pauli master equations

One of the earliest efforts to describe relaxation processes in quantum mechanical settings was the Pauli master equation (Van Hove, 1962). In essence, the master equation reduces the full density matrix equation to a system of coupled equations for the diagonal density matrix elements alone. Over the years, numerous attempts have been made to justify this reduction. While it is mathematically possible to obtain the Pauli master equation as the double limit of long times and weak coupling (Van Hove, 1962), it is not physically compelling to do so. Here, we show how it sits naturally in the stochastic Schrödinger equation context. Moreover, in addition to equations coupling the diagonal matrix elements alone, there is a companion system of equations for the off-diagonal matrix elements, the *coherences*. These extra equations prove valuable in the theory for spectral line shapes (Faid, 1986).

Two versions of the Pauli master equation will be derived below: one for the microcanonical equilibrium and one for the canonical equilibrium. In the first instance we make the Q_{ijkl} in (1.5.4) satisfy the special condition

$$Q_{ijkl} = Q_{ijkl}(\delta_{ik}\delta_{jl} + \delta_{il}\delta_{jk}). \tag{1.6.1}$$

Physically, this says that transitions between different pairs of subsystem states induced by the reservoir are statistically independent. Define the averaged probability to be in the ith eigenstate of $\mathbf{H}_0$ by

$$P_i \equiv \langle i|\langle \rho(t)\rangle|i\rangle = \langle \rho_{ii}(t)\rangle. \tag{1.6.2}$$

Using (1.6.1) in (1.5.9) and (1.5.10) yields

$$\frac{\mathrm{d}}{\mathrm{d}t}P_i = \sum_j W_{ij}P_j - \sum_j W_{ij}P_i, \tag{1.6.3}$$

where

$$W_{ij} \equiv \frac{2}{\hbar^2} Q_{ijji} \geqslant 0. \tag{1.6.4}$$

No summation is intended in (1.6.4). The positivity of W_{ij}, as well as its index symmetry, follows from (1.5.4). Equations (1.6.3) and (1.6.4) are precisely the Pauli master equation, and the equilibrium created is the microcanonical equilibrium. However, this is not all one gets from (1.6.1) and (1.5.9). Define the averaged coherence by

$$C_{kl}(t) \equiv \langle \rho_{kl}(t)\rangle = \langle k|\langle \rho(t)\rangle|l\rangle. \tag{1.6.5}$$

In addition to (1.6.3), one also gets

$$\frac{\mathrm{d}}{\mathrm{d}t}C_{kl}(t) = -\mathrm{i}\omega_{kl}C_{kl}(t) - \frac{1}{2}\sum_j (W_{kj} + W_{lj})C_{kl}(t) + \frac{2}{\hbar^2} Q_{klkl}C_{lk}(t), \tag{1.6.6}$$

in which $\omega_{kl} \equiv (E_k - E_l)/\hbar$. Note that C_{kl} is coupled to itself and transpose only.

While (1.6.3) describes the relaxation of the system, (1.6.6) is directly related to the response of the system to an optical spectroscopic probe.

We may also obtain a Pauli master equation which relates the system to the canonical equilibrium if we use the temperature-dependent method introduced in the preceding section. This requires imposing the special condition

$$Q^k_{ijmn} = Q^k_{ijji}\delta_{in}\delta_{jm} \tag{1.6.7}$$

on the Q^k_{ijmn} of (1.5.21), as well as retaining the restrictions in (1.5.33). The physical interpretation is the same as for (1.6.1) above, although (1.5.33) is responsible for the difference between (1.6.1) and (1.6.7). Putting all of this into (1.5.35) yields

$$\begin{aligned} R_{ijmn} = -\frac{1}{\hbar^2}\Bigg\{ &\delta_{jn}\delta_{im}\Bigg[\Bigg(Q_{illi}\frac{\theta(\omega_{li})}{\exp(\beta\hbar\omega_{li})-1} \\ &+ Q_{jllj}\frac{\theta(\omega_{lj})}{\exp(\beta\hbar\omega_{lj})-1}\Bigg) \\ &+\Bigg(Q_{liil}\frac{\theta(\omega_{il})\exp(\beta\hbar\omega_{il})}{\exp(\beta\hbar\omega_{il})-1} + Q_{ljjl}\frac{\theta(\omega_{jl})\exp(\beta\hbar\omega_{jl})}{\exp(\beta\hbar\omega_{jl})-1}\Bigg)\Bigg] \\ &- 2\delta_{nm}\delta_{ij}\Bigg(Q_{niin}\frac{\theta(\omega_{in})}{\exp(\beta\hbar\omega_{in})-1} \\ &+ Q_{inni}\frac{\theta(\omega_{ni})\exp(\beta\hbar\omega_{ni})}{\exp(\beta\hbar\omega_{ni})-1}\Bigg)\Bigg\}, \end{aligned} \tag{1.6.8}$$

in which $\theta(x)$ is the Heaviside function which equals one for $x > 0$ and is zero otherwise, and in which we have dropped the superscript k on the Q's since the k-dependence has been summed out in accord with (1.5.33). It is now easily seen that R couples diagonal density matrix elements to other diagonal elements only. Using the definition

$$P_i = \mathrm{Trace}_{\mathrm{R}}\langle i|\langle\rho(t)\rangle|i\rangle = \mathrm{Trace}_{\mathrm{R}}\langle\rho_{ii}(t)\rangle = \langle\!\langle\rho_{ii}(t)\rangle\!\rangle \tag{1.6.9}$$

we obtain

$$\frac{\mathrm{d}}{\mathrm{d}t}P_i(t) = \sum_j W_{ij}P_j(t) - \sum_j W_{ij}P_i(t) \tag{1.6.10}$$

in which W_{ij} is now given by

$$W_{ij} \equiv \frac{1}{\hbar^2}\left(Q_{jiij}\frac{\theta(\omega_{ij})}{\exp(\beta\hbar\omega_{ij})-1} + Q_{ijji}\frac{\theta(\omega_{ji})\exp(\beta\hbar\omega_{ji})}{\exp(\beta\hbar\omega_{ji})-1}\right). \tag{1.6.11}$$

Clearly, for fixed i and j only one of the two expressions in (1.6.11) will contribute because of the Heaviside functions. Moreover, this implies that W_{ij} is not symmetric, as in the microcanonical case, but instead satisfies the *detailed balancing* relation

$$W_{ij}\exp[-\beta E_j] = W_{ji}\exp[-\beta E_i] \tag{1.6.12}$$

as follows from (1.6.11). This means that the equilibrium is the canonical equilibrium.

We also get equations for the coherences, $C_{kl}(t)$, defined by

$$C_{kt}(t) \equiv \langle\langle \rho_{kl}(t) \rangle\rangle. \tag{1.6.13}$$

These equations are

$$\frac{d}{dt} C_{kl}(t) = -i\omega_{kl} C_{kl}(t) - \frac{1}{2} \sum_j (W_{kj} + W_{lj}) C_{kl}(t). \tag{1.6.14}$$

This looks similar to (1.6.6), but differs in two ways: it contains no C_{lk} and the W_{kj} is defined by (1.6.11) instead of by (1.6.4).

All of the results given in this section have been for white noise. They may be generalized to colored noise (Faid, 1986) which greatly complicates the expressions, but in a relatively straightforward way.

1.7 Magnetic relaxation and Redfield's equation

Magnetic relaxation provides an especially nice, relatively simple physical example for the preceding considerations. With it, we can exhibit the importance of the equations for the coherences and the consequences of colored noise in place of white noise.

The Hamiltonian for a spin-$\frac{1}{2}$ magnetic moment in a magnetic field is

$$\mathbf{H} = -\frac{e\hbar}{2mc} \vec{\sigma} \cdot \vec{B}, \tag{1.7.1}$$

in which e is the charge of the electron, m is the spin's mass, c is the speed of light, $\vec{B}$ is the magnetic field

$$\vec{B} = B\hat{k} + \tilde{\vec{B}}(t), \tag{1.7.2}$$

which has a constant part, $B\vec{k}$, and a stochastic part, $\tilde{\vec{B}}(t)$, and $\vec{\sigma}$ is the Pauli matrix vector

$$\vec{\sigma} = \hat{i}\sigma_x + \hat{j}\sigma_y + \hat{k}\sigma_z \tag{1.7.3}$$

expressed in terms of the conventional spin-$\frac{1}{2}$ Pauli matrices. If the fluctuating magnetic field is isotropic, Gaussian, white noise, then

$$\langle \tilde{B}_i(t) \rangle = 0 \quad i = x, y, z \tag{1.7.4}$$

and

$$\langle \tilde{B}_i(t) \tilde{B}_j(s) \rangle = 2Q_{ij}\delta(t-s). \tag{1.7.5}$$

We also assume that different cartesian components of the fluctuating magnetic field are statistically independent, although still isotropic, i.e.

$$Q_{ij} = Q\delta_{ij}. \tag{1.7.6}$$

Introduce the Rabi frequency, Ω, by

$$\Omega \equiv \frac{eB}{mc}. \tag{1.7.7}$$

The averaged density matrix equation turns out to be

$$i\hbar\frac{\partial}{\partial t}\langle\rho\rangle = \left[-\frac{\hbar}{2}\Omega\sigma_z, \langle\rho\rangle\right] + i\hbar\mathbf{R}\langle\rho\rangle, \tag{1.7.8}$$

in which the relaxation operator, $\mathbf{R}$, is given by

$$\mathbf{R} \equiv -\left(\frac{e}{2mc}\right)^2 Q\{6\bullet - 2\sigma_i\bullet\sigma_i\}, \tag{1.7.9}$$

in which there is a summation over i. Equation (1.7.8) is essentially Redfield's equation (Fox, 1978; Redfield, 1965). Since $\langle\rho\rangle$ is 2×2 in this special case, and it is Hermitian, we may expand it in terms of Pauli matrices:

$$\langle\rho\rangle = \tfrac{1}{2}\sigma_0 + \vec{M}(t)\cdot\vec{\sigma}, \tag{1.7.10}$$

in which σ_0 is the 2×2 identity and $\vec{M}(t)$ is the averaged time-dependent magnetic moment vector. With this rewriting of $\langle\rho\rangle$, (1.7.8) turns into

$$\frac{\mathrm{d}}{\mathrm{d}t}\begin{pmatrix} M_x \\ M_y \\ M_z \end{pmatrix} = \begin{pmatrix} 0 & \Omega & 0 \\ -\Omega & 0 & 0 \\ 0 & 0 & 0 \end{pmatrix}\begin{pmatrix} M_x \\ M_y \\ M_z \end{pmatrix} - \frac{1}{T}\begin{pmatrix} 1 & 0 & 0 \\ 0 & 1 & 0 \\ 0 & 0 & 1 \end{pmatrix}\begin{pmatrix} M_x \\ M_y \\ M_z \end{pmatrix}, \tag{1.7.11}$$

in which $(1/T) \equiv 2Q(e/mc)^2$ is the relaxation time. This is just Bloch's description (Bloch, 1932) of magnetic relaxation giving both the longitudinal (M_z) relaxation and the transverse (M_x, M_y) relaxation. (We have here the most simple case of the high temperature limit with identical longitudinal and transverse relaxation times, T.)

Pauli's master equation, (1.6.3), yields only the $(\mathrm{d}/\mathrm{d}t)(M_z)$ part of (1.7.11). The corresponding equation for the coherences, (1.6.6), yields the (M_x, M_y) part of (1.7.11). This shows clearly that the van Hove type approach to the Pauli master equation would totally eliminate the transverse magnetic relaxation in this context. Since the transverse relaxation is readily observed, this strengthens the view that our stochastic density matrix approach is richer.

It is possible to study this example with respect to the influence of colored noise in place of white noise. Replace (1.7.5) and (1.7.6) with

$$\langle\tilde{B}_i(t)\tilde{B}_j(s)\rangle = \frac{Q}{\tau}\exp\left[-\frac{|t-s|}{\tau}\right]\delta_{ij}. \tag{1.7.12}$$

The limit $\tau\to 0$ restores the white noise situation just analyzed above. In order to obtain the averaged density matrix equation which replaces (1.7.8) we use the time ordered cumulant method. If we approximate the cumulant expansion with its second term (this is no longer exact for $\tau \neq 0$, even though our process is Gaussian), then we get

$$i\hbar\frac{\partial}{\partial t}\langle\rho\rangle=\left[-\frac{\hbar}{2}\Omega\left(1+\frac{\tau e^2Q}{m^2c^2(1+\Omega^2\tau^2)}\right)\sigma_z,\langle\rho\rangle\right]$$
$$+i\hbar\mathbf{R}\langle\rho\rangle, \tag{1.7.13}$$

in which

$$\mathbf{R}=-\left(\frac{e}{2mc}\right)^2Q\left\{\left(\frac{4}{1+\Omega^2\tau^2}+2\right)\bullet-\frac{2}{1+\Omega^2\tau^2}\right.$$
$$\left.\times(\sigma_x\bullet\sigma_x+\sigma_y\bullet\sigma_y)-2\sigma_z\bullet\sigma_z\right\}, \tag{1.7.14}$$

which clearly reduces to (1.7.9) when $\tau=0$. The remarkable result here is that colored noise not only modifies the relaxation times, but that it also modifies the commutator term in (1.7.13). This amounts to the creation of energy shifts in the eigenenergies of the non-stochastic part of the Hamiltonian, and can be interpreted as the creation of an anomalous magnetic moment

$$\frac{\Delta\mu}{\mu}=\frac{\tau e^2Q}{m^2c^2(1+\Omega^2\tau^2)}. \tag{1.7.15}$$

This turns out to be indicative of the general situation: colored noise creates energy shifts added to the unperturbed eigenenergies of the non-stochastic Hamiltonian, H_0; whereas relaxation is already created by white noise. This means that a spectroscopic measurement not only yields the strength of the fluctuations, through the width parameter of the spectrum, but also yields the magnitude of the coloring, through the energy shifts.

The results in this section represent the temperature-independent case. The more complicated consequences of temperature-dependence created by the model described in Section 1.5 have been worked out in detail by Faid (1986). His studies cover temperature-dependence, anisotropy, and colored noise.

1.8 A variety of physical applications

The limits of space prohibit a detailed account of the applications which follow. In each case much more can be said than will be here. The reader is urged to see the references for additional detail.

Spectral line shapes

Faid (1986) and Faid and Fox (1986) have used the formalism presented here to derive spectral line shapes for the interaction of light with a relaxing molecule. This has been done both with and without temperature-dependence. It provides a generalization of earlier work by Mukamel (1982). This work is likely to have the most direct contact with experiments. The important feature of the analysis which emerges is that, while relaxation is described nicely by the Pauli master equations for the state probabilities, it is the coherences which determine the spectral line shape. The relaxation parameters in the coherences and in the probabilities are not necessarily the same. It becomes important to distinguish life-times and line-widths, which are identical only for special cases.

Free induction decay (FID)

FID has been studied (Berman and Brewer, 1985; Schenzle, Mitsumaga, DeVoe and Brewer, 1984) in the context of the optical Bloch equations. The fluctuations in this case only appear in the diagonal elements of the Hamiltonian, i.e.

$$\tilde{H}_{ij}(t) \equiv 0 \text{ unless } i = j. \tag{1.8.1}$$

This creates *no relaxation* of the probabilities, but it still does yield non-trivial behavior in the coherences. The formalism presented here applies very naturally to FID and again underscores the important distinction between probability relaxation and coherence relaxation. The earlier literature appears to be devoid of a proper discussion of this distinction. Thus, FID may prove to be another useful place to explore contact with experiment.

Laser fluctuations

Contact with experiment has already been made in the study of laser fluctuations in pumped dye lasers (Fox, James and Roy, 1984). This work utilizes the averaged density matrix formalism presented here. It generalizes and extends earlier work on lasers and quantum relaxation (Louisell, 1973; Sargent, Scully and Lamb, 1974). Spontaneous emission is modeled by additive fluctuations, whereas pump fluctuations are multiplicative. Using first passage time techniques, it is possible to determine relatively easily (Roy, Yu and Zhu, 1985) all of the noise parameters for both the additive and multiplicative fluctuations. S. Zhu (1987) has just completed a PhD dissertation with Roy which contains a detailed account of this work and the extensive earlier work of others upon which it is based.

Quantum Langevin equations

An alternative approach to these problems begins with the Heisenberg operator equations rather than with the Schrödinger equation. Usually, a quantum oscillator in a blackbody radiation field is studied (Ford, Kac and Mazur, 1965; Ford, Lewis and O'Connell, 1985). This approach yields a quantum Langevin equation, i.e. an additive stochastic process instead of the multiplicative density operator equations presented here. Very beautiful exact results regarding the free energy of the oscillator in a radiation field have been obtained (Ford, Lewis and O'Connell, 1985). However, the method is limited to a kind of linearity of interaction in order that the Heisenberg equations obtained remain tractable. Additional results along these lines (requiring some kind of linearity) have been obtained by Lax (1968) and have been extended very elegantly by Gardiner (1983). Gardiner's work makes use of coherent states (Glauber, 1983a, b) in a very effective way. An alternative approach to both of these efforts, but still along similar lines, was initiated by Mori (1965) and utilizes the projection operator method, again yielding Langevin-like equations. This approach has been utilized by many researchers since its inception. All three of these approaches attempt to analyze an exact many body problem rather than a stochastic model. In that sense, they differ from our approach which is phenomenological. Nevertheless, many of the results have parallel developments in the stochastic models. This has been shown in detail for phonon and photon reservoirs by Fox (1978, 1986c). That approach shows that density operators, time order cumulants, and characteristics functionals provide a unifying methodology for the study of these exact model systems.

Acknowledgement

This work was partially supported by NSF grant PHY-8603729.

References

Anderson, P. W. 1954. *J. Phys. Soc. Jpn.* **9**, 316.
Arnold, L. 1974. *Stochastic Differential Equations.* New York: John Wiley.
Berman, R. P. and Brewer, R. G. 1985. *Phys. Rev. A* **32**, 2784.
Berne, B. J. and Pecora, R. 1976. *Dynamic Light Scattering* New York: John Wiley.
Bloch, F. 1932. *Z. Physik* **74**, 295.
Faid, K. 1986. PhD Thesis. Georgia Institute of Technology. Available from University Microfilms, Ann Arbor, Michigan.
Faid, K. and Fox, R. F. 1986. *Phys. Rev. A* **34**, 4286.
Faid, K. and Fox, R. F. 1987. *Phys. Rev. A* **35**, 2684.
Ford, G. W., Kac, M. and Mazur, P. 1965. *J. Math. Phys.* **6**, 504.
Ford, G. W., Lewis, J. T. and O'Connell, R. F. 1985. *Phys. Rev. Lett.* **55**, 2273.
Fox, R. F. 1978. *Phys. Rep.* **48**, 179.
Fox, R. F. 1979. *J. Math. Phys.* **20**, 2467.
Fox, R. F. 1986a. *Phys. Rev. A* **33**, 467.
Fox, R. F. 1986b. *Phys. Rev. A* **34**, 4525.
Fox, R. F. 1986c. In *Probability, Statistical Mechanics, and Number Theory* (G-C. Rota, ed.), Advances in Mathematics Supplemental Studies, vol. 9, p. 125. New York: Academic Press.
Fox, R. F. and Uhlenbeck, G. E. 1970. *Phys. Fluids* **13**, pp. 1893, 2881.
Fox, R. F., James, G. E. and Roy, R. 1984. *Phys. Rev. A* **30**, 2482.
Gardiner, C. W. 1983. *Handbook of Stochastic Methods*, chapter 10. Berlin: Springer-Verlag.
Glauber, R. J. 1983a. *Phys. Rev.* **130**, 2529.
Glauber, R. J. 1963b. *Phys. Rev.* **131**, 2766.

Grabert, H., Hänggi, P. and Talkner, P. 1979. *Phys. Rev. A* **19**, 2440.

Kubo, R. 1954. *J. Phys. Soc. Jpn.* **9**, 935.

Kubo, R., Toda, M. and Nashitsume, N. 1985. *Statistical Physics II*, chapter 2. New York: Springer-Verlag.

Lax, M. 1968. *Phys. Rev.* **172**, 350.

Lichtenberg, A. J. and Lieberman, M. A. 1983. *Regular and Stochastic Motion.* New York: Springer-Verlag.

Louisell, W. H. 1973. *Quantum Statistical Properties of Radiation.* New York: John Wiley.

McQuarrie, D. A. and Keizer, J. E. 1981. *Theoretical Chemistry: Advances and Perspectives*, vol. *6A*, pp. 165–213. New York: Academic Press.

Mori, H. 1965. *Prog. Theor. Phys.* **34**, 399.

Mukamel, S. 1982. *Phys. Rep.* **93**, 1.

Nelson, E. 1966. *Phys. Rev.* **150**, 1079.

Onsager, L. and Machlup, S. 1953, *Phys. Rev.* **91**, pp. 1505, 1512.

Redfield, A. 1965. In *Advances in Magnetic Resonance* (J. S. Waugh, ed.), vol. 1, pp. 1–32. New York: Academic Press.

Roy, R., Yu, A. W. and Zhu, S. 1985. *Phys. Rev. Lett.* **55**, 2794.

Sakurai, J. J. 1967. *Advanced Quantum Mechanics.* Reading, MA: Addison-Wesley.

Sargent, M., Scully, M. O. and Lamb, W. E. 1974. *Laser Physics.* Reading, MA: Addison-Wesley.

Schenzle, A., Mitsumaga, M., DeVoe, R. G. and Brewer, R. G. (1984). *Phys. Rev. A* **30**, 325.

Steiger, U. R. and Fox, R. F. 1982. *J. Math. Phys.* **23**, 1678.

van Hove, L. 1962. In *Fundamental Problems in Statistical Mechanics* (E. G. D. Cohen, ed.), chapter 6, pp. 157–72. Amsterdam: North-Holland.

Zaslavsky, G. M. 1981. *Phys. Rep.* **80**, 157–250.

Zhu, S. 1987. PhD Thesis, Georgia Institute of Technology. Available from University Microfilms, Ann Arbor, Michigan.

2 Self-diffusion in non-Markovian condensed-matter systems

TOYONORI MUNAKATA

2.1 Introduction

Atomic diffusion in condensed-matter systems has important implications in solid state physics and chemistry (Murch and Nowick, 1984). Over the last fifty years there have been developed many theoretical models, which may be classified in various ways, for example stochastic (mesoscopic) or atomistic (microscopic), discrete or continuous, and classical (thermal activation) or quantal (tunnelling) (Flynn, 1972).

From the very nature of diffusion, a stochastic point of view is one of the essential ingredients in the theory for mass transport in solids (Dieterich, Fulde and Peschel, 1980). Actually, a hopping or random walk model, in which the mean residence time τ_r of a diffusing atom on an atomic site is assumed to be much longer than the time of flight τ_f, has long been used to study atomic diffusion. The hopping rate $\Gamma = 1/\tau_r$, an important parameter in this model, is calculated microscopically based on, e.g., a transition-state (TS) theory (Vineyard, 1957) or a dynamic theory (Flynn, 1968, 1972). We note that a diffusing atom is restricted to discrete sites and thus an oscillatory behavior during its stay on an atomic site or motion associated with barrier-crossing cannot be described within the hopping model.

To retrieve these features of a diffusion process, a continuous diffusion model (CDM) was proposed, in which an impurity moving through interstitial sites is assumed to perform Brownian motion in a periodic potential $V_p(X)$ produced by a host crystal (Fulde, Pietronero, Schneider and Strassler, 1975). The CDM is concisely expressed by the following Langevin equation (2.1.1) with the fluctuation–dissipation theorem (FDT) (2.1.2):

$$\dot{P} = \mathrm{d}P/\mathrm{d}t = -\int_0^t \mathrm{d}s\, K(t-s)P(s) - \mathrm{d}V_p(X)/\mathrm{d}X + F(t), \qquad (2.1.1)$$

$$\langle F(t)F \rangle = Mk_B TK(t), \qquad (2.1.2)$$

where $P = M\dot{X}$, M denoting the mass of an impurity. On account of its simplicity and its close relationship to the hopping model in the case of large friction at low temperature, the CDM has been studied intensively (Dieterich, Fulde and Peschel, 1980; Risken, 1984).

In the first half of this chapter, we are concerned with a stochastic process governed by (2.1.1) and we calculate the hopping rate with emphasis put on non-Markovian effects in an atomic migration process (see Section 2.2). Here it is shown that a kind of resonant activation is possible when dissipation is weak and the characteristic frequency of the random force $F(t)$ is near the natural frequency defined at the bottom of the periodic potential.

From a microscopic viewpoint, atomic jump events result from complex interplay between the diffusing atom and the surrounding crystalline atoms, and it is highly desirable if one could derive a Langevin equation like (2.1.1) from first principles. In the latter half of this article we are concerned with microscopic (or statistical mechanical) aspects of diffusion and try to obtain what is called a reduced description by projecting or integrating out irrelevant variables, i.e. phonon-modes in our problem. With use of the projection operator (PO) formalism (Mori, 1965) we derive the Langevin equation (2.1.1) with (2.1.2) starting from a Hamiltonian system (see Section 2.3). It is also shown that coupling of impurity to (optic) phonon-modes can lead to a non-Markovian noise $F(t)$ with a characteristic frequency of the phonon-modes, thus giving a microscopic motivation for our study in Section 2.2.

We note that our approach stated above is based on classical statistical mechanics and (2.1.1) can describe only a thermal activation process. It is well known that as the temperature of the system decreases, tunnelling begins to dominate thermal activation in hopping kinetics. In order to investigate the crossover from thermal to quantum hopping, we next employ another method to eliminate phonon-modes, the technique of path-integration (PI) (Feynman and Hibbs, 1965), and obtain a hopping rate formula which bridges between the low and the high temperature regions (Section 2.4). The final section contains some remarks.

2.2 Resonant activation in non-Markovian processes

2.2.1 *A hopping rate in a sinusoidal potential*

Let us consider a diffusion process governed by the non-Markovian Langevin equation (NMLE) (2.1.1) with the FDT (2.1.2). The periodic potential $V_p(X)$ represents the field produced by the host lattice and is assumed here to be sinusoidal of amplitude E_b,

$$V_p(X) = (E_b/2)[1 + \cos(2\pi X/a)], \tag{2.2.1}$$

with a denoting the lattice constant. In connection with diffusion in solids, the Josephson tunnelling junction (Büttiker, Harris and Landauer, 1983), dynamics of polar molecules in an electric field (Wyllie, 1980), etc., the model of Brownian motion in a periodic potential has long been studied by many

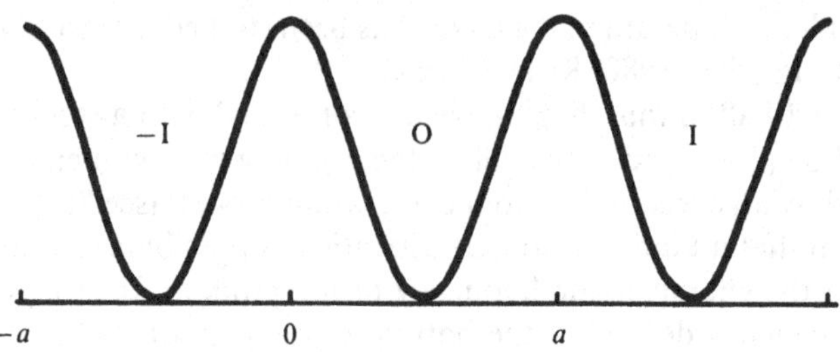

Figure 2.1. Periodic potential $V_p(X)$.

authors. However, most of the studies centre on the Markovian case where the random force $F(t)$ is white,

$$K(t) = 2\zeta\delta(t). \tag{2.2.2}$$

In this section we consider non-Markovian effects on the hopping rate from an interstitial site O to the neighbouring site −I or I (Figure 2.1). Before proceeding to our subject, we briefly survey for later convenience the theory of the thermal activation rate and give some expressions for the rate to be used for a sinusoidal potential.

In 1940 Kramers found famous formulas for the rate based on the Markovian CDM. The set of equations (2.1.1), (2.1.2) and (2.2.2) are equivalent to the Fokker–Planck equation (FPE) for the distribution function $F(X, P = M\dot{X}, t)$,

$$\partial F/\partial t + (F, P^2/2M + V_p(X)) = \zeta\partial/\partial P[(P + Mk_B T\partial/\partial P)F], \tag{2.2.3}$$

where $(\ldots,\ldots)$ denotes the Poisson bracket. For intermediate and large values of ζ, the rate Γ was given as

$$\Gamma_{M,I} = 2(\omega_0/2\pi\omega_b)\{[(\zeta/2)^2 + \omega_b^2]^{1/2} - \zeta/2\}\exp(-E_b/k_B T), \tag{2.2.4}$$

where the suffices M and I on Γ denote Markov and intermediate, respectively. The frequencies ω_0 and ω_b are associated with the second derivative of the potential $V_p(X)$ at the bottom and the top, respectively, and in our case (2.2.1)

$$\omega_0 = \omega_b = (2\pi^2 E_b/Ma^2)^{1/2}. \tag{2.2.5}$$

The factor of two in front of the right hand side (rhs) of (2.2.4) comes from the fact that we have two exits at $X = 0$ and a. If we take formally the limit $\zeta \to 0$ in (2.2.4), the transition-state result is obtained;

$$\Gamma_{TS} = (\omega_0/\pi)\exp(-E_b/k_B T). \tag{2.2.6}$$

In the case of low friction the energy E $(= M\dot{X}^2/2 + V_p(X))$ or the action $J = \oint P\,dX/2\pi$ is a slowly varying variable and Kramers (1940) derived first an FPE for J from (2.2.3),

$$\partial F(J,t)/\partial t = \frac{\partial}{\partial J}\{[\zeta J/\omega(E)](\omega(E) + k_B T\partial/\partial J)F\}, \tag{2.2.7}$$

and obtained

$$\Gamma_{M,L} = \omega_0(\zeta J_b/k_B T)\exp(-E_b/k_B T), \tag{2.2.8}$$

where $dJ/dE = 1/\omega(E)$ and $J_b = J(E = E_b)$.

Recently Kramers' results have been refined and/or generalized in various ways (for a review see Hänggi, 1986). Referring to Chapter 3 of this volume for the refinements achieved in the underdamped case, we focus on the generalization to the non-Markovian case of our main concern. The rate $\Gamma_{NM,I}$, which reduces to (2.2.4) in the Markovian limit, is obtained with a variety of methods (Grote and Hynes, 1980; Hänggi and Mojtabai, 1982, 1983; Munakata, 1985) as

$$\Gamma_{NM,I} = (s/\pi)\exp(-E_b/k_B T), \tag{2.2.9}$$

where s is the (largest) positive root of

$$s^2 + s\int_0^\infty dt\, K(t)\exp(-st) - \omega_0^2 = 0. \tag{2.2.10}$$

Equation (2.2.9) will be mentioned in Section 2.4 where we study the high temperature limit of a quantum hopping rate. For the low friction case the starting point is the non-Markovian generalization of the FPE (2.2.7),

$$\partial F/\partial t = \frac{\partial}{\partial J}\{\varepsilon(J)[\omega(E) + k_B T\partial/\partial J]F\}, \tag{2.2.11}$$

which is derived from the NMLE (2.1.1) (Carmeli and Nitzan, 1983; Grote and Hynes, 1982). The action diffusion constant $\varepsilon(J)$ is concisely expressed as

$$\varepsilon(J) = (M/\omega^2(E))\int_0^\infty dt\, K(t)\langle \dot{X}(0)\dot{X}(t)\rangle, \tag{2.2.12}$$

where the velocity correlation function is defined for the deterministic motion with $\langle\cdots\rangle$ meaning the average over the initial phase. Grote and Hynes (1982) show that $\varepsilon(J)$ is well approximated by the following equation, which we employ later in numerical calculations:

$$\varepsilon(J) = [J/\omega(E)]\int_0^\infty dt\, K(t)\cos[\omega(E)t]. \tag{2.2.13}$$

Since $J = M\oint X\, dX/2\pi = M\langle \dot{X}^2\rangle/\omega(E)$, it is readily seen from (2.2.2) and (2.2.12) that the FPE (2.2.11) becomes (2.2.7) in the Markovian limit. From the comparison of (2.2.11) with (2.2.7) the natural extension of $\Gamma_{M,L}$, (2.2.8), is given by $\Gamma^{(1)}_{NM,L} = \omega_0[\omega(E_b)\varepsilon(J_b)/k_B T]\exp(-E_b/k_B T)$. A refined rate formula, which reduces to $\Gamma^{(1)}_{NM,L}$ for $E_b/k_B T \gg 1$, is given from (2.2.11) by (Grote and Hynes, 1982)

$$\Gamma^{(2)}_{NM,L} = [(k_B T)^{-1}\int_0^{J_b} dJ\exp[E(J)/k_B T]\varepsilon^{-1}(J)\int_0^J dJ' \times\exp[-E(J')/k_B T]]^{-1} = \tau^{-1}(J_b), \tag{2.2.14}$$

where $\tau(J_b)$ is the mean time for the impurity to gain energy E_b with the reflecting wall at $E = 0$ (see, e.g., Gardiner, 1983). Recently a bridging formula, which connects $\Gamma_{NM,I}$ and $\Gamma^{(2)}_{NM,L}$, was derived by Carmeli and Nitzan (1984),

$$\Gamma_{NM,B} = 1/[\tau(J_1) + S/\Gamma_{NM,I}]. \tag{2.2.15}$$

The action $J_1 (J_b > J_1 > 0)$ marks a transition point, below which an energy-accumulation process is rate-determining while for $J > J_1$ diffusion in X-space plays a decisive role for barrier-crossing. For the expressions for J_1 and S see Carmeli and Nitzan (1984). Here we only note that, as dissipation becomes weak, J_1 goes to J_b and from $\tau(J_b) \gg S/\Gamma_{NM,I}$ we recover $\Gamma_{NM,B} = \Gamma^{(2)}_{NM,L}$, while in the opposite limit J_1 and S go, respectively, to zero and one to ensure $\Gamma_{NM,B} = \Gamma_{NM,I}$. The $\Gamma_{NM,B}$ turns out to be useful when one cannot estimate the magnitude of dissipation beforehand.

2.2.2 *Resonant activation under a non-Markovian noise*

When an internal time scale of the system of interest is shorter than or comparable to that of the surrounding heat bath, non-Markovian properties of the random force in (2.1.1) have to be fully taken into account. In some cases a spectrum of the random force $F(t)$ may have considerable weight around a characteristic frequency, to be denoted as Ω. For example, Ω may be given by the phonon frequency ω_p of optic longitudinal modes in case of impurity diffusion in solids (Section 2.3). From these considerations, it is of some interest to investigate effects of a noise with a characteristic frequency on the hopping rate (Munakata, 1986a; Munakata and Kawakatsu, 1985).

As a model simulating such a random force, we choose the process described by

$$\ddot{F} = -\Omega^2 F - \gamma\dot{F} + f(t) \quad \text{and} \quad \langle f(t)f(t')\rangle = 2\alpha\delta(t-t'). \tag{2.2.16}$$

The kernel $K(t)$ is easily calculated from (2.1.2) and (2.2.16) as

$$K(t) = (\Omega^2\zeta/\gamma)e^{-\gamma t/2}[\cos(\omega_1 t) + (\gamma/2\omega_1)\sin(\omega_1 t)] \tag{2.2.17}$$

with $\omega_1^2 \equiv (\Omega^2 - \gamma^2/4)$ and the Markovian (low-frequency) friction is defined by $\zeta \equiv \int_0^\infty \mathrm{d}t\, K(t) = \alpha/(Mk_B T\Omega^4)$.

In order to see some characteristic features of the hopping process, we show in Figure 2.2 arbitrarily chosen sample trajectories of $F(t)$, $\tilde{X}(t) = X(t) - a/2$ and the energy $E(t) = M\dot{X}^2/2 + V_p(X)$, obtained by solving the set of equations (2.1.1), (2.1.2) and (2.2.16) numerically. While $F(t)$ behaves rather irregularly with varying amplitude and frequency, $X(t)$ oscillates with an amplitude changing slowly in time. Four bumps are observed in energy variation $E(t)$ with the last one around $\omega_0 t \simeq 115$ when a barrier-crossing has occurred. A close inspection of Figure 2.2 shows that the energy $E(t)$ increases when $F(t)$ and $\dot{X}(t)$ (i.e. the gradient of the quantity plotted vary in phase. This is easily understood from the equation for energy variation $\mathrm{d}E/\mathrm{d}t = F(t)\dot{X}(t)$,

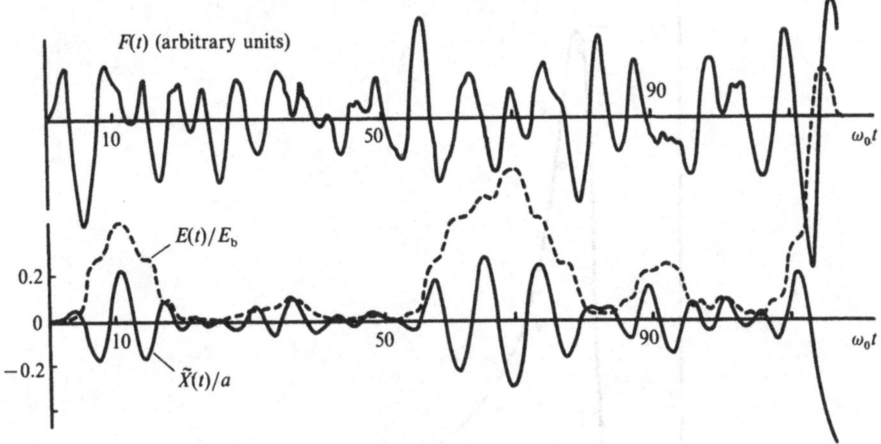

Figure 2.2. Sample trajectories of $\tilde{X}(t)$, $E(t)$ and $F(t)$ for $k_B T/E_b = 0.5$, $\zeta/\omega_0 = 0.01$, $\gamma/\omega_0 = 0.2$ and $\Omega/\omega_0 = 1.0$ (Munakata, 1986a).

derivable from (2.1.1) with the neglect of the memory term, which is expected to be unimportant for low friction and long memory. In the final stage of barrier-crossing, which starts at $\omega_0 t \simeq 106$, the impurity gains energy very efficiently from the bath mainly due to the coincidence of the phases of $F(t)$ and $\dot{X}(t)$ as explained above. The possibility that $F(t)$ and $\dot{X}(t)$ oscillate in phase for some period of time results from the fact that in the case treated here the frequency of oscillation of $X(t)$ is nearly equal to that of the noise $F(t)$ and furthermore the correlation time $2/\gamma$ of the noise is not small in comparison with the period $2\pi/\Omega$. When these conditions are met the impurity has a high probability of being resonantly activated over the barrier.

From the considerations above we infer that the rate Γ, as a function of Ω, takes its maximum around $\Omega/\omega_0 = 1$. That this is the case is shown in Figure 2.3, in which the experimental results for the rate Γ_{exp} are plotted; Γ_{exp} is obtained as follows. We put a particle at time $t = 0$ at the bottom of the potential well ($X = a/2$) and follow the particle until it escapes over one of the barriers at $X = 0$ and a at time $t = \tau$. Then the rate is calculated from the relation $\Gamma_{\text{exp}} = 1/\langle\tau\rangle$, where $\langle\tau\rangle$ is the average over 500 escape events. The resonance manifests itself as a rather sharp peak near $\Omega_r/\omega_0 = 0.85$.

We now turn to a theoretical interpretation of the results. It is seen from Figure 2.2 that the energy-accumulation process is intimately related to barrier-crossing. The efficiency of energy exchange between the system (impurity) and the bath is measured by the energy (action) diffusion constant $\varepsilon(J)$, (2.2.12), in the FPE (2.2.11). In Figure 2.4 is depicted the average $\langle\varepsilon\rangle = \int_0^{E_b} dE\ \varepsilon(J(E))/E_b$ as a function of Ω/ω_0. We see a sharp peak near Ω_r, which is well-correlated to the peak of Γ_{exp}. If we employ a slightly different definition for the average, namely the average of $\varepsilon(J)$ over the equilibrium distribution

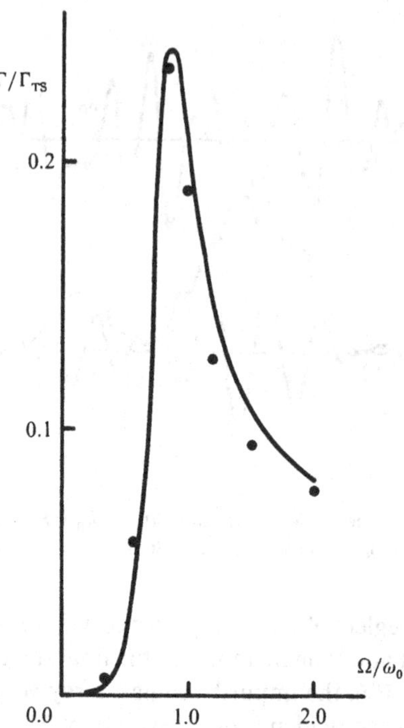

Figure 2.3. The hopping rate as a function of the noise frequency Ω for $k_B T/E_b = 0.5$, $\zeta/\omega_0 = 0.01$ and $\gamma/\omega_0 = 0.2$. The simulation result, Γ_{exp}, is shown by black circles. The curve represents the theoretical result obtained from the bridging formula (2.2.15) (Munakata, 1986a).

$\exp(-E(J)/k_B T)$, the main feature of $\langle \varepsilon \rangle$ turns out to be conserved. The difference between Ω_r and the natural frequency ω_0 results from the sinusoidal potential (2.2.1) for which $\omega(E)$ is a decreasing function of E and $\omega(E=0) = \omega_0$. Based on $\Gamma_{NM,B}$, (2.2.15), with necessary modifications of the original expression formulated for the one-exit problem, we calculated the rate numerically (Figure 2.3). It is seen that the overall features of Γ_{exp} are well reproduced by $\Gamma_{NM,B}$. For $\zeta/\omega_0 = 0.05$ with other parameter values held fixed, the rate nearly doubles and a similar (but slightly poorer) agreement between Γ_{exp} and $\Gamma_{NM,B}$ was observed. At this point, it is to be noted that the FPE (2.2.11) is valid when the following condition is satisfied (Carmeli and Nitzan, 1983):

$$\left| \int_0^\infty dt\, K(t) \exp(in\omega_0 t) \right| (n; \text{integer}) \ll \gamma \ll \omega_0. \tag{2.2.18}$$

As ζ becomes large and the left inequality is violated, the energy $E(t)$ changes rapidly compared with the random force $F(t)$, and the Markovian description in energy space such as (2.2.11) is no longer valid. In this case we should resort

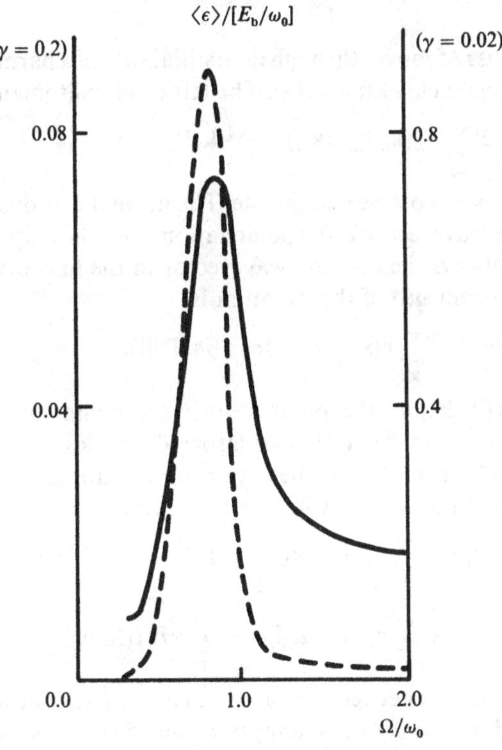

Figure 2.4. The action diffusion constant $\langle\varepsilon\rangle$ as a function of the noise frequency Ω for the case $k_B T/E_b = 0.5$ and $\zeta/\omega_0 = 0.01$. The full curve is for $\gamma/\omega_0 = 0.2$, and the broken one is for $\gamma/\omega_0 = 0.02$ (Munakata and Kawakatsu, 1985).

to $\Gamma_{NM,I}$, (2.2.9), which does not describe the resonant activation (Munakata, 1985).

We note that the resonance we found results from an internal noise, in contrast to common resonance phenomena which are caused by external forces. With regard to this point, we remark that a resonant activation in presence of both an external periodic force and a fluctuating one was recently found and discussed by Carmeli and Nitzan (1985) in connection with an activation process in Josephson junctions.

2.3 Self-diffusion and the generalized Langevin equation (GLE)

In this and the subsequent sections we consider some microscopic aspects of impurity diffusion in solids, starting from the following model.

2.3.1 Model

An impurity with mass M moves through interstitial sites of a harmonic lattice, composed of N atoms each with mass m. The lattice Hamiltonian is given by

$$H_L = (1/2)\sum_{\mathbf{k},j}\{\dot{Q}(\mathbf{k},j)\dot{Q}^*(\mathbf{k},j) + \omega^2(\mathbf{k},j)Q(\mathbf{k},j)Q^*(\mathbf{k},j)\}, \tag{2.3.1}$$

where the asterisk means complex conjugate. The normal coordinate $Q(\mathbf{k},j)$ is characterized by the wavevector $\mathbf{k}$, the polarization j and the frequency $\omega(\mathbf{k},j)$. We remark that k always denotes the wavevector in the first Brillouin zone (FBZ). The displacement $\mathbf{u}(\mathbf{l})$ of the $\mathbf{l}$th atom is

$$\mathbf{u}(\mathbf{l}) = (Nm)^{-1/2}\sum_{\mathbf{k},j}\mathbf{e}(\mathbf{k},j)Q(\mathbf{k},j)\exp(\mathrm{i}\mathbf{k}\cdot\mathbf{R}(\mathbf{l})), \tag{2.3.2}$$

where $\mathbf{e}(\mathbf{k},j)$ and $\mathbf{R}(\mathbf{l})$ denote the polarization vector and the equilibrium position of the $\mathbf{l}$th atom, respectively (for lattice dynamics, see Maradudin, Montroll, Weiss and Ipatova, 1971). The impurity Hamiltonian consists of the kinetic energy and the interaction with the host lattice, thus

$$\begin{aligned} H_\mathrm{I} &= M\dot{X}^2/2 + \sum_{\mathbf{l}} v(|\mathbf{X}-\mathbf{R}(\mathbf{l})-\mathbf{u}(\mathbf{l})|) \\ &\cong M\dot{X}^2/2 + \sum_{\mathbf{l}} v(|\mathbf{X}-\mathbf{R}(\mathbf{l})|) + \sum_{\mathbf{l}}\mathbf{u}(\mathbf{l})\cdot\mathbf{f}(\mathbf{X};\mathbf{l}), \end{aligned} \tag{2.3.3}$$

where $v(r)$ is the interaction between the impurity and a lattice atom. In (2.3.3) we have introduced a simplifying assumption that displacements of lattice atoms are small enough to allow the linear approximation. The $\mathbf{f}(\mathbf{X};\mathbf{l})$ denotes the force on the impurity exerted by the $\mathbf{l}$th atom at its equilibrium position, $\mathbf{f}(\mathbf{X};\mathbf{l}) = -\nabla_\mathbf{X} v(|\mathbf{X}-\mathbf{R}(\mathbf{l})|)$. From (2.3.2) and the Fourier transform $\tilde{v}(q) = \int \mathrm{d}\mathbf{r}\, v(r)\exp(\mathrm{i}\mathbf{q}\cdot\mathbf{r})$, $\mathbf{H}_\mathrm{I}$ is expressed as

$$H_\mathrm{I} = M\dot{\mathbf{X}}^2/2 + V_\mathrm{p}(\mathbf{X}) + \sum_{\mathbf{k},j} Q(\mathbf{k},j)F(\mathbf{X};\mathbf{k},j), \tag{2.3.4}$$

where the periodic potential $V_\mathrm{p}(\mathbf{X})$ and $F(\mathbf{X};\mathbf{k},j)$ are given by

$$V_\mathrm{p}(\mathbf{X}) = n\sum_{\mathbf{G}}\tilde{v}(\mathbf{G})\exp(-\mathrm{i}\mathbf{G}\cdot\mathbf{X}), \tag{2.3.5a}$$

$$\begin{aligned} F(\mathbf{X};\mathbf{k},j) &= n\mathrm{i}(Nm)^{-1/2}\sum_{\mathbf{G}_0}\mathbf{e}(\mathbf{k},j)\cdot[-\mathbf{k}+\mathbf{G}_0]\tilde{v}(|-\mathbf{k}+\mathbf{G}_0|) \\ &\quad\times\exp[\mathrm{i}(\mathbf{k}-\mathbf{G}_0)\cdot\mathbf{X}], \end{aligned} \tag{2.3.6a}$$

with n denoting the number density of the host lattice. While $\mathbf{G}_0$ denotes a general reciprocal lattice vector (RLV) including zero vector, $\mathbf{G}$ will be specifically used for a non-zero RLV. The total Hamiltonian of our system is thus

$$H = H_L + M\dot{\mathbf{X}}^2/2 + V_\mathrm{p}(\mathbf{X}) + \sum_{\mathbf{k},j} Q(\mathbf{k},j)F(\mathbf{X};\mathbf{k},j). \tag{2.3.7}$$

As for the interaction $v(r)$ we assume that $v(r)$ is characterized by the amplitude V_0 and the range σ. More explicitly, in numerical calculations to be given later we take the repulsive Gaussian interaction $v(r) = V_0 \exp(-r^2/\sigma^2)$. Since $v(q) \sim \exp[-(q\sigma/2)^2]$, if the range σ satisfies the relation $\pi^2\sigma^2/a^2 \gg 1$, the periodic potential $V_p(\mathbf{X})$, (2.3.5a), is approximated by

$$V_p(\mathbf{X}) = n \sum_{\mathbf{G}}' \tilde{v}(\mathbf{G}) \exp(-i\mathbf{G}\cdot\mathbf{X}), \tag{2.3.5b}$$

where the prime on $\sum$ denotes the sum over the RLVs with the smallest amplitude. Similarly, since $v(k) \gg v(|\mathbf{k}+\mathbf{G}|)$ except for the zone-boundary wavevector $\mathbf{k}$, $\mathbf{F}(\mathbf{X};\mathbf{k},j)$ is simplified considerably as

$$F(\mathbf{X};\mathbf{k},j) = -in(Nm)^{-1/2}\mathbf{e}(\mathbf{k},j)\cdot\mathbf{k}\tilde{v}(k)\exp(i\mathbf{k}\cdot\mathbf{X}) \tag{2.3.6b}$$

In Section 2.4, where the system (2.3.7) is treated quantum mechanically, we employ (2.3.5b) and (2.3.6b) from the outset. However, in this section we derive a GLE in a general framework and introduce the condition $\pi^2\sigma^2/a^2 \gg 1$ at the end of the calculations.

2.3.2 Generalized Langevin equation (GLE)

The following exact GLE holds for a set of dynamical variables $\mathbf{A}$ (Mori, 1965):

$$\dot{\mathbf{A}} = i\hat{\omega}\cdot\mathbf{A} - \int_0^t ds\, \hat{\phi}(t-s)\cdot\mathbf{A}(s) + \mathbf{F}(t). \tag{2.3.8}$$

On the basis of a physical picture that the impurity moves in a periodic potential produced by the host lattice under an action of phonons, we choose as $\mathbf{A}$

$$\mathbf{A} = (\mathbf{P}, \exp[i\mathbf{G}\cdot\mathbf{X}] - \gamma(\mathbf{G}))^\dagger \equiv (\mathbf{P}, A_{\mathbf{G}})^\dagger, \tag{2.3.9}$$

where $\gamma(\mathbf{G}) = \langle \exp(i\mathbf{G}\cdot\mathbf{X})\rangle$ and $\mathbf{A}$ is a $(3+\infty)$-dimensional column vector with $\dagger$ denoting the transpose. We note first that we are mainly interested in the GLE for $\mathbf{P} = M\dot{\mathbf{X}}$ and $A_{\mathbf{G}}(\mathbf{G}\in\text{RLV})$ included in $\mathbf{A}$ serves to extract coupling of the impurity to the periodic structure of the host lattice. Secondly, since the exact calculation of the frequency matrix $i\hat{\omega} = (\dot{\mathbf{A}}, \mathbf{A})(\mathbf{A}, \mathbf{A})^{-1}$ and the damping matrix $\hat{\phi}(t) = (\mathbf{F}(t), \mathbf{F})(\mathbf{A}, \mathbf{A})^{-1}$ with $\mathbf{F}(t) = \dot{\mathbf{A}} - i\hat{\omega}\cdot\mathbf{A} = (\mathbf{F}_p, \mathbf{F}_{\mathbf{G}})^\dagger$ is inhibitively difficult, we assume that the interaction $v(r)$ is weak compared to the coupling among the lattice atoms. This assumption is also implied in the linear approximation (2.3.3). Consequently our task is to obtain $i\hat{\omega}$ and $\hat{\phi}(t)$ to order v^2. Since the details of the calculations are given elsewhere (Munakata, 1986b*), we only give an outline of the derivation of (2.1.1) and (2.1.2).

* The erroneous factor $\frac{1}{2}$ on the rhs of (39) in this paper should be removed. Necessary corrections are given in this section.

The upper three components of $i\hat{\omega}\cdot\mathbf{A}$, which we denote by $[i\hat{\omega}\cdot\mathbf{A}]$, is given by

$$[i\hat{\omega}\cdot\mathbf{A}] = -in\sum_{\mathbf{G}}\mathbf{G}\left\{\tilde{v}(G) - (n/Nm)\sum_{\mathbf{G},\mathbf{k},j} H(\mathbf{k},j|\mathbf{G}_0,\mathbf{G}_0+\mathbf{G})\right\}$$
$$\times\exp(i\mathbf{G}\cdot\mathbf{X}) \equiv -\nabla_X \bar{V}_p(\mathbf{X}), \qquad (2.3.10)$$

where, with $\mathbf{q}_0 \equiv \mathbf{k} + \mathbf{G}_0$ and $\mathbf{q}_0' \equiv \mathbf{k} + \mathbf{G}_0'$,

$$H(\mathbf{k},j|\mathbf{G}_0,\mathbf{G}_0') = \tilde{v}(q_0)\tilde{v}(q_0')\mathbf{e}(\mathbf{k},j)\cdot(\mathbf{q}_0)\mathbf{e}(\mathbf{k},j)$$
$$\cdot(\mathbf{q}_0')/[2\omega^2(\mathbf{k},j)]. \qquad (2.3.11)$$

It is easily confirmed that (2.3.10), the systematic part of the force on the impurity, expresses the average force with the impurity fixed at the position **X**. The $\bar{V}_p(\mathbf{X})$ in (2.3.10) and $V_p(\mathbf{X})$, (2.3.5), differ by a term of order v^2.

For the damping matrix we see that

$$[\hat{\phi}(t-s)\cdot\mathbf{A}(s)] = \hat{\phi}_p(t-s)\cdot\mathbf{P}(s)$$
$$-2\sum_{G}\boldsymbol{\phi}_{\mathbf{G}}(t-s)\sin[\mathbf{G}\cdot\mathbf{X}(s)], \qquad (2.3.12)$$

where

$$\hat{\phi}_p(t) = (Mk_BT)^{-1}(\mathbf{F}_p(t),\mathbf{F}_p)$$
$$= (Mk_BT)^{-1}\sum_{\mathbf{k},j}\langle Q_0(\mathbf{k},j,t)Q(-\mathbf{k},j)\rangle_0$$
$$\times\langle\nabla_X F(\mathbf{X}_0(t);\mathbf{k},j)\nabla_X F(\mathbf{X};-\mathbf{k},j)\rangle_0. \qquad (2.3.13)$$

The first term on the rhs of (2.3.12) represents damping effects due to dynamic coupling to phonons, while the second term stands for a retarded periodic force of order v^2. In (2.3.13) the average $\langle\cdots\rangle_0$ is over the canonical distribution $z_0\exp(-\beta H_0)$ with $H_0 = H_L + M\dot{\mathbf{X}}^2/2$ and the time evolution of $Q_0(\mathbf{k},j,t)$ and $\mathbf{X}_0(t)$ is governed by H_0. Since the periodic potential $\bar{V}_p(\mathbf{X})$ contains the dominant contribution of order v, namely $V_p(\mathbf{X})$, (2.3.5), we neglect the terms of order v^2 in $\bar{V}_p(\mathbf{X})$ and the retarded periodic force in (2.3.12).

Putting the above results together we obtain the GLE,

$$d\mathbf{P}/dt = -\nabla_{X(t)}V_p(\mathbf{X}(t)) - \int_0^t ds\,\hat{\phi}_p(t-s)\cdot\mathbf{P}(s) + \mathbf{F}_p(t), \qquad (2.3.14)$$

where $\hat{\phi}_p(t)$ and the random force $\mathbf{F}_p(t)$ are related through the FDT, (2.3.13).

The results obtained above, (2.3.14) with (2.3.13), are rather formal. We now calculate the damping matrix explicitly for an isotropic Debye model with the polarization vector $\mathbf{e}(\mathbf{k},L)$ for a longitudinal mode parallel to **k**, for which transverse modes do not contribute to damping (see (2.3.6b) and (2.3.7)). We take $\omega(\mathbf{k},L) = c_L k$ for an acoustic system and $\omega(\mathbf{k},L) = \omega_p$ (plasma frequency) for an optic one. As the latter system we have in mind a one-component plasma (solid phase) (Baus and Hansen, 1980) or a simplified model of a superionic conductor (Perram, 1983; Vashishta, Mundy and Shenoy, 1979). If we further specify the interaction $v(r)$ to be the repulsive Gaussian and employ the condition $\pi^2\sigma^2/a^2 \gg 1$, we find from (2.3.13) that

$$\hat{\phi}_p(t) = [n^2/3MmN]\sum_{\mathbf{k}}[v(k)]^2\psi_{s,0}(k,t)k^4$$
$$\times \cos[\omega(\mathbf{k},L)t]\hat{I}/\omega^2(\mathbf{k},L) = K(t)\hat{I}, \qquad (2.3.15)$$

where $\psi_{s,0}(k,t) = \exp(-k_BTk^2t^2/2M)$ and $\hat{I}$ denote the self-correlation function of the ideal gas (Hansen and McDonald, 1976) and the (3×3) unit matrix, respectively. The kernel $K(t)$ defined in (2.3.15) is explicitly obtained after integration in k-space as

$$K_A(t) = (B\pi^{1/2}/8c_L^2)\exp(-c_L^2t^2/4\tau)[3\tau^{-5/2}$$
$$-3c_L^2t^2\tau^{-7/2} + c_L^4t^4\tau^{-9/2}/4], \qquad (2.3.16a)$$

$$K_O(t) = (15B\pi^{1/2}/16\omega_p^2)\tau^{-7/2}\cos[\omega_p t], \qquad (2.3.16b)$$

where the subscripts A and O refer to acoustic and optic, respectively, and $\tau = (\sigma^2 + k_BTt^2/M)/2$ and $B = \pi n V_0^2\sigma^6/(6Mm)$. Finally, if we take a simple cubic host lattice for simplicity, we obtain from (2.3.5b)

$$V_p(\mathbf{X}) = (E_b/2)[\cos(2\pi X/a) + \cos(2\pi Y/a) + \cos(2\pi Z/a)] \qquad (2.3.17)$$

with $E_b = 4n\tilde{v}(G = 2\pi/a)$.

2.3.3 Kernel and the hopping rate

In Figures 2.5 and 2.6 we show, for illustrative purposes, the memory function $K(t)$ for the case $\delta = M/m = 1/3$ and 3, respectively, with the following values for the system parameters: $a = 3\times 10^{-8}$ cm, $\sigma = a/\sqrt{3}$, $c_L = 10^5$ cm s^{-1}, $\omega_p = c_L k_{max} = 1.3\times 10^{13}$ s^{-1}, $V_0 = 6.9\times 10^{-13}$ erg and $m = 100$ amu $= 1.66\times 10^{-22}$ g, with k_{max} denoting the radius of the spherical FBZ. We note that $K_O(t)$ exhibits an oscillatory behavior for a long time compared to $K_A(t)$, which decays to zero rapidly after one overdamped oscillation. This tendency becomes more and more pronounced as the temperature becomes low and the ratio δ becomes large. The rapid decay of $K_A(t)$ results from the relation $\omega(\mathbf{k},L) = c_L k$.

The hopping rate Γ is now discussed based on the GLE (2.1.1) with the kernel (2.3.16a, b). For the acoustic case one can adopt the Markovian approximation (2.2.2) with ζ given by

$$\zeta = (\pi n V_0^2\sigma^2/6mc_L^2)(2\pi/k_BTM)^{1/2}\exp(-Mc_L^2/2k_BT). \qquad (2.3.18)$$

The important parameter which measures the strength of damping effects is $\bar{\zeta} \equiv \zeta/\omega_0$, where ω_0 is given by (2.2.5). Since we consider the interaction $v(r)$ to be weak, we regard ζ to be small and employ the Kramers' formula (2.2.8) to

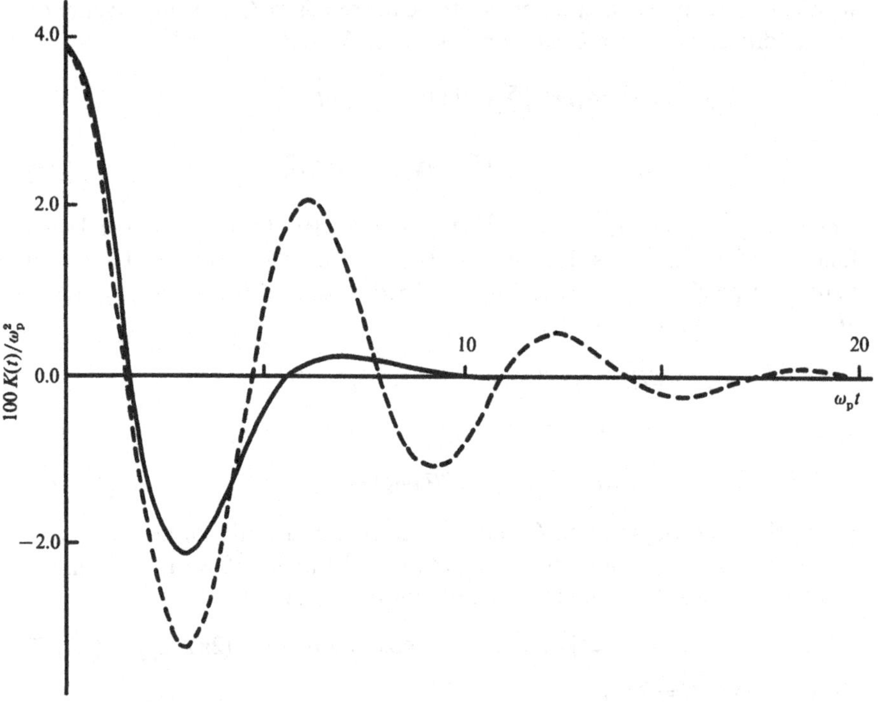

Figure 2.5. Memory function $K(t)$ for $\delta = M/m = \frac{1}{3}$. The solid and dashed curves represent $K_A(t)$ and $K_O(t)$, respectively, for $T = 100\,\mathrm{K}$. For other parameter values see the text (Munakata, 1986b).

obtain

$$\Gamma_A = (2\pi/k_B TM)^{1/2}(2E_b V_0^2 \sigma^2/3a^3 mc_L^2 k_B T) \times \exp(-E_a/k_B T), \qquad (2.3.19)$$

where we note that the activation energy $E_a = E_b + Mc_L^2/2$ has the extra contribution $Mc_L^2/2$ from dynamic coupling to the host lattice. The prefactor depends on the impurity mass as $M^{-1/2}$. This dependence is the same as the TS theory predicts. It is seen from (2.3.19) that as c_L becomes large, the rate or the diffusion constant $D(=\Gamma a^2/2)$ becomes small. Similar results were obtained by Kleppmann and Zeyher (1980) based on a mode-coupling approximation.

Next we consider the optic case, (2.3.16b). Since the kernel $K_O(t)$ generally oscillates in time, we can no longer use the approximation (2.2.2). For our underdamped situation we may apply (2.2.14) to calculate the rate. From (2.2.13) and (2.3.16b) we obtain

$$\varepsilon(J) = W\{|\omega(E) - \omega_p|^3 K_3[(\sigma^2 M/k_B T)^{1/2}|\omega(E) - \omega_p|] + (\omega_p \to -\omega_p)\}, \qquad (2.3.20)$$

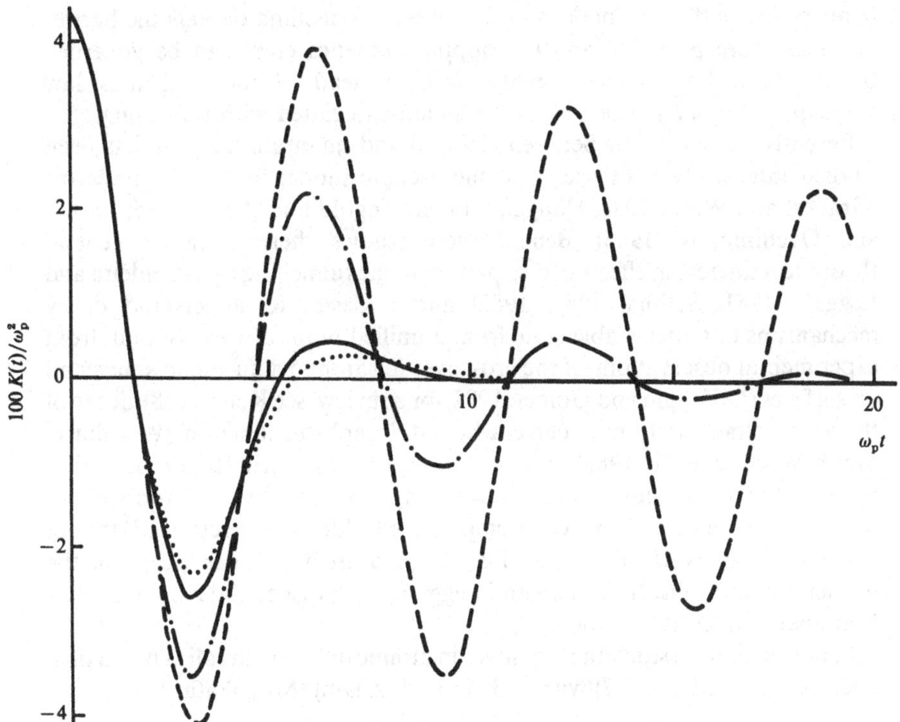

Figure 2.6. Memory function $K(t)$ for $\delta = 3$. The solid and dotted curves represent $K_A(t)$ for $T = 100$ and 1000 K, respectively. The dashed and dash-dotted curves represent $K_O(T)$ for $T = 100$ and 1000 K, respectively (Munakata, 1986b).

where $W \equiv 15\pi BJM^2/[16\sqrt{2}\Gamma(7/2)\omega(E)\omega_p^2\sigma^3(k_B T)^2]$ with B defined as previously, just after (2.3.16b). $\Gamma(x)$ and $K_3(x)$ denote, respectively, the Gamma function and the modified Bessel function of the third order (Abramowitz and Stegun, 1970). We find a sharp increase of Γ_O as δ becomes small (Munakata, 1986b). This is partly explained by the δ-dependence of $\varepsilon(J)$, (2.3.20). First we note that $K_3(x)$ decreases exponentially for large x. As δ becomes small, $\omega(E)$ becomes large and comparable in order of magnitude with ω_p, resulting in large $\varepsilon(J)$. This reminds us of the resonant activation discussed in Section 2.2, where it was investigated based on a simple stochastic model (2.1.1), (2.1.2) and (2.2.17).

2.4 Crossover from thermal to quantum hopping

So far, we have been concerned with diffusion in a high temperature region where thermal activation *over* a potential barrier dominates the hopping process and the quantum mechanical corrections can be neglected. As the temperature of the system decreases, however, tunnelling *through* the barrier becomes more probable and the hopping kinetics comes to be governed by the tunnelling rate $\Gamma \sim \exp[-S_B/\hbar]$ instead of the Arrhenius law $\Gamma \sim \exp[-E_a/k_B T]$, where S_B is the action associated with tunnelling.

Recently the crossover between classical and quantum hopping has been studied intensively with use of a microscopic model for a heat reservoir (Grabert and Weiss, 1984; Hänggi, Grabert, Ingold and Weiss, 1985; Larkin and Ovchinnikov, 1985). Behind these studies there is rather general theoretical interest in effects of dissipation on the tunnelling rate (Caldeira and Leggett, 1981; Sethna, 1981, 1982) and a desire to understand decay mechanisms of a metastable state from a unified point of view. Stimuli from experimental observations of the crossover for atomic diffusion in solids and on surfaces (DiFoggio and Gomer, 1982; for a review, see Kehr, 1978), decay of the zero-voltage state in a current-biased Josephson junction (Washburn, Webb, Voss and Faris, 1985), etc., should also be mentioned. Here we note that most of the works, investigating tunnelling and quantum diffusion under dissipative influences of reservoirs, employ the Caldeira–Leggett (1981) type of reservoir which is Markovian at high temperature and linear except for the barrier potential $V(X)$ (Caldeira and Leggett, 1983; Fisher and Zwerger, 1985; Marianer and Deutsch, 1985).

In this section we study the crossover in atomic diffusion in solids based on a microscopic model, (2.3.7) with (2.3.5b) and (2.3.6b) (Munakata, 1987).

2.4.1 The effective action $S[X]$

We introduce real phonon-variables $q_\lambda(\mathbf{k},j)(\lambda = 1, 2 k_z > 0)$ by $Q(\mathbf{k},j) = [q_1(\mathbf{k},j) + iq_2(\mathbf{k},j)]/\sqrt{2}$, $Q(-\mathbf{k},j) = [q_1(\mathbf{k},j) - iq_2(\mathbf{k},j)]/\sqrt{2}$ for convenience of path-integration. The Hamiltonian H, (2.3.7), is now expressed in terms of $q_\lambda(\mathbf{k},j)$ as

$$H = H_L + H_s + H_{int}, \tag{2.4.1}$$

where $H_L = \frac{1}{2}\sum'_{\mathbf{k},j,\lambda}[\dot{q}_\lambda^2(\mathbf{k},j) + \omega^2(\mathbf{k},j)q_\lambda^2(\mathbf{k},j)]$ with the prime on $\sum$ denoting the sum in the half of the FBZ, $k_z > 0$, and $H_s = M\dot{\mathbf{X}}^2/2 + V_p(\mathbf{X})$. The H_{int} is obtained from (2.3.6b) and (2.3.7) as

$$H_{int}(X, \{q_\lambda(\mathbf{k},j)\}) = V_1 \sum_{\mathbf{k}}{}'[q_1(\mathbf{k},L)\sin(\mathbf{k}\cdot\mathbf{X}) + q_2(\mathbf{k},L)\cos(\mathbf{k}\cdot\mathbf{X})]k\exp(-\sigma^2k^2/4) \tag{2.4.2}$$

for the isotropic Debye lattice with the Gaussian interaction $v(r) = V_0\exp(-r^2/\sigma^2)$. Here V_1 in (2.4.2) is given by $V_1 = (2\pi^3/Nm)^{1/2}V_0(\sigma/a)^3$. For a one-dimensional (1D) system, the prime in (2.4.2) means the sum in the 1D FBZ with $k > 0$ and $V_1 = (2\pi/Nm)^{1/2}V_0(\sigma/a)$ (1D).

Let us first consider a 1D system with the periodic potential $V_p(X)$ given by (2.2.1) (see Figure 2.1). The density matrix $K(X_f, \{q_\lambda(k,j)\}_f; X_i, \{q_\lambda(k,j)\}_i; \beta\hbar)$ is expressed in a path-integral form as

$$K = \int \mathrm{D}X \int \mathrm{D}\{q_\lambda(k,j)\} \exp\left(- \int_0^{\beta\hbar} \mathrm{d}uH/\hbar \right), \tag{2.4.3}$$

where $\int \mathrm{D}X$ denotes integral over the path $X(u)$ which connects two points, $X(0)=X_i$ and $X(\beta\hbar)=X_f$. Following Feynman and Hibbs (1965), we integrate over phonon-variables to obtain $K(X_f; X_i; \beta\hbar) = \int \mathrm{D}X \exp(-S[X]/\hbar)$, where the effective action $S[X]$ is

$$S[X] = \int_0^{\beta\hbar} \mathrm{d}u\, H - (V_1^2/4) \int_0^{\beta\hbar} \mathrm{d}u \int_0^{\beta\hbar} \mathrm{d}u' \sum_k{}' k^2 g(k; u-u') \times \cos[k(X(u) - X(u'))] \exp(-\sigma^2 k^2/2)/\omega(k), \tag{2.4.4}$$

with

$$g(k;u) = \mathrm{e}^{-\omega(k)|u|} + [1 + \mathrm{e}^{-\beta\hbar\omega(k)}] \cos \mathrm{h}[\omega(k)u]/\sin \mathrm{h}[\beta\hbar\omega(k)]. \tag{2.4.5}$$

The impurity is initially trapped in an interstitial site O and hops to the neighboring site $-$I or I after some time τ. As is stated below, the rate $\Gamma = 1/\langle\tau\rangle$ is related to the imaginary part of the free energy $F = -k_B T \ln Z$ with $Z = \mathrm{Trace} \exp(-H/k_B T) = \int \mathrm{d}X_0 \int \mathrm{D}X \exp(-S[X]/\hbar)$, $(X_i = X_f = X_0)$. Thus the task is to perform the path-integration to obtain the imaginary part of F, $\mathrm{Im}\, F$. For the purpose we employ the so-called bounce technique (Schulman, 1981) and a perturbation method (Grabert and Weiss, 1984).

2.4.2 *The hopping rate*

(i) The case $T > T_c$

At high temperatures $\mathrm{Im}\, F$ is dominated by the paths in the neighbourhood of the saddle points $X_s(u) = a$ and $X_s(u) = 0$. From the usual WKB procedure (Schulman, 1981) it is seen that

$$\mathrm{Im}\, F = k_B T \prod_1^\infty [\lambda_{n,\mathrm{eq}}/\lambda_{n,\mathrm{s}}] \exp(-\beta E_b), \tag{2.4.6}$$

where $\lambda_{n,\mathrm{eq(s)}} = \nu^2 n^2 \pm \omega_0^2 + \bar{\zeta}_n^{(4)} (\nu \equiv 2\pi/\beta\hbar)$ and $\bar{\zeta}_n^{(m)}$ is defined as $\bar{\zeta}_n^{(m)} = \zeta_n^{(m)} - \zeta_0^{(m)}$ with

$$\left.\begin{aligned} \zeta_n^{(m)} &= \int_0^{\beta\hbar} \mathrm{d}u\, \kappa^{(m)}(u) \exp(-\mathrm{i}\nu n u)/M \\ \kappa^{(m)}(u) &= -(V_1^2/2) \sum{}' k^m \exp(-\sigma^2 k^2/2) g(k;u)/\omega(k) \\ &\equiv \sum{}' \kappa^{(m)}(k;u). \end{aligned}\right\} \tag{2.4.7}$$

From (2.4.5) and (2.4.7) it is readily shown that $0 < \bar{\zeta}_n^{(4)} < \bar{\zeta}_m^{(4)}$ $(n < m)$ and that $\bar{\zeta}_1^{(4)}$ is expressed as $\bar{\zeta}_1^{(4)} = \nu \int_0^\infty \mathrm{d}t\, \bar{K}(t) \exp(-\nu t)$ with

$$\bar{K}(t) = (n^2/mMN) \sum [v(k)]^2 k^4 \cos[\omega(k)t]/\omega^2(k). \tag{2.4.8}$$

For the rate in the region $T > T_c$ we employ the equation $\Gamma = (2/\hbar)(\beta/\beta_c)\mathrm{Im}\,F \equiv \Gamma_H$ (Affleck, 1981; Grabert and Weiss, 1984). As T decreases $\lambda_{n,s}$ $(n = 1,\ldots)$ also decreases with $\lambda_{1,s}$, the smallest of them, taking zero value at some temperature which we denote by T_c. We note from (2.4.6) that as T approaches T_c from above Γ_H diverges as $(T - T_c)^{-1}$. This unphysical divergence of Γ_H, which will be removed later, results from our use of the WKB calculus outside the range of its validity.

We next consider the high-T limit of Γ_H. For $T \gg T_c$, it holds that $\lambda_{n,eq}/\lambda_{n,s} \cong 1$ and Γ_H $(T \gg T_c) = (2/\hbar\beta_c)\exp(-\beta E_b) = (\nu_c/\pi)\exp(-\beta E_b) = (\omega_0^2 - \bar{\zeta}_{1,c}^{(4)})^{1/2}\Gamma_{TS}/\omega_0$ (see (2.2.6) for Γ_{TS}). It is interesting to note that Γ_H $(T \gg T_c)$ is obtained also from the following non-Markovian GLE

$$M\ddot{X} = M\int_0^t \mathrm{d}s\,\bar{K}(t-s)\dot{X}(s) - \mathrm{d}V_p(X)/\mathrm{d}X + F(t) \tag{2.4.9}$$

together with the rate formula (2.2.9), $\Gamma_{NM,I}$. Here $\bar{K}(t)$ is the kernel defined by (2.4.8). Comparing the kernel $\bar{K}(t)$, derived from path-integration, with the 1D version of the kernel $K(t)$, (2.3.15), derived from the projection operator method, we notice that $\bar{K}(t)$ coincides with $K(t)$ if we set $\psi_{s,0}(k, t)$ in (2.3.15) equal to one. It is to be noted that the equation-of-motion approach by Cortes, Wert and Lindenberg (1985) also gives the kernel $\bar{K}(t)$.

(*ii*) *The case* $T < T_c$

At low temperatures below T_c we have two bounce solutions $X_B(u)$ to the equation of motion $\delta S[X_B]/\delta X_B(u) = 0$. We concentrate on the contribution from the paths near $X = 0$ and double it for the rate. Introducing the fluctuation $y(u)$ by $X(u) = X_B(u) + y(u)$, we expand $S[X]$ as follows:

$$\begin{aligned} S[X] &= S[X_B] + (1/2)\delta^2 S[y] + (1/3!)\delta^3 S[y] \\ &\quad + (1/4!)\delta^4 S[y] + \cdots = S_B + \Delta S. \end{aligned} \tag{2.4.10}$$

Terms of order y^3 and higher, usually neglected in the WKB calculus, are retained for latter convenience.

$X_B(u)$ is determined from

$$\begin{aligned} &M\ddot{X}_B + (E_b\pi/a)\sin(2\pi X_B/a) \\ &\quad + \sum{}' \int_0^{\beta\hbar} \mathrm{d}u'\,\kappa_3(k; u - u')\sin\{k[X_B(u) - X_B(u')]\} = 0. \end{aligned} \tag{2.4.11}$$

Since we are mainly interested in crossover behavior near T_c, we can develop a perturbation theory with $\varepsilon = (T_c - T)/T_c$ as a smallness parameter. For $X_B(u) = \Sigma X_n \exp(\mathrm{i}\nu n u)$ $(X_n = X_{-n})$ we find

$$X_1^2 = -\lambda_{1,s}/B, \quad B \equiv (2\pi^2/a^2)\omega_0^2 + [\bar{\zeta}_2^{(6)}/2 - 2\bar{\zeta}_1^{(6)}] \tag{2.4.12}$$

and the other Fourier coefficients turn out to be higher order in ε. We will assume that B is positive near T_c. From (2.4.4), (2.4.5) and (2.4.12),

$$S_B = S[X_B] = E_b\beta\hbar - M\beta\hbar|\lambda_{1,s}|^2/(2B). \tag{2.4.13}$$

The $\delta^2 S[y]/2$ is obtained from $y(u) = \sum Y_n \exp(i\nu nu)$ as

$$\delta^2 S[y]/2 = (M\beta\hbar/2)\left[\sum_{n\neq\pm1} \lambda_{n,s}|Y_n|^2 + BX_1^2(Y_1 + Y_{-1})^2\right], \tag{2.4.14}$$

yielding after integration over all $\{Y_n\}$

$$\Gamma = (2/\hbar)\mathrm{Im}\,F = \lambda_{1,\mathrm{eq}} \prod_{n=2}^{\infty} (\lambda_{n,\mathrm{eq}}/\lambda_{n,s})(8\pi M/\beta\hbar^2 B)$$
$$\times \exp(-S_B/\hbar) \equiv \Gamma_L. \tag{2.4.15}$$

(*iii*) *The case* $T \cong T_c$

We briefly look back on our treatment of the case $T < T_c$ in (ii). The bounce solution (2.4.12) of (2.4.11) was obtained and the magnitude of the fluctuation Y_1 about X_1 is measured by $\langle |Y_1|^2 \rangle = 1/(M\beta|\lambda_{1,s}|)$. Since $X_1 = 0$ (i.e. $X_B = 0$) gives an extra stationary point of $S[X]$, we cannot use the method of steepest descent (or method of Laplace) if $\langle |Y_1|^2 \rangle > X_1^2$, or if the ratio $r \equiv |\lambda_{1,s}|/(B/M\beta)^{1/2}$ is smaller than one. In this region we have to take effects of higher (than second) order terms in (2.4.10) into account. Following Grabert and Weiss (1984), we include terms up to order y^4 in (2.4.10) to obtain

$$\Delta S = (M\beta\hbar/2)\left[\sum_{n\neq\pm1} \lambda_{n,s}|Y_n|^2 + B(X_1(Y_1 + Y_{-1}) + Y_1 Y_{-1})^2\right], \tag{2.4.16}$$

and after integration over fluctuations we see that

$$\Gamma = \Gamma_L \,\mathrm{erfc}\,[(M\beta B/2)^{1/2} X_1^2]/2 \equiv \Gamma_{\mathrm{int}}, \tag{2.4.17}$$

where $\mathrm{erfc}(x) = 2\pi^{-1/2} \int_{-\infty}^{x} dt \exp(-t^2)$. It is easily seen that Γ_{int} bridges between Γ_H and Γ_L smoothly.

For a 3D system with a periodic potential (2.3.17), the rate can be calculated in a similar way to the 1D case. Only the points to be altered are given here. First we multiply Γ given above by three since three times more paths are available for hopping. Secondly k^4 and k^6 in the summands of $\kappa^{(4)}(u)$ and $\kappa^{(6)}(u)$ in (2.4.7) should be replaced by $k^2k_x^2$ and $k^2k_x^4$, respectively.

In Figure 2.7 we show temperature (T)-dependence of the rate, calculated from Γ_H, Γ_L and Γ_{int} for a 3D acoustic system with parameter values given in Section 2.3.3. Γ_{int} is used when the ratio r, defined above, is smaller than one. It is observed that high-T behaviour is well described by an Arrhenius law with obvious deviation therefrom at low T, $T < T_c$.

The impurity-mass dependence of T_c is most roughly estimated as follows: at extremely low and high temperature it holds that $\Gamma_L \sim \exp(-S_B/\hbar)$

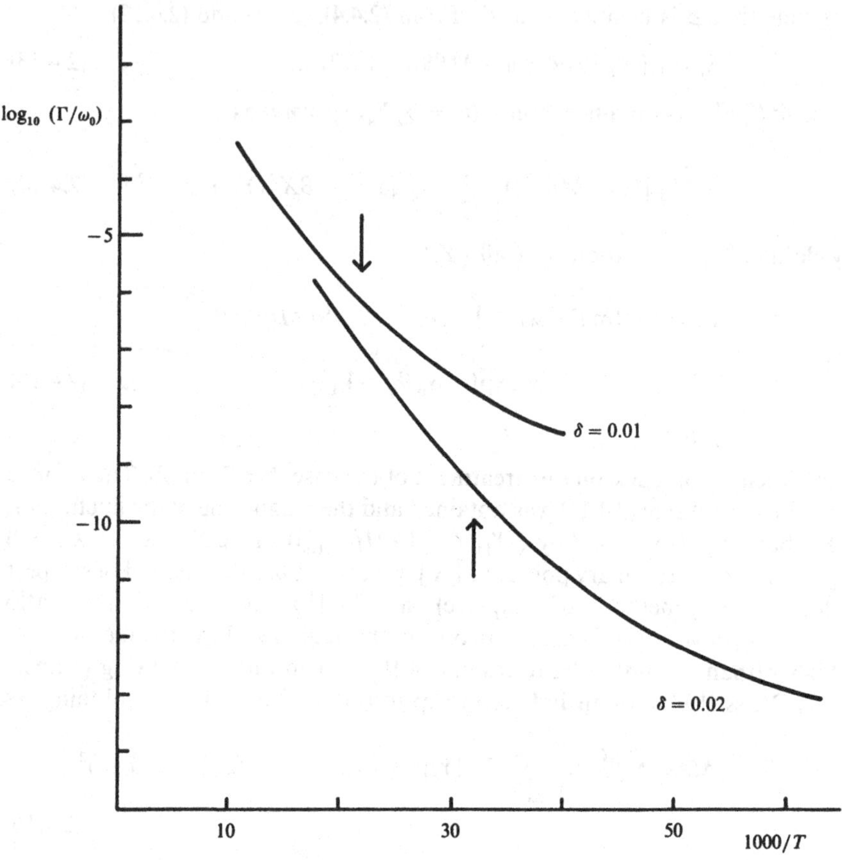

Figure 2.7. Hopping rate as a function of 1000/*T*. The arrows indicate the crossover temperature T_c, and δ denotes the mass ratio M/m.

and $\Gamma_H \sim \exp(-E_a/k_B T)$, respectively. If we equate them and note $S_B = \int dX|P| \propto M^{1/2}$, the relation $T_c \propto M^{-1/2}$ follows. Such dependence is obtained also based on the results of Section 2.4.2 (Munakata, 1987).

2.5 Summary and remarks

In this chapter elementary processes associated with atomic diffusion in solids are considered with emphasis put on non-Markovian effects on the rate Γ. As stated in Section 2.2.1 considerable progresses have been achieved recently for the rate of non-Markovian processes, and in Section 2.2.2 we utilized the outcome to study effects of noise with a characteristic frequency. The motivation behind this study is given in Section 2.3 where the CDM, (2.1.1) and (2.1.2), was *derived* microscopically with the aid of the projection operator formalism. In Section 2.4 we lowered the temperature of the system with the

Hamiltonian (2.3.7) and studied quantum aspects of impurity diffusion. By taking the high-T limit of our results, we again encountered the Langevin equation (2.4.9) and the kernel (2.4.8). As a byproduct the crossover temperature T_c was related to the activation rate of a non-Markovian process, (2.4.9), in a similar way as was found by Grabert and Weiss (1984), Hänggi *et al.* (1985) and Wolynes (1981). Here we make the following remark: in spite of the fact that no assumption was introduced explicitly as to the magnitude of the damping (friction) effect, $\Gamma_H(T \gg T_c)$ was related to $\Gamma_{NM,I}$, (2.2.9), which is to be used for the case of relatively large (effective) friction. It is supposed that the relation is caused by our (quasi-stationary) assumption of thermal equilibrium prevailing throughout the interior of the cell O (Figure 2.1), thus neglecting effects of finite relaxation time to the equilibrium distribution (Waxman and Leggett, 1985). Quantum treatment of the hopping rate in case of low friction (both Markovian and non-Markovian) seems to remain as a challenging problem.

A final remark is concerned with the relation between Γ and the diffusion constant D. For discussion of D we invoked a simple (and approximate) relation between Γ and D, $D(\Gamma) = \Gamma a^2/2$. To be exact one could start from the time-correlation-function (TCF) expression (Hansen and McDonald, 1976) $D(\mathrm{TCF}) = \int_0^\infty \mathrm{d}t \langle \dot{X}(t)\dot{X} \rangle$ (Dieterich, Fulde and Peschel, 1980; Risken, 1984). It seems that the relation between $D(\Gamma)$ and $D(\mathrm{TCF})$ is not clearly settled. At present we know that as the effective friction becomes large $D(\Gamma)$ approaches $D(\mathrm{TCF})$ (A. Igarashi and T. Munakata, manuscript in preparation) and that when the friction goes to zero, $D(\mathrm{TCF})$ diverges and $D(\Gamma)$ vanishes.

Acknowledgements

The author expresses his sincere gratitude to Professor A. Ueda for useful discussions and comments.

References

Abramowitz, H. and Stegun, I. A. 1970. *Handbook of Mathematical Functions.* New York: Dover Publications.
Affleck, I. 1981. *Phys. Rev. Lett.* **46**, 388–91.
Baus, M. and Hansen, J. P. 1980. *Phys. Rep.* **59**, 1–94.
Büttiker, M., Harris, E. P. and Landauer, R. 1983. **28**, 1268–75.
Caldeira, A. O. and Leggett, A. J. 1981. *Phys. Rev. Lett.* **46**, 211–14.
Caldeira, A. O. and Leggett, A. J. 1983. *Physica* **121A**, 587–616.
Carmeli, B. and Nitzan, A. 1983. *J, Chem. Phys.* **79**, 393–404.
Carmeli, B. and Nitzan, A. 1984. *Phys. Rev. A* **29**, 1481–95.
Carmeli, B. and Nitzan, A. 1985. *Phys. Rev. A* **32**, 2439–54.
Cortes, E., Wert, B. J. and Lindenberg, K. 1985. *J. Chem. Phys.* **82**, 2708–17.
Dieterich, W., Fulde, P. and Peschel, I. 1980. *Adv. Phys.* **29**, 527–605.
DiFoggio, R. and Gomer, R. 1982. *Phys. Rev. B* **25**, 3490–511.
Feynman, R. P. and Hibbs, A. R. 1965. *Quantum Mechanics and Path Integral.* New York: McGraw-Hill.

Fisher, M. P. A. and Zwerger, W. 1985. *Phys. Rev. B* **32**, 6190–206.
Flynn, C. P. 1968. *Phys. Rev.* **171**, 682–98.
Flynn, C. P. 1972. *Point Defects and Diffusion.* Oxford: Clarendon Press.
Fulde, P., Pietronero, L., Schneider, W. R. and Strassler, S. 1975. *Phys. Rev. Lett.* **35**, 1776–9.
Gardiner, C. W. 1983. *Handbook of Stochastic Methods.* Berlin: Springer.
Grabert, H. and Weiss, U. 1984. *Phys. Rev. Lett.* **53**, 1787–90.
Grote, R. F. and Hynes, J. T. 1980. *J. Chem. Phys.* **73**, 2715–32.
Grote, R. F. and Hynes, J. T. 1982. *J. Chem. Phys.* **77**, 3736–43.
Hänggi, P. 1986. *J. Stat. Phys.* **42**, 105–48.
Hänggi, P. and Mojtabai, F. 1982. *Phys. Rev. A* **26**, 1168–70.
Hänggi, P. and Mojtabai, F. 1983. *J. Stat. Phys.* **30**, 401–12.
Hänggi, P., Grabert, H., Ingold, G. L. and Weiss, U. 1985. *Phys. Rev. Lett.* **55**, 761–4.
Hansen, J. P. and McDonald, I. A. 1976. *Theory of Simple Liquids.* New York: Academic Press.
Kehr, K. W. 1978. In *Hydrogen in Metals* (G. Alefeld and J. Volkl, eds.), vol. 1, chapter 8. Berlin: Springer Verlag.
Kleppmann, W. G. and Zeyher, R. 1980. *Phys. Rev. B* **22**, 6044–64.
Kramers, H. A. 1940. *Physica* **7**, 284–304.
Larkin, A. L. and Ovchinnikov, Y. N. 1985. *J. Stat. Phys.* **41**, 425–43.
Maradudin, A. A., Montroll, E. W., Weiss, C. H. and Ipatova, I. P. 1971. *Theory of Lattice Dynamics in the Linear Approximation.* New York: Academic Press.
Marianer, S. and Deutsch, J. M. 1985. *Phys. Rev. B* **31**, 7478–81.
Mori, H. 1965. *Prog. Theor. Phys.* **33**, 423–55.
Munakata, T. 1985. *Prog. Theor. Phys.* **73**, 826–9.
Munakata, T. 1986a. *Prog. Theor. Phys.* **75**, 747–50.
Munakata, T. 1986b. *Phys. Rev. B* **33**, 8016–26.
Munakata, T. 1987. *Prog. Theor. Phys.* **77**, 1–6.
Munakata, T. and Kawakatsu, T. 1985. *Prog. Theor. Phys.* **74**, 262–71.
Murch, G. E. and Nowick, A. S. (eds.) 1984. *Diffusion in Crystalline Solids.* New York: Academic Press.
Perram, J. W. 1983. *The Physics of Superionic Conductors and Electrode Materials.* New York: Plenum.
Risken, H. 1984. *The Fokker–Planck Equation.* Berlin: Springer.
Schulman, L. S. 1981. *Techniques and Applications of Path Integration.* New York: John Wiley.
Sethna, J. P. 1981. *Phys. Rev. B* **24**, 698–713.
Sethna, J. P. 1982. *Phys. Rev. B.* **25**, 5050–63.
Vashishta, P., Mundy, J. N. and Shenoy, G. K. 1979. *Fast Ion Transport in Solids.* Amsterdam: North-Holland.
Vineyard, G. H. 1957. *J. Phys. Chem. Solids* **3**, 121–7.
Washburn, S., Webb, R. A., Voss, R. F. and Faris, S. H. 1985. *Phys. Rev. Lett.* **54**, 2712–15.
Waxman, D. and Leggett, A. J. 1985. *Phys. Rev. B* **32**, 4450–68.
Wolynes, P. G. 1981. *Phys. Rev. Lett.* **47**, 968–71.
Wyllie, G. 1980. *Phys. Rep.* **61**, 327–76.

3 Escape from the underdamped potential well

M. BÜTTIKER

3.1 Introduction

In this chapter we discuss a modest refinement of Kramers (1940) result for the noise activated escape out of an underdamped potential well. At very low damping, motion in a potential well (see Figure 3.1) is almost conservative, and the escape up in energy, out of the well, becomes difficult. This was appreciated by Kramers, who derived a result for extremely low damping. For a particle subject to thermal noise and damping proportional to the momentum p with a relaxation rate η, Kramers found an escape rate from the metastable state which is proportional to η. In the low friction limit it is diffusion up along the energy coordinate which limits the escape rate. On the other hand for large friction, the escape out of the potential well is limited by the diffusion along the reaction coordinate. For large friction the escape rate is proportional to η^{-1}. Thus the escape rate as a function of the friction constant peaks at an intermediate value of the damping constant as shown in Figure 3.2. Kramers noted in his paper that it is the smaller of the two rates which applies. At very low damping, as the friction increases, the escape rate must eventually deviate from the linear behavior predicted by Kramers to merge at large η into the heavy damping behavior. It is this departure from the linear behavior for small damping which is the subject of this chapter. The refinement which we present, based on a simple approach due to Büttiker, Harris and Landauer (1983), clearly points to the essential physical facts. We find that the first correction term to the linear behavior found by Kramers is proportional to $\eta^{3/2}$, both for the escape out of a single well (Figure 3.1) and for the hopping rate in a double well potential (Büttiker, 1984), in agreement with Risken and Voigtlaender (1985). The results of Risken and Voigtlaender are based on a genuinely two dimensional solution for the distribution function (see chapter 6, Volume 1). Despite this, we emphasize here the earlier approach of Büttiker, Harris and Landauer (1983), which provides physical insight, and as a consequence permits a more ready appraisal of other approaches. It is furthermore easy to generalize to differing physical situations (Hänggi and Weiss, 1984; Rips and Jortner, 1986). To further elucidate the physics we study

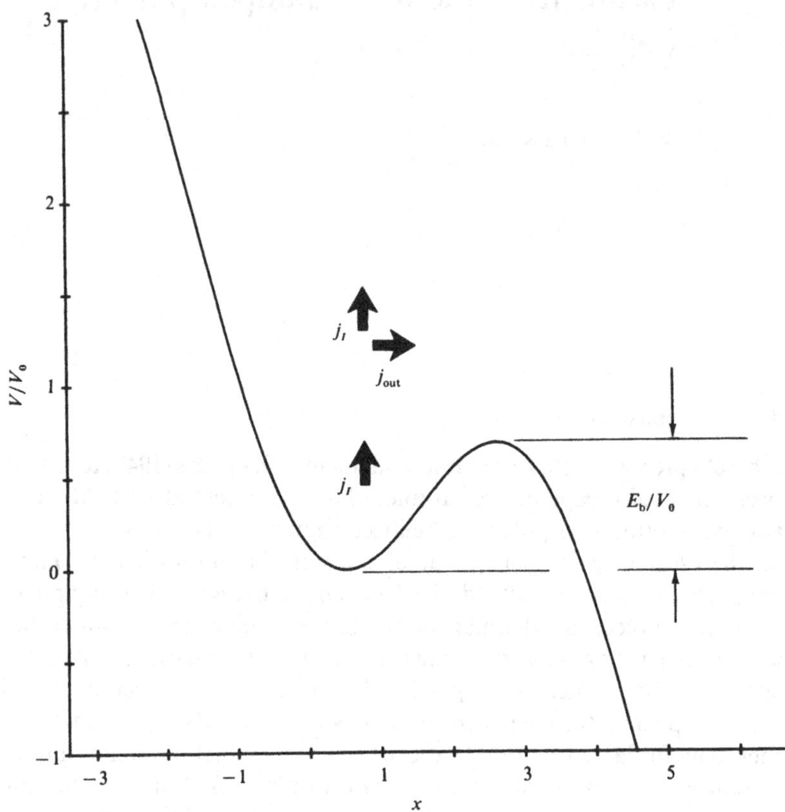

Figure 3.1. Metastable state of the potential $V(x) = V_0(1 - \cos(x)) - Fx$. In the strongly underdamped case, there is a diffusive flux j_I up along the energy coordinate. For energies above the barrier peak energy E_b we have, in addition to this vertical current, a horizontal current j_{out}, giving the flux out of the well. After Büttiker, Harris and Landauer (1983).

the energy distribution of the particles as they escape over the potential barrier (Büttiker and Landauer, 1984a, b). The low friction result of Kramers results in an average escape energy which is zero. For moderate and large friction the average escape energy at the barrier is of the order of kT. Our approach interpolates between these two results and finds an average escape energy which increases proportional to $\eta^{1/2}$ for low friction and tends to kT as the friction becomes large.

The motivation of BHL (abbreviation for Büttiker, Harris and Landauer, 1983) for a refinement of the Kramers low damping result was twofold. First, especially among physicists, it seemed that Kramers' original discussion of the extremely underdamped case was often not appreciated. As discussed by Landauer (Chapter 1, Volume 1) some of the most important treatments of the noise activated escape from the metastable state do not reflect Kramers'

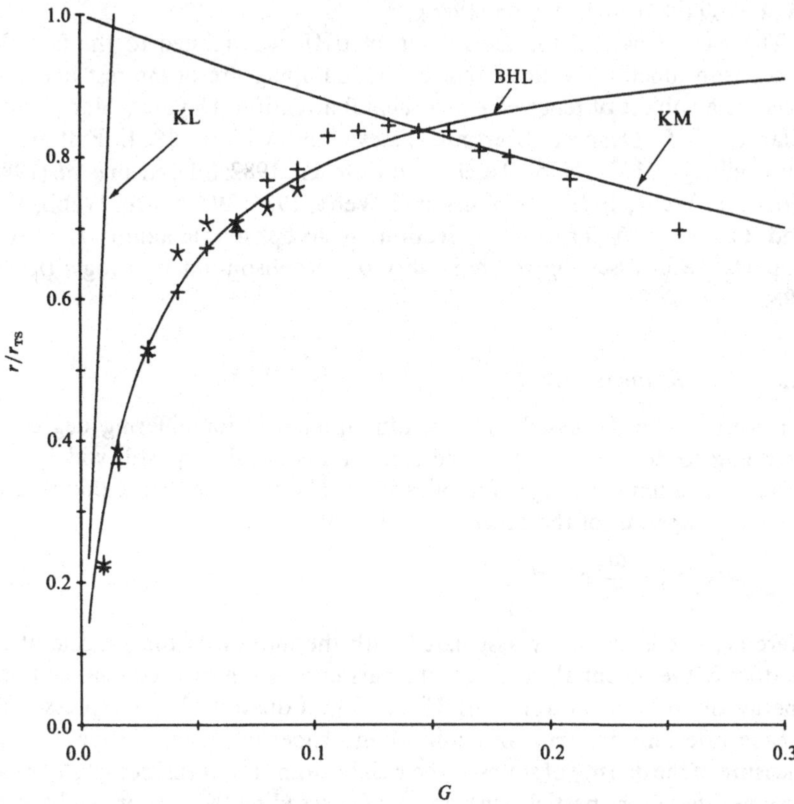

Figure 3.2. Escape rate r, in units of the transition-state theory result r_{TS}, as a function of the dimensionless damping constant $G = \eta(m/V_0)^{1/2}$. The solid lines are Kramers' moderate damping result (KM), Kramers' low damping result (KL), and the refined low damping result of Büttiker, Harris and Landauer (1983) (BHL). These results are compared with computer simulations using two different techniques (+, *). After Büttiker, Harris and Landauer (1983).

understanding of the very lightly damped case. There are a number of notable exceptions to this situation. Landauer and Swanson (1961), in a paper extending Kramers' discussion of the heavy damping limit to a multidimensional situation, also discuss the underdamped limit. Kramers' approach to the extremely underdamped potential well was applied to Josephson junctions by Lee (1971). An attempt to interpolate between the extremely underdamped case and the strongly damped case was made by Visscher (1976). The papers of Iche and Noizieres (1976) and Leuthäusser (1981) focus on the desorption of adatoms in a lightly damped well. As emphasized, our criticism primarily addresses the physicists. Chemistry has had a different history in this subject and we cite here only the work of Grote and Hynes (1982) and Larson and Kostin (1980) and refer the reader to the review articles by Frauenfelder and

Wolynes (1985), and Hänggi (1986).

The second part of the motivation of BHL was related to the fact that Josephson junctions with extremely low damping are of interest and have become a subject of renewed experimental attention (Devoret, Martinis and Clarke, 1985; Devoret, Martinis, Esteve and Clarke, 1984; Fulton and Dunkelberger, 1974; Naor, Tesche and Ketchen, 1982; Silvestrini *et al.* (1988); Voss and Webb, 1981; Washburn and Webb, 1986; Washburn, Webb, Voss and Farris, 1985. Thermal activation in Josephson junctions is also of importance in assessing the reliability of Josephson-junction logic (Raver, 1982).

3.2 Kramers' results

Kramers (1940) discussed three results applicable for differing degrees of damping to demonstrate that the transition-state theory, still widely used today, has a limited range of applicability. The transition-state theory gives for the escape out of the potential well a rate

$$r_{\mathrm{TS}} = \frac{\omega_A}{2\pi} \mathrm{e}^{-E_b/kT}. \tag{3.2.1}$$

Here ω_A is the frequency associated with the motion of the particle at the bottom of the potential well, E_b is the barrier height measured relative to the energy of the well bottom (see Figure 3.1). Equation (3.2.1) expresses the escape rate only in terms of equilibrium properties of the system. ω_A is a measure of the density of states at the well bottom. The transition-state theory assumes that every particle with an energy exceeding the barrier height and a velocity at the barrier peak describing motion away from the well leaves the well.

In the limit of heavy damping, as Kramers has pointed out, the particle in the well exhibits highly diffusive behavior. A density gradient develops across the barrier peak. As a consequence, the escape rate in the presence of heavy damping is smaller than that predicted by (3.2.1) and is given by

$$r_{\mathrm{KH}} = \frac{|\omega_B|}{\eta} r_{\mathrm{TS}}. \tag{3.2.2}$$

Here, ω_B is the imaginary frequency associated with the unstable potential curvature at the barrier. Equation (3.2.2) applies, according to Kramers, for $\eta \gg |\omega_B|$. In the heavy damping limit, relaxation of the energy of the particle is so fast that the velocity distribution of the particles is always given by the Maxwell distribution $\exp(-mv^2/2kT)$ even for particles near the barrier peak. The behavior of the distribution at the barrier can be described by a density gradient in x alone. For moderate damping this is not the case, and inertial effects become important. The velocity distribution of carriers near the barrier peak is no longer Maxwellian. As shown in Figure 3.3 we approach thermal

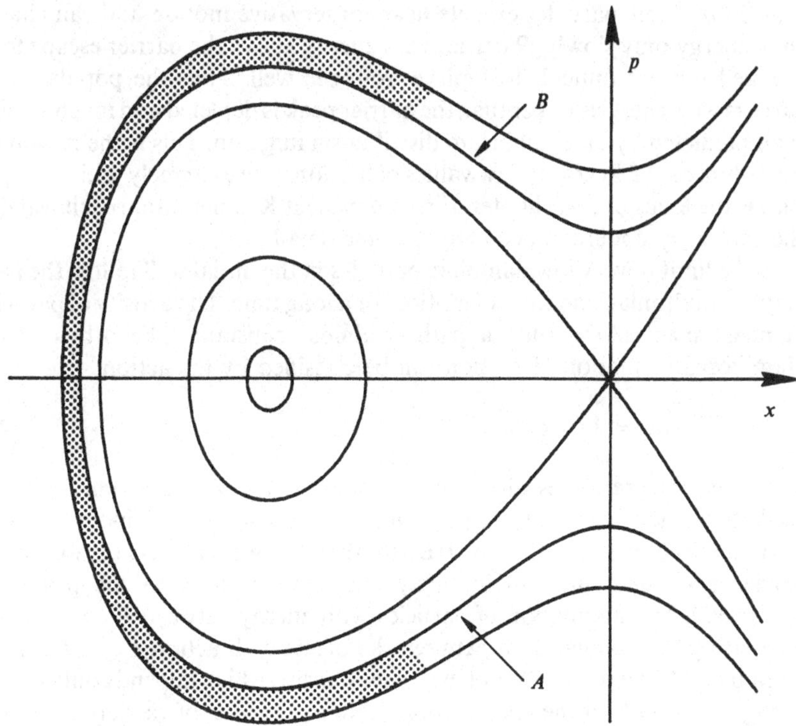

Figure 3.3. Phase plane (p, x) with constant energy contours. The shaded region shows the range over which Kramers' discussion of the moderate damping rate assumes thermal equilibrium. In the underdamped regime no particles enter the initial well at A. The population in the annulus is built up by equilibration with the filled well at lower energies as we move along the annulus toward point B. After Büttiker and Landauer (1984a).

equilibrium only as we move away from the potential well. Kramers presented an approach which takes such inertial effects into account and found

$$r_{\mathrm{KM}} = \frac{1}{|\omega_B|}\left[\left(\frac{1}{4}\eta^2 + |\omega_B|^2\right)^{1/2} - \frac{\eta}{2}\right] r_{\mathrm{TS}}. \tag{3.2.3}$$

For large damping, $\eta \gg |\omega_B|$, (3.2.3) reduces to (3.2.2). For weak damping (3.2.3) yields the transition-state result (3.2.1). Chandrasekhar (1943), in his otherwise authoritative review paper, claims that (3.2.3) is valid for all degrees of damping. This is astonishing since Kramers presents in his paper a third result which is valid for extremely low damping and differs from (3.2.3).

The derivation of (3.2.3) only takes into account the difficulty involved in crossing the barrier, and assumes that the population, even for energies above the barrier, is an equilibrium distribution away from the barrier (see Figure 3.3). The real situation, in the limit of small friction, is different from that. In the well, particles execute near conservative motion and can change their energy only slowly. Particles with energy above the barrier escape from the well but are difficult to replace from the well. Thus the population of carriers with energies larger than the barrier peak is depleted and much smaller than predicted by an equilibrium distribution function. This is the reason for the failure of (3.2.3) at very low values of friction. The extremely underdamped case is the focus of this chapter. First we present Kramers' approach, valid in the extremely underdamped case, in some detail.

In the limit of very low damping, particles in the initial well follow the path of the Hamiltonian equation of motion for a long time. The density of particles in phase-space along such a path is almost constant. The orbits of the Hamiltonian equation of motion can be classified by the action

$$I(E) = \int p\,\mathrm{d}x, \tag{3.2.4}$$

where the integration is along the orbit. Near the well bottom, for small oscillations, the action is proportional to the energy and increases with increasing energy to a value $I_b \equiv I(E_b)$ for the orbit which enters the well at the barrier peak and after an infinitely long time returns to the peak (see Figure 3.3). The population of particles with energy between E and $E + \mathrm{d}E$, populating the annular area between the orbits with action I and $I + \mathrm{d}I$, is given by $\rho \mathrm{d}I$, where $\rho(E)$ is a phase-space density which depends only on the energy E but not on the spatial coordinate x. The flux of carriers along the action coordinate,

$$j_I = -\eta I(\rho + kT\partial\rho/\partial E) \tag{3.2.5}$$

has two terms. The first term describes the relaxation in energy due to the friction η. A carrier, completing a revolution, loses on average, due to friction, an energy ηI. The second term is a consequence of the thermal fluctuations and describes the diffusive process in energy which permits carriers to move away from equilibrium.

In the absence of a current, (3.2.5) yields the thermal equilibrium distribution

$$\rho_{\mathrm{eq}}(E) = \rho_0 \mathrm{e}^{-E/kT}. \tag{3.2.6}$$

The normalization constant is determined by $\int \rho_{\mathrm{eq}}(E)\,\mathrm{d}x\,\mathrm{d}p = N$, where N is the total number of particles in the well. This yields $\rho_0 = \omega_A N/2\pi kT$.

The simplest way to find the escape rate is to consider a steady-state problem. We imagine that we feed a current j_I into the bottom of the valley, replacing the carriers which escape over the barrier. We then have to find a solution of (3.2.5) for a steady-state current j_I. The distribution function $\rho(E)$ describing transport deviates most strongly from the equilibrium solution $\rho_{\mathrm{eq}}(E)$, where the equilibrium solution is small. It is convenient to express the transport solution in the form

$$\rho(E) = \beta(E)\rho_{\mathrm{eq}}(E), \tag{3.2.7}$$

with a correction factor $\beta(E)$. Thus β is almost constant and equal to unity except near the barrier peak where the equilibrium distribution is small and a strong variation of β is needed to maintain the current j_I. Using (3.2.7) in (3.2.5) yields

$$j_I = -\eta I k T \rho_{eq}(E)\partial\beta/\partial E. \tag{3.2.8}$$

In the steady-state j_I is independent of E and integration of (3.2.8) from $E = E_1 \simeq kT$ to $E = E_b$ yields

$$j_I = \eta k T \frac{\beta(E_1) - \beta(E_b)}{\int_{E_1}^{E_b} (1/I)(1/\rho_0)e^{E/kT}\,dE}. \tag{3.2.9}$$

The lower limit of (3.2.9) must be taken with care. At the bottom of the well the action I is proportional to E. The integral in (3.2.9) diverges if we take $E = 0$ to be the lower limit. This divergence is compensated by a divergence of β at low energies. If the current j_I is fed into the well at the very bottom the correction factor diverges logarithmically. These complications can be avoided if we assume that the current j_I is not fed into the well at the very bottom but at an energy $E_1 \simeq kT$ above the barrier bottom. In fact the exact energy at which we inject the current j_I is unimportant as long as this is done several kT below the barrier peak energy. Thus we can safely take $\beta(E_1) = 1$. The integral in (3.2.9) is then controlled by the upper limit and we find

$$j_I = \eta I_b(1 - \beta(E_b))\rho_0 e^{-E_b/kT}. \tag{3.2.10}$$

Here I_b is the action of the orbit at the barrier peak. This describes a particle which starts with zero velocity at the barrier peak, makes an excursion into the well, and returns to the barrier peak approaching the starting point with zero velocity.

Kramers (1940) now makes the additional assumption that every particle with energy above the barrier leaves the well. He assumes that the population at these energies is completely depleted setting $\rho(E_b) = 0$. Using (3.2.7) this implies that $\beta(E_b) = 0$, and hence for low friction he finds from (3.2.10) an escape rate

$$r_{KL} = j_I/N = (\eta I_b/kT)r_{TS}. \tag{3.2.11}$$

Kramers' result for low friction, (3.2.11), is shown in Figure 3.2 labeled KL together with the result for moderate friction, (3.2.3), labeled KM. In the next section we present an extension of Kramers' result (3.2.11), valid for extremely low friction, to a wider range of friction.

3.3 Refined low damping result

The key assumption of Kramers giving rise to (3.2.11) is the complete depletion of the particle population for energies above E_b. Clearly as the friction increases this cannot be correct. A small population of carriers, much smaller than the equilibrium population assumed in the derivation of (3.2.3), will exist. Due to thermal noise, carriers at an energy $E \leqslant E_b$ in one round trip can gain an energy ΔE determined by the diffusion constant of (3.2.5), $\langle(\Delta E)^2\rangle \propto 2\eta I_b kT$. Carriers in the potential well populate an energy range up to $E_b + \Delta E$. Thus for $E > E_b$ we still have a flux along the action coordinate given by (3.2.5). Due to the narrow range of energies under consideration we can take the action in (3.2.5) to be independent of energy and equal to I_b. However, for energies above E_b, carriers which reach the barrier peak with a positive velocity will escape. Thus in addition to the vertical flux j_I we also have a horizontal flux j_{out} describing the flow of particles out of the potential well (see Figure 3.1). For a metastable state with a potential as shown in Figure 3.1 there is, at energies larger than the barrier peak, no flow of carriers into the potential well. In the annular area of Figure 3.3 connecting points A and B the population is zero at A and builds up as we follow the annulus into the potential well. Clearly for energies $E > E_b$ the population along an annular area is nonuniform. Thus $\beta = \rho/\rho_{eq}$ is no longer constant along an energy contour. However, (3.2.8) still applies but now for an effective correction factor β averaged along an orbit of constant energy. The number of carriers exiting at B is determined by the actual density in phase space at B and not by the averaged phase-space density. The phase-space density builds up along an orbit of constant energy and at the exit point is larger than the effective β. This can be taken into account by introducing an additional correction factor α multiplying β to yield the correct phase-space density at the barrier peak. The rate at which particles leave at the barrier peak $x = x_b$ in the energy range from E to $E + \mathrm{d}E$ is given by

$$j_{out}\mathrm{d}E = \rho(E, x_b)v(E, x_b)\mathrm{d}p = \rho(E, x_b)\mathrm{d}E = \alpha\beta(E)\rho_{eq}(E)\mathrm{d}E. \quad (3.3.1)$$

In the steady-state the outflow diminishes the vertical flux j_I. The decrease of the vertical flow must compensate the outflow of carriers, $\mathrm{d}j_I = -j_{out}\,\mathrm{d}E$, or

$$\frac{\mathrm{d}j_I}{\mathrm{d}E} = -j_{out} = -\alpha\beta\rho_{eq}. \quad (3.3.2)$$

The divergence of the vertical flux is determined by the rate at which particles leave the potential well. Using (3.2.8) and taking into account that $\rho_{eq}^{-1}(\mathrm{d}\rho_{eq}/\mathrm{d}E) = -1/kT$ we find from (3.3.2) an equation for the effective correction factor,

$$\eta I_b kT\frac{\mathrm{d}^2\beta}{\mathrm{d}E^2} - \eta I_b\frac{\mathrm{d}\beta}{\mathrm{d}E} - \alpha\beta = 0. \quad (3.3.3)$$

The linear homogeneous differential equation (3.3.3) has solutions of the form $\beta = \mathrm{e}^{sE/kT}$. Inserting this into (3.3.3) yields

$$s_{\pm} = 1/2[1 \pm [1 + (4\alpha kT/\eta I_b)]^{1/2}]. \quad (3.3.4)$$

s_+ yields a solution which increases exponentially with energy and is not relevant in our problem. s_- yields a solution which decreases exponentially with increasing energy. It accentuates the drop-off present in equilibrium. It is the physical solution which we are looking for. For severely underdamped systems, $kT \gg \eta I_b$ the exponential drop-off is fast and determined by $s_- \propto -(\alpha kT/\eta I_b)^{1/2}$. Thus $\rho \propto \rho_{eq} e^{s_- E/kT}$ decreases much faster than ρ_{eq}. However, if the damping increases and $\eta I_b \gg kT$ the correction factor becomes unimportant.

To find the escape rate we must match the probability density for $E > E_b$ and the probability density for $E < E_b$. We require continuity of the density and its derivative at $E = E_b$ or equivalently continuity of the density and the vertical current at $E = E_b$. For $E < E_b$ the density is given by (3.2.7). For $E > E_b$ the density is given by $\rho(E) = A \exp(s_- E/kT)\rho_{eq}(E)$. Continuity of the two densities at E_b determines the constant $A = \beta(E_b)\exp(-s_- E_b/kT)$. Continuity of the derivative of the two distribution functions yields $d\beta/dE|_{E=E_b} = s_- \beta(E_b)/kT$. The derivative of β can be obtained from (3.2.8) by eliminating the current with the help of (3.2.10). This yields $1 - \beta(E_b) = -s_- \beta(E_b)$ or

$$\beta(E_b) = 1/(1 - s_-) = 1/s_+ = 2/([1 + (4\alpha kT/\eta I_b)]^{1/2} + 1). \qquad (3.3.5)$$

In the extremely underdamped limit, $\eta I_b \ll kT$, the correction factor β at the barrier energy is proportional to $(\eta I_b/kT)^{1/2}$ and increases with increasing friction to the equilibrium value $\beta = 1$. The correction factor for energies $E > E_b$ multiplying the equilibrium distribution function is given by

$$\beta(E) = (1/s_+)e^{s_-(E - E_b)/kT}. \qquad (3.3.6)$$

Inserting β given by (3.3.5) into (3.2.10) yields the refined version of Kramers' low damping result obtained by BHL,

$$r = j_I/N = \frac{[1 + (4\alpha kT/\eta I_b)]^{1/2} - 1}{[1 + (4\alpha kT/\eta I_b)]^{1/2} + 1}\left(\frac{\eta I_b}{kT}\right) r_{TS}. \qquad (3.3.7)$$

As η tends to zero, this result tends to the Kramers low friction result (3.2.11). As η becomes large (3.3.7) yields an escape rate $r = \alpha r_{TS}$, where r_{TS} is given by (3.2.1).

Equation (3.3.7) is plotted in Figure 3.2 and is compared with computer simulations for the escape rate out of a local minimum of the potential $V = V_0(1 - \cos x) - Fx$. For the computer simulations the field is $F = 0.985 V_0$ and the temperature is $V_0 = 1135.9\, kT$ corresponding to a ratio $E_b/kT = 3.938$ at this field. Two different techniques were used to compute escape rates. In a conventional simulation (+ in Figure 3.2) the Langevin equation was integrated with the particles initially started at the well bottom. Such a simulation is not very economical since the particle spends most of the time near the well bottom. On the other hand, we know that the phase-space density near the well bottom is simply the equilibrium distribution. To speed up the simulations BHL adapted an approach by Bennett (1977). This approach uses the analytical solution near the well bottom and uses computer simulation only for energies above a certain threshold, still a few kT below the barrier peak. The computational results for the phase-space density and the analytical solution are then matched at the threshold energy. The results of this technique are also shown in Figure 3.2 (data points shown as *). It was found that $\alpha = 1$ yields a reasonable fit of (3.3.7) to the computer simulations. On theoretical grounds, as discussed above, we expected to find $\alpha > 1$. Below we return to the discussion of appropriate values for α. Equation (3.3.7) was used by Devoret, Martinis and Clarke (1985) to analyze the escape rates of underdamped Josephson junctions.

Subsequently the reasoning of BHL leading to (3.3.7) was also applied to a symmetric double well potential (Büttiker, 1984). In such a potential the rate for particles to hop from one well to the other is

$$r = \frac{1}{2} \frac{[1 + (8\alpha_d kT/\eta I_b)]^{1/2} - 1}{[1 + (8\alpha_d kT/\eta I_b)]^{1/2} + 1} \left(\frac{\eta I_b}{kT} \right) r_{TS}. \tag{3.3.8}$$

Here α_d plays the same role as α in the single well potential. Note that (3.3.8), in the limit $\eta \to 0$ yields an escape rate which is only half as large as Kramers' result (3.2.11) for the escape out of a single underdamped potential well. (The action I_b, in (3.3.8), also measures the phase-space of one well only.) Particles with energy $E > E_b$ have a probability to return to their original well. It is this back flow of particles which leads to an additional reduction of the escape rate.

To compare these results with other related work it is useful to expand (3.3.7) in powers of $\eta^{1/2}$. From (3.3.7) we find

$$r = [1 - (\eta I_b/\alpha kT)^{1/2}](\eta I_b/kT) r_{TS}. \tag{3.3.9}$$

Thus the first correction term to the Kramers low damping result is proportional to $\eta^{3/2}$. Precisely a result of this form was found by Risken and Vollmer (1985, unpublished calculations) and Risken and Voigtlaender (1985). To order $\eta^{1/2}$ they take the spatial variation of the distribution function for energies near the top of the barrier into account. To this order their solution for the distribution function is genuinely two dimensional and allows them to determine the parameter α. Comparison (see Risken and Voigtlaender, 1985) of their result with (3.3.8) yields $\alpha = 1.47$ in the case of a single metastable potential well and $\alpha_d = 2.146$ in the case of a double well potential. (A discussion of additional related work is the subject of Section 3.6.)

Clearly, these values of α, determined for very low friction, should not be expected to provide the best fit of (3.3.7) to experimental or computational data over a wider range of damping constants. The two dimensional character of the distribution function depends on the temperature and the friction. Hence α should be considered to be a function of damping and temperature with the properties (for a single metastable well) $\alpha \to 1.47$ as $\eta I_b/kT \to 0$ and $\alpha \to 1$ as $\eta I_b/kT \to \infty$. The second limit requires the escape rate to approach the transition-state result for large damping. This might then also explain why a value $\alpha \simeq 1$ provides a good overall fit to the computational results in Figure 3.2.

For $kT \gg \eta I_b$ the BHL result (3.3.7) approaches the transition-state result. From (3.3.7) we find an escape rate, $r = (1 - \alpha kT/\eta I_b)\alpha r_{TS}$ which is reduced compared to the transition-state result by a correction factor proportional to $1/\eta$. That the BHL escape rate approaches the transition-state result, and not Kramers' moderate damping result (3.2.3) is due to the differing mechanisms taken into account. The BHL approach completely neglects the difficulty in crossing the barrier due to diffusion along the reaction coordinate which becomes important for moderate damping. Slow diffusive motion leads with increasing damping to a return flux from the barrier into the well. This is not taken into account in (3.3.7) and (3.3.8). Thus, as we cross from the regime of low damping into the regime of moderate damping, it is eventually diffusion along the reaction coordinate, incorporated in (3.2.3) which becomes the physical process limiting the escape from the well.

3.4 Escape energies

In this section we study the average energy which an escaping particle has when it crosses the barrier peak. This quantity is of interest in the desorption of adatoms (Leuthäusser, 1981). Our original motivation (Büttiker and Landauer, 1984a, b) for the study of the average escape energy was to have an additional quantity, other than the escape rate, to distinguish between competing theories (see Section 3.6).

Consider, first, the transition-state theory. This approach assumes that the distribution function is, for all energies, given by the equilibrium distribution function (3.2.6). The flux of particles out of the well with energies between E and $E + \mathrm{d}E$ is given by $\mathrm{d}j = \rho_{eq}(E)v\,\mathrm{d}p$. Thus the probability $p(E)$ for an escaping carrier to have energy between E and $E + \mathrm{d}E$ is $p(E)\mathrm{d}E = \mathrm{d}j/j$, where $j = \int_0^\infty \rho_{eq}(E)v\mathrm{d}p$ is the total flux out of the well. The excess energy $\Delta E = E - E_b = mv^2/2$, at the barrier, is purely kinetic. Hence the average energy of an escaping particle, at the barrier peak, is given by

$$\langle \Delta E \rangle = \int_{E_b}^{\infty} \Delta E p(E) \mathrm{d}E. \tag{3.4.1}$$

For the transition state theory we find $p(E) = (1/kT)\mathrm{e}^{-\Delta E/kT}$ and evaluation of (3.4.1) yields

$$\langle \Delta E \rangle = kT. \tag{3.4.2}$$

Consider next Kramers' treatment of the moderate damping regime with an escape rate given by (3.2.3). Kramers (1940) derives for this regime a distribution function (his equation (23)) which can be used to find the average escape energy. A calculation (Büttiker and Landauer, 1984b) yields

$$\langle \Delta E \rangle = [1 + \tfrac{1}{2}(1 - \eta/a)^{1/2}\{1 - (1 - \eta/a)^{1/2}\}]kT, \tag{3.4.3}$$

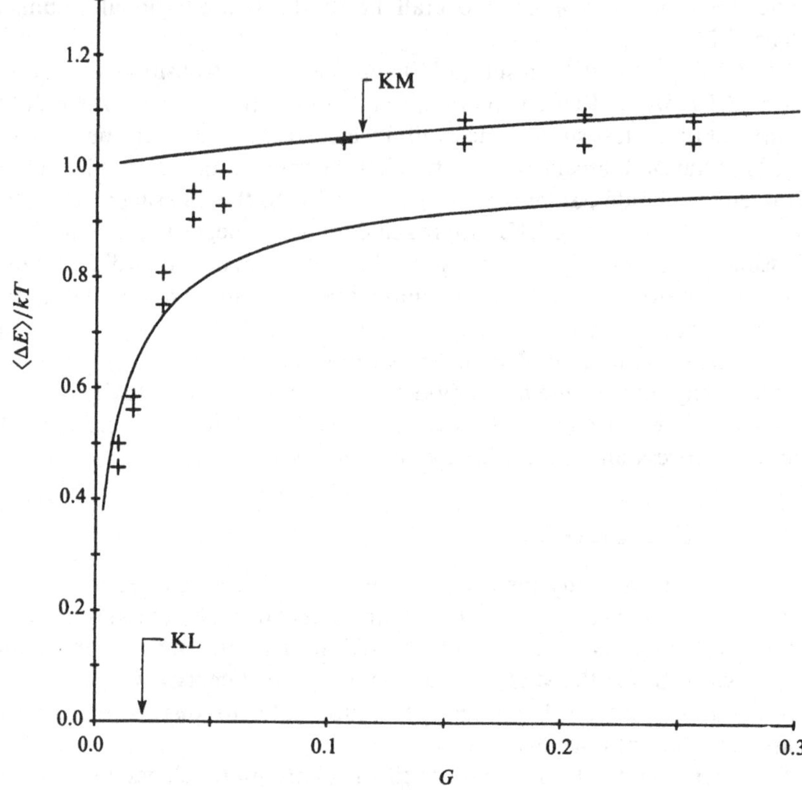

Figure 3.4. Average kinetic energy $\Delta E = \frac{1}{2}\langle mv^2 \rangle$ of escaping particles as a function of the dimensionless damping constant G. The computational data (+) are compared with (3.4.4) for $\alpha = 1$ and with Kramers' moderate (KM) damping approach, (3.4.3). Kramers' discussion of the extremely underdamped case (KL), yields $\langle \Delta E \rangle = 0$ independent of the damping constant. After Büttiker and Landauer (1984b).

where

$$a = \tfrac{1}{2}\eta + (\tfrac{1}{4}\eta^2 + |\omega_B|^2)^{1/2}.$$

Equation (3.4.3) is plotted in Figure 3.4 and labeled KM. For $\eta = 0$ (3.4.3) gives the transition-state theory result (3.4.2). Equation (3.4.3) yields a maximum for the average escape energy $\langle \Delta E \rangle = 1.125\,kT$ for $\eta/|\omega_B| = 1/2$ and again approaches kT for large η where $\langle \Delta E \rangle = kT(1 + |\omega_B|^2/2\eta^2)$.

The discussion by Kramers of the weakly damped case, leading to (3.2.11), assumes a complete depletion of the population above the barrier peak, $\rho = 0$ for $E > E_b$. All escaping particles have zero kinetic energy, and hence $\langle \Delta E \rangle = 0$. This result is also shown in Figure 3.4, labeled KL.

Let us next apply the approach of Section 3.3 to find the average escape energy for low damping. The flow of particles out of the well in the energy range between E and $E + \mathrm{d}E$ is $\mathrm{d}j = -j_{\text{out}}\,\mathrm{d}E = \alpha\beta\rho_{\text{eq}}\,\mathrm{d}E$ with β determined by (3.3.6). This gives a probability $p(E) = (s_+/kT)\exp(-s_+\Delta E/kT)$ for a particle to escape with energy E. This yields an average energy

$$\langle\Delta E\rangle = \frac{2kT}{(1 + 4\alpha kT/\eta I_{\text{b}})^{1/2} + 1}. \tag{3.4.4}$$

For large damping (3.4.4) predicts an average escape energy of kT and for small damping the average escape energy is $\langle\Delta E\rangle = (\eta I_{\text{b}}kT/\alpha)^{1/2}$. Equation (3.4.4) is plotted in Figure 3.4 (for $\alpha = 1$) together with computational data for the same potential and parameters as in Figure 3.2. For each value of the dimensionless damping parameter $G = \eta(m/V_0)^{1/2}$ two data points, each obtained by simulating 200 escape events, are shown corresponding to two different starting points of the random generator. The BHL approach, discussed in Section 3.3, fits these data rather well.

3.5 Non-Markovian damping and quantum corrections

The approach of BHL has been extended in two directions. First, all the results which we have presented so far are for thermal (white) noise. In recent years there has been an interest in problems with memory dependent damping and colored noise. An extension of the underdamped case, as discussed in Section 3.3, to the case of non-Markovian dissipation and fluctuations, has been given by Hänggi and Weiss (1984). Furthermore, over the recent years, the quantum escape out of a metastable state in the presence of dissipative events has been the subject of intense theoretical (Caldeira and Leggett, 1983; Hänggi, 1986) and experimental investigations (Devoret, Martinis and Clarke, 1985; Devoret *et al.*, 1984; Voss and Webb, 1981; Washburn and Webb, 1986; Washburn *et al.*, 1985). The techniques most widely applied to this problem invoke a thermal equilibrium population in the well, and yield a crossover from the quantum regime at low temperatures to the Kramers moderate damping result, (3.2.3), at high temperatures (Grabert, Olschowski and Weiss, 1985; Larkin, Likharev and Ovchinnikov, 1984). In the limit of very low friction this is, of course, incorrect. As shown by Rips and Jortner (1986), for weak damping the crossover can be described, at least for a certain range of parameters, by a simple modification of the purely classical treatment of BHL. Below we discuss this modification in more detail.

Consider the case where the damping constant and the temperature are high enough such that a particle in the well, but near the top of the barrier, during one revolution, loses phase memory. In this case the dynamics in the well can be described classically. The quantum escape can be taken into account by modifying the horizontal flow, (3.3.1). If we can neglect quantum coherence in the well, then the quantum penetration of the barrier can be described by the probability $T(E)$ for transmission through the barrier for a particle incident on the barrier with an energy E. Thus the flux $\alpha\beta(E)\rho_{\text{eq}}(E)$ incident on the barrier has probability T for transmission and probability $R = 1 - T$ for reflection back into the well. Hence, the horizontal flux, (3.3.1), is now given by

$$j_{\text{out}} = \alpha\beta(E)\rho_{\text{eq}}(E)T(E). \tag{3.5.1}$$

Note that in the presence of quantum tunneling we have a horizontal flux out of the well even for energies below the barrier peak energy. Here, we have deliberately included α to account for the nonuniformity of the distribution along a path of constant energy, as discussed in connection with (3.3.1). Using (3.5.1) and $\mathrm{d}j_I/\mathrm{d}E = -j_{\text{out}}$ we find, instead of (3.3.3),

$$\eta I_{\text{b}} kT \frac{\mathrm{d}^2\beta}{\mathrm{d}E^2} - \eta I_{\text{b}} \frac{\mathrm{d}\beta}{\mathrm{d}E} - \alpha\beta T(E) = 0. \tag{3.5.2}$$

Equation (3.5.2), in contrast to (3.3.3), is valid at all energies. Again, we are concerned with a narrow energy interval around the barrier peak energy E_{b} and hence we can again neglect the energy dependence of the action I. On the other hand, the transmission probability near the top of the barrier,

$$T(E) = [1 + \exp(-2\pi\Delta E/\hbar|\omega_B|)]^{-1}, \tag{3.5.3}$$

where $\Delta E = E - E_{\text{b}}$, varies exponentially with energy. The physical solution of (3.5.2) obeys the boundary conditions, $\beta(E) \to 1$ for $E \to 0$ and $\beta(E) \to 0$ for $E \to \infty$. This yields an escape rate (Rips and Jortner, 1986)

$$r = \frac{\eta I_{\text{b}}}{kT}\left(\frac{-s_-}{s_+}\right)\frac{[\Gamma(1 + s_+/t)]^2}{[\Gamma(1 - s_-/t)]^2}\frac{\pi/t}{\sin(\pi/t)[\Gamma(1 + 1/t)]^2} r_{\text{TS}}. \tag{3.5.4}$$

Here, $s_\pm$ are given by (3.3.4), and $t \equiv 2\pi kT/\hbar|\omega_B|$. When the thermal energy is high compared to the barrier frequency, $t \gg 1$, the quantum corrections become unimportant. In this limit the arguments of the gamma functions tend to one and (3.5.4) yields $r = (\eta I_{\text{b}}/kT)(-s_-/s_+)r_{\text{TS}}$ which is (3.3.7). Rips and Jortner (1986) have taken $\alpha = 1$. But clearly even in the presence of quantum penetration the distribution function in the well is nonuniform along the contours of constant energies. This effect can be expected to be small at energies for which the transmission probability is small but cannot be neglected at energies for which the transmission probability is substantial.

Let us briefly discuss the limits of such an approach. Equation (3.5.4) treats the escape of carriers near the top of the barrier. This requires that $\mathrm{e}^{-E_{\text{b}}/kT}$ is large compared to the quantum tunneling rate $\propto \mathrm{e}^{-2I_0/\hbar}$ for escape out of the ground state of the well. Here I_0 is the action associated with the WKB path for escape out of the ground state. Comparison of the two exponentials yields a lower bound for the temperature, $kT > \hbar E_{\text{b}}/2I_0$. Quantum effects near the barrier top are most pronounced if the temperature is comparable to or smaller than $\hbar|\omega_B|$. Hence the strongest deviations from the classical result (3.3.7) can be expected for temperatures in the range $\hbar E_{\text{b}}/2I_0 < kT < \hbar|\omega_B|$. Equation (3.5.4) is more limited and applicable only if kT exceeds $\hbar|\omega_B|$. (The rate given by (3.5.4) diverges for $t = 1$, i.e. $2\pi kT = \hbar|\omega_B|$.) Equation (3.5.4) assumes that the dynamics in the well can be treated classically, i.e. that the quantum phase of the carriers is randomized during one round trip. For this to be the case, the coupling of the particle to the bath needs to be sufficiently strong. Phase randomization occurs if the energy of the particle suffers an energy change during one cycle comparable to the separation of the energy levels Δ in the well. Both the drift and the fluctuations contribute to energy exchange and this yields $\eta I_b > \Delta$ or $\eta I_b kT > \Delta^2$. This sets a lower bound for the value of the friction constant. Note that it is the energy separation of the eigenstates in a range kT below E_b which counts, and in this range Δ is very small. In the discussion leading to (3.5.4) the exponential suppression of the tunneling rate due to inelastic events (Caldeira and Leggett, 1981) has been neglected. However, we can assume that inelastic events in the tunneling process are only important if the energy dissipated during barrier traversal is of the order of $|\hbar\omega_B|$. To estimate the dissipated energy, we follow Büttiker and Landauer (1982) and use the absolute value of the momentum $|p| = (2m)^{1/2}(V(x) - E)^{1/2}$ which determines the exponential decay of the wave function in the barrier. The dissipated energy is then found from the classical expression $E_d = \int \eta |p| \, dv$. For a carrier incident with energy $E = E_b - \hbar|\omega_B|$ this estimate yields $E_d = \pi\hbar\eta$. Thus in the range of damping constants for which the underdamped case is of interest, $\eta \ll |\omega_B|$, neglecting the inelastic events during the tunneling process is well justified. For a discussion of the underdamped quantum escape problem in the limit where the discreteness of the energy levels in the well is relevant we refer to the work of Larkin and Ovchinnikov (1986).

3.6 Discussion of related work

We have shown that the first correction (see (3.3.9)) to the Kramers low damping result is proportional to $\eta^{3/2}$. Furthermore, we have investigated the average energy of the escaping particles and found that for small damping this energy is proportional to $\eta^{1/2}$. These two results provide simple criteria to distinguish the approach of BHL from other work. We have already stressed the agreement of these results with those of Risken (see Chapter 6, Volume 1). Therefore, we turn now to the discussion of the work of other authors.

A number of attempts (Carmeli and Nitzan, 1983; Dekker, 1985, 1986a, b; Matkowsky, Schuss and Tier, 1984; Visscher, 1976) have been made at interpolating between the Kramers low friction result and Kramers' high friction result. It is the nature of such interpolation attempts that they have the correct limiting behavior built in and easily achieve reasonable results for the escape rate. Visscher's interpolation formula and the results of Carmeli and Nitzan (1983) and Matkowsky, Schuss and Tier (1984) fail, however, to exhibit the $\eta^{3/2}$-correction term to the Kramers low damping result. The approach of Carmeli and Nitzan invokes the distribution of the Kramers low damping approach inside the potential well and matches it to the distribution function of Kramers' approach to the moderate damping result to obtain the phase space density near the barrier peak. Kramers' moderate damping approach, as discussed in Section 3.4, yields an escape energy of the order of kT for all damping constants (see Figure 3.4). Thus the approach of Carmeli and Nitzan yields an escape energy of order kT independent of the degree of damping. Matkowsky *et al.* use a mean first passage time approach to discuss the escape of particles. There is no simple way to calculate an escape energy from this approach. A set of more stringent conditions for the validity of such an approach has recently been discussed by Ryter (1988). Dekker derives a variational functional for the distribution function. The quality of the results obtained by variation hinges on the properties of the test functions. Dekker's results at low damping do not reduce to the Kramers low friction result, (3.2.4).

The work of Mel'nikov and Meshkov (1986) presents results for the escape rate and the escape energy which for low friction agree with (3.3.7) and (3.4.4). Comparison of their result for the escape rate yields $\alpha = 1.49$ which is close to the value $\alpha = 1.47$ found by Risken and Voigtlaender (1985). This is astonishing, since, as we explain below, the origin of this factor in the work of Mel'nikov and Meshkov is different from that of Risken and Voigtlaender. Mel'nikov and Meshkov (1986) characterize the earlier work as based on 'unjustified assumptions' and '*ad hoc*' interpolations. Their technique, however, in contrast to those of Risken and Voigtlaender (1985) and BHL, is even more drastically limited to one dimension, i.e. to motion along the energy coordinate. They point out that for $\eta I_b \gg kT$ their results differ from (3.3.7) and (3.4.4), without an explanation indicating which result is the more plausible one. They find for $\eta I_b \gg kT$ an escape rate and an average escape energy which exhibit only exponentially small deviations from the transition-state results. In contrast, (3.3.7) and (3.4.4) exhibit in this limit corrections to the transition-state results which are proportional to $1/\eta$. We will go on to show why the slow approach to the transition-state results, given by (3.3.7) and (3.4.4), is in fact to be expected. First, we present a physical argument, and later we go on to pinpoint the source of the difference between the BHL results and those of Mel'nikov and Meshkov (1986).

The following argument has been pointed out to us by Landauer. For $\eta I_b \gg kT$ both for the approach of Mel'nikov and Meshkov (1986) and that of BHL the escape current is to first order fixed by the transition-state value $j_{TS} = N r_{TS}$. To get this current up into the region $E > E_b$ the population in this energy range must be depleted below the equilibrium value. Using $J_I = j_{TS}$ in (3.2.8) yields for the correction factor $d\beta(E)/dE = -1/\eta I_b$. Hence the population at an energy ΔE above E_b is only a fraction $(1 - \Delta E/\eta I_b)$ of the equilibrium population. But the depletion of the population for energies $E > E_b$ also reduces the escape current. Since the escape rate is proportional to the population above E_b we see that the resulting escape current must exhibit a deviation from the transition-state result proportional to $1/\eta$. Therefore, it is the BHL result which is the plausible one.

Let us now investigate in more detail the difference between the approach of Mel'nikov and Meshkov (1986) and that of BHL. Mel'nikov and Meshkov invoke a formalism advanced by Iche and Noizieres (1976) and Leuthäusser (1981). The only physical ingredient in this formalism is the probability $W(E, E')\,\mathrm{d}E$ that a particle subject to drift and fluctuations which initially has energy E' *after one round trip between two turning points* acquires an energy between E and $E + \mathrm{d}E$. The distribution function $\rho(E)$ describing the escape out of the well obeys the simple integral equation

$$\rho(E) = \int_0^{E_b} \mathrm{d}E'\, W(E, E')\rho(E'). \tag{3.6.1}$$

The lower limit of the integral corresponds to the well bottom. The upper limit of the integral reflects the fact that carriers which exceed the barrier peak energy eventually escape out of the well. Again, we need only to consider a narrow effective range near the barrier top. In this case, W becomes a function of the energy difference $\varepsilon = E - E'$ only. Below we show that (3.6.1), apart from neglecting the spatially nonuniform character of the phase-space density for energies above E_b, is correct and with the appropriate hopping probability W leads to the results discussed in Sections 3.3 and 3.4. Mel'nikov and Meshkov (1986) find a probability $W(\varepsilon)$ which is a displaced Gaussian $W(\varepsilon) \propto \exp[-(\varepsilon + \eta I)^2/4\eta IkT]$. Below, we show that the Gaussian hopping probability $W(\varepsilon)$ is not appropriate for the integral equation (3.6.1). They determine W as the solution describing diffusion of a population initially concentrated at energy E' along the energy coordinate. This solution allows a particle to make many revolutions to arrive at an energy E.

To determine the hopping probability $W(E - E')$ let us assume that a current $\delta(E - E')\rho_{eq}(E')v(E')\,\mathrm{d}p$ is injected into the well at energy E'. To obtain a steady-state problem, we remove the particles after one revolution from the well (even if their energy is below the barrier peak energy). Thus at each energy we have a horizontal flux $\rho(E)v(E)\,\mathrm{d}p$ out of the well, and at E' we have both an incoming flux and an outgoing flux (particles whose energy has not changed during one round trip). Here $\rho(E)$ is the particle distribution which remains to be determined. Let us for this discussion introduce a correction factor $g = \rho(E)/\rho_{eq}(E)$. The divergence of the vertical flux is

$$\mathrm{d}j_i/\mathrm{d}E = (\delta(E - E') - g)\rho_{eq}. \tag{3.6.2}$$

Note that we have taken $\alpha = 1$. We neglect completely the possible deviations from uniformity of the distribution $\rho(E)$ along a path of constant energy. Using (3.2.8) with g instead of β, and (3.6.2) we find

$$\eta IkT\frac{\mathrm{d}^2 g}{\mathrm{d}E^2} - \eta I\frac{\mathrm{d}g}{\mathrm{d}E} - g = \delta(E - E'). \tag{3.6.3}$$

The physical solution $g(E-E')$ of (3.6.3) obeys the boundary conditions $g \to 0$ for $E \to \pm\infty$. With the help of this solution the probability $W\,\mathrm{d}E$ for a particle to hop, during one revolution, from energy E' into the energy range $E, E+\mathrm{d}E$ is determined by the flux $\mathrm{d}j$ in the energy range $E, E+\mathrm{d}E$ normalized by the total incident flux at E', $W\,\mathrm{d}E = \mathrm{d}j/j$. The flux in the interval E, $E+\mathrm{d}E$ is $\mathrm{d}j = g(E-E')\rho_{\mathrm{eq}}(E)\mathrm{d}E$ and the total flux injected into the well is $j = \rho_{\mathrm{eq}}(E')$. We find

$$W(E-E') = g(E-E')\rho_{\mathrm{eq}}(E)/\rho_{\mathrm{eq}}(E')$$

$$= \frac{1}{\eta I}\frac{1}{s_+ - s_-}\mathrm{e}^{-s_\pm \varepsilon/kT}, \tag{3.6.4}$$

where $s_\pm$ are given by (3.3.4) with the action I_b replaced by the action $I = I(E')$ and with $\alpha = 1$. In (3.6.4) the upper sign in the exponent applies for $\varepsilon = E - E' > 0$ and the lower sign applies for $\varepsilon < 0$. The probability W is normalized, $\int \mathrm{d}\varepsilon\, W(\varepsilon) = 1$, and obeys detailed balance $W(\varepsilon) = \exp(-\varepsilon/kT)\cdot W(-\varepsilon)$ since $s_+ + s_- = 1$. The first moment of the probability distribution is determined by the average energy loss during one cycle,

$$\langle \varepsilon \rangle = \eta I, \tag{3.6.5}$$

and the second moment is given by

$$\langle \varepsilon^2 \rangle = 2[(\eta I)^2 + \eta I kT]. \tag{3.6.6}$$

The hopping probability, (3.6.4), has been used by Leuthäusser (1981) to treat the desorption of lightly damped adatoms. To show that the distribution (3.6.1) determined with the help of (3.6.4) is the same as that found by BHL and discussed in Section 3.3, we notice the following: (3.3.3), valid for $E > E_\mathrm{b}$, can be extended to be valid at all energies. For $E < E_\mathrm{b}$, $\beta(E)$ obeys $\mathrm{d}j_I/\mathrm{d}E = 0$ with j_I given by (3.2.8). This yields (3.3.3) without the term $(-\alpha\beta)$ describing the flow out of the well. Hence, the extension of (3.3.3) to all energies is

$$\eta I kT \frac{\mathrm{d}^2\beta}{\mathrm{d}E^2} - \eta I \frac{\mathrm{d}\beta}{\mathrm{d}E} - \beta = -\theta(E_\mathrm{b} - E)\beta(E), \tag{3.6.7}$$

where θ is the step function and $\alpha = 1$. The solution $g(E-E')$ of (3.6.3) is just the Green's function which allows an integral representation of the solution of (3.6.7). We find $\beta(E) = \int \mathrm{d}E'\, g(E-E')\theta(E_\mathrm{b} - E')\beta(E')$. Multiplying this equation on both sides by $\rho_{\mathrm{eq}}(E)$ and taking into account that $W(E-E') = g(E-E')\rho_{\mathrm{eq}}(E)\rho_{\mathrm{eq}}^{-1}(E')$ transforms this expression into (3.6.1). Thus (3.6.1) with the hopping probability (3.6.4) yields the distribution function of BHL discussed in Sections 3.3 and 3.4, except that now $\alpha = 1$. Note that (3.6.4) is different from the Gaussian hopping probability invoked by Mel'nikov and Meshkov (1986). However, for extremely low damping the first and second moments of the Gaussian hopping probability invoked by Mel'nikov and Meshkov (1986) agree with (3.6.5) and (3.6.6), and their results are thus identical to ours in this limit. But with increasing damping the Gaussian probability becomes increasingly flat as a function of ε enhancing the escape out of the well and leading to a fast approach to the transition-state result. In contrast, (3.6.4) for strong damping, $\eta I \gg kT$, and positive energies, behaves like the equilibrium distribution $W(\varepsilon) \propto \exp(-\varepsilon/kT)\exp(-\varepsilon/\eta I)$ limiting the extent of the positive energy tail of W. This accounts for the slower approach to the transition-state result with a correction term proportional to $1/\eta$ predicted by BHL and Büttiker and Landauer (1984a, b). For strong damping and negative ε the hopping rate is proportional to $\exp(\varepsilon/\eta I_b)$.

We have mentioned in the introduction to this chapter (see also Chapter 1, Volume 1) that the appreciation of Kramers' discussion of the low damping limit was extremely slow. We believe that, with the modest refinement and extension of this result, we have contributed to an acceleration of this process. This is demonstrated by the number of competing theories (Carmeli and Nitzan, 1983; Dekker, 1986b; Matkowsky, Schuss and Tier, 1984; Mel'nikov and Meshkov, 1986; Risken and Voigtlaender, 1985) which have appeared on this subject since 1983. It is further demonstrated by the rapidity with which the BHL approach has been extended and generalized (Hänggi and Weiss, 1984; Rips and Jortner, 1986).

The general physical insight which we gain from investigating the escape rate out of a well in the presence of friction is that inelastic processes can enhance transport. There is an intermediate value of damping for which the escape rate reaches a maximum (see Figure 3.2). This phenomenon is common to a variety of transport processes. Localization of electrons in a thin disordered wire is one example: for weak inelastic scattering, electrons in the wire are effectively localized and the resistance of the wire is very high. Increasing inelastic scattering permits carriers to escape from the localized states and decreases the resistance. Strong inelastic scattering, on the other hand, impedes the motion of the carriers and the resistance increases again (Büttiker, 1986; Thouless, 1981). There is thus an intermediate inelastic scattering time for which the resistance reaches a minimum (the conductance reaches a maximum). Other examples are current transport in superlattices (Büttiker and Thomas, 1979; Esaki and Tsu, 1970) or current transport in tiny normal metal loops driven by a time-dependent flux through the hole of the loop (Landauer and Büttiker, 1985). Of all the examples the escape from a well is perhaps the simplest and most illustrative.

References

Bennett, C. H. 1977. In *Algorithms for Chemical Computations* (R. E. Christoffersen, ed.), pp. 63–97. Washington, DC: American Chemical Society.

Büttiker, M. 1984. Proceedings of the *17th International Conference on Low Temperature Physics* (U. Eckern, A. Schmidt, W. Weber and H. Wühl, eds.), pp. 1155–6. Amsterdam: North-Holland.

Büttiker, M. 1986. *Phys. Rev. B* **33**, 3020–6.

Büttiker, M., Harris, E. P. and Landauer, R. 1983. *Phys. Rev. B* **28**, 1268–75.
Büttiker, M. and Landauer, R. 1982. *Phys. Rev. Lett.* **49**, 1739–42.
Büttiker, M. and Landauer, R. 1984a. *Phys. Rev. Lett.* **52**, 1250.
Büttiker, M. and Landauer, R. 1984b. *Phys. Rev. B* **30**, 1551–3.
Büttiker, M. and Thomas, H. 1979. *Z. Phys. B* **34**, 301–11.
Caldeira, A. O. and Leggett, T. 1981. *Phys. Rev. Lett.* **46**, 211–14.
Caldeira, A. O. and Leggett, T. 1983. *Ann. Phys. (NY)* **149**, 374–456.
Carmeli, B. and Nitzan, A. 1983. *Phys. Rev. Lett.* **51**, 233–6.
Chandrasekhar, S. 1943. *Rev. Mod. Phys.* **15**, 1–89.
Dekker, H. 1985. *Phys. Lett.* **113A**, 193–6.
Dekker, H. 1986a. *Physica* **135A**, 80–104.
Dekker, H. 1986b. *Physica* **136A**, 124–46.
Devoret, M. H., Martinis, J. M. and Clarke, J. 1985. *Phys. Rev. Lett.* **55**, 1908–11.
Devoret, M. H., Martinis, J. M., Esteve, D. and Clarke, J. 1984. *Phys. Rev. Lett.* **53**, 1260–3.
Esaki, L. and Tsu, R. 1970. *IBM J. Res. Dev.* **14**, 61–5.
Frauenfelder, H. and Wolynes, P. G. 1985. *Science* **229**, 337–45.
Fulton, T. A. and Dunkelberger, L. N. 1974. *Phys. Rev. B* **9**, 4760–8.
Grabert, H., Olschowski, P. and Weiss, U. 1985. *Phys. Rev. B* **32**, 3348–50.
Grote, R. F. and Hynes, J. T. 1982. *J. Chem. Phys.* **77**, 3736–43.
Hänggi, H. 1986. *J. Stat. Phys.* **42**, 105–40.
Hänggi, H. and Weiss, U. 1984. *Phys. Rev. A* **29**, 2265–7.
Iche, G. and Noizieres, P. 1976. *J. Physique* **37**, 1313–23.
Kramers, H. A. 1940. *Physica* **7**, 284–304.
Landauer, R. and Büttiker, M. 1985. *Phys. Rev. Lett.* **54**, 2049–52.
Landauer, R. and Swanson, J. A. 1961. *Phys. Rev.* **121**, 1668–74.
Larkin, A. I., Likharev, K. K. and Ovchinnikov, Y. N. 1984. *Physica* **126B**, 414–22.
Larkin, A. I. and Ovchinnikov, Y. N. 1986. *Sov. Phys. JETP* **64**, 185–9.
Larson, R. S. and Kostin, M. D. 1980. *J. Chem. Phys.* **72**, 1392–400.
Lee, P. A. 1971. *J. Appl. Phys.* **42**, 325–34.
Leuthäusser, U. 1981. *Z. Phys.* **44**, 101–8.
Matkowsky, B. J., Schuss, Z. and Tier, C. 1984. *J. Stat. Phys.* **35**, 443–56.
Mel'nikov, V. I. and Meshkov, S. V. 1986. *J. Chem. Phys.* **85**, 1018–27.
Naor, M., Tesche, C. D. and Ketchen, M. B. 1982. *Appl. Phys. Lett.* **41**, 202–4.
Raver, N. 1982. *IEEE J. Solid State Circuits* **17**, 932–7.
Rips, I. and Jortner, J. 1986. *Phys. Rev. B* **34**, 233–9.
Risken, H. and Voigtlaender, K. 1985. *J. Stat. Phys.* **41**, 825–63.
Ryter, D. 1988. *J. Stat. Phys.* **49**, 751–65.
Silvestrini, P., Pagano, S., Cristiano, R., Liengme, O. and Gray, K. E. 1988. *Phys. Rev. Lett.* **60**, 844–7.
Thouless, D. J. 1981. In *Physics in One Dimension* (J. Bernasconi and T. Schneider, eds.), pp. 306–9. Heidelberg: Springer.
Visscher, P. B. 1976. *Phys. Rev. B* **14**, 347–53.
Voss, R. F. and Webb, R. A. 1981. *Phys. Rev. Lett.* **47**, 265–8.
Washburn, S. and Webb, R. A. 1986. *Ann. NY Acad. Sci.* **480**, 66–77.
Washburn, S., Webb, R. A., Voss, R. F. and Farris, S. M. 1985. *Phys. Rev. Lett.* **54**, 2712–15.

4 Effect of noise on discrete dynamical systems with multiple attractors

EDGAR KNOBLOCH and JEFFREY B. WEISS

4.1 Introduction

The isolation of a physical system from its environment is a frequent approximation which leads to the idea of a deterministic conservative system. All physical systems are, however, coupled to the outside world, giving rise to the related phenomena of fluctuations and dissipation. In a dissipative dynamical system volumes in phase space contract onto attractors. Fluctuations allow the system to escape from attractors, rendering all attractors metastable. The long time behavior of a noisy dissipative system is thus intermittent, consisting of motion near the various attractors of the system alternating with transitions between attractors. In the limit of small noise, the times spent on the attractors become longer, and the transitions rarer. In this chapter we describe some recent work on stochastic dynamical systems with multiple attractors, with particular emphasis on systems possessing multiple limit cycles.

Although there has been much interest in noisy iterated maps, to our knowledge no one has actually derived a noisy map from a stochastic differential equation. In Section 4.2 we discuss some of the considerations that must go into any such derivation. As an example, we examine how noise affects a driven oscillator in both the phase-locked and unlocked regimes. The details of the noisy dynamics play an important role in determining how noise must affect the resulting map. Section 4.3 contains a formal derivation of a noisy iterated map from a linearized stochastic differential equation. Once a map is obtained one can analyze the dynamics to understand the time evolution of either a single system or an ensemble of systems. In Section 4.4 we discuss the dynamics of noisy iterated maps. We outline a derivation of the moments of the metastable distribution around a single attractor, and examine how the global structure of the basins of attraction influences the dynamics. These considerations are then used in Section 4.5 to study the dynamics of a one-dimensional cubic map perturbed by additive Gaussian white noise.

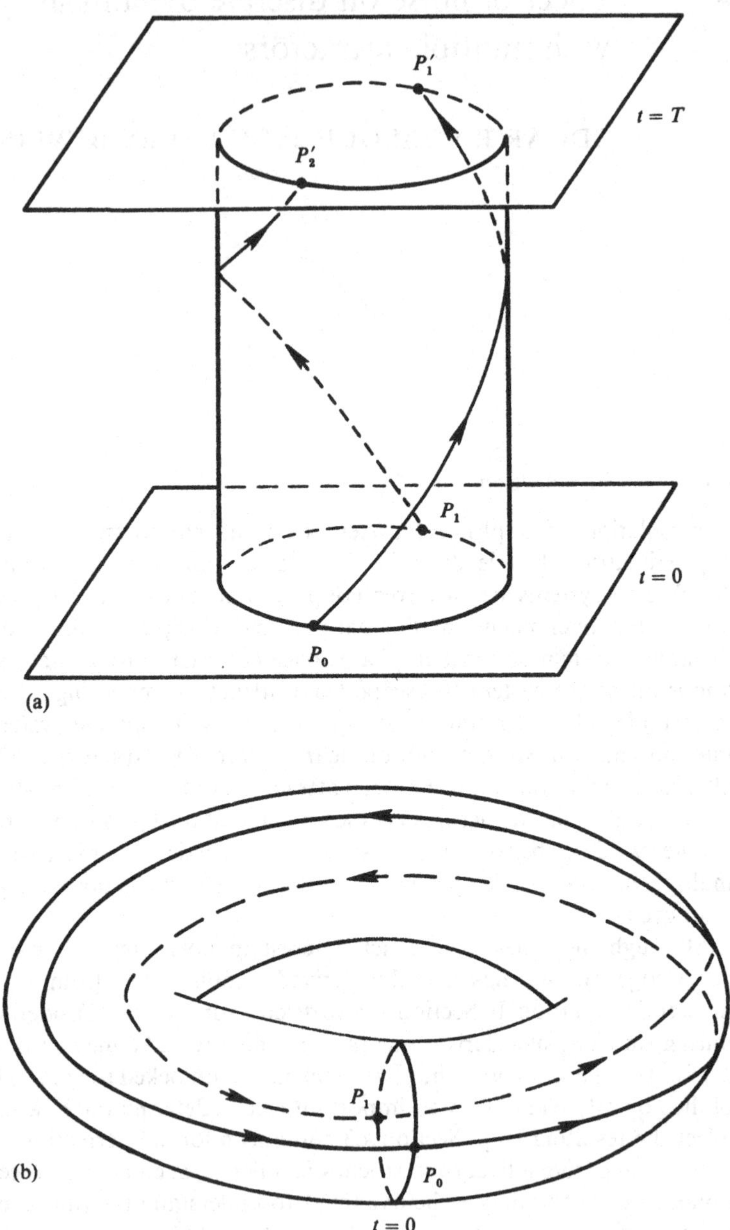

4.2 Stochastic differential equations and iterated maps

It is convenient to discuss dynamical systems with time-dependent attractors in terms of iterated maps. For autonomous systems of differential equations such maps may be obtained through the technique of the Poincaré surface of

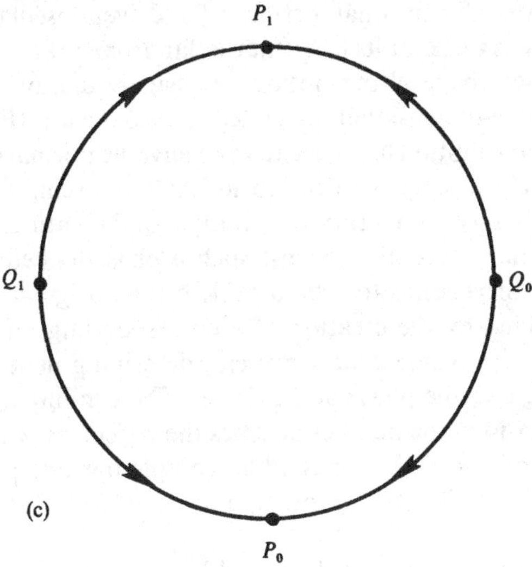

Figure 4.1. (a) A 2-torus in extended phase space $\mathbb{R}^2 \times \mathbb{R}$ showing a period-two orbit. Because of the periodicity in t the points P'_1 and P_1 are identified, as are P'_2 and P_0. (b) The 2-torus of (a) shown in the phase space $\mathbb{R}^2 \times S^1$. (c) The section $t = 0$ through the 2-torus showing a stable period-two orbit (P_0, P_1) and an unstable period-two orbit (Q_0, Q_1).

section (Guckenheimer and Holmes, 1983). First, an oriented surface in phase space is chosen transverse to the flow. The position on the surface is noted each time the continuous trajectory intersects the surface in a prescribed direction. The action of the resulting map is to take each such intersection into the succeeding one. The surface of section technique thus reduces the dimension of the dynamical system by one, but at the expense of replacing the continuous system with a discrete one. When the continuous system contains a stable limit cycle the Poincaré map will contain a stable periodic attractor. A different map, a time-T map, can be constructed for T-periodic nonautonomous systems by plotting the position in phase space of the trajectory at times $t = nT$, for integer n. A time-T map will reduce a continuous nonautonomous system to a discrete but autonomous system. It is important to stress that unless small parameters are present in the dynamical system these maps cannot in general be constructed analytically. They can, however, be obtained numerically or experimentally. Often they take the form of *noninvertible* maps $x_{n+1} = f(x_n)$. Strictly speaking, such maps cannot arise from a dynamical system. Indeed, as demonstrated by Hénon (1976), these maps contain an uncountable number of 'leaves', typically spaced so closely that they may not be resolvable. Nonetheless, if such a 'map' is approximated by a map of the form $x_{n+1} = f(x_n)$ the resulting dynamics are reflected in the dynamics of the underlying dynamical system. Because discrete dynamical systems are easier to simulate, such maps have seen extensive numerical investigation.

Consider now a one-dimensional periodically driven oscillator. If the strength of the forcing is weak, or its frequency is 'far' from a rational multiple of the natural frequency, the resulting motion is quasiperiodic, and takes place on a 2-torus in three-dimensional extended phase space (Figure 4.1a). Attracting quasiperiodic motion has a negative eigenvalue normal to the torus and two zero eigenvalues tangent to the torus. As the forcing frequency is brought towards resonance with a rational multiple of the natural frequency, locking of the frequencies typically ensues. Such a phase-locked solution is singly periodic, and corresponds to a limit cycle on the original 2-torus. Its creation is accompanied by the creation of a corresponding unstable limit cycle. Only one zero eigenvalue is now present, describing neutral stability with respect to changes in the phase of the driver. The winding number of a periodic orbit is the ratio of the number of times the trajectory winds around the two angular directions of the torus while completing one period. This number determines the period of the corresponding attractor in the time-T map (Figure 4.1c).

The phase-locking process may be described by the nonlinear circle map

$$\theta_{n+1} = \theta_n + \alpha + \kappa \sin \theta_n \tag{4.2.1}$$

relating the relative phases θ at $t = nT$ and $t = (n+1)T$. Here $\alpha = 2\pi\omega_{\mathrm{n}}/\omega_{\mathrm{f}}$, where ω_{n}, ω_{f} are, respectively, the natural and forcing frequencies, and κ denotes the strength of the forcing. When $\omega_{\mathrm{n}}/\omega_{\mathrm{f}}$ is irrational and $\kappa = 0$ the motion described by (4.2.1) is quasiperiodic. As κ increases locking occurs in thin wedges, called Arnol'd tongues, emanating from each rational number and opening with increasing κ (Bohr, Bak and Jensen, 1984; Jensen, Bak and Bohr, 1984). A continuous version of (4.2.1) is the equation

$$\frac{\mathrm{d}\theta}{\mathrm{d}t} = A - K \sin \theta, \quad A, K > 0, \tag{4.2.2}$$

which may be thought of as describing motion of an overdamped particle in the 'washboard' potential $V(\theta) = -A\theta - K\cos\theta$ (Adler, 1946; Büttiker and Landauer, 1981). When $A > K$ the phases are unlocked; when A falls below K the relative phase locks at $\theta_1 = \sin^{-1}(A/K) + 2n\pi$ with unstable phase-locked states at $\theta_2 = \pi - \theta_1$. Here n is an integer. Equation (4.2.2) can only be derived using the method of averaging, and hence requires the presence of a small parameter. There exist phenomenological models of both resistively shunted Josephson junctions (Ambegaokar and Halperin, 1969; Lindelof, 1981) and charge density wave conduction (Grüner, Zawadowski and Chaikin, 1981; Grüner and Zettl, 1985) which are analogous to the periodically driven oscillator and exhibit phase locking (Bohr, Bak and Jensen, 1984; Jensen, Bak and Bohr, 1984). Both the nonlinear circle map and the washboard potential have thus been important in understanding the dynamics in the above solid state systems.

Fluctuations are characterized by their correlation time, or equivalently by their frequency spectrum. Processes with finite correlation time have a peaked frequency spectrum, and hence are called colored noise. When the correlation time of the noise is much shorter than any timescale of the deterministic system it is appropriate to take the limit of zero correlation time, resulting in white noise and a flat frequency spectrum. Fluctuations can be divided into two classes: additive noise and multiplicative noise. In the case of additive noise, stochastic terms are added to the deterministic equations of motion, while multiplicative noise refers to the case where the parameters entering the deterministic equations are themselves fluctuating.

The effect of noise on a system with multiple stable equilibria is twofold. An ensemble of systems starting on a given equilibrium will, on a short timescale, evolve to a metastable distribution; on a longer timescale fluctuations will drive the system from one equilibrium to another. These transitions will, for small noise, occur almost exclusively over saddle points on the basin boundary. The deterministic flow near an equilibrium is reflected in its eigenvalues. Hence, fluctuations in directions with different eigenvalues will affect the dynamics differently (Knobloch and Wiesenfeld, 1983). Fluctuations parallel to eigenvectors with negative eigenvalues will equilibrate to a metastable state in which the contraction is balanced by the noise-induced expansion. Directions with positive eigenvalues indicate that trajectories leave the neighborhood of the equilibrium, i.e. the equilibrium is either unstable or a saddle; noise in these directions may slow the escape, or even drive the system back across the equilibrium. As saddle points often separate basins of attraction, fluctuations which drive a system over a saddle can cause qualitative changes in the dynamics. Fluctuations along neutrally stable eigenvectors are similar to an unbiased random walk, and will result in diffusion. Of course, as the eigenvalues get smaller and the noise strength larger the above distinctions become blurred, and noise-induced diffusion dominates.

These considerations enter into a discussion of the effects that fluctuations have on an attracting limit cycle embedded in a 2-torus. Fluctuations normal to the 2-torus will be countered by contraction onto the torus. The effect of fluctuations tangent to the 2-torus but normal to the flow depends on whether the system is phase-locked. If the system is not phase-locked the limit cycle covers the torus densely and these fluctuations will continuously drive the system onto nearby parts of the *same* limit cycle, but with different phases. The result is that the system wanders over the whole torus, continuously making finite jumps in phase. In the phase-locked case, however, different sections of the stable limit cycle are separated by an unstable limit cycle, and the system is locally stable against small fluctuations normal to the flow. Fluctuations normal to the flow may still cause a finite phase shift, although now they must drive the system across an unstable orbit. Finally, fluctuations tangent to the flow will cause the system to diffuse in phase along the trajectory. It must be remembered that these directions are determined by a coordinate system comoving with the instantaneous deterministic trajectory. Experiments on charge density waves in $NbSe_3$ indicate that the broadband voltage noise is greatly reduced in the phase-locked state (Sherwin and Zettl, 1985). Wiesenfeld and Satija (1987) use the noisy circle map to show that this noise reduction is a general property of phase-locked dynamics.

Fluctuations may not necessarily occur in all directions of phase space. An important example is provided by the distinction between a noisy oscillator coupled to a deterministic driver, and two weakly coupled noisy oscillators. Both deterministic motions take place on a 2-torus and exhibit phase-locking, but the phase space for the coupled oscillators is $\mathbb{R}^4$. Since the noise in the driven oscillator does not affect the driver, the fluctuations only occur in the two-dimensional phase space of the oscillator itself. This corresponds to fluctuations occurring in planes which are normal to the axis of the torus of Figure 4.1(a). Thus, the driven oscillator does not have fluctuations in a direction tangent to the flow and there is no phase diffusion. For the coupled oscillators, however, the fluctuations affect both oscillators and thus contain components in all directions of $\mathbb{R}^4$. In the phase-locked state both systems will exhibit small fluctuations in relative phase, with occasional transitions across an unstable orbit. In addition, the coupled oscillators will diffuse in absolute phase due to fluctuations tangent to the limit cycle.

Continuous systems with noise are described by stochastic differential equations. Although the trajectories of a stochastic differential equation are not differentiable, they are continuous, and thus can in principle be described by an iterated map. One would like to approximate the exact map in the following manner: some 'average' map deterministically takes the system from one intersection with the surface to the next, followed by a fluctuation which moves the system to a new point on the surface. This approximation results in a deterministic map perturbed at each iteration by additive noise. Studies of such noisy maps have been undertaken by many authors, generally under the assumption of a simple noise process. While we shall follow this approach here, we pause to point out that the relationship between the noise in a stochastic differential equation and that appearing in the resulting map is by no means trivial. The situation is particularly complex for autonomous systems of coupled oscillators, where the diffusion in absolute phase presents new difficulties for the construction of a noisy map. In the deterministic case the Poincaré map may contain a periodic attractor, corresponding to a stable limit cycle. The different points of the attractor are visited sequentially. The addition of small amplitude noise to the coupled oscillators has two effects. Owing to the noise the system will now make occasional discrete jumps in phase due to transitions across unstable periodic orbits. In the Poincaré map these phase shifts will break the sequential visiting of neighborhoods of the periodic points of the corresponding attractor. This effect can be described by a noisy Poincaré map. The diffusion in absolute phase will also break the sequential visiting of the periodic points. Consider a trajectory which has just pierced the surface of section in the appropriate direction. A fluctuation directed against the deterministic flow may drive the system back through the surface, only to intersect it again as the flow carries the system forward. This sequence of events would appear in the resulting map as the system remaining near a single periodic point for two iterations, and cannot be described by a simple Poincaré map. Phase diffusion causes an ensemble of systems starting in phase to evolve to a stationary distribution in phase. An ensemble of maps distributed in phase with such a distribution may be used to model the stationary distribution on the attractor.

4.3 Reduction to a stochastic map

We now examine in more detail the derivation of a map for a randomly perturbed nonlinear oscillator which we write in the form

$$\dot{x} = f(x) + \varepsilon\xi(t), \quad \varepsilon \ll 1, \quad x \in R^m, \tag{4.3.1}$$

where $\xi(t)$ indicates the noise, and ε its strength. We assume that the deterministic system ($\varepsilon = 0$) has an attracting limit cycle γ with period T:

$$x = x_0(t), \quad x_0(t+T) = x_0(t). \tag{4.3.2}$$

We introduce the new variable $y = x - x_0$ and linearize (4.3.1) to obtain

$$\dot{y} = M(t)y(t) + \varepsilon\xi(t), \tag{4.3.3}$$

where $M \equiv \mathrm{D}f(x_0(t))$ is a T-periodic matrix. This linearization enables us to formally construct both a time-T map relating $y(0)$ to $y(T)$, and a surface of section map. In the calculation care must be taken to distinguish between motion towards the limit cycle, i.e. in the 'amplitude' direction, and motion in the tangential or 'phase' direction, since the former is strongly influenced by the stability characteristics of the limit cycle, while the latter is influenced by the neutral stability relative to phase changes. We therefore define the unit vector

$$v(t) = \frac{f(x_0(t))}{|f(x_0(t))|}, \quad v(t+T) = v(t), \tag{4.3.4}$$

tangent to γ at time t, and use it to construct the projections

$$\left.\begin{aligned} y^{\parallel} &= (v(t)\cdot y(t))v(t) \\ y^{\perp} &= y(t) - (v(t))\cdot y(t))v(t). \end{aligned}\right\} \tag{4.3.5}$$

We refer to these as the parallel and perpendicular components, respectively.

These components satisfy the differential equations

$$\left.\begin{aligned}\frac{d}{dt}y_i^{\parallel}(t) &= L_{ij}^{\parallel}(t)y_j^{\parallel}(t) + L_{ij}^{\parallel}(t)y_j^{\perp}(t) + \varepsilon\xi_i^{\parallel}(t)\\ \frac{d}{dt}y_i^{\perp}(t) &= L_{ij}^{\perp}(t)y_j^{\parallel}(t) + L_{ij}^{\perp}(t)y_j^{\perp}(t) + \varepsilon\xi_i^{\perp}(t),\end{aligned}\right\} \tag{4.3.6}$$

where

$$\left.\begin{aligned}L_{ij}^{\parallel}(t) &= v_i(t)v_k(t)M_{kj}(t) + v_i(t)\dot{v}_j(t) + \dot{v}_i(t)v_j(t)\\ L_{ij}^{\perp}(t) &= M_{ij}(t) - L_{ij}^{\parallel}(t).\end{aligned}\right\} \tag{4.3.7}$$

The matrices $L^{\parallel}$, $L^{\perp}$ are therefore also T-periodic, and depend only on the deterministic limit cycle γ. In general, the matrix $L^{\perp}$ will be invertible, while $L^{\parallel}$ will not be. It is therefore simplest to introduce the Green operators

$$\left.\begin{aligned}A_{ij}^{\parallel}(t,t') &= \Theta(t-t')\exp\int_{t'}^{t} L_{ij}^{\parallel}(t'')\,dt''\\ A_{ij}^{\perp}(t,t') &= \Theta(t-t')\exp\int_{t'}^{t} L_{ij}^{\perp}(t'')\,dt'',\end{aligned}\right\} \tag{4.3.8}$$

where $\Theta(t-t')$ is the step function and time-ordered exponentials are understood, and use these to solve (4.3.6):

$$\left.\begin{aligned}y_i^{\parallel}(t) &= A_{ij}^{\parallel}(t,0)y_j^{\parallel}(0) + \int_0^{\infty} A_{ij}^{\parallel}(t,t')\{L_{jk}^{\parallel}(t')y_k^{\perp}(t') + \varepsilon\xi_j^{\parallel}(t')\}\,dt'\\ y_i^{\perp}(t) &= A_{ij}^{\perp}(t,0)y_j^{\perp}(0) + \int_0^{\infty} A_{ij}^{\perp}(t,t')\{L_{jk}^{\perp}(t')y_k^{\parallel}(t') + \varepsilon\xi_j^{\perp}(t')\}\,dt'.\end{aligned}\right\} \tag{4.3.9}$$

The initial conditions $y^{\parallel}(0)$, $y^{\perp}(0)$ are nonstochastic. The quantities $y^{\parallel}$, $y^{\perp}$ can now be eliminated between these equations to obtain a closed stochastic integral equation for either component. Since these equations are linear, the solutions can be written in the form

$$\left.\begin{aligned}y^{\parallel}(t) &= y_0^{\parallel}(t) + \varepsilon\zeta^{\parallel}(t)\\ y^{\perp}(t) &= y_0^{\perp}(t) + \varepsilon\zeta^{\perp}(t),\end{aligned}\right\} \tag{4.3.10}$$

where $y_0^{\parallel}(t), y_0^{\perp}(t)$ satisfy the deterministic part of (4.3.9), i.e. (4.3.9) with $\varepsilon = 0$, and $\zeta^{\parallel}(t)$, $\zeta^{\perp}(t)$ satisfy the stochastic part of (4.3.9):

$$\left.\begin{aligned}\zeta_i^{\parallel}(t) = {}& \int_0^{\infty} dt' \int_0^{\infty} dt''\, A_{ij}^{\parallel}(t,t')L_{jk}^{\parallel}(t')A_{kl}^{\perp}(t',t'')L_{lm}^{\perp}(t'')\zeta_m^{\parallel}(t'')\\ &+ \int_0^{\infty} dt'\, A_{ij}^{\parallel}(t,t')\left\{\xi_j^{\parallel}(t') + L_{jk}^{\parallel}(t')\int_0^{\infty} dt''\, A_{kl}^{\perp}(t',t'')\xi_l^{\perp}(t'')\right\}\\ \zeta_i^{\perp}(t) = {}& \int_0^{\infty} dt' \int_0^{\infty} dt''\, A_{ij}^{\perp}(t,t')L_{jk}^{\perp}(t')A_{kl}^{\parallel}(t',t'')L_{lm}^{\parallel}(t'')\zeta_m^{\perp}(t'')\\ &+ \int_0^{\infty} dt'\, A_{ij}^{\perp}(t,t')\left\{\xi_j^{\perp}(t') + L_{jk}^{\perp}(t')\int_0^{\infty} dt''\, A_{kl}^{\parallel}(t',t'')\xi_l^{\parallel}(t'')\right\}.\end{aligned}\right\} \tag{4.3.11}$$

The quantities $y_0^{\parallel}$, $y_0^{\perp}$ describe the deterministic response of the system to a perturbation at $t = 0$, while $\zeta^{\parallel}, \zeta^{\perp}$ describe the cumulative effect of the noise since $t = 0$.

At this point we still have the phase of $x_0(t)$ at our disposal. In the deterministic case when γ is a *periodic attractor* we are guaranteed by a theorem of Hirsch and Smale (1974) the existence of a point $x_0(0)$ on γ such that $|y_0(t)| \to 0$ as $t \to \infty$. This point, which we denote by $z_0(0)$, is the best 'tracking' point for an initial point $y_0(0)$. It allows us to define a *linear* deterministic time-T map whose fixed point is $z_0(0)$. If we did not make this choice of $x_0(0)$ the asymptotic phase shift would prevent $|y_0^{\parallel}(t)|$ and $|y_0^{\perp}(t)|$ from tending to zero with increasing time.

In the stochastic problem we may retain $z_0(0)$ as the tracking point. Then, because of phase diffusion, the relative phase between $x(t)$ and $z_0(0)$ will slip, leading to a mean secular growth in $y(t)$ with the number of iterations of the time-T map. The consequences of this are twofold. First, the linearization (4.3.3) becomes invalid even though the stochastic trajectory remains in the neighborhood of γ. Second, the system will be prevented from returning every period to a neighborhood of $z_0(0)$ as $t \to \infty$. Consequently, points along the entire limit cycle will appear in the map as the phase diffuses, i.e. the map will be in $\mathbb{R}^m$ with no reduction in dimension. Typically, the distance the system diffuses in time t is of order $\varepsilon t^{1/2}$; the linearization will thus remain valid for times much less than order $(2\pi/\varepsilon)^2$.

A second possibility is to choose a new point z_0 for each iteration. This eliminates the difficulty with phase diffusion but because $z_0 = z_0(y)$ the resulting map will be formally *nonlinear*. This is so even though we start with linearized equations around an attracting limit cycle, and reflects the fact that owing to phase diffusion the growth of $y(t)$ is bounded only by the size of γ.

Similarly, one can formally construct a return map. Due to phase diffusion the time for the trajectory to intersect the surface of section is now a stochastic quantity. Thus, the time t to be used in (4.3.9) and (4.3.11) is a function of the noise realization. The return map, by definition, only contains points in an $(m-1)$-dimensional surface. The use of a stochastic time, however, means that the parameters entering the map will depend on the noise, i.e. we obtain a map with multiplicative noise.

4.4 Noisy maps

Several authors have examined the bifurcation sequence of noisy maps as a control parameter is varied (e.g. Crutchfield, Farmer and Huberman, 1982; Mayer-Kress and Haken, 1981; Napiórkowski, 1985; Shraiman, Wayne and Martin, 1981). The main result is that high period orbits get washed out by the noise. Thus, a period-doubling cascade is truncated and higher period windows disappear. The lower period attractors, however, remain, and it is on these remaining attractors that we wish to study the dynamics. Consider the one-dimensional map

$$x_{n+1} = f(x_n, a) + \xi_n, \quad x \in \mathbb{R}, \tag{4.4.1}$$

where a is a parameter and ξ_n is an additive noise characterized by the probability distribution $P_\xi(\xi_n)\mathrm{d}\xi_n$. For example, when ξ_n is a Gaussian white noise process

$$P_\xi(\xi_n)\mathrm{d}\xi_n = \frac{1}{(2\pi\sigma^2)^{1/2}} \exp\left(-\frac{\xi_n^2}{2\sigma^2}\right)\mathrm{d}\xi_n, \tag{4.4.2}$$

where σ^2 measures the mean-square strength of the noise.

The dynamics of the map (4.4.1) depend on the parameter a. In the following we shall assume that a is chosen such that the deterministic system $x_{n+1} = f(x_n, a)$ has multiple periodic attractors. In order to understand the effect of noise on such a system it is essential to first understand the structure of the basins of attraction. Let $\{\omega_i\}_{i=0}^{N-1}$ be the periodic points of a period N attractor: $f(\omega_i) = \omega_{i+1 \,(\mathrm{mod}\, N)}$, $f^N(\omega_i) = \omega_i$. Then the modulus of the eigenvalue $\lambda = f'(\omega_0)f'(\omega_1)\ldots f'(\omega_{N-1})$ is less than unity. The basin of the attractor can be divided into N pieces, corresponding to the N distinct phases of motion on the attractor. We shall refer to these subsets of the full basin as phase basins. Under the action of the map, the system sequentially visits each phase basin. If one looks at the Nth iterate of the map, each phase point of the attractor becomes a fixed point, and each phase basin becomes the full basin of attraction of its corresponding fixed point. Surrounding each periodic point of an attractor is a connected region in the phase basin of that point, bounded by an unstable periodic point and its preimage. We shall refer to this region as the local phase basin. The union of the local phase basins of an attractor will be called the local basin of attraction. Adjoining each local phase basin is a region in either the basin of another attractor or another phase basin of the same attractor. In one dimension the boundaries of the local basin are the 'accessible points' (Alligood and Yorke, 1987; Grebogi, Ott and Yorke, 1986) of the basin boundary – those points on the boundary which can be reached by a continuous finite path which starts on the attractor and only intersects the boundary at the point in question. Each basin of a period N attractor is thus the union of N disjoint phase basins. Each phase basin has a connected subset, the local phase basin, containing the periodic point of the attractor of the corresponding phase.

As a period-one attractor progresses through a period-doubling bifurcation, its basin evolves into a complex structure of phase basins. At the first period-doubling bifurcation the basin of the period-one point splits into the two phase basins of the new period-two attractor. The now unstable fixed point plays a key role in the phase basin structure: it separates the two local phase basins of the attractor, and its preimages separate the pieces of the global phase basins. There is a countable infinity of preimages of the above unstable point accumulating at each point on the basin boundary of the old period-one

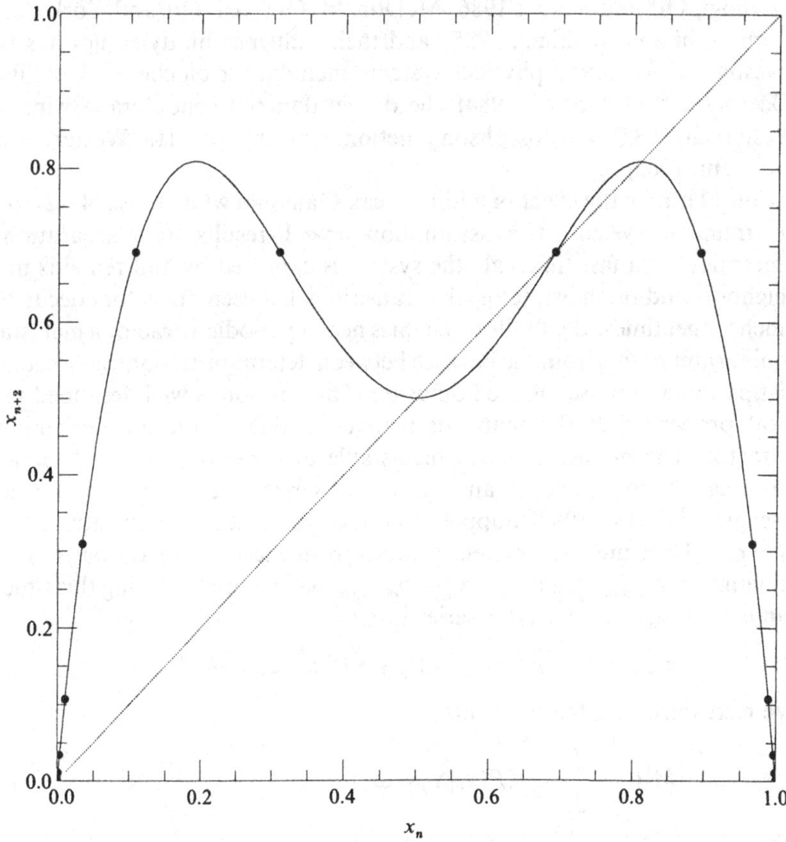

Figure 4.2. Two iterations of the logistic map $x_{n+2} = f(f(x_n))$, $f(x) = ax(1-x)$ at the superstable period-two point $a = 3.236068$. The solid dots are the unstable fixed point and its preimages, i.e. phase basin boundaries.

attractor. In these regions there are an infinite number of disconnected pieces of the two phase basins, whose widths go to zero as the boundary is approached. The situation is illustrated in Figure 4.2 which depicts the logistic map $x_{n+1} = ax_n(1-x_n)$ at the superstable period-two-point $a = 3.236\ldots$. At every period-doubling bifurcation the process repeats: each of 2^n newly unstable periodic points separates a pair of the 2^{n+1} local phase basins of the new period-2^{n+1} attractor, and the points which previously separated the 2^n phase basins become accumulation points for the new phase basin boundaries. Thus, Figure 4.2 can also be thought of as picturing the old local phase basin of a period-2^n attractor which, through a period-doubling bifurcation, has split into the two phase basins of the new period-2^{n+1} attractor. By the time one reaches high period attractors, the basin of attraction has become shredded into an extremely complicated set of phase basins. The existence and bifurcations of fractal basin boundaries have been studied by several authors

(Grebogi, Ott and Yorke, 1986; McDonald, Grebogi, Ott and Yorke, 1985; Yamaguchi and Mishima, 1985), and their influence on dynamics has been investigated for many physical systems including a biochemical oscillator (Decroly and Goldbeter, 1984), the driven damped pendulum (Gwinn and Westervelt, 1985), and Josephson junctions (Iansiti, Qing Hu, Westervelt and Tinkham, 1985).

Consider now the effect of adding weak Gaussian white noise (4.4.2) to the deterministic system. The assumption $\sigma^2 \ll 1$ results in a separation of timescales: on a fast timescale the system is captured by and remains in the neighborhood of an attractor; the transitions between attractors occur on a much longer timescale. While a system is near a periodic attractor a metastable equilibrium results from the balance between deterministic contraction due to dissipation and noise-induced diffusion. This motion is well described using local properties of the map – derivatives evaluated on the deterministic attractor. The moments of the metastable distribution near each periodic point can be expressed as an asymptotic series in the mean-square noise strength σ^2 (Weiss, 1987). Suppose that the system starts on the attractor, i.e. $x_0 = \omega_0$. Then until the system escapes from the local phase basin x_n will remain near $\omega_{n(\mathrm{mod}\,N)}$, and $\varepsilon_n = x_n - \omega_{n(\mathrm{mod}\,N)}$ will be small. During this time we can express ε_{n+1} as a Taylor series in ε_n:

$$\varepsilon_{n+1} = \xi_n + f'(\omega_{n(\mathrm{mod}\,N)})\varepsilon_n + \tfrac{1}{2}f''(\omega_{n(\mathrm{mod}\,N)})\varepsilon_n^2 + \cdots. \tag{4.4.3}$$

We next introduce the moments

$$M^k(t) = \int \mathrm{d}x_n\, \varepsilon_n^k P(x_n | x_0 = \omega_0), \tag{4.4.4}$$

where $P(x_n|x_0 = \omega_0)$, the conditional probability, can be expressed in terms of the probability distribution for ξ_n given by (4.4.2). Iterating (4.4.3) N times and averaging over the noise in (4.4.4) results in a map for the moments:

$$\left.\begin{aligned} M^1((m+1)N) &= \lambda M^1(mN) + \tfrac{1}{2}M^2(mN)L_0 + \sigma^2 B + O(\sigma^4), \\ M^2((m+1)N) &= \lambda^2 M^2(mN) + \sigma^2 C + O(\sigma^4), \end{aligned}\right\} \tag{4.4.5}$$

where

$$\left.\begin{aligned} L_n &= \sum_{m=n}^{N-1} f'(\omega_{N-1})\ldots f'(\omega_{m+1})f''(\omega_m) \\ &\quad \times (f'(\omega_{m-1})\ldots f'(\omega_n))^2, \\ B &= \sum_{n=1}^{N-1} L_{N-n}, \\ C &= 1 + \sum_{n=1}^{N-1} (f'(\omega_{N-1})\ldots f'(\omega_{N-n}))^2. \end{aligned}\right\} \tag{4.4.6}$$

In general the moments depend on the noise strength as $M^{2n-1} \sim M^{2n} \sim O(\sigma^{2n})$. The moments of the metastable distribution are given by the fixed points of (4.4.5),

$$\left.\begin{aligned} M_{\text{eq}}^{1} &= \frac{\sigma^2 B}{1-\lambda} + \frac{\sigma^2 L_0 C}{1-\lambda^2} + O(\sigma^4), \\ M_{\text{eq}}^{2} &= \frac{\sigma^2 C}{1-\lambda^2} + O(\sigma^4). \end{aligned}\right\} \tag{4.4.7}$$

The condition for stability of the attractor, $|\lambda| < 1$, is identical to the condition for the stability of the fixed points M_{eq}^1, M_{eq}^2. Thus, the moments of the metastable distribution within each local phase basin can be calculated from local properties of the deterministic map.

There are two types of transition phenomena associated with periodic attractors; the system can either escape to a different attractor, or it can remain on the same attractor with a phase shift. In the case of a phase shift the system jumps to a phase basin other than the one given by the deterministic map. Both types of transitions will be referred to as escape events as they involve crossing an unstable orbit. In order to discuss the first passage time for escape one must define a region of phase space from which to escape. The local basin of attraction provides a natural choice for this region. If the basin boundaries are simple the system will quickly relax into the neighborhood of an attractor. When the basin boundaries are fractal, however, the system may exhibit transient chaotic motion before relaxing onto an attractor. The presence of noise acts to smooth out the basin structure. When the system is in a region of phase space with finely interwoven basins, fluctuations will cause frequent transitions between basins.

The final equilibrium distribution is primarily composed of the metastable distributions around each attractor weighted by a factor depending on both the escape time from and the return time to that attractor. In addition, there will be a small contribution to the final equilibrium distribution arising from the transients occurring between times spent near an attractor. As the noise strength increases these transients become more frequent and make a more significant contribution to the equilibrium distribution. One property of interest is the 'most stable' attractor, i.e. the attractor near which a system spends the most time. As emphasized by Landauer (1979), determining the relative stability of two attractors can be quite complicated. The attractor with the longest escape time may not be the attractor most often occupied; a very long return time can cause the system to spend a small fraction of its time near this attractor. The average return time will in general depend on the details of the global basin structure.

4.5 The cubic map

The simplest map with multiple attractors is a one-dimensional cubic map. We shall study the cubic map

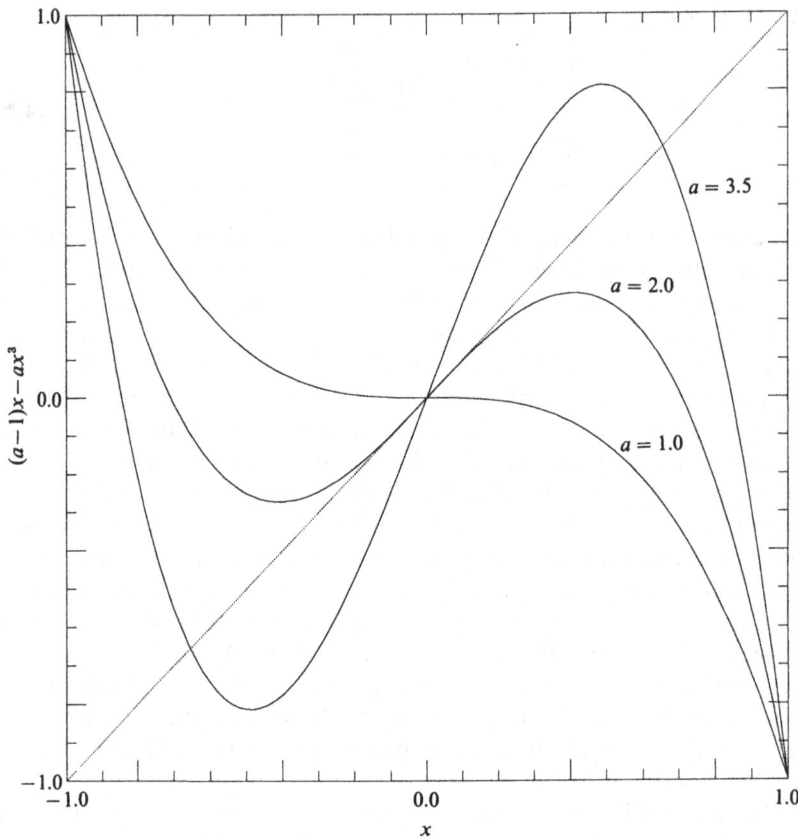

Figure 4.3. The cubic map $f(x) = (a-1)x - ax^3$.

$$x_{n+1} = f(x_n) + \xi_n, \quad f(x) = (a-1)x - ax^3, \tag{4.5.1}$$

$$\langle \xi_n \rangle = 0, \quad \langle \xi_m \xi_n \rangle = \sigma^2 \delta_{m,n}, \quad \sigma^2 \ll 1, \tag{4.5.2}$$

which contains two symmetric attractors over the parameter range $2.0 < a < 4.0$. Arecchi, Badii and Politi (1984) have studied the power spectrum and Lyapunov exponent in this map as the noise strength varies. May (1979) studied the bifurcation structure of a different cubic map ($x_{n+1} = -f(x_n)$) in the context of population genetics, and Testa and Held (1983) went on to examine the periodic windows and the structure of the basins of attraction for a period-eight attractor of the same map. A cubic map with a quadratic term to break the symmetry has been studied in both the deterministic (Fraser and Kapral, 1982), and the noisy case (Celarier, Fraser and Kapral, 1983; Fraser, Celarier and Kapral, 1983). Our interest lies in the dynamics of (4.5.1) at a fixed

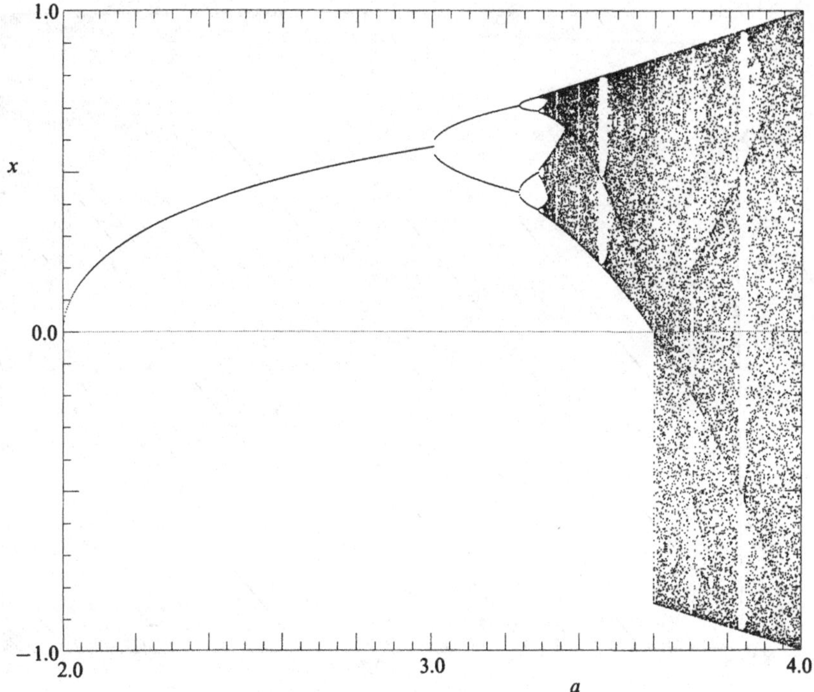

Figure 4.4. Bifurcation diagram for the cubic map following one of two attractors. The second attractor is obtained by reflection in $x = 0$.

parameter value for which the deterministic system has two periodic attractors.

The deterministic cubic map $x_{n+1} = f(x_n) = (a-1)x_n - ax_n^3$ is a bounded dynamical system with phase space $x \in [-1, 1]$ for parameter values $0 \leqslant a \leqslant 4$. Figure 4.3 shows the cubic map for various parameter values. Figure 4.4 is the bifurcation diagram following one of the two attractors. The second attractor may be obtained by symmetry. For $a < 2$ the origin is a stable fixed point. A pitchfork bifurcation occurs at $a = 2.0$ creating a symmetric pair of fixed points. As a increases each attractor follows a period-doubling route to chaos. At $a_{\text{crisis}} = 3\sqrt{3}/2 + 1$ the critical points get mapped to a root of f and there is a crisis bifurcation (Grebogi, Ott and Yorke, 1983a, b): the chaotic attractor collides with the unstable fixed point at $x = 0$. Prior to this crisis the two attractors lie on opposite sides of the origin, while afterward the attractors become intertwined. Periodic windows occur throughout the chaotic regime. We shall pay particular attention to the dynamics within the period-three window containing $a = 3.981798$, a window which is too narrow to be visible in Figure 4.4. At $a = 3.981798$ the period-three attractor has an eigenvalue near zero ($\lambda = -0.0118$) and the three stable periodic points are $\omega_1 = 0.993175\ldots, \omega_2 = -0.939379\ldots$, and $\omega_3 = 0.499633\ldots$. At $a = 4.0$ there is a

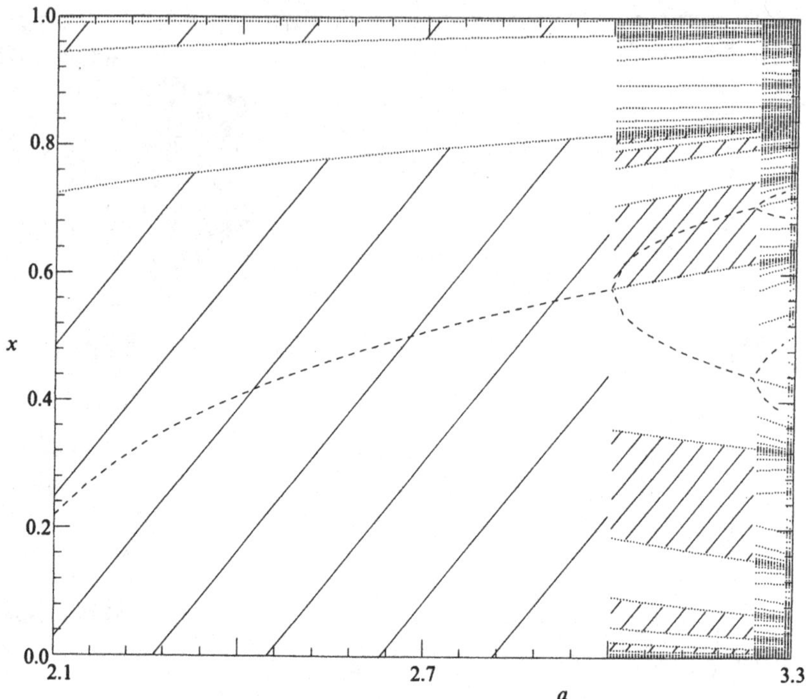

Figure 4.5. Phase basin boundaries for the cubic map as a function of a. The dashed line is the bifurcation diagram for one of the two attractors. In the period-one regime the basin of attraction of the first attractor is cross-hatched, while the basin of the second attractor is white. In the period-two regime one phase basin of the first attractor is cross-hatched while the other phase basin of the same attractor is white.

crisis in which the chaotic attractors collide with the unstable period-two points at $x = \pm 1$. For $a > 4$ the motion is no longer bounded.

As a increases through the period-doubling cascade the phase basins of each attractor become increasingly complex as described in the previous section (Figure 4.5). For $a < a_{\text{crisis}}$ the critical points of f are mapped into the same basin in which they reside and the local phase basins of a given attractor are surrounded by regions in other phase basins of the same attractor. Thus, most escape events involve phase shifts, with an extraordinarily large fluctuation required to produce a jump between attractors. Above a_{crisis} the two attractors become intertwined and the situation is very different. We use the period-three window as an example. Immediately outside each local phase basin the basins of attraction are very finely interwoven. Escape from the local phase basin results in transient chaotic motion followed by capture by an attractor.

Numerically, we approximate each Gaussian fluctuation as the sum of twenty evenly distributed pseudorandom numbers. The convergence of the sum to a Gaussian distributed random variable is given by the central limit theorem (Khinchin, 1949). Technically, the addition of noise renders the cubic map unbounded; fluctuations can drive the system across the unstable period-two points at ± 1. These events are extremely rare, however, occurring on a timescale much longer than any other timescale of the dynamics, and shall be ignored.

A noisy system starting on a deterministic attractor will quickly relax to motion within the local phase basin described by the metastable distribution. On a longer timescale the system will experience a fortuitous sequence of fluctuations and escape the local phase basin. This separation of timescales has a consequence for numerical simulations. The decay to a metastable distribution is easily computed numerically, while transitions between attractors are much slower and thus require more computing time. As a result, we have limited ourselves to studying escape events for a single parameter in the cubic map: the period-three window at $a = 3.981798$. The first passage time for escape from an attractor is a stochastic quantity. It is thus the mean first passage time which is often considered to be of interest. The mean of a stochastic quantity is only representative of individual events if their distribution is sharply peaked. If the distribution is broad the rms deviation can be as large as the mean and large deviations from the mean are likely. A simple example is given by the exponential distribution

$$\rho(t)\,\mathrm{d}t = \frac{1}{\tau}\mathrm{e}^{-t/\tau}\,\mathrm{d}t. \tag{4.5.3}$$

The exponential distribution is not sharply peaked, and the mean and the rms deviation are both equal to τ. The numerically measured first passage time distribution for several noise strengths is shown in Figure 4.6. The distribution rises sharply from zero, has a peak at small times, and then slowly decays. Notice that the mean first passage time is quite different from the most probable first passage time. For rms noise strengths 1.2×10^{-5} or less (mean first passage times greater than 200) a least-squares fit of the first passage time distribution to an exponential distribution yields a mean first passage time that averages only 5% too high with a correlation of 0.94. The distribution's fit to an exponential deteriorates with higher noise strengths, although it still has a large rms first passage time.

In periodic windows the local phase basin is bounded on one side by the unstable periodic point created with the stable periodic point at the tangent bifurcation, and on the opposite side by a preimage of the unstable periodic point. The size of the local phase basin varies from one phase to another, and the likelihood of escape is greatest in the narrowest local phase basin. For $a = 3.981798$ the sizes of the three local phase basins l_i are roughly an order of magnitude apart: $l_1 = 7.63 \times 10^{-5}, l_2 = 6.71 \times 10^{-4}$, and $l_3 = 5.07 \times 10^{-3}$. Escape is most likely when the system is in the neighborhood of the periodic point with the smallest local phase basin, ω_1. We find numerically that the fraction of escapes from each local phase basin is independent of which

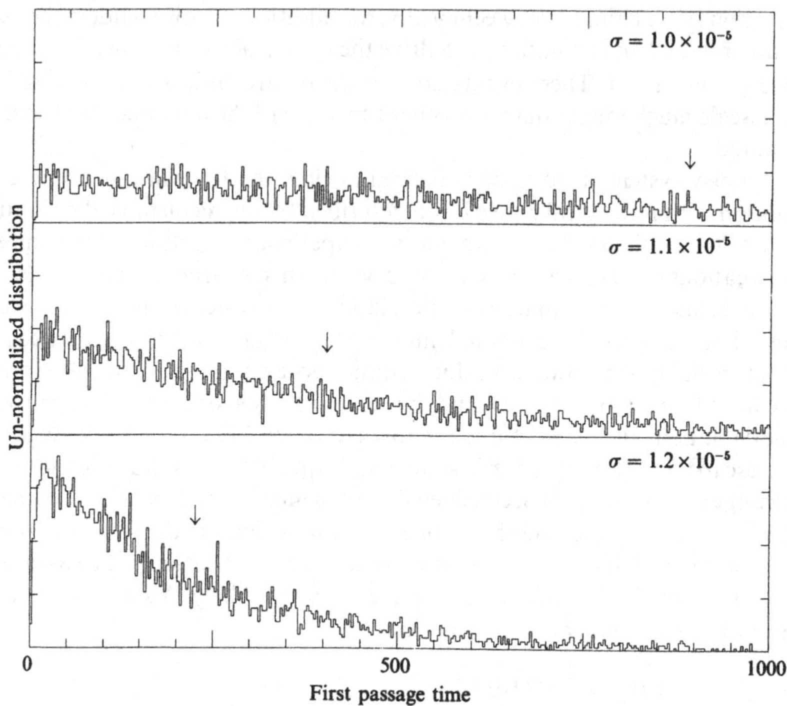

Figure 4.6. First passage time distribution for escape from the period-three attractor of the noisy cubic map at $a = 3.981798$. Arrows indicate the mean first passage time, and σ is the rms noise strength.

periodic point is used as an initial condition, and is roughly constant for rms noise strengths below 2.0×10^{-5}, corresponding to mean first passage times above twenty-five iterations. The variation of the escape fraction with noise strength is of the same size as the fluctuation due to different noise realizations. Averaging over noise strength and initial phase we find that 90.8% of the escapes are from the narrowest basin, 8.5% are from the intermediate size basin, and 0.7% are from the largest basin.

The question arises as to whether escape from the local basin is the result of a small number of relatively large fluctuations, or is due to a long sequence of smaller fluctuations. We can answer this question by examining a probability distribution conditioned on escape in the future, instead of a distribution conditioned in the past. The time for such a distribution to relax backwards in time to the metastable distribution gives an idea of the timescale of escape events. Figure 4.7 compares this future-conditioned distribution with the Gaussian approximation to the metastable distribution given by (4.4.7). Due to the action of the map, most escapes occur at the boundary composed of the unstable point. If the system approaches the preimage of the unstable point, one iteration of the map brings the system to the side of the attractor bounded

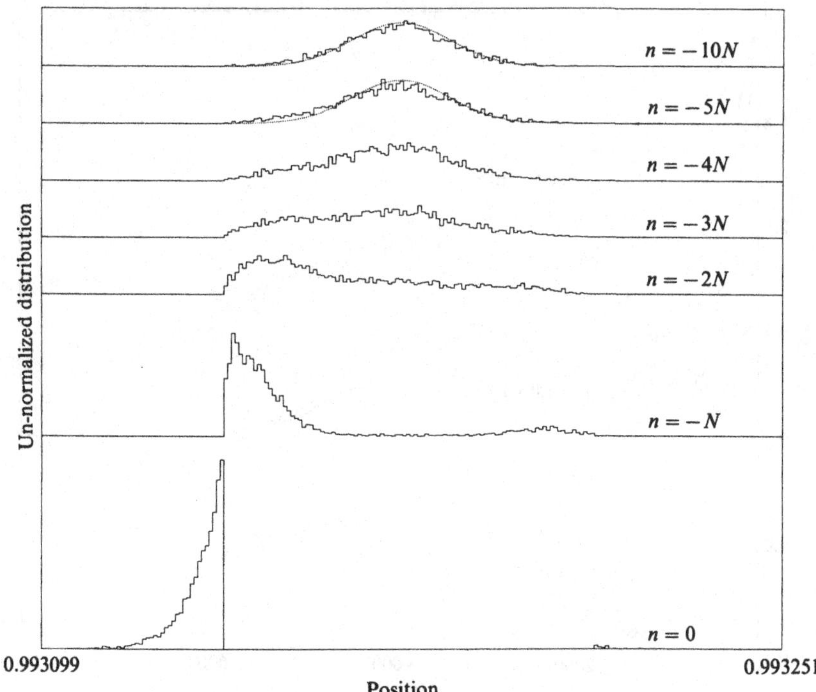

Figure 4.7. Distribution of position for an ensemble of systems which escape the period-three attractor of the noisy cubic map ($a = 3.981798$) at $n = 0$. The dotted line is the analytically calculated metastable distribution. N denotes the period of the attractor. The rms noise strength is $\sigma = 1.0 \times 10^{-5}$.

by the unstable periodic point. We see that the distribution conditioned on escape approximates the metastable distribution only five periods before escape, although the tail is slightly larger and the peak somewhat smaller. These discrepancies slowly diminish as we look further backwards in time. The conclusion is that most escape events are short time processes, occurring within a small number of periods. Individual systems do not slowly diffuse out of the attractor; there is a relatively sharp transition between motion within the metastable distribution and motion outside the local basin.

Once a system has escaped from the local phase basin it wanders on a transient chaotic trajectory before being captured by an attractor (Grebogi, Ott and Yorke, 1983a, b). The region of phase space covered by the chaotic transient is also the region with finely interwoven basins of attraction. The noise thus makes it impossible to predict which attractor will eventually capture the system. The average time spent in the transient appears to be insensitive to the noise strength, and like the escape time discussed above it has an rms deviation approximately equal to the mean.

An ensemble of systems starting in phase on a single attractor will slowly

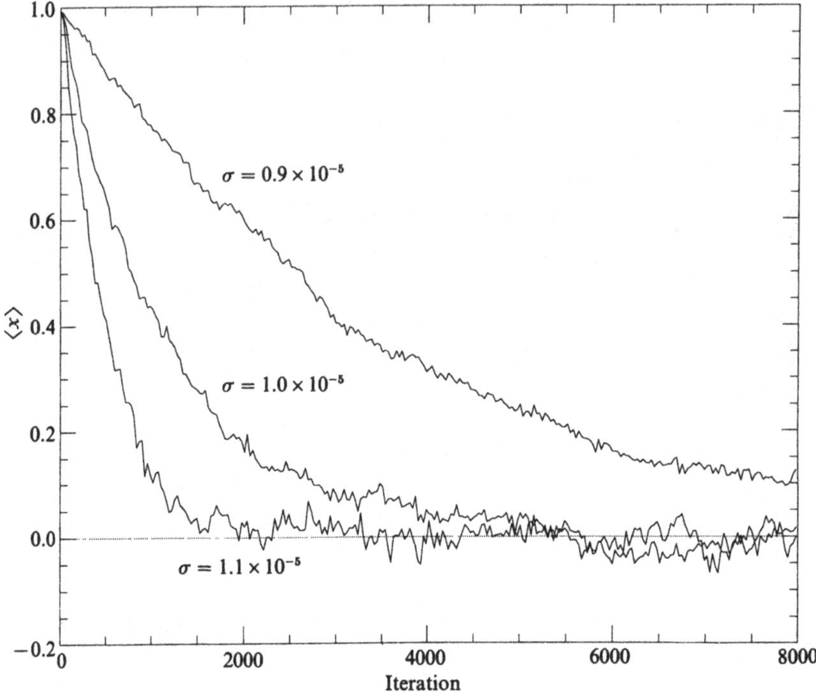

Figure 4.8. Mean position as a function of iterations of the noisy cubic map ($a = 3.981798$) for an ensemble of systems starting on the stable period-three point $x = 0.993175\ldots$. The position is plotted for iterations which are an integer number of periods. The rms noise strength is denoted by σ.

diffuse in phase and onto the other attractor due to escape events. The probability distribution of such an ensemble will evolve to a stable equilibrium on a slow timescale dependent on the noise strength. Although individual systems experience relatively sudden escapes from the metastable distribution, the large rms first passage time means that an ensemble of systems will slowly diffuse out of an attractor. This diffusion can be characterized by studying the time dependence of the mean position. For the cubic map, the final equilibrium distribution will contain equal probability of being in the metastable distribution of the two symmetric attractors, and equal probability for all phases. The average position is thus zero. Figure 4.8 shows the long time evolution of the mean position for an ensemble of systems starting on the attracting point $x = \omega_1 = 0.993175\ldots$ of the period-three attractor at $a = 3.981798$. The diffusion is linear at first, and then slows down as the final equilibrium is reached. Fluctuations about the final equilibrium are due to the finite size ensemble.

4.6 Conclusions

In this chapter we have discussed several aspects of noisy iterated maps with multiple attractors. One expects such maps to arise in systems described by stochastic differential equations with locally stable limit cycles. The phenomenon of phase diffusion, the cumulative effect of fluctuations in directions with zero eigenvalues, plays an important role in the dynamics. As the phase diffuses, the system no longer returns to the same region of phase space every period. Hence a time-T map contains points along the entire limit cycle and there is no reduction of dimension. Alternatively, the stochastic time required for the system to return to a given surface in phase space implies that the surface of section map contains multiplicative noise. As the distance between a stochastic system and its deterministic counterpart grows via diffusion, any linearization using that distance as a small parameter must break down even though the stochastic system may remain close to the limit cycle. Systems with fluctuations in neutrally stable directions are thus poorly suited for modelling with iterated maps perturbed by additive noise. On the other hand, if fluctuations only occur in locally stable directions the dynamics may be adequately represented by such a map, as is the case for a noisy oscillator which is phase-locked to a deterministic periodic driver.

The dynamics of an iterated map with small amplitude additive noise depend sensitively on the structure of the basins of attraction of the various attractors. The division of each basin of attraction into phase basins and local phase basins allows one to determine which transitions are likely; most transitions will be from a local basin of attraction to the basin adjoining it. At sufficiently low periods the local phase basins are surrounded by regions in other phase basins of the same attractor; consequently most transitions consist of phase shifts. As an attractor progresses through a period-doubling cascade the phase basin structure becomes exceedingly complicated. In periodic windows in the chaotic regime the basins outside the local basin of attraction are sufficiently intertwined to prevent prediction of the probable transitions.

We have numerically investigated the dynamics of escape events for a cubic map in a period-three window. For small noise the first passage time distribution is well approximated by an exponential distribution. Despite the long escape time, the actual transition between motion within the metastable equilibrium distribution and motion outside the local basin takes place in a small number of iterations. The broad distribution of escape times results in an ensemble of systems slowly diffusing out of the local phase basin, onto both other attractors and other phase basins of the same attractor as determined by the basin structure.

Acknowledgements

This work was supported in part by the California Space Institute and the IBM Distributed Academic Computing Environment Grant.

References

Adler, R. 1946. *Proc. IRE* **34**, 351.

Alligood, K. T., and Yorke, J. A., 1987. Preprint.

Ambegaokar, V. and Halperin, B. I. 1969. *Phys. Rev. Lett.* **22**, 1364.

Arecchi, F. T., Badii, R. and Politi, A. 1984. *Phys. Rev. A* **29**, 1006.

Bohr, T., Bak, P. and Jensen, M. H. 1984. *Phys. Rev. A* **30**, 1970.

Büttiker, M. and Landauer, R. 1981. In *Nonlinear Phenomena at Phase Transitions and Instabilities* (T. Riste, ed.), p. 111, New York: Plenum Press.

Celarier, E. A., Fraser, S. and Kapral, R. 1983. *Phys. Lett.* **94A**, 247.

Crutchfield, J. P., Farmer, J. D. and Huberman, B. A. 1982. *Phys. Rep.* **92**, 45.

Decroly, O. and Goldbeter, A. 1984. *Phys. Lett.* **105A**, 259.

Fraser, S., Celarier, E. and Kapral, R. 1983. *J. Stat. Phys.* **33**, 341.

Fraser, S. and Kapral, R. 1982. *Phys. Rev. A* **25**, 3223.

Grebogi, C., Ott, E. and Yorke, J. A. 1983a. *Physica* **7D**, 181.

Grebogi, C., Ott, E. and Yorke, J. A. 1983b. *Phys. Rev. Lett.* **50**, 935.

Grebogi, C., Ott, E. and Yorke, J. A. 1986. *Phys. Rev. Lett.* **56**, 1011.

Grüner, G., Zawadowski, A. and Chaikin, P. M. 1981. *Phys. Rev. Lett.* **46**, 511.

Grüner, G. and Zettl, A. 1985. *Phys. Rep.* **119**, 117.

Guckenheimer, J. and Holmes, P. 1983. *Nonlinear Oscillations, Dynamical Systems, and Bifurcations of Vector Fields.* New York: Springer-Verlag.

Gwinn, E. G. and Westervelt, R. M. 1985. *Phys. Rev. Lett.* **54**, 1613.

Hénon, M. 1976. *Commun. Math. Phys.* **50**, 69.

Hirsch, M. W. and Smale, S. 1974. *Differential Equations, Dynamical Systems, and Linear Algebra.* New York: Academic Press.

Iansiti, M., Qing Hu, Westervelt, R. M. and Tinkham, M. 1985. *Phys. Rev. Lett.* **55**, 746.

Jensen, M. H., Bak, P. and Bohr, T. 1984. *Phys. Rev. A* **30**, 1960.

Khinchin, A. I. 1949. *Mathematical Foundations of Statistical Mechanics.* New York: Dover Publications.

Knobloch, E. and Wiesenfeld, K. A. 1983. *J. Stat. Phys.* **33**, 611.

Landauer, R. 1979. *Ann. NY Acad. Sci.* **316**, 433.

Lindelof, P. E. 1981. *Rep. Prog. Phys.* **44**, 949.

McDonald, S. W., Grebogi, C., Ott, E. and Yorke, J. A. 1985. *Physica* **17D**, 125.

May, R. M. 1979. *Ann. NY Acad. Sci.* **316**, 517.

Mayer-Kress, G. and Haken, H. 1981. *J. Stat. Phys.* **26**, 149.

Napiórkowski, M. 1985. *Phys. Lett.* **112A**, 357.

Sherwin, M. S. and Zettl, A. 1985. *Phys. Rev. B* **32**, 5536.

Shraiman, B., Wayne, C. E. and Martin, P. C. 1981. *Phys. Rev. Lett.* **46**, 935.

Testa, J. and Held, G. A. 1983. *Phys. Rev. A* **28**, 3085.

Wiesenfeld, K., and Satija, I. 1987. Preprint.

Weiss, J. B. 1987. *Phys. Rev. A* **35**, 879.

Yamaguchi, Y. and Mishima, N. 1985. *Phys. Lett.* **109A**, 196.

5 Discrete dynamics perturbed by weak noise

PETER TALKNER and PETER HÄNGGI

5.1 Perspectives

In Chapter 9 of Volume 1, one of us elaborated on noisy dynamical flows described by the set of first-order differential equations

$$\dot{x}_\alpha = f_\alpha(x, \lambda) + \sum_i q_{\alpha i}(x, \lambda)\xi_i(t). \tag{5.1.1}$$

Such systems exhibit already for a small number of coupled state variables an overwhelming complexity which generally defies an analytic description. It turns out, however, that the complexity of such systems can be extracted from a stroboscope-like discretization in time of a single trajectory $x(t)$. With strong dissipation even a multidimensional flow of a huge number of coupled degrees of freedom can be described in terms of a one-dimensional discrete iterative mapping. That is, one can predict in a causal manner with surprisingly good accuracy, e.g., the maximum of a state variable $x_\alpha(t_{n+1})$ of a complex dynamics if we only know the previous maximum $x_\alpha(t_n)$; i.e. $x(t_{n+1}) = F(x(t_n); \lambda)$, where λ denotes a set of control parameters. Likewise, the set of values $\{x_\alpha(t_0), x_\alpha(t_0 + \tau), \ldots, x_\alpha(t_0 + n\tau)\}$ of a sequence of observations at constant time intervals, τ, also obeys the same law. $F(x, \lambda)$ often exhibits turning points, or maxima and minima. This stroboscope-like procedure then yields an approximation to the stochastic flow in (5.1.1) which takes on the form of a noisy discrete dynamics

$$x_{n+1} = F(x_n, \lambda) + \xi_n. \tag{5.1.2}$$

Without the noisy term ξ_n, equations such as (5.1.2) have been studied extensively in the context of 'deterministic chaos' (e.g. see Coullet and Tresser, 1978a, b; Feigenbaum, 1978a, b, 1979; Grossman and Thomae, 1977; for a review, see Schuster, 1984). In the following we shall not comment on the many interesting results of deterministic chaos, but rather put our focus on the effects of weak random perturbations ξ_n. The importance of residual noise on the discrete dynamics in (5.1.1) has been recognized in numerical studies (for a review see Crutchfield, Farmer and Huberman, 1982; some recent papers are Linz and Lücke, 1986; B. Morris and F. Moss, 1986, and 1988, unpublished).

Our goal in this study of discrete dynamical systems is the (predominantly analytical) calculation of stationary probabilities, mean first passage times, and related quantities, in the *limit of weak noise* (Talkner, Hänggi, Freidkin and Trautmann, 1987).

5.2 Discrete dynamics driven by white noise

In the following we shall restrict the discussion to discrete systems driven by white noise only. In other words, we shall consider the map dynamics in (5.1.1) with ξ_n being independent and identically distributed noise. It should be stressed, however, that ξ_n generally will depend on the state x_n, or more precisely it will depend on the iterated state $F(x_n)$; i.e.

$$x_{n+1} = F(x_n) + \xi_n(F(x_n)). \tag{5.2.1}$$

$\xi_n(x)$ is specified by its probability density $\rho(\xi, x)$ for finding $\xi_n(x)$ in the interval $(\xi, \xi + \mathrm{d}\xi)$

$$P(\xi(x) \in [\xi, \xi + \mathrm{d}\xi]) = \rho(\xi, x)\,\mathrm{d}\xi. \tag{5.2.2}$$

With these conditions for ξ_n, (5.2.1) defines a Markov process; put more precisely, $\{x_n\}$ in (5.2.1) defines a Markov chain (Feller, 1966).

The probability density $W_n(x)$ for finding x_n in the interval $[x, x + \mathrm{d}x]$ then obeys the master equation

$$W_{n+1}(x) = \int P(x\,|\,y)\,W_n(y)\,\mathrm{d}y, \tag{5.2.3}$$

where the transition probability $P(x\,|\,y)$ is given by

$$P(x\,|\,y) = \rho(x - F(y), F(y)). \tag{5.2.4}$$

Often one assumes for ξ_n a *multiplicative* noise structure of the form

$$\xi_n(x) = g(x_n)\zeta_n, \tag{5.2.5}$$

with ζ_n being white noise with an x-independent probability $\varphi(\zeta)$. The noise ξ_n is termed *additive* if the coupling function $g(x)$ is independent of the state variable x. The probability density for $\xi_n(x)$ in (5.2.5) reads (Z: normalization)

$$\rho(\xi, x) = |g(x)|^{-1}\varphi(\xi/g(x))/Z. \tag{5.2.6}$$

In many other cases, the physics dictates a process x_n that is restricted to an *a priori* fixed interval $I = [x^{(1)}, x^{(2)}]$, for all times n. Then the noise ξ_n cannot be additive. With multiplicative noise, $g(x)$ then must vanish at the boundary points $x^{(1)}, x^{(2)}$, in order to prevent an eventual escape out of the interval. This, however, implies that there cannot be any fluctuations at the boundary itself; i.e.

$$\rho(\xi, x^{(\mathrm{B})}) = 0, \quad x^{(\mathrm{B})} = x^{(1)}, x^{(2)}. \tag{5.2.7}$$

Depending on the problem under consideration the condition in (5.2.7) might not be appropriate. As an illustrative example let us take a random number ξ from an ensemble with density $\varphi(y)$; this will be taken as an allowed realization of $\xi_n(x)$ if $(\xi + x_n)$ is contained in the interval. If not true, we must select another random number. This procedure implies a state-dependent fluctuating force with density (Haken and Mayer-Kress, 1981)

$$\rho(\xi, x) = \begin{cases} \varphi(\xi) \Big/ \int_{x^{(1)}}^{x^{(2)}} \varphi(\xi' - x)\mathrm{d}\xi', & \text{if } (x + \xi) \in I \\ 0, & \text{otherwise.} \end{cases} \tag{5.2.8}$$

Other examples that also cannot be modeled by multiplicative noise are processes driven by additive noise with *periodic* or reflecting boundary conditions (b.c.). In the case of a periodic b.c., $(x + \xi)$ is taken modulo the length of the interval, if not contained in I; likewise, for a reflecting b.c., $(x + \xi)$ is mirrored at the adjacent boundary as many times until the image falls into the interval. Thus, we find for the density ρ_{P} and ρ_{R} for periodic b.c. and reflecting b.c., respectively

$$\rho_{\mathrm{P,R}}(\xi, x) = \sum_{n=-\infty}^{\infty} \varphi(S_n^{\mathrm{P,R}}(\xi + x) - x) \tag{5.2.9a}$$

with

$$S_n^{\mathrm{P}}(x) = 2n(x^{(2)} - x^{(1)}) + x \tag{5.2.9b}$$

and

$$S_n^{\mathrm{R}}(x) = 2n(x^{(2)} - x^{(1)}) + (-1)^n x. \tag{5.2.9.c}$$

Probability densities for the random force in the presence of more general b.c., such as, e.g., a sticking b.c., etc., are constructed analogously. In conclusion, the multiplicative form in (5.2.5) does not present the most general possible noise structure.

5.3 Stationary probability for weak Gaussian noise

Next we shall consider one-dimensional maps which are weakly disturbed (strength measured by the parameter $\varepsilon < 1$) by white Gaussian noise; i.e.

$$x_{n+1} = F(x_n) + \varepsilon^{1/2}\xi_n, \tag{5.3.1}$$

where $\varepsilon > 0$ and ξ_n has a density

$$\rho(\xi) = (2\pi)^{-1/2} \exp(-\tfrac{1}{2}\xi^2). \tag{5.3.2}$$

If

$$F'_{\pm\infty} \equiv \lim_{x \to \pm\infty} F'(x) \quad \text{exists, and if } |F'_{\pm\infty}| < 1,$$

there is a uniquely defined stationary probability $W(x), x \in (-\infty, \infty)$, of the process defined by (5.3.1) and (5.3.2). This invariant probability (solution with eigenvalue 1) obeys the master equation

$$W(x)=(2\pi\varepsilon)^{-1/2}\int_{-\infty}^{\infty}\exp[-(x-F(y))^2/2\varepsilon]\,W(y)\,\mathrm{d}y. \tag{5.3.3}$$

In order to solve (5.3.3) for weak noise we use a WKB-type ansatz

$$W(x)=Z(x)\exp(-\Phi(x)/\varepsilon), \tag{5.3.4}$$

wherein both $\Phi(x)$ and $Z(x)$ shall not depend on ε. This procedure is well-known in the study of continuous-time Markov processes (Graham, 1981; Kubo, Matsuo and Kitahara, 1973; Ludwig, 1975; Matsuo, 1977). Inserting this ansatz into (5.3.3), one finds

$$Z(x)=(2\pi\varepsilon)^{-1/2}\int_{-\infty}^{\infty}Z(y)\exp(-Y^2(x,y)/2\varepsilon)\,\mathrm{d}y, \tag{5.3.5}$$

where we have defined

$$Y^2(x,y)=2[\Phi(y)-\Phi(x)]+[x-F(y)]^2. \tag{5.3.6}$$

Our goal is to determine Φ in such a way that

$$Y^2(x,y)\geqslant 0, \tag{5.3.7}$$

as indicated already by the notation.

In the following we shall work out a seemingly strange chain of arguments involving the quantity $Y^2(x,y)$. This reasoning, however, will result in (5.3.13) which provides us the key to find the potential function $\Phi(x)$ occurring in (5.3.4).

First we note that the root of Y^2, i.e. $Y(x,y)$, defines for each x an invertible transformation of the old coordinate, y, to the new one, Y. Thus, it is sufficient and necessary that the derivative of $Y(x,y)$ with respect to y is not vanishing; i.e.

$$\frac{\partial Y(x,y)}{\partial y}\neq 0. \tag{5.3.8}$$

For the moment, let us assume that we know a function Φ, defining a function $Y^2(x,y)$, (see(5.3.6)), obeying the properties (5.3.7) and (5.3.8). Then we can perform in the integral (5.3.5) a change of coördinates from y to Y; i.e.

$$Z(x)=(2\pi\varepsilon)^{-1/2}\int_{-\infty}^{\infty}Z\{y(Y,x)\}\times\exp(-Y^2/2\varepsilon)\frac{\mathrm{d}Y}{|\partial Y(x,y)/\partial y|_{y=y(Y,x)}}. \tag{5.3.9}$$

Here, $y(Y,x)$ is the unique solution of (5.3.6) for y. Further, if we assume that the quantity

$$Z\{y(Y,x)\}/|\partial Y/\partial y|_{y=y(Y,x)}$$

is a smoothly varying function of Y in a sufficiently large neighborhood around $Y=0$, say $Y<10\,\varepsilon^{1/2}$, we can evaluate the integral (5.3.9) for small ε. We find up to order $O(\varepsilon^0)$

$$Z(x)=Z\{y(Y=0,x)\}/|\partial Y/\partial y|_{y=y(Y=0,x)}. \tag{5.3.10}$$

This constitutes a linear functional equation in the unknown function $Z(x)$; it can be solved iteratively.

The positivity condition on Y^2, (see(5.3.7)), has still another important consequence. Setting $x=F(y)$ in (5.3.6), we observe that $\Phi(x)$ *decreases* along the deterministic trajectory $x_{n+1}=F(x_n)$; i.e.

$$\Phi(F(y))-\Phi(y)=-\tfrac{1}{2}Y^2(x,y)\leqslant 0. \tag{5.3.11}$$

To put it differently, $\Phi(x)$ is a *Lyapunov function* of the deterministic dynamics $x_{n+1}=F(x_n)$. Now we shall establish an equation for $\Phi(x)$ itself. In order for (5.3.8) to hold true at each pair of points (x,y) at which $\partial Y^2(x,y)/\partial y=0$, $Y^2(x,y)$ must vanish too. First we obtain from (5.3.6) that for every y with $F'(y)\neq 0$ (prime indicates differentiation) there exists an x-value such that $\partial Y^2/\partial y$ vanishes; i.e.

$$x=(\Phi'(y)/F'(y))+F(y). \tag{5.3.12}$$

Now, however, in order that $\partial Y/\partial y\neq 0$ holds, (see (5.3.8)), for all x and y, $Y^2(x,y)$ must vanish for those x given by (5.3.12). With (5.3.6) this yields the desired equation for $\Phi(x)$ itself

$$\tfrac{1}{2}(\Phi'(y)/F'(y))^2+\Phi(y)-\Phi\left(\frac{\Phi'(y)}{F'(y)}+F(y)\right)=0. \tag{5.3.13}$$

This is a nonlinear functional differential equation for the potential function Φ. At first sight, this equation looks even more complicated than the linear integral equation (5.3.3), which was our starting point. Equation (5.3.13), however, can be related to a two-dimensional Hamiltonian system with discrete time, which can be solved iteratively (P. Talkner, manuscript in preparation).

5.3.1 Examples

Rather than developing the general theory for the solution of (5.3.13) we shall now discuss two examples. First, let us check the theory for a linear map with a stable fixed point at $x=0$; i.e.

Example 1

$$F(x)=Ax,\quad |A|<1. \tag{5.3.14}$$

From (5.3.13) we find

$$(\Phi'(y))^2+2A^2\Phi(y)-2A^2\Phi\left(Ay+\frac{\Phi'(y)}{A}\right)=0. \tag{5.3.15}$$

Making the ansatz $\Phi(y) = By^2$ we find from (5.3.15) three values for B, namely

$$B_1 = \tfrac{1}{2}(1 - A^2), \quad B_2 = -\tfrac{1}{2}A^2, \quad B_3 = 0. \tag{5.3.16}$$

The solution B_1 yields with (5.3.6) a positive Y^2:

$$Y^2(x, y) = (y - Ax)^2. \tag{5.3.17}$$

Equation (5.3.10) for $Z(x)$ becomes

$$Z(x) = Z(Ax),$$

which implies a constant; i.e. $Z(x) = \text{const}$. The solution B_2 yields via (5.3.6) a negative $Y^2(x, y)$ for some (x, y)-pairs. Thus, it must be excluded. The solution B_3 implies the trivial solution $\Phi(x) = 0$; therefore it must be excluded, too. Hence, the invariant density for a linear map in the presence of Gaussian noise reads

$$W(x) = \left(\frac{1 - A^2}{2\pi\varepsilon}\right)^{1/2} \exp[-(1 - A^2)x^2/2\varepsilon]. \tag{5.3.18}$$

Actually, this is the exact solution for (5.3.3) and (5.3.14); it has been obtained previously by other means (see, e.g., Haken and Wunderlin, 1982).

Example 2

In our example we consider weakly nonlinear maps of the form

$$F(x) = x - aU'(x), \quad a > 0, \tag{5.3.19}$$

where a is a small positive parameter, and $U(x)$ is a smooth potential. In this case (5.3.13) reads

$$\tfrac{1}{2}(\Phi'(y))^2 + (1 - aU''(y))^2\Phi(y)$$
$$- (1 - aU''(y))^2\Phi\left(y - aU'(y) + \frac{\Phi'(y)}{1 - aU''(y)}\right) = 0.$$

If we set $\Phi(y) = a\varphi(y)$ we find in leading order in a the following a-independent equation for the scaled potential φ:

$$-\tfrac{1}{2}(\varphi'(y))^2 + U'(y)\varphi'(y) = 0.$$

Again we disregard the trivial solution $\varphi'(y) = 0$ and obtain up to an arbitrary constant

$$\varphi(y) = 2U(y). \tag{5.3.20}$$

From (5.3.6) and (5.3.20) we obtain a positive $Y^2(x, y)$, yielding to leading order in a

$$Y(x, y) = y - x - a[U'(y) - 2(U(y) - U(x))/(y - x)].$$

At $Y = 0$, we find

$$y(Y = 0, x) = x - aU'(x) + O(a^2).$$

and

$$\left.\frac{\partial Y(y,x)}{\partial y}\right|_{y=y(Y=0,x)} = 1 + O(a^2).$$

Equation (5.3.10) thus gives a prefactor

$$Z(x) = Z(x - aU'(x))(1 + O(a^2)). \tag{5.3.21}$$

Hence, up to corrections of order $O(a^2)$, $Z(x)$ is a constant.

Combining (5.3.4), (5.3.20) and (5.3.21) the stationary probability at weak noise (small ε) thus reads (note that a/ε may be large)

$$W(x) = N\exp(-2aU(x)/\varepsilon). \tag{5.3.22}$$

This result agrees with a previous treatment (Talkner *et al.*, 1987) of the same class of map functions specified in (5.3.19). Corrections to the leading order result (5.3.22) will be discussed elsewhere (P. Talkner, manuscript in preparation).

5.4 Circle map: lifetime of metastable states

In this section we shall elaborate on the noise-induced escape in a periodically continued map (circle map) $F(n + x) = n + F(x)$. Specifically we take the climbing sine map

$$x_{n+1} = x_n + a\sin(2\pi x_n) + \varepsilon^{1/2}\xi_n, \tag{5.4.1}$$

with ξ_n being independent, Gaussian distributed noise (5.3.2). The strength of the sine-force is denoted by $a > 0$, which will be assumed to be small, $a < (2\pi)^{-1}$. For $\varepsilon = 0$, the deterministic map has unstable fixed points at $x^u = n$, and stable fixed points at $x_n^s = (2n + 1)/2$, $n = 0, \pm 1, \pm 2, \ldots$. A trajectory which starts in the interior of the interval $I = [0, 1]$ will be attracted by the stable point $x^s = \frac{1}{2}$, and never leaves the interval I. For arbitrarily small ε, however, the interval will be left eventually. A quantitative measure for the occurrence of these rare events is the mean first passage time (MFPT), i.e. the mean number of steps after which a random walker starting at $x \in [0, 1]$ reaches the exterior of $[0, 1]$ for the first time. For a discrete dynamics the MFPT obeys the inhomogeneous backward equation (Haken and Wunderlin, 1982; Talkner *et al.*, 1987)

$$t(x) - 1 = \int_0^1 P(y|x)t(y)\,dy, \quad x \text{ in } [0, 1]$$

$$t(x) = 0, \quad x \text{ outside } [0, 1]. \tag{5.4.2}$$

With (5.2.4) and (5.3.2) we find

$$t(x) - 1 = (2\pi\varepsilon)^{-1/2} \int_0^1 t(y)\exp[-(y - F(x))^2/2\varepsilon]\,dy,$$

$$x \in I = [0, 1]. \tag{5.4.3}$$

For physical reasons this equation has a unique, bounded solution for all $\varepsilon > 0$. Due to the symmetry of the corresponding potential $U(x) = \cos(2\pi x)/2\pi$ in (5.4.1) (see (5.3.19)) around $x_s = \frac{1}{2}$, $t(x)$ itself becomes a symmetric function about the stable fixed point $x_s = \frac{1}{2}$, and attains its maximal value at $x = x_s = \frac{1}{2}$. Setting $t(\frac{1}{2}) = T$, it follows from (5.4.3) that

$$T - 1 = (2\pi\varepsilon)^{-1/2} \int_0^1 t(y) \exp(-(y-\tfrac{1}{2})^2/2\varepsilon)\,dy$$

$$\leqslant T(2\pi\varepsilon)^{-1/2} \int_0^1 \exp(-(y-\tfrac{1}{2})^2/2\varepsilon)\,dy. \tag{5.4.4}$$

Hence, the MFPT does obey the inequality

$$t(x) \leqslant T \leqslant \left[\operatorname{erfc}\left(\frac{1}{2(2\varepsilon)^{1/2}} \right) \right]^{-1}, \tag{5.4.5}$$

with erfc (x) denoting the complementary error function. This estimate just happens to agree with the approximation by Arecchi, Badii and Politi (1985). In contrast to the case of Markovian Fokker–Planck processes we shall see that $t(x)$ possesses jumps at the exit boundaries $x = 0$ and $x = I^*$. If we start at a boundary, say $x = 0$, the trajectory can return into the interior of the interval $[0, 1]$ with finite probability

$$p = (2\pi\varepsilon)^{-1/2} \int_0^1 \exp(-y^2/2\varepsilon)\,dy;$$

For weak noise, $\varepsilon \ll 1$, p almost equals 1/2. This finite return probability p thus implies a non-zero jump $t(0) > 0$. A more precise estimate for the jump can be devised from (5.4.3); i.e

$$t(0) - 1 = (2\pi\varepsilon)^{-1/2} \int_0^1 t(y) \exp(-y^2/2\varepsilon)\,dy, \tag{5.4.6}$$

or

$$t(0) = CT + 1. \tag{5.4.7}$$

In terms of the form function $\tilde{h}(x)$

$$t(x) = T\tilde{h}(x), \quad \tilde{h}(x) \leqslant 1, \tag{5.4.8}$$

the constant C in (5.4.7) is given by

$$C = (2\pi\varepsilon)^{-1/2} \int_0^1 \tilde{h}(y) \exp(-y^2/2\varepsilon)\,dy < \tfrac{1}{2}. \tag{5.4.9}$$

It then follows from the behavior of $h(y)$ at weak noise (see (5.4.11) and (5.4.12)

* Similar jumps for the MFPT occur for continuous time processes driven by colored noise (Hänggi and Talkner, 1985) or white non-Gaussian noise sources (Knessl, Matkowsky, Schuss and Tier, 1986; Troe, 1977; Weiss and Szabo, 1983).

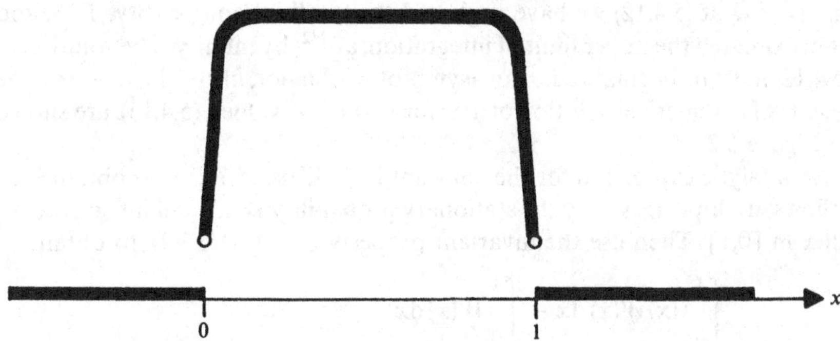

Figure 5.1. Qualitative sketch of the MFPT $t(x)$ versus x. Illustrated are the absorbing lines outside the domain of attraction, as well as the characteristic boundary jumps at the exit points.

below) that the constant C is of the order $O(\varepsilon^0)$; in other words the jump does not approach zero as $\varepsilon \to 0$. In Figure 5.1 we depict the qualitative behavior of the MFPT $t(x)$.

5.4.1 *Weak noise analysis of the MFPT*

At a weak noise level the trajectory remains for most of the time near the stable fixed point, and only rarely will it make a large excursion. Thus, the MFPT attains a very large value inside the internal $[0, 1]$ which deviates only little from its maximal value; i.e. $t(x) \simeq T$. Significant deviations from the constant T occur only near the exit boundaries. Thus $\tilde{h}(x)$ defined in (5.4.8) becomes a boundary layer function which deviates from the value $\tilde{h}(x) \simeq 1$ only in a small neighborhood near the boundaries. The width of the boundary layer function will turn out to be of order $O(\varepsilon^{1/2})$.
We now insert the ansatz (5.4.8) into (5.4.3) to find

$$h(x) - T^{-1} = (2\pi\varepsilon)^{-1/2} \int_0^1 h(y) \exp(-(y - F(x))^2/2\varepsilon)\, dy. \quad (5.4.10)$$

Because the integral kernel is sharply peaked around $y = F(x)$ we can for x near zero linearize $F(x)$ around $x = 0$. In terms of the scaled boundary layer function $h(x)$

$$h(x) = \tilde{h}((2\varepsilon)^{1/2} x) \quad (5.4.11)$$

we thus obtain the following integral equation:

$$h(x) = \pi^{-1/2} \int_0^\infty h(y) \exp(-(y - Ax)^2)\, dy, \quad (5.4.12)$$

where

$$A = F'(0) = 1 + 2\pi a. \quad (5.4.13)$$

In arriving at (5.4.12) we have neglected the small inhomogeneity, T^{-1}, and approximated the upper limit of integration, $\varepsilon^{-1/2}$, by infinity. The solution of (5.4.12) must be normalized to an asymptotic behavior, $h(x) \to 1$ as $x \to \infty$. The results of a numerical solution of $h(x)$ for various A-values, (5.4.13), are shown in Figure 5.2.

An analytic expression for the constant large lifetime T can be obtained as follows: multiply (5.4.3) by the stationary probability $W(x)$, and integrate over all x in $[0,1]$. Then use the invariant property of $W(x)$, (5.3.3), to obtain

$$\begin{aligned}\int_0^1 t(x)W(x)\,dx - \int_0^1 W(x)\,dx &= \int_0^1 t(y)W(y)dy - (2\pi\varepsilon)^{-1/2}\left\{\int_{-\infty}^0 + \int_1^\infty\right\}dx \\ &\quad \times \left\{\int_0^1 t(y)\exp[-(y-F(x))^2/2\varepsilon]\,dy\right\}W(x).\end{aligned} \tag{5.4.14}$$

Utilizing (5.4.8) we thus find the central result

$$T^{-1} = 2(2\pi\varepsilon)^{-1/2}\left[\frac{\displaystyle\int_{-\infty}^0 dx\,W(x)\int_0^1 dy\,\tilde{h}(y)\exp(-(y-F(x))^2/2\varepsilon)}{\displaystyle\int_0^1 W(x)\,dx}\right]. \tag{5.4.15}$$

Hereby we made use of the symmetry about $x^s = \frac{1}{2}$. The result in (5.4.15) is an *exact* expression for $t(\frac{1}{2}) = T$. At weak noise it can be simplified further: the invariant probability, $W(x) = N\exp(-\Phi(x)/\varepsilon)$, is sharply peaked at $x^s = \frac{1}{2}$. For small nonlinearity, a, we can now use the results of Example 2 in Section 5.3; i.e. with (5.3.19), (5.3.22) and (5.4.1) we have

$$\Phi(x) = \frac{a}{\pi}\cos(2\pi x). \tag{5.4.16}$$

With a steepest descent approximation the denominator of (5.4.15) thus becomes

$$\int_0^1 W(x)\,dx \simeq N\frac{1}{2}\left(\frac{\varepsilon}{a}\right)^{1/2}\exp\left(\frac{a}{\pi\varepsilon}\right).$$

The numerator of (5.4.15) simplifies for weak noise as well. The map $F(x)$ can be linearized around $x^u = 0$, and the invariant probability can be approximated around $x^u = 0$ by

$$W(x) \simeq N\exp\left(-\frac{a}{\pi\varepsilon}\right)\exp(4\pi a x^2).$$

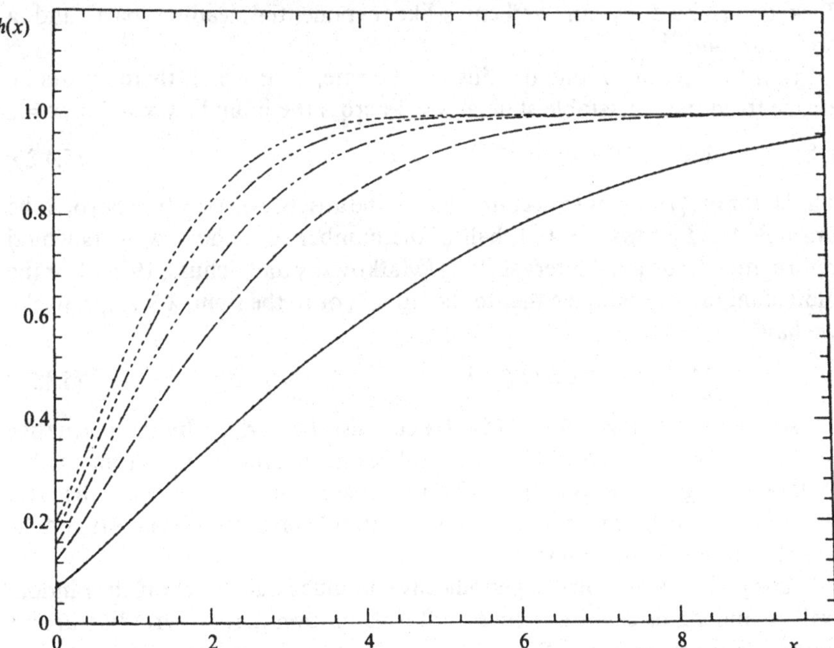

Figure 5.2. Numerical solution of the scaled boundary layer function $h(x)$ for the following A-values: --------- $A = 1.09$; ------- $A = 1.07$; ---- $A = 1.05$; —— $A = 1.03$; ——— $A = 1.01$.

Collecting everything thus yields for (5.4.15)

$$T = \frac{1}{2a^{1/2}R(A)} \exp\left(\frac{2a}{\pi\varepsilon}\right), \tag{5.4.17a}$$

where ($A = 1 + 2\pi a$, see (5.4.13))

$$R(A) = \int_0^\infty \operatorname{erfc}(Ay)h(y)\exp[(A^2 - 1)y^2]\,dy. \tag{5.4.17b}$$

The quantity $R(A)$ is an ε-*independent* function which can be evaluated numerically. For small positive a, one finds within numerical accuracy

$$R(A) = \left(\frac{A^2 - 1}{\pi}\right)^{1/2} + O(A^2 - 1). \tag{5.4.18}$$

Combining (5.4.18) with (5.4.17a) we find for the lifetime T at weak noise the result

$$T = \frac{1}{4a} \exp\left(\frac{2a}{\pi\varepsilon}\right). \tag{5.4.19}$$

T is determined by an Arrhenius-like exponential leading part and a prefactor, $(4a)^{-1}$.

From the lifetime T one obtains for the rate, λ, at which there occurs an escape from the metastable state at $x=\frac{1}{2}$ across the boundary $x=0$ or $x=1$

$$\lambda=\tfrac{1}{2}T^{-1}=\lambda^{+}+\lambda^{-}. \tag{5.4.20}$$

The factor of $(\frac{1}{2})$ takes into account that, in the absence of a capture beyond the unstable fixed points, $x=0, 1$, half of the number of random walkers would return into the original interval $[0, 1]$ (Matkowsky and Schuss, 1979). For the individual rates of escape either to the left, λ^{-}, or to the right, λ^{+}, respectively, we have

$$\lambda^{+}=\lambda^{-}=\tfrac{1}{2}\lambda=(4T)^{-1}. \tag{5.4.21}$$

These rate results (5.4.20) and (5.4.21) can also be derived by an alternative method (Talkner *et al.*, 1987) which utilizes ideas underlying Kramers' flux method (Hänggi, 1986; Kramers, 1940); i.e. one evaluates the rates as the ratio of a nonvanishing, stationary probability current across the exit points and the population inside the interval.

In our present situation of a periodically continued map function, the random walker undergoes a noise-induced diffusive motion across periodic barriers with a diffusion constant D

$$\langle (x_n-\langle x_n\rangle)^2\rangle \to 2Dn \quad \text{as} \quad n\to\infty. \tag{5.4.22}$$

D itself is determined by the forward and backward hopping rates λ^{+}, λ^{-}, and the step size $L=1$; i.e.

$$D=\tfrac{1}{2}(\lambda^{+}+\lambda^{-})L^2=(4T)^{-1}. \tag{5.4.23}$$

References

Arecchi, F. T., Badii, R. and Politi, A. 1985. *Phys. Rev. A* **32**, 402.

Coullet, P. and Tresser, J. 1978a, *C.R. Acad. Sci.* **287**, 577.

Coullet, P. and Tresser, J. 1978b. *J. Phys. (Paris)* **C5**, 25.

Crutchfield, J. R., Farmer, J. D. and Huberman, B. A. 1982. *Phys. Rep.* **92**, 45.

Feigenbaum, M. J. 1978a, *J. Stat. Phys.* **19**, 25.

Feigenbaum, M. J. 1978b. *J. Stat. Phys.* **21**, 669.

Feigenbaum, M. J. 1979. *Physica* **7D**, 16.

Feller, W. 1966. *Introduction to Probability and its Applications*, vols. I, II. New York: Wiley.

Graham, R. 1981. In *Stochastic Nonlinear Systems* (L. Arnold and R. Lefever, eds.), p. 202. New York: Springer.

Grossmann, S. and Thomae, S. 1977. *Z. Naturforschung* **32A**, 1353.

Haken, H. and Mayer-Kress, G. 1981. *Physik B* **43**, 185.

Haken, H. and Wunderlin, A. 1982, *Z. Physik B* **46**, 181.

Hänggi, P. and Talkner P. 1985. *Phys. Rev. A* **32**, 1934.

Hänggi, P. 1986, *J. Stat. Phys.* **42**, 105 and 1003 (addendum).

Knessl, C., Matkowsky, B. J., Schuss, Z. and Tier, C. 1986. *J. Stat. Phys.* **42**, 169.
Kramers, H. A. 1940. *Physica* **7**, 284.
Kubo, R., Matsuo, K. and Kitahara, K. 1973. *J. Stat. Phys.* **9**, 51.
Linz, S. J. and Lücke, M. 1986. *Phys. Rev. A* **33**, 2694.
Ludwig, D. 1975. *SIAM Rev.* **17**, 605.
Matkowsky, B. J. and Schuss, Z. 1979. *SIAM J. Appl. Math.* **35**, 604.
Matsuo, K. 1977. *J. Stat. Phys.* **16**, 169.
Morris, B. and Moss, F. 1986. *Phys. Lett.* **118A**, 117.
Schuster, H. G. 1984. *Deterministic Chaos* Weinheim: VCH Publishers.
Talkner, P., Hänggi, P., Freidkin, E. and Trautmann, D. 1987. *J. Stat. Phys.* **48**, 231.
Troe, J. 1977. *J. Chem. Phys.* **66**, 4745.
Weiss, G. H. and Szabo, A. 1983. *Physica* **119A**, 569.

6 Bifurcation behavior under modulated control parameters

M. LÜCKE

6.1 Introduction

Temporal modulation of a control parameter of nonlinear systems can induce interesting behavior in particular near instabilities. Examples are mechanical and electrical systems subject to parametric modulation (Andronov, Vitt and Khaiken, 1966; Hayashi, 1964; Minorsky, 1962; Nayfeh and Mook, 1979) but also systems undergoing pattern forming instabilities (a review of the effect of parametric modulation on various hydrodynamical instabilities is given by Davis, 1976; for additional references see, e.g., Ahlers, Hohenberg and Lücke 1985a, b, and Kumar, Bhattacharjee and Banerjee, 1986) and other physical and mathematical systems (for reviews see, e.g., Horsthemke and Kondepudi, 1984 and Horsthemke and Lefever, 1984).

Here we consider only simple model systems that show in the absence of modulation a supercritical pitchfork bifurcation, a hysteric subcritical bifurcation, or a transcritical bifurcation. On the one hand, these systems are used to elucidate in detail mathematical methods which, for more complicated systems, allow us to determine the behavior, under parametric modulation in the vicinity of an instability, of a known basic state. On the other hand, is the response of these systems interesting enough by itself to warrant detailed studies?

In the first part of this work we investigate the behavior of a degree of freedom, $x(t)$, that within a mechanical picture is described as a classical particle of mass m moving in the presence of friction γ in a potential $V(x; \varepsilon(t))$ the form of which is modulated in time via a time-dependent control parameter $\varepsilon(t)$

$$m\ddot{x} + m\gamma\dot{x} = -\frac{\partial V(x; \varepsilon(t))}{\partial x}. \tag{6.1.1}$$

In fact, the range of applications of such differential equations seems to be rather wide. They do not only appear in mechanical systems and electronic circuits but also as model equations to field equations that describe systems with pattern forming instabilities. Therein x represents the spatial mode of the order parameter field that grows first, e.g., out of a basic homogeneous state represented by $x = 0$ when the latter becomes unstable.

We investigate here only potentials that vary $\sim x^2$ at the origin so that the force (6.1.1) vanishes for $x = 0$. Thus the trivial state, $x = 0$, is always a fixed point solution of (6.1.1) and the bifurcation of another solution, $x(t)$, from it is a sharp one. This is in contrast to time-dependent 'additive' forcing where there is no stationary state and hence no perfect bifurcation from $x = 0$ so that bifurcation diagrams of moments, e.g., are rounded.

We consider not only systems with inertia but also the overdamped limit case, $m \to 0$ such that $m\gamma \to 1$, yielding a purely relaxational dynamics. We study in particular the effect of parametric modulation (i) on the bifurcation threshold where the basic state, $x = 0$, loses its stability and a time-dependent solution bifurcates and (ii) on the nonlinear behavior in the vicinity of the bifurcation threshold. We find that in the presence of inertia any modulation process with small amplitude stabilizes the basic state, $x = 0$. In systems (6.1.1) with a pure relaxation dynamics ($m \to 0; m\gamma \to 1$), on the other hand, the bifurcation threshold is not shifted by switching on the modulation.

We elucidate a generalized version (Lücke and Schank, 1985) of a Poincaré–Lindstedt expansion (Davis and Rosenblat, 1977) of the bifurcating orbit, $x(t)$, close to threshold around the basic state and we discuss its quality and validity in comparison (Lazarov, 1987) with other methods and with numerical experiments. We determine the response, $x(t)$, to parametric modulation itself and its first moments, i.e. quantities that are directly accessible in experiments rather than probability distributions of x. (For a review of approaches to solve Fokker–Planck equations see, e.g., Horsthemke and Kondepudi, 1984; Horsthemke and Lefever, 1984; Risken, 1984.)

In the last part of this work we investigate the behavior (Linz and Lücke, 1986) of the nonlinear discrete system

$$x_{n+1} = r(1 + \Delta\xi_n)x_n(1 - x_n) \tag{6.1.2}$$

under 'temporal' modulation of the control parameter r by a stochastic process ξ_n. While additive noise destroys the transcritical bifurcation of the unmodulated map at $r = 1$ (trajectories are driven to $-\infty$) multiplicative noise leaves this first bifurcation from the $x = 0$ fixed point sharp. The bifurcation threshold, $r_c(\Delta)$, is determined exactly and shows that any small-amplitude multiplicative forcing stabilizes the basic fixed point $x = 0$.

We determine the statistical dynamics of the bifurcating orbit numerically and for small noise amplitude also with perturbation theory. The latter allows us to derive a condition under which additive and multiplicative noise of small amplitude causes similar statistical response behavior for r sufficiently beyond the threshold $r_c(\Delta)$. Finally we discuss the effect of noise on the first period-doubling pitchfork bifurcation of the unforced map at $r = 3$. There the order parameter $\langle (x_{n+1} - x_n)^2 \rangle$ appropriate to this time translational symmetry breaking bifurcation shows, in the presence of noise, a rounded, imperfect

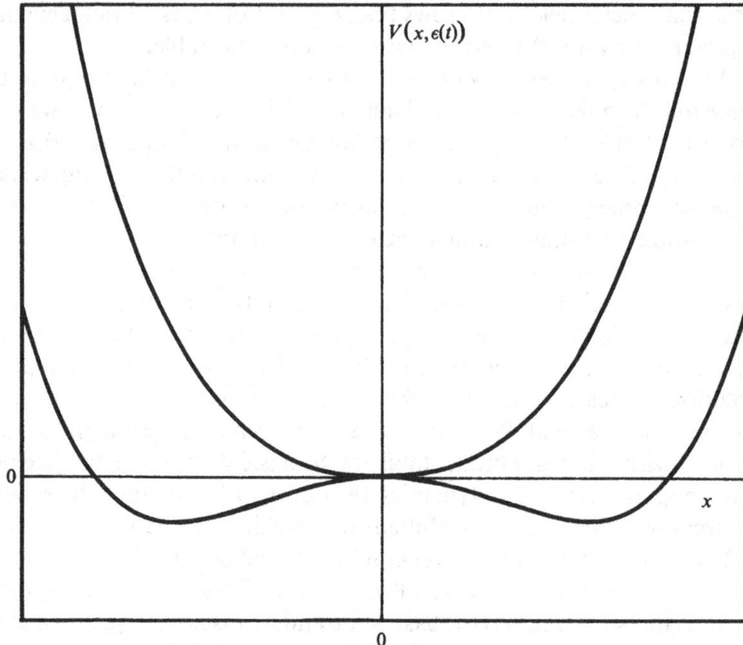

Figure 6.1. Configurations of the potential $V(x;\varepsilon(t)) = -\varepsilon(t)x^2/2 + x^4/4$ for $\varepsilon(t) < 0$ (upper curve) and for $\varepsilon(t) > 0$ (lower curve).

bifurcation with averages varying nonanalytically with growing noise amplitude Δ.

6.2 Parametrically driven Duffing oscillator

Since the Duffing oscillator is a representative example for many instabilities, and since many methods for investigating the response against modulated control parameters can be well demonstrated for this system, we think it is worthwhile to study it in some detail.

6.2.1 *The system*

Here we shall investigate the dynamics of the parametrically driven Duffing oscillator

$$m\ddot{x} + m\gamma\dot{x} = \varepsilon(t)x - x^3. \tag{6.2.1}$$

The associated potential

$$V(x: \varepsilon(t)) = -\tfrac{1}{2}\varepsilon(t)x^2 + \tfrac{1}{4}x^4 \tag{6.2.2}$$

is shown in Figure 6.1. For $\varepsilon(t) < 0$ it has a single-well form and for $\varepsilon(t) > 0$ it consists of two symmetric wells separated by a central barrier.

Let the control parameter

$$\varepsilon(t) = \varepsilon + \Delta\zeta(t) \tag{6.2.3}$$

be modulated with amplitude Δ around a mean

$$\varepsilon = \langle \varepsilon(t) \rangle. \tag{6.2.4}$$

Hence we shall take $\zeta(t)$ to represent a real, stationary process with zero mean, $\langle \zeta(t) \rangle = 0$, and some arbitrary two-point correlation

$$\langle \zeta(t)\zeta(t') \rangle = D(|t - t'|) \tag{6.2.5}$$

depending on the time distance only with a spectrum

$$D(\omega) = \int_{-\infty}^{\infty} \mathrm{d}t\, \mathrm{e}^{\mathrm{i}\omega t} D(t) = \int_{-\infty}^{\infty} \frac{\mathrm{d}\omega'}{2\pi} \langle \zeta(\omega)\zeta(\omega') \rangle \tag{6.2.6}$$

that is positive definite and symmetric in ω. It might be monochromatic, containing in the case of a periodic modulation

$$\zeta(t) = \cos \Omega t \tag{6.2.7a}$$

just one line

$$D(\omega) = \frac{\pi}{2}[\delta(\omega - \Omega) + \delta(\omega + \Omega)]; \tag{6.2.7b}$$

it might contain several lines, or it might be continuous. In the two former situations the angular brackets may be thought of as time averages. In the latter case of a stochastic driving they may also be interpreted as ensemble averages with the statistical weight given by the path probability distribution of the process.

6.2.2 *Bifurcation without modulation*

Before we investigate the response of $x(t)$ to the modulation (6.2.3) we briefly recall the bifurcation properties of (6.2.1) in the absence of modulation, $\Delta = 0$. In this static case $x(t)$ runs for $\varepsilon \leqslant 0$ into the fixed point $x = 0$. While this equilibrium solution exists for all ε it loses its stability at the critical value, $\varepsilon_c(\Delta = 0) = 0$, of the control parameter. For $\varepsilon > 0$ two additional stable equilibrium solutions, $x = \pm \varepsilon^{1/2}$, develop in the two minima of the double-well potential. The bifurcation diagram of the square of these equilibrium solutions versus control parameter ε is shown in Figure 6.2 by thin lines. There we also show schematically the bifurcation behavior of the mean squared amplitude $\langle x^2 \rangle$ under modulation (6.2.3) with small amplitude (see Sections 6.3 and 6.4 for further discussion).

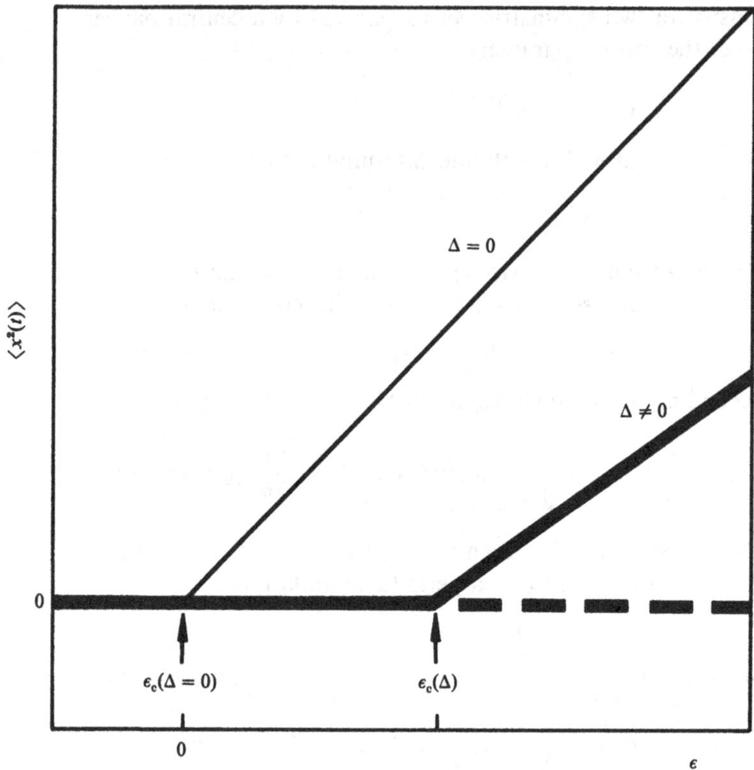

Figure 6.2. Schematic bifurcation diagram. The mean square amplitude $\langle x^2 \rangle$ of the Duffing oscillator, (6.2.1)–(6.2.6) subject to modulation with small amplitude Δ is shown with thick lines as a function of ε (see Sections 6.3 and 6.4 for a discussion). The $x = 0$ solution is stable for $\varepsilon < \varepsilon_c$ (full horizontal line) and unstable for $\varepsilon > \varepsilon_c$ (dashed horizontal line). In contradistinction to additive forcing, in the case of parametric modulation, $x = 0$ is always a solution and the bifurcation from it is a perfect one. The unmodulated case, $\Delta = 0$, is represented by thin lines that partly coincide with the thick ones.

6.2.3 *Overdamped limit*

We shall see that the stabilization of the $x = 0$ fixed point solution under small-amplitude modulation, i.e. the shift of the bifurcation threshold $\varepsilon_c(\Delta)$ towards positive values and the decrease of $\langle x^2 \rangle$ when the modulation is turned on, is purely a result of inertia. In the overdamped limit, $m \to 0$ and $m\gamma \to 1$, leading to the modulated Landau equation

$$\dot{x} = \varepsilon(t)x - x^3, \tag{6.2.8a}$$

the mean squared amplitude has – for any stationary modulation process – the same bifurcation behavior

$$\langle x^2 \rangle = \langle \varepsilon(t) \rangle = \varepsilon \tag{6.2.8b}$$

with the same bifurcation threshold

$$\varepsilon_c(\Delta) = 0 \tag{6.2.8c}$$

as in the static case, $\Delta = 0$. This exact result follows from taking the average of the equation of motion (6.2.8a), $x^2 = \varepsilon(t) - \dot{x}/x$, above threshold where x is nonzero – if x becomes zero it remains so according to (6.2.8a). Also other moments $\langle x^n \rangle$ become finite at $\varepsilon = 0$ but in contradistinction to the second moment their values and ε-dependence differ in general from the analogous $\Delta = 0$ quantities. And thus the stationary probability distribution of x may also for some small $\varepsilon > 0$ still be peaked at $x = 0$ which may be interpreted as a modulation-induced stabilization of the most probable value of x at zero. However, the mean squared amplitude $\langle x^2 \rangle$ (6.2.8b) is not depressed by the modulation.

6.3 Linear stability analysis of the basic state

The stability threshold $\varepsilon_c(\Delta)$ of the $x = 0$ fixed point of (6.2.1) is defined by the requirement that the solution x_c of the linearized equation

$$[m\partial_t^2 + m\gamma\partial_t - \varepsilon_c(\Delta) - \Delta\zeta(t)]x_c(t; \Delta) = 0 \tag{6.3.1}$$

describing infinitesimal deviations from the fixed point is marginal, i.e. shows neither overall growth nor decay. For the monochromatic driving, $\zeta(t) = \cos \Omega t$, x_c is periodic, while for a general stationary driving process ζ the solution x_c of (6.3.1) should represent a stationary process with finite moments. For the sake of convenience we use the above notation $\varepsilon_c = \varepsilon_c(\Delta)$ and $x_c = x_c(t; \Delta)$. However, it should be kept in mind that the stability threshold ε_c and the marginal solution x_c depend not only on Δ but also on the dynamical and statistical properties of $\zeta(t)$.

6.3.1 Periodic modulation

Here we briefly recall the stability behavior of the $x = 0$ fixed point in the presence of a periodic modulation, $\zeta(t) = \cos \Omega t$. In Figure 6.3 we show the stability domain of the $x = 0$ solution in the ε-Δ plane for $\gamma/\Omega = 1$ as a representative example.

The boundary curves (a) and (c) are both harmonic bifurcation thresholds. There the marginal solution x_c has the same periodicity as the driving force. Curve (b) is a subharmonic bifurcation threshold where x_c has twice the periodicity of the driving. With decreasing (increasing) γ/Ω the tongue-shaped boundaries (b) and (c) move closer to (further away from) the ε-axis and become sharper. The form of the stability boundaries in the limit $\gamma \to 0$ with curves (b) and (c) touching the ε-axis at $-m\Omega^4/4$ and $-m\Omega^2$, respectively, is shown – often in an upside-down or rotated version of Figure 6.3 – in many textbooks.

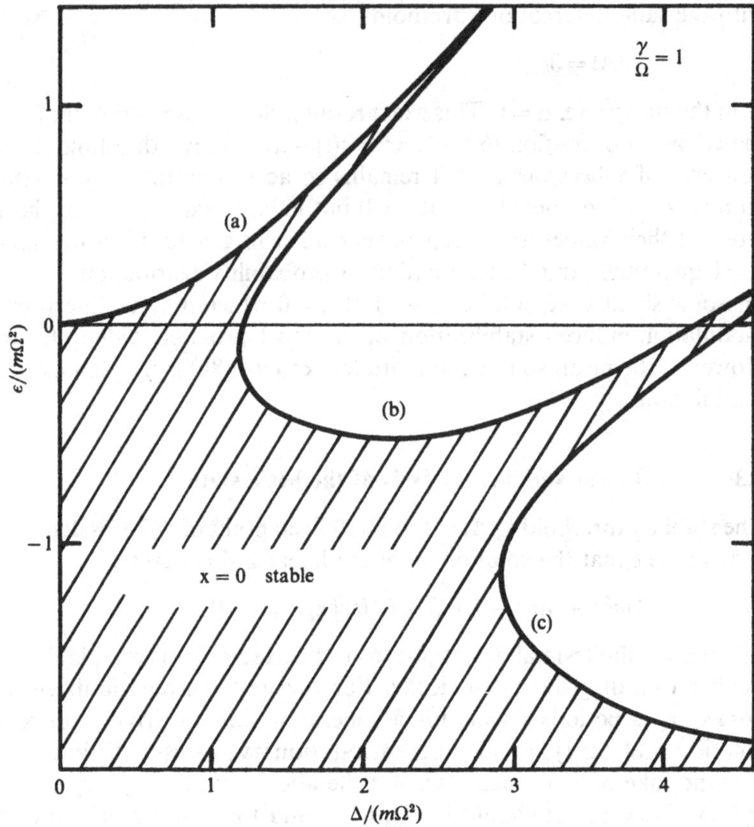

Figure 6.3. Stability domain of the $x = 0$ fixed point of the Duffing oscillator (6.2.1) with modulation $\varepsilon(t) = \varepsilon + \Delta \cos \Omega t$ for $\gamma/\Omega = 1$ as a representative example. Thick lines are the bifurcation thresholds ε_c where $x = 0$ becomes unstable. Curves (a) and (c) are both harmonic bifurcation thresholds–there the period of the marginal solution x_c of (6.3.1) is the same as that of the modulation. Curve (b) is a subharmonic bifurcation threshold, and here the period of x_c is twice that of the modulation. The thin horizontal line marks the value, $\varepsilon_c(\Delta = 0) = 0$, of the bifurcation threshold in the absence of modulation, $\Delta = 0$. See the text for further discussion.

Curve (a), on the other hand, always starts at the origin, $\varepsilon = \Delta = 0$. Then it bends upwards with a curvature increasing to a maximal value at $\gamma = 0$. Thus periodic modulation with small amplitude Δ always stabilizes the $x = 0$ fixed point beyond the static threshold $\varepsilon_c(\Delta = 0) = 0$ up to a positive value ε_c given by curve (a) in Figure 6.3. With larger modulation amplitudes the stability behavior of $x = 0$ is more complicated. For example, resonance amplification occurs when crossing curves (b) or (c) at sufficient negative ε where the potential is always curved upwards.

The stability thresholds in Figure 6.3 were obtained by numerically evaluating the characteristic exponents of the Floquet solutions of

$$[m\partial_t^2 + m\gamma\partial_t - \varepsilon - \Delta\cos\Omega t]x(t) = 0. \tag{6.3.2}$$

To that end one transforms (6.3.2) into the standard form of the Mathieu equation

$$[\partial_\tau^2 + a - 2q\cos 2\tau]y(\tau) = 0 \tag{6.3.3a}$$

with

$$y(\tau) = e^{\tau\gamma/\Omega}x(t); \quad \tau = \Omega t/2 \tag{6.3.3b}$$

$$a = -\frac{4}{m}\frac{\varepsilon}{\Omega^2} - \frac{\Gamma^2}{\Omega^2}; \quad q = \frac{2}{m}\frac{\Delta}{\Omega^2}. \tag{6.3.3c}$$

Then one evaluates the Floquet exponent $\mu(a,q)$ of y with an integration procedure described, e.g., by Blanch (1965). Since the Floquet solution of (6.3.2) reads

$$x(t) = e^{(\mu - \gamma/\Omega)\tau}P(\tau) \tag{6.3.4}$$

with $P(\tau)$ being periodic, the stability boundary of $x = 0$ is given by

$$\operatorname{Re}\mu(a,q) = \gamma/\Omega. \tag{6.3.5}$$

Bifurcation thresholds obtained by this method for values of $\gamma/\Omega \neq 1$ are shown in a recent paper by Ahlers, Hohenberg and Lücke, (1985a).

6.3.2 *Analytic theory for arbitrary small-amplitude modulation*

Here we present a perturbation analysis of the linear stability problem (6.3.1). This theory yields for small modulation amplitude Δ analytical results for the marginal solution $x_c(t;\Delta)$ of (6.3.1) and for the stability boundary $\varepsilon_c(\Delta)$ which connects at $\Delta = 0$ to the threshold value $\varepsilon_c(\Delta = 0) = 0$ of the unmodulated system. In particular for the cosine modulation it yields the small-Δ asymptotics (Ahlers, Hohenberg and Lücke, 1985a) of branch (a) in Figure 6.3. However, this method is not restricted to periodic modulation (Lücke and Schank, 1985).

Furthermore, it may be applied to systems with many degrees of freedom. Venezian (1969) determined with this method the stability of the conductive state of a fluid layer heated from below in the presence of periodic modulation of the temperature of the horizontal boundaries by using a complete Galerkin expansion of the hydrodynamic fields. Also few-mode truncations of hydrodynamic field equations have been analyzed with this perturbation method to determine the stability of the conductive basic state of the modulated Bénard problem (Ahlers, Hohenberg and Lücke, 1985a; Kumar, Bhattacharjee and Banerjee, 1986) and the stability of circular Couette flow in the annulus between concentric cylinders when the rotation rate of the inner one is modulated periodically (Kuhlmann, 1985; Kumar, Bhattacharjee and Banerjee, 1986).

The expansion

The perturbation method consists of expanding both the marginal solution

$$x_c(t;\Delta) = x_c^{(0)}[1 + \Delta x_c^{(1)}(t) + \Delta^2 x_c^{(2)}(t) + \cdots] \tag{6.3.6a}$$

and the bifurcation threshold

$$\varepsilon_c(\Delta) = \Delta\varepsilon_c^{(1)} + \Delta^2 \varepsilon_c^{(2)} + \cdots. \tag{6.3.6b}$$

Thereby (6.3.1) is transformed into a sequence of inhomogeneous differential equations with constant coefficients. The requirement that in each order $x_c^{(n)}(t)$ has to be marginal allows the determination of both expansion 'coefficients', $\varepsilon_c^{(n)}$ and $x_c^{(n)}(t)$, simultaneously.

In (6.3.6a, b) we have explicitly displayed the fact that in the absence of modulation the bifurcation threshold is zero, $\varepsilon_c^{(0)} = \varepsilon_c(\Delta = 0) = 0$, and that the marginality condition on the solution of the lowest order equation,

$$\mathscr{L}_0 x_c^{(0)} = m\partial_t(\partial_t + \gamma)x_c^{(0)} = 0 \tag{6.3.7}$$

requires $x_c^{(0)}$ to be an (undetermined) constant. To first order in Δ one obtains the equation

$$m\partial_t(\partial_t + \gamma)x_c^{(1)}(t) = \varepsilon_c^{(1)} + \zeta(t). \tag{6.3.8a}$$

The condition that $x_c^{(1)}(t)$ is marginal, i.e. that the time average of the left side of (6.3.8a) vanishes, enforces with $\langle\zeta(t)\rangle = 0$

$$\varepsilon_c^{(1)} = 0. \tag{6.3.8b}$$

In fact, for a driving $\zeta(t)$ that is distributed symmetrically around zero, $\varepsilon_c(\Delta)$ is an even function of Δ since the latter enters only via the combination $\Delta\zeta(t)$ into (6.3.1). Hence every $\varepsilon_c^{(2n+1)} = 0$.

Marginality and Fredholm solvability condition

As an aside we remark that the requirement of marginality is nothing but a Fredholm solvability condition imposed on each inhomogeneity $h^{(n)}(t)$ that appears in the nth order equation

$$\mathscr{L}_0 x_c^{(n)}(t) = h^{(n)}(t). \tag{6.3.9}$$

The Fredholm alternative demands that the constant, marginal solution $y_c^{(0)}$ of the equation

$$\mathscr{L}_0^+ y_c^{(0)} = m\partial_t(\partial_t - \gamma)y_c^{(0)} = 0 \tag{6.3.10}$$

that is adjoint to (6.3.7) with a scalar product

$$(f(t)|g(t)) = \langle f(t)g(t)\rangle \tag{6.3.11}$$

defined by the average has to be orthogonal to the inhomogeneity $h^{(n)}(t)$

$$0 = (h^{(n)}(t)|y_c^{(0)}) = \langle h^{(n)}(t)\rangle y_c^{(0)} \tag{6.3.12}$$

since

$$0 = (\mathcal{L}_0^+ y_c^{(0)}|x_c^{(n)}(t)) = (y_c^{(0)}|\mathcal{L}_0 x_c^{(n)}(t)) = y_c^{(0)}\langle \mathcal{L}_0 x_c^{(n)}(t)\rangle. \quad (6.3.13)$$

First order marginal solution

The solution of (6.3.8) in frequency space reads

$$x_c^{(1)}(\omega) = \chi(\omega)\zeta(\omega) \quad (6.3.14)$$

with

$$\chi(\omega) = \frac{-1/m}{\omega(\omega + i\gamma)}. \quad (6.3.15)$$

For the special case of a cosine modulation, $\zeta(t) = \cos \Omega t$, one has

$$x_c^{(1)}(t) = \mathrm{Re}[\chi(\Omega)e^{-i\Omega t}]. \quad (6.3.16)$$

Here the dynamical susceptibility,

$$\chi(\omega) = \lim_{\omega_0 \to 0} \frac{-1/m}{\omega^2 - \omega_0^2 + i\gamma\omega}, \quad (6.3.17)$$

of a harmonic oscillator towards additive forcing enters in the limit of vanishing restoring force $\omega_0^2 = -\varepsilon/m \to 0$ since the perturbation expansion (6.3.6a, b) proceeds around the flat potential configuration at $\varepsilon = 0$. This is the basic reason for the appearance of infrared divergencies in the expansion (6.3.6a, b). The first one is most clearly seen in (6.3.14–6.3.15): in the absence of a restoring force, $\omega_0^2 = 0$, the lowest order response $x_c^{(1)}$ towards the driving ζ diverges if the latter has a finite zero-frequency spectral component. We shall discuss this point further later on. For the moment let us restrict ourselves to driving processes for which $\zeta(\omega \to 0)$ vanishes sufficiently fast.

Stability threshold

To second order in the small-Δ expansion (6.3.6a, b) the equation of motion (6.3.1) reads

$$\mathcal{L}_0 x_c^{(2)}(t) = \varepsilon_c^{(2)} + \zeta(t)x_c^{(1)}(t). \quad (6.3.18)$$

Marginality of $x_c^{(2)}(t)$, i.e. the solvability condition, requires

$$\varepsilon_c^{(2)} = -\langle \zeta(t)x_c^{(1)}(t)\rangle = -\int_{-\infty}^{\infty} \frac{d\omega}{2\pi}\chi(\omega)D(\omega). \quad (6.3.19)$$

Here we have used (6.3.14) and (6.2.6) to express the equal-time correlation between driving $\zeta(t)$ and lowest order response, $x_c^{(1)}(t)$, in terms of a spectral integral. Since the spectrum $D(\omega)$ is symmetric in ω, only the even, negative real part of $\chi(\omega)$ enters into (6.3.19). Thus, to second order in Δ, the stability threshold

$$\varepsilon_c(\Delta) = \Delta^2 \frac{1}{m}\int_0^{\infty} \frac{d\omega}{\pi}\frac{D(\omega)}{\omega^2 + \gamma^2} + O(\Delta^4). \quad (6.3.20)$$

is positive since $D(\omega) \geqslant 0$.

Any stationary small-amplitude modulation stabilizes the trivial solution, $x = 0$, into a driving domain, $\varepsilon_c(\Delta) > 0$, in which the time averaged potential has a maximum at the origin. Each frequency mode ω of the modulation contributes additively to the stabilization, with a weight given by the spectral intensity $D(\omega)$ of the driving and the real part of the dynamical response function $\chi(\omega)$ (6.3.17) of a harmonic oscillator with vanishing restoring force.

To second order in Δ, the threshold $\varepsilon_c(\Delta)$ is determined by $D(\omega)$, the two-point correlation spectrum of the modulation, only. Higher order correlations of the modulation enter only into higher orders of Δ – a four-point correlation of ζ enters, for example, into the next nonvanishing coefficient, $\varepsilon_c^{(4)}$.

Stabilization by inertia

The dynamical stabilization by small-amplitude modulation is caused by inertia. Approaching the pure relaxation dynamics with $m \to 0$ and $m\gamma \to 1$ the stability threshold $\varepsilon_c(\Delta)$ (6.3.20) decreases linearly with m towards $\varepsilon_c = 0$, i.e. the bifurcation point for static driving. Thus the small-Δ expansion reproduces this exact result discussed already in Section 6.2.3.

Inertia delays the particle's 'rolling' down from the vicinity of the mountain position after the potential has developed a downwards curvature at $x = 0$. Inertia also lets the particle that has started to move inwards, towards $x = 0$ in an upwards curved potential, continue its motion into the rather flat central potential region close to $x = 0$ for a while even after $x = 0$ is no longer a minimum. It is easy and fascinating to observe this inertia-induced asymmetry in the dynamics of the motion towards and away from $x = 0$ in the $x - \dot{x}$ phase plane on a computer screen, say, for a periodic modulation.

The stabilizing effect of modulation via inertia may also be explained with analytical arguments that are particularly simple for the special case of a modulation, $\varepsilon(t)$, that is piecewise constant as a function of time, taking values, ε_t, that are positive or negative. The characteristic exponents describing the motion during the time intervals of constant, small driving ε_t are

$$\lambda_1 = \frac{\varepsilon_t}{m\gamma} \tag{6.3.21a}$$

$$\lambda_2 = -\gamma\left(1 + \frac{\varepsilon_t}{m\gamma^2}\right), \tag{6.3.21b}$$

while in the overdamped case, $m \to 0, m\gamma \to 1$, there is only one exponent, $\lambda = \varepsilon_t$. In the latter situation growth ($\varepsilon_t > 0$) and decay ($\varepsilon_t < 0$) rates of $x(t)$ are symmetrically distributed around zero if $\varepsilon = \langle \varepsilon_t \rangle = 0$ and a modulation with $\varepsilon = \langle \varepsilon_t \rangle > 0$ leads to an overall growth. For finite mass m, however, λ_2 (6.3.21b) is always negative. Thus, with only one exponent, λ_1, changing sign during the modulation the motion towards $x = 0$ is enhanced in comparison with the motion away from $x = 0$.

Comparison with other authors; special cases

The formula (6.3.20) for the threshold $\varepsilon_c(\Delta)$ holds for arbitrary small-amplitude modulation. It contains and reproduces the result that Graham and Schenzle (1982) have derived with a completely different technique for a Gaussian process with a Brownian spectrum $D(\omega) = D_0\Gamma^2/(\omega^2 + \Gamma^2)$. Note that in contrast to (6.3.14) the threshold formula (6.3.20) is not plagued by an infrared divergency as long as γ is finite. In Section 6.4.3 we shall reproduce (6.3.20) by expanding the solution of the nonlinear problem (6.2.1) for small Δ around the $\Delta = 0$ fixed point $x^2 = \varepsilon$ at finite $\varepsilon > 0$. The upwards shift, $\varepsilon_c(\Delta)$, of the stability boundary of $x = 0$ is also consistent with the variation of Lyapunov exponents with ε and Δ investigated by Arnold, Papanicolaou and Wihstutz (1986).

However, Knobloch and Wiesenfeld (1983) and Seshadri, West and Lindenberg (1981) reported results that disagree with ours. From the work of Knobloch and Wiesenfeld one infers that the stationary average $\langle (x + \dot{x}/\gamma)^2 \rangle$, i.e. $\langle x^2 \rangle$ itself, becomes finite already at negative $\varepsilon > \varepsilon_c^{KW} = -\Delta^2 D_0 5/(2m\gamma)$, where D_0 is the spectral strength of a white noise Gaussian process. Seshadri, West and Lindenberg derive results according to which $\langle x^2 \rangle$ is finite for any– negative and positive – ε and any strength of Gaussian white noise multiplicative forcing. We have numerically approximated such a process by adding twenty-four random numbers and integrating the equation of motion (see Section 6.4.3 for details). In this way we saw what is physically quite obvious: for small Δ and large negative ε the force

$$-|\varepsilon|\left[1 - \frac{\Delta}{|\varepsilon|}\xi\right]x(1 + x^2)$$

on the particle is directed practically always towards $x = 0$ since the square bracket is mostly positive due to the smallness of $\Delta/|\varepsilon|$. Thus amplitudes decrease basically with the damping rate $\gamma/2$ to the fixed point $x = 0$. For example, for $m = \gamma = D_0 = 1$, $\Delta = 0.7$ for which $\varepsilon_c^{KW} = -1.225$, the quantity $x^2 + \dot{x}^2$ drops for $\varepsilon = -1$ from 1 to 10^{-40} within a time of about 40 ln 10.

For a periodic modulation, $\zeta(t) = \cos \Omega t$ with the line spectrum $D(\omega) = (\pi/2)[\delta(\omega - \Omega) + \delta(\omega + \Omega)]$ (6.3.20) yields the small-Δ asymptotics

$$\varepsilon_c(\Delta) = \Delta^2 \frac{1}{2m} \frac{1}{\Omega^2 + \gamma^2} + O(\Delta^4) \tag{6.3.22}$$

of branch (a) of the stability boundary shown in Figure 6.3. Thus, right at the bifurcation threshold, $\varepsilon = \varepsilon_c(\Delta)$, the control parameter $\varepsilon(t) = \varepsilon_c + \Delta \cos \Omega t$ is negative for a fraction

$$\frac{1}{2} - \frac{1}{\pi} \frac{\varepsilon_c(\Delta)}{\Delta}$$

of the modulation period $2\pi/\Omega$. In view of the inertia-induced asymmetry between the decay rates towards $x=0$ and the growth rate away from the fixed point, it is not surprising that the stabilization increases with decreasing modulation frequency since the particle has more and more time to approach $x=0$.

Formula (6.3.22) makes it very clear that the dynamical stabilization of the $x=0$ state is not a friction effect but an inertia effect. The threshold shift $\varepsilon_c(\Delta)$ is largest in the absence of friction while it vanishes for $\gamma\to\infty$, both with finite m and also in the inertia free limiting case of an overdamped dynamics described by $m\to 0$ such that $m\gamma\to$ const.

Standing pendulum

We finally mention the closely related phenomenon of dynamically stabilizing a pendulum in its statically unstable upwards position by shaking, i.e. modulating, its point of support vertically. For small deviations of the angle α from the upwards position, $\alpha=0$, the equation of motion for the pendulum reads

$$\left[\partial_t^2-\frac{g}{l}+\frac{A}{l}\Omega^2\cos\Omega t\right]\alpha(t)=0. \tag{6.3.23}$$

Here g is the gravitational acceleration, l is the length of the pendulum, Ω is the frequency of the vertical motion of the point of support, and A is its amplitude. Obviously one has to identify

$$\alpha\leftrightarrow x;\quad g/l\leftrightarrow\varepsilon/m;\quad \Omega^2 A/l\leftrightarrow -\Delta/m \tag{6.3.24}$$

to map the pendulum problem onto the modulated potential problem (6.3.3a–c). Since (6.3.22) with $\gamma=0$ implies stabilization if

$$\Delta^2>2m\Omega^2\varepsilon, \tag{6.3.25a}$$

one infers that the upwards position is stabilized by sinusoidally shaking the point of support up and down with frequency Ω and amplitude A such that

$$A\Omega>(2gl)^{1/2}. \tag{6.3.25b}$$

Landau and Lifshitz (1960) derive in problem 1 of § 30 the same requirement, however, with a different method.

We conclude this section by pointing out that small-amplitude parametric modulation does not always, and in any system, stabilize the basic state.

6.4 The bifurcating solution

Here we investigate the nonlinear behavior beyond the bifurcation threshold by using a variant of a method originally developed by Lindstedt, Poincaré and Hopf (see, e.g., the notes of chapter II of Joseph, 1976). We also compare its result with a straightforward perturbation expansion of the solution above threshold for small Δ and in addition with numerical experiments.

6.4.1 Poincaré–Lindstedt expansion

The method consists of expanding the bifurcating solution close to the bifurcation threshold around the basic solution. In our model system of the modulated Duffing oscillator the basic solution is $x = 0$, which loses stability at the threshold value $\varepsilon = \varepsilon_c(\Delta)$. Note, however, that this technique can be applied also to more complicated systems of differential equations (Ahlers, Hohenberg and Lücke, 1985a) and to partial differential equations (S. Schmitt, 1988, unpublished diploma thesis, Universität des Saarlandes) if the basic solution from which the bifurcation occurs is known. Furthermore, the Poincaré–Lindstedt method seems to work not only for static control parameters and periodic parametric modulation but also for more general modulation. Lacking rigorous mathematical proofs we come to this conclusion by comparison with other methods and with numerical experiments (Lazarov, 1987). Lastly we point out that the bifurcating solution to be approximated by the expansion does not have to be a stable one. In Section 6.5 we discuss the subcritical bifurcation of a periodic repeller.

In the following, we demonstrate this versatile technique for the modulated Duffing oscillator (6.2.1) as a concrete example. In this special case one can perform at $\varepsilon > 0$ also a straightforward small-Δ expansion around the 'nontrivial', i.e. bifurcating, solution $x = \pm\varepsilon^{1/2}$ for stationary control parameter (see Section 6.4.3). However, for more complicated systems the stationary bifurcating solution may not be available analytically – the stationary convective state in a fluid layer heated from below or the stationary Taylor vortex flow state in a rotating Couette system are examples. In such cases a Poincaré–Lindstedt expansion around the basic state is still possible.

Let us assume that the solution $x(t)$ of (6.2.1) that bifurcates at the threshold value ε_c of the control parameter ε from the basic solution, $x = 0$, varies continuously with the deviation, $\varepsilon - \varepsilon_c$, from criticality. Then we expand

$$x(t) = \lambda x_1(t) + \lambda^2 x_2(t) + \cdots \tag{6.4.1a}$$

$$\varepsilon - \varepsilon_c = \lambda\varepsilon_1 + \lambda^2\varepsilon_2 + \cdots. \tag{6.4.1b}$$

Equations (6.4.1a, b) and all others in this section apply also to the overdamped limit case, $m \to 0$ with $m\gamma \to 1$, where, e.g., $\varepsilon_c = 0$.

The expansion parameter λ may be thought of as measuring the mean amplitude of the bifurcating solution – to lowest order $\lambda \sim \langle x^2 \rangle^{1/2}$ if $\langle x_1^2 \rangle$ is $O(1)$. Equivalently, λ can be regarded as measuring an appropriate power of $\varepsilon - \varepsilon_c$. In our case of small-amplitude modulation we shall see that $\varepsilon_1 = 0$ so that $\lambda \sim (\varepsilon - \varepsilon_c)^{1/2}$. Note, however, that the expansion parameter is only an ordering device. It will be eliminated between (6.4.1a) and (6.4.1b) later on to yield – e.g. by solving (6.4.1b) for λ and inserting the result into (6.4.1a) – a series of x in powers of $(\varepsilon - \varepsilon_c)^{1/2}$ in our case. In Sections 6.5.3 and 6.5.4, on the other hand, we shall solve (6.4.1a) for λ and insert the result into (6.4.1b) to obtain $\varepsilon - \varepsilon_c$ expressed as a power series in x.

In the following we show how to determine successively all unknown expansion coefficients $x_n(t)$ and ε_n. Inserting (6.4.1) into (6.2.1) one finds in order λ the linearized equation of motion at threshold, $\varepsilon = \varepsilon_c$,

$$\mathscr{L}x_1(t) = 0 \tag{6.4.2a}$$

$$\mathscr{L} = m\partial_t^2 + m\gamma\partial_t - \varepsilon_c(\Delta) - \Delta\zeta(t). \tag{6.4.2b}$$

This equation is solved by the marginal solution

$$x_1(t) = x_c(t; \Delta), \tag{6.4.2c}$$

which can be determined with the small-Δ perturbation theory of Section 6.3.2. In addition, we have to solve the problem

$$\mathscr{L}^+ y_1(t) = 0 \tag{6.4.3a}$$

$$\mathscr{L}^+ = m\partial_t^2 - m\gamma\partial_t - \varepsilon_c(\Delta) - \Delta\zeta(t) \tag{6.4.3b}$$

that is adjoint to (6.4.2a, b) with the scalar product (6.3.11). Its solution is the adjoint marginal solution at threshold, $\varepsilon = \varepsilon_c(\Delta)$,

$$y_1(t) = y_c(t, \Delta). \tag{6.4.3c}$$

It can also be obtained by the small-Δ perturbation expansion described in Section 6.3.2.

To second order in λ one obtains

$$\mathscr{L}x_2(t) = \varepsilon_1 x_1(t) \tag{6.4.4a}$$

and

$$\varepsilon_1 = 0. \tag{6.4.4b}$$

The latter follows from the solvability condition

$$0 = (y_1 | \mathscr{L}x_2) = \varepsilon_1 (y_1 | x_1) \tag{6.4.4c}$$

requiring the inhomogeneity on the right hand side of (6.4.4a) to be orthogonal to the eigenvector y_1 of $\mathscr{L}^+$ with zero eigenvalue. Note that $(y_1 | x_1)$ is finite as may be seen directly from the small-Δ expansion of x_1 and y_1 (see Section 6.4.2).

Since the nonlinearity in (6.2.1) is cubic, x^3, and since ε_1 vanishes, the first nonlinear term in the solution $x(t)$ of (6.2.1) to be generated by the equation of motion should be $\lambda^3 x_3(t)$. Therefore it is consistent to solve (6.4.4a, b) by

$$x_2(t) = 0. \tag{6.4.4d}$$

Then the third order equation in λ reads

$$\mathscr{L}x_3(t) = \varepsilon_2 x_1(t) - x_1^3(t), \tag{6.4.5a}$$

and the above described Fredholm solvability condition gives

$$\varepsilon_2 = \frac{\langle y_1 x_1^3 \rangle}{\langle y_1 x_1 \rangle}. \tag{6.4.5b}$$

Thus with the known functions $x_1(t)$ and $y_1(t)$ one can evaluate ε_2 and then determine the solution $x_3(t)$ of (6.4.5a), e.g., via a small-Δ expansion.

To fourth order in λ one obtains

$$\varepsilon_3 = 0 \tag{6.4.6a}$$

$$x_4 = 0, \tag{6.4.6b}$$

and the Fredholm condition of order λ^5 yields

$$\varepsilon_4 = 3\frac{\langle y_1 x_1^2 x_3 \rangle}{\langle y_1 x_1 \rangle} - \varepsilon_2 \frac{\langle y_1 x_3 \rangle}{\langle y_1 x_1 \rangle}. \tag{6.4.7}$$

This suffices to make the logic of the Poincaré–Lindstedt procedure clear. Having determined $x_1, \ldots, x_{n-1}$ and $\varepsilon_1, \ldots, \varepsilon_{n-1}$ one evaluates ε_n with the solvability condition of order λ^n and then one can solve the nth order equation for x_n.

The next step is to eliminate λ from the expansions (6.4.1a, b) which reduce in our case to

$$x(t) = \lambda[x_1(t) + \lambda^2 x_3(t) + \cdots] \tag{6.4.8a}$$

$$\varepsilon - \varepsilon_c = \lambda^2[\varepsilon_2 + \lambda^2 \varepsilon_4 + \cdots]. \tag{6.4.8b}$$

Inverting (6.4.8b) and inserting the result into (6.4.8a) one finds that (6.4.8a, b) correspond to an expansion

$$x(t) = (\varepsilon - \varepsilon_c)^{1/2}[\tilde{x}_1(t) + (\varepsilon - \varepsilon_c)\tilde{x}_3(t) + O(\varepsilon - \varepsilon_c)^2] \tag{6.4.9}$$

in powers of the square root of the distance of the control parameter from the threshold. The first coefficient functions are

$$\tilde{x}_1(t) = \frac{x_c(t)}{(\varepsilon_2)^{1/2}} = \left(\frac{\langle y_c(t) x_c(t) \rangle}{\langle y_c(t) x_c^3(t) \rangle}\right)^{1/2} x_c(t) \tag{6.4.10}$$

$$\tilde{x}_3(t) = \varepsilon_2^{-3/2}\left[x_3(t) - \frac{1}{2}\frac{\varepsilon_4}{\varepsilon_2} x_c(t)\right]. \tag{6.4.11}$$

Here we have used (6.4.2c), $x_1(t) = x_c(t)$. Note that (6.4.9)–(6.4.11) are the final and central results of the Poincaré–Lindstedt expansion. In the next section we show how to obtain the functions $\tilde{x}_n(t)$ explicitly for small modulation amplitudes Δ. However, one should keep in mind that for general modulation processes with arbitrary amplitudes one will not be able to calculate the functions $\tilde{x}_n(t)$ and the threshold $\varepsilon_c(\Delta)$ analytically. Thus, the interesting bifurcation behavior close to the thresholds (b) and (c) of Figure 6.3, for example, seem to be accessible only by numerical means (Ahlers, Hohenberg and Lücke, 1985a).

A remark concerning the solution of (6.4.4a, b) should be made. We could have solved these equations also by $x_2 = \alpha x_1$ with a finite undetermined constant α instead of $\alpha = 0$ since the linear differential equation (6.4.4a, b) for

x_2 is identical to the one for x_1, (6.4.2a). However, that does not change the final result (6.4.9). While a finite α generates odd coefficients ε_{2n+1} that are related to the previous even ones, e.g. $\varepsilon_3 = 2\alpha\varepsilon_2$, and even coefficient functions $x_{2n}(t)$ that are related to the previous odd ones, e.g. $x_4(t) = 3\alpha x_3(t)$, they do not contribute to the final result for $x(t)$.

6.4.2 *Small-Δ expansion of the coefficients of the Poincaré–Lindstedt expansion*

It is instructive to write down the first functions $\tilde{x}_1(t)$ and $\tilde{x}_3(t)$ entering the Poincaré–Lindstedt expansion explicitly for small Δ. To that end we make a second, additional expansion up to second order in Δ of $x_c(t;\Delta)$, $y_c(t;\Delta)$, and $x_3(t;\Delta)$ with the direct perturbation method described in Section 6.3.2. Doing that we, of course, assume that these three functions entering into (6.4.10) and (6.4.11) allow a Taylor expansion around $\Delta = 0$.

We begin with the marginal solutions of the linear equations (6.4.2a–c) and (6.4.3a–c) at threshold, $\varepsilon = \varepsilon_c$,

$$x_c(t;\Delta) = x_c^{(0)}[1 + \Delta x_c^{(1)}(t) + \Delta^2 x_c^{(2)}(t) + O(\Delta^3)] \tag{6.4.12a}$$

$$y_c(t;\Delta) = y_c^{(0)}[1 + \Delta y_c^{(1)}(t) + \Delta^2 y_c^{(2)}(t) + O(\Delta^3)]. \tag{6.4.12b}$$

First note that the undetermined lowest order constants $x_c^{(0)}$ and $y_c^{(0)}$ drop out of all expressions for observable quantities like (6.4.10) and (6.4.11). The first order functions are most conveniently given in frequency space

$$x_c^{(1)}(\omega) = \chi(\omega)\zeta(\omega) \tag{6.4.13a}$$

$$y_c^{(1)}(\omega) = \chi^*(\omega)\zeta(\omega), \tag{6.4.13b}$$

where $\chi^*(\omega)$ is the complex conjugate of the susceptibility $\chi(\omega)$ (6.3.15). These quantities are the inverses in frequency space of the operators $\mathscr{L}_0^+ = \mathscr{L}^+(\Delta = 0) = m(\partial_t^2 - \gamma\partial_t)$ and of $\mathscr{L}_0 = \mathscr{L}(\Delta = 0) = m(\partial_t^2 + \gamma\partial_t)$, respectively.

The second order coefficients $x_c^{(2)}$ and $y_c^{(2)}$ follow from (6.3.18) and the adjoint equations

$$x_c^{(2)}(\omega) = \chi(\omega)\int_{-\infty}^{\infty} \mathrm{d}t\, \mathrm{e}^{\mathrm{i}\omega t}[\zeta(t)x_c^{(1)}(t) - \langle\zeta(t)x_c^{(1)}(t)\rangle] \tag{6.4.14a}$$

$$y_c^{(2)}(\omega) = \chi^*(\omega)\int_{-\infty}^{\infty} \mathrm{d}t\, \mathrm{e}^{\mathrm{i}\omega t}[\zeta(t)y_c^{(1)}(t) - \langle\zeta(t)y_c^{(1)}(t)\rangle]. \tag{6.4.14b}$$

Here we have used (6.3.19), which implies that

$$-\varepsilon_c^{(2)} = \langle\zeta x_c^{(1)}\rangle = \langle\zeta y_c^{(1)}\rangle = \int_{-\infty}^{\infty} \frac{\mathrm{d}\omega}{2\pi} \operatorname{Re}\chi(\omega)D(\omega).$$

To order Δ^2, however, $y_c^{(2)}$ does not enter into (6.4.10) and (6.4.11) since $\langle y_c^{(2)}(t)\rangle = 0$.

Inserting (6.4.12a, b) into (6.4.10) one finally finds the small-Δ representation of the first coefficient function of (6.4.9)

$$\tilde{x}_1(t) = 1 + \Delta x_c^{(1)}(t) + \Delta^2[x_c^{(2)}(t) - \langle y_c^{(1)}(t)x_c^{(1)}(t) + \tfrac{3}{2}(x_c^{(1)}(t))^2\rangle] + O(\Delta^3) \tag{6.4.15}$$

with $x_c^{(1)}, y_c^{(1)}, x_c^{(2)}$ being given by (6.4.13a, b) and (6.4.14a, b) in terms of the modulation ζ. The averages in the square bracket of (6.4.15) arise from expanding the prefactor $1/(\varepsilon_2)^{1/2}$ in (6.4.10) according to

$$\varepsilon_2 = \frac{\langle y_c x_c^3\rangle}{\langle y_c x_c\rangle} = (x_c^{(0)})^2\left[\frac{1 + \Delta^2 3\langle y_c^{(1)}x_c^{(1)} + (x_c^{(1)})^2\rangle + O(\Delta^4)}{1 + \Delta^2\langle y_c^{(1)}x_c^{(1)}\rangle + O(\Delta^4)}\right]. \tag{6.4.16}$$

Note that *both* of these averages

$$\langle y_c^{(1)}x_c^{(1)}\rangle = \frac{1}{m^2}\int_{-\infty}^{\infty}\frac{d\omega}{2\pi}\frac{D(\omega)}{\omega^2}\frac{\omega^2 - \gamma^2}{(\omega^2 + \gamma^2)^2} \tag{6.4.17a}$$

and

$$\langle (x_c^{(1)})^2\rangle = \frac{1}{m^2}\int_{-\infty}^{\infty}\frac{d\omega}{2\pi}\frac{D(\omega)}{\omega^2}\frac{1}{\omega^2 + \gamma^2} \tag{6.4.17b}$$

show infrared divergencies for modulation spectra $D(\omega)$ that contain nonzero spectral weight at $\omega = 0$. We shall discuss this breakdown of the small-Δ expansion of the coefficient ε_2 of the Poincaré–Lindstedt expansion further later on.

To determine the next coefficient function $x_3(t;\Delta)$, (6.4.11) of the bifurcating solution (6.4.9) we need, in addition to the marginal solutions $x_c(t;\Delta)$ and $y_c(t;\Delta)$ of the linearized problem at threshold, the solution $x_3(t;\Delta)$ of the nonlinear problem in order λ^3, (6.4.5a, b). Its small-Δ expansion reads

$$x_3(t;\Delta) = (x_c^{(0)})^3[\Delta x_3^{(1)}(t) + \Delta^2 x_3^{(2)}(t) + O(\Delta^3)]. \tag{6.4.18}$$

Here we used the fact that $x_3(t;\Delta = 0) = 0$. The Poincaré–Lindstedt expansions (6.4.1a, b) reproduce for $\Delta = 0$ the static fixed point $x = \varepsilon^{1/2}$ with $x_{2n+1} = \varepsilon_{2n} = 0$ for all $n \geqslant 1$. Using (6.4.18) together with (6.4.12a, b), (6.4.14a, b) and (6.4.16) one finds

$$\varepsilon_4 = (x_c^{(0)})^4\Delta^2 2\langle (x_c^{(1)} + y_c^{(1)})x_3^{(1)}\rangle + O(\Delta^4) \tag{6.4.19}$$

and thus

$$\tilde{x}_3(t;\Delta) = \Delta x_3^{(1)}(t) + \Delta^2[x_3^{(2)}(t) - \langle (3x_c^{(1)} + y_c^{(1)})x_3^{(1)}\rangle] + O(\Delta^3). \tag{6.4.20}$$

The functions $x_3^{(1)}(t)$ and $x_3^{(2)}(t)$ are the solutions of (6.4.5a, b) to first and second order in Δ, respectively,

$$x_3^{(1)}(\omega) = -2\chi(\omega)x_c^{(1)}(\omega) \tag{6.4.21}$$

$$x_3^{(2)}(\omega)=\chi(\omega)h(\omega)-2x_c^{(2)}(\omega) \tag{6.4.22a}$$

$$h(t)=x_3^{(1)}(t)\zeta(t)-\langle x_3^{(1)}\zeta\rangle-3[(x_c^{(1)}(t))^2-\langle(x_c^{(1)})^2\rangle]. \tag{6.4.22b}$$

6.4.3 *Mean square of the bifurcating solution*

In this section we discuss as a representative example the second moment $\langle x^2(t)\rangle$ of the bifurcating solution in more detail. We compare the prediction of the Poincaré–Lindstedt expansion with numerical experiments and a direct small-Δ perturbation expansion.

The Poincaré–Lindstedt result

From (6.4.9) one obtains the following expression for the second moment of the bifurcating solution up to order $(\varepsilon-\varepsilon_c)^2$:

$$\langle x^2(t)\rangle=(\varepsilon-\varepsilon_c)s(\Delta)+(\varepsilon-\varepsilon_c)^2\tau(\Delta)+O(\varepsilon-\varepsilon_c)^3. \tag{6.4.23}$$

The initial slope of $\langle x^2\rangle$ at threshold

$$s(\Delta)=\langle\tilde{x}_1^2(t;\Delta)\rangle=\frac{\langle y_c(t;\Delta)x_c(t;\Delta)\rangle\langle x_c^2(t;\Delta)\rangle}{\langle y_c(t;\Delta)x_c^3(t;\Delta)\rangle} \tag{6.4.24}$$

is determined by equal-time correlations of the marginal functions x_c and y_c. The prefactor of the quadratic term in (6.4.23) reads

$$\tau(\Delta)=2\langle\tilde{x}_1(t;\Delta)\tilde{x}_3(t;\Delta)\rangle. \tag{6.4.25}$$

Up to second order in Δ these quantities are given by

$$s(\Delta)=1-\Delta^2 2\langle x_c^{(1)}(x_c^{(1)}+y_c^{(1)})\rangle+O(\Delta^4) \tag{6.4.26a}$$

$$=1-\Delta^2\frac{4}{m^2}\int_{-\infty}^{\infty}\frac{d\omega}{2\pi}\frac{D(\omega)}{(\omega^2+\gamma^2)^2}+O(\Delta^4) \tag{6.4.26b}$$

and by

$$\tau(\Delta)=-\Delta^2 2\langle x_3^{(1)}(y_c^{(1)}+2x_c^{(1)})\rangle+O(\Delta^4) \tag{6.4.27a}$$

$$=-\Delta^2\frac{4}{m^3}\int_{-\infty}^{\infty}\frac{d\omega}{2\pi}\frac{D(\omega)}{\omega^2}\frac{3\omega^2-\gamma^2}{(\omega^2+\gamma^2)^3}+O(\Delta^4). \tag{6.4.27b}$$

Thus the initial slope $s(\Delta)$ of the mean squared solution (6.4.23) close to threshold $\varepsilon_c(\Delta)$ is in the presence of small-amplitude modulation and inertia smaller than without modulation, $\Delta=0$. In the overdamped limit case, however, neither the mean squared amplitude itself nor its slope is suppressed by modulation: the exact result $\langle x^2\rangle=\varepsilon$ of Section 6.2.3 is reproduced to the order Δ^2 considered here since $s\to1$ and $\tau\to0$ when $m\to0$ with $m\gamma\to1$. The bifurcation behavior of $\langle x^2\rangle$ in the presence of modulation and inertia is shown schematically in Figure 6.2 (thick lines) in comparison with the bifurcation diagram for the unmodulated case (thin lines).

It is important to keep the range of validity of (6.4.26a, b) and (6.4.27a, b) in mind. Whereas the formula for the bifurcation threshold $\varepsilon_c(\Delta)$ obtained with a small-Δ expansion, (6.3.20), seems to be correct for *any* modulation spectrum $D(\omega)$, this is neither the case for the small-Δ expressions (6.4.27a, b) of the initial curvature τ nor for the Δ^2 correction to the initial slope $s(\Delta = 0) = 1$ of $\langle x^2 \rangle$ induced by modulation. To derive the small-Δ formulas (6.4.26a, b) and (6.4.27a, b) for $s(\Delta)$, (6.4.24), and $\tau(\Delta)$, (6.4.25), we used the Δ-expansion of Section 6.4.2 for the functions x_c, y_c and x_3 entering into the equal-time correlations of (6.4.24) and (6.4.25). Then there appear in order Δ^2 infrared divergencies, e.g. in the expressions (6.4.17a, b) for $\langle y_c^{(1)} x_c^{(1)} \rangle$ and $\langle (x_c^{(1)})^2 \rangle$ if the modulation spectrum $D(\omega)$ has a finite weight at zero frequency. While such a divergence manifestly can appear in (6.4.27b) it cancels in (6.4.26a, b). Therefore it has been speculated (Lücke and Schank, 1985) that (6.4.26a, b) might hold also for modulation spectra with $D(\omega = 0) \neq 0$. That, unfortunately, is wrong, as we shall see further below. The Δ^2-term in the expansions (6.4.26a, b) of the slope $s(\Delta)$ seems to be correct only for modulation processes with spectra such that $D(\omega = 0) = 0$ (Lazarov, 1987) and thus formulas (12) and (13) in the work by Lücke and Schank (1985) apply only for the above described spectra. One arrives at this conclusion by comparing with a direct expansion of $x(t)$ for small Δ and with numerical experiments.

Direct small-Δ expansion of the bifurcating solution

The relative simplicity of the parametrically modulated Duffing oscillator allows us to expand the full nonlinear bifurcating solution of (6.2.1) directly around the corresponding $\Delta = 0$ supercritical solution

$$x(t; \Delta = 0) = \varepsilon^{1/2} \tag{6.4.28a}$$

according to

$$x(t) = \varepsilon^{1/2} + \Delta X_1(t; \varepsilon) + \Delta^2 X_2(t; \varepsilon) + \cdots . \tag{6.4.28b}$$

We use capital letters for the coefficients of the direct expansion (6.4.28a, b) to avoid confusion with the coefficients $x_n(t; \Delta)$ appearing in the Poincaré–Lindstedt expansion (6.4.1a, b) around the basic solution, $x = 0$, for small $\varepsilon - \varepsilon_c$. The expansion will work particularly well for large ε. Then the potential (6.2.2) has deep minima at positions $x^2 = \varepsilon[1 + (\Delta/\varepsilon)\zeta(t)]$ that differ for $\Delta/\varepsilon \ll 1$ only slightly from the minima at $x = \pm\varepsilon^{1/2}$ for the static, $\Delta = 0$, case. However, we shall consider (6.4.28a, b) also for small ε.

Inserting (6.4.28a, b) into (6.2.1) one finds

$$X_1(\omega) = \varepsilon^{1/2} R(\omega) \zeta(\omega) \tag{6.4.29}$$

$$X_2(\omega) = R(\omega) \int_{-\infty}^{\infty} \mathrm{d}t\, \mathrm{e}^{\mathrm{i}\omega t} X_1(t)[\zeta(t) - 3\varepsilon^{1/2} X_1(t)], \tag{6.4.30}$$

with

$$R(\omega)=\frac{-1/m}{\omega^2-2\varepsilon/m+\mathrm{i}\gamma\omega} \tag{6.4.31}$$

denoting the dynamical response function of a damped harmonic oscillator with a characteristic frequency $(2\varepsilon/m)^{1/2}$ determined by the mean curvature of the potential side minima at $\pm\varepsilon^{1/2}$. The above functions (6.4.29)–(6.4.31) still depend on ε, which will be shown explicitly in the list of arguments only when needed.

Note that $X_1(t)$, $X_2(t)$ and hence also $x(t)$, (6.4.28a, b), evaluated up to order Δ^2, do not vanish at the threshold value $\varepsilon=\varepsilon_c$ where the correct nonlinear solution of (6.2.1) bifurcates from the trivial solution $x=0$. However, the direct expansion (6.4.28a, b) is meaningful and very useful for $\varepsilon\geqslant\varepsilon_c$ in particular for a comparison of the second moment $\langle x^2\rangle$ of (6.4.28a, b) with that of the Poincaré–Lindstedt expansion.

To order Δ^2 one obtains for the former

$$\langle x^2\rangle=\varepsilon+\Delta^2\langle X_1^2+2\varepsilon^{1/2}X_2\rangle+O(\Delta^4). \tag{6.4.32a}$$

Here $\langle X_2\rangle$ follows from taking the average of the equation of motion at order Δ^2

$$2\varepsilon\langle X_2\rangle=\langle X_1\zeta-3\varepsilon^{1/2}X_1^2\rangle. \tag{6.4.32b}$$

Expressing the equal-time correlations by spectral integrals one finds

$$\langle x^2\rangle=\varepsilon-\Delta^2\frac{1}{m}\int_{-\infty}^{\infty}\frac{\mathrm{d}\omega}{2\pi}\frac{\omega^2 D(\omega)}{(\omega^2-2\varepsilon/m)^2+\gamma^2\omega^2}+O(\Delta^4). \tag{6.4.32c}$$

Note that the expression (6.4.32c) for $\langle x^2\rangle$ evaluated up to order Δ^2 becomes positive at the threshold value

$$\varepsilon_c=\Delta^2\frac{1}{m}\int_{-\infty}^{\infty}\frac{\mathrm{d}\omega}{2\pi}\frac{D(\omega)}{\omega^2+\gamma^2}+O(\Delta^4) \tag{6.4.33}$$

that was shown in Section 6.3.2 to be the bifurcation threshold. In this sense the direct small-Δ expansion of the second moment of the *bifurcating* solution is consistent with and reproduces the correct stability threshold of the *basic* solution $x=0$. This is in complete analogy to the behavior of the moments of a modulated discrete dynamical system (see Section 6.6.3) studied by Linz and Lücke (1986).

In order to arrive at (6.4.33) one has to evaluate the spectral integral in (6.4.32c) at $\varepsilon=\varepsilon_c(\Delta=0)=0$. While this implies extending the Δ-expansion down to $\varepsilon=0$, it does not cause mathematical difficulties in (6.4.32c) since the integrand is well behaved near $\varepsilon=0$. The situation for the slope $S(\varepsilon;\Delta)=\partial\langle x^2\rangle/\partial\varepsilon$

$$S(\varepsilon;\Delta)=1-\Delta^2\frac{4}{m^2}\int_{-\infty}^{\infty}\frac{\mathrm{d}\omega}{2\pi}\omega^2 D(\omega)\frac{\omega^2-2\varepsilon/m}{[(\omega^2-2\varepsilon/m)^2+\gamma^2\omega^2]^2}+O(\Delta^4) \tag{6.4.34}$$

of the second moment as a function of ε, on the other hand, is more subtle as we shall see now.

Since the Poincaré–Lindstedt expansion proceeds around $\varepsilon = \varepsilon_c(\Delta)$ one has to identify

$$S(\varepsilon = \varepsilon_c(\Delta), \Delta) \leftrightarrow s(\Delta) \tag{6.4.35}$$

in order to compare the result of the direct expansion method for the initial slope of $\langle x^2 \rangle$ at threshold $\varepsilon = \varepsilon_c$ with that one of the Poincaré–Lindstedt expansion. Then, if one wants to compare the secondary expansion of $s(\Delta)$ (6.4.26a, b) to order Δ^2 with the corresponding order of $S(\varepsilon_c, \Delta)$, one has to evaluate the spectral integral in (6.4.34) for $\varepsilon \to 0$ since $\varepsilon_c \sim \Delta^2$. However, in this limit the square bracket in the denominator of the integrand in (6.4.34) contains a double zero at $\omega = 0$. To handle it properly one should factorize

$$(\omega^2 - 2\varepsilon/m)^2 + \gamma^2\omega^2 = (\omega^2 + \alpha^2)(\omega^2 + \eta^2) \tag{6.4.36}$$

and observe that for small ε

$$\alpha = \frac{\gamma}{2} + \left(\frac{\gamma^2}{4} - \frac{2\varepsilon}{m}\right)^{1/2} = \gamma + O(\varepsilon) \tag{6.4.37a}$$

$$\eta = \frac{\gamma}{2} - \left(\frac{\gamma^2}{4} - \frac{2\varepsilon}{m}\right)^{1/2} = \frac{2\varepsilon}{m\gamma} + O(\varepsilon^2). \tag{6.4.37b}$$

Then one finds

$$S(\varepsilon_c, \Delta) = 1 - \Delta^2 \frac{4}{m^2} \int_{-\infty}^{\infty} \frac{d\omega}{2\pi} \frac{D(\omega)}{(\omega^2 + \gamma^2)^2} \left[1 - \gamma \frac{\pi}{2} \delta(\omega)\right] + O(\Delta^4). \tag{6.4.38}$$

Here the Dirac δ-function arises from the second term in the integrand of (6.4.34) since

$$\frac{\omega^2 \eta}{(\omega^2 + \eta^2)^2} \to -\omega \frac{\pi}{2} \frac{\partial}{\partial \omega} \delta(\omega).$$

A partial integration then leads to the second term in (6.4.38). Thus the initial slope $S(\varepsilon_c, \Delta)$, (6.4.38), obtained from the direct small-Δ expansion is larger than the slope $s(\Delta)$, (6.4.26a, b), given by the secondary small-Δ expansion of the primary Poincaré–Lindstedt expansion by the positive amount

$$S(\varepsilon_c, \Delta) - s(\Delta) = \Delta^2 \frac{D(\omega = 0)}{m^2 \gamma^3} + O(\Delta^4), \tag{6.4.39}$$

which vanishes only when the modulation spectrum has no weight at zero frequency.

For example, for a white noise spectrum, $D(\omega) = D(\omega = 0) = D_0$, one finds $S = 1 + O(\Delta^4)$ while $s = 1 - \Delta^2 D_0/(m^2\gamma^3)$. In fact, for this special spectrum,

(6.4.32c) and (6.4.34) show that the slope is $S = 1 + O(\Delta^4)$ for all ε – the spectral integral in (6.4.32c) reduces to (6.4.33) and that of (6.4.34) vanishes. Thus the direct small-Δ expansion gives to order Δ^2 for the second moment above threshold, $\varepsilon_c = \Delta^2 D_0/(2\gamma m)$,

$$\langle x^2 \rangle = \varepsilon - \varepsilon_c. \tag{6.4.40}$$

Numerical experiments

Here we describe some of the numerical experiments (Lazarov, 1987) that were performed to compare with theory. To integrate (6.2.1) we used a one-step forward difference method. For the case

$$m = \gamma = 1 \tag{6.4.41a}$$

that will be discussed in the remainder of this section the discretized version of (6.2.1) used by us was

$$x_{n+1} = x_n + \mathrm{d}t\, v_n \tag{6.4.41b}$$

$$v_{n+1} = v_n + \mathrm{d}t[-v_n + (\varepsilon + \Delta\zeta_n - x_n^2)x_n]. \tag{6.4.41c}$$

The index n identifies the nth time step of size $\mathrm{d}t = 0.005$.

A white-noise process $\zeta_w(t)$ with a spectrum

$$D(\omega) = D_0 \qquad \text{(W)} \tag{6.4.42}$$

was approximated by

$$\zeta_w(t_n) = (12 D_0/\mathrm{d}t)^{1/2}\, \xi_n. \tag{6.4.43a}$$

Here ξ_n are computer generated uncorrelated pseudorandom numbers such that

$$\langle \xi_n \rangle = 0; \quad \langle \xi_n \xi_m \rangle = \tfrac{1}{12}\delta_{n,m}. \tag{6.4.43b}$$

Since ξ was distributed with equal weight over the interval $(-\tfrac{1}{2}, \tfrac{1}{2})$ we did not simulate a Gaussian process. In addition, processes with spectra

$$D(\omega) = \frac{D_0}{(\omega^2 - \Omega^2)^2 + \Gamma^2\omega^2} \qquad \text{(A)} \tag{6.4.44}$$

$$D(\omega) = \frac{\omega^2 D_0}{(\omega^2 - \Omega^2)^2 + \Gamma^2\omega^2} \qquad \text{(B)} \tag{6.4.45}$$

$$D(\omega) = \frac{D_0}{\omega^2 + \Gamma^2} \qquad \text{(C)} \tag{6.4.46}$$

were generated by solving the finite difference version of

$$(\partial_t^2 - \Omega^2 + \Gamma\partial_t)\zeta(t) = \zeta_w(t) \tag{6.4.47}$$

for $\zeta(t)$ (A) or $\dot{\zeta}(t)$ (B). The solution of

$$(\partial_t + \Gamma)\zeta(t) = \zeta_w(t) \tag{6.4.48}$$

yields a process with a spectrum (C). To facilitate a comparison with the monochromatic spectrum (6.2.7b) of the periodic modulation (6.2.7a) the total spectral weight of (A)–(C) was enforced to be $\frac{1}{2}$:

$$\int_{-\infty}^{\infty} \frac{d\omega}{2\pi} D(\omega) = 1/2 \tag{6.4.49}$$

by choosing D_0 as

$$D_0 = \Gamma\Omega^2 \quad (A); \quad D_0 = \Gamma \quad (B) \text{ and } (C). \tag{6.4.50}$$

All time averages were taken over an interval of 10^4. The errors involved in obtaining $\langle x^2 \rangle$ from the solution of (6.4.41a–c) were estimated on the basis of an error analysis of the numerical results for the moments of (6.4.47) and (6.4.48). In the range of ε and Δ values discussed below they were about 0.4% (Lazarov, 1987). We found that already the second moment, $\langle \xi_n^2 \rangle$, when evaluated over different sequences of random numbers of an HP9000 computer varied by about 0.2% for the typical sequence lengths of 2×10^6 numbers used in the numerical experiments.

Comparisons

We shall concentrate here on the supercritical behavior of the mean squared amplitude $\langle x^2(t; \varepsilon, \Delta) \rangle$ of the bifurcated solution. Extrapolating this quantity down to zero as a function of ε yields the bifurcation threshold $\varepsilon_c(\Delta)$. In such a way it was found that the theoretical prediction (6.3.20) agreed perfectly (Lazarov, 1987) with all numerically determined thresholds for all small-amplitude modulation processes (cos, A, B, C, W) that were investigated.

To compare the theoretical predictions (6.4.23)–(6.4.27a, b) and (6.4.32c) for the small-ε and small-Δ dependence of $\langle x^2 \rangle$ with each other and with numerical experiment we discuss the depression

$$d(\varepsilon, \Delta) = x^2(\varepsilon, \Delta = 0) - \langle x^2(t; \varepsilon, \Delta) \rangle \tag{6.4.51}$$

of the square of the bifurcated solution, $x^2(\varepsilon, \Delta = 0) = \varepsilon$, caused by modulation of amplitude Δ. This is also an appropriate quantity to determine experimentally: for example, at a fixed ε one could switch on the modulation and see how $x^2 = \varepsilon$ for $\Delta = 0$ is changed to $x^2(t)$ in the presence of modulation.

In Figure 6.4 we show the depression d, (6.4.51), as a function of ε for fixed $\Delta = 0.2$ for a cosine modulation and a stochastic modulation process with spectrum (B). Note that in each case $D(\omega)$ vanishes at $\omega = 0$ so that the small-Δ expansion of the Poincaré–Lindstedt result discussed earlier (shown by full lines in Figure 6.4) agrees with the direct Δ expansion. We have continued the full lines in Figure 6.4 all the way down to $\varepsilon = 0$ although $\langle x^2 \rangle$ is nonzero only above ε_c. This quantity, being very small for the modulation process of Figure 6.4, may be read off from the ordinate by extrapolation, e.g. $\varepsilon_c = 4 \times 10^{-3}$ for the cosine modulation and $\varepsilon_c = 3.3 \times 10^{-3}$ for modulation (B). The depression of $\langle x^2 \rangle$ is larger for the cosine modulation than for process

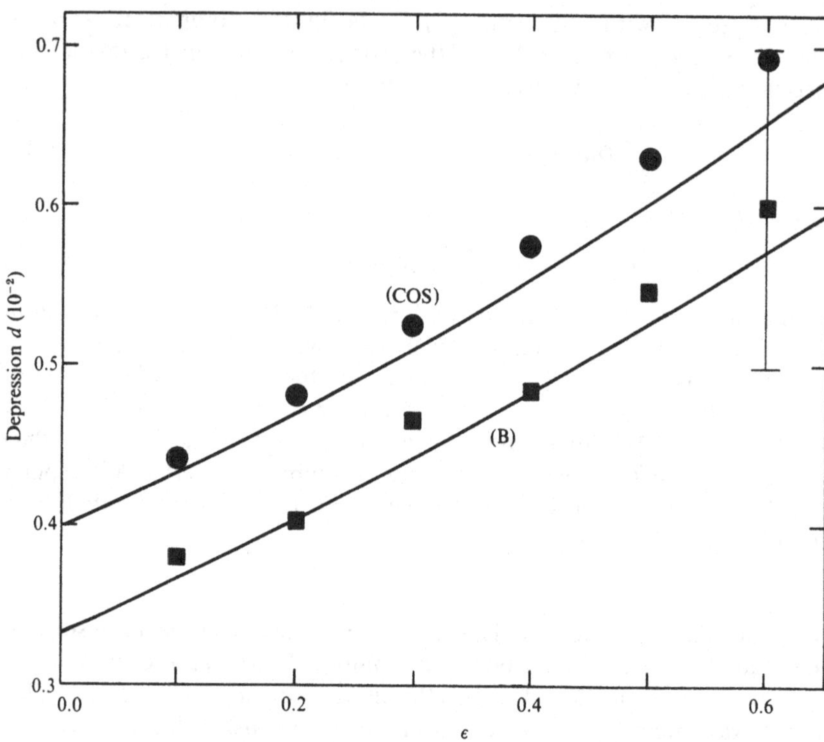

Figure 6.4. Modulation-induced depression of the mean squared amplitude $\langle x^2 \rangle$ of the Duffing oscillator ($m = 1 = \gamma$) versus ε for fixed amplitude $\Delta = 0.2$. Modulation processes were a cosine of frequency $\Omega = 2$, and a stochastic process with spectrum (B) ($\Omega = 2$, $\Gamma = 1$) peaked at $\Omega \simeq 2$. The symbols show numerical results, and the lines represent the Poincaré–Lindstedt expansion (6.4.23) up to $(\varepsilon - \varepsilon_c)^2$ with the secondary expansion of ε_c (6.3.20), s (6.4.26b), τ (6.4.27b) up to Δ^2. The result (6.4.32c) of the direct Δ expansion almost coincides with the full lines. The error estimate is explained in the text.

(B) although both have the same total spectral weight. However, the factor $1/(\omega^2 + \gamma^2)$ entering into the various spectral integrals has for the parameter combinations considered here ($\gamma = \Gamma = 1, \Omega = 2$) a larger overall spectral overlap with the line spectrum at $\Omega = 2$ than with the broadened spectrum (B). In Figure 6.5 we show, again for $\Delta = 0.2$, the depression d for a white-noise modulation with $D_0 = 1$ and for the spectrum (A), (6.4.44), showing with $\Gamma = 1$, $\Omega = 2$ a peak near $\Omega = 2$ of height $D_A(\Omega) = 1$ and a zero-frequency weight of $D_A(0) = \frac{1}{4}$. For both of these spectra the small-Δ expansion of the Poincaré–Lindstedt coefficient $\tau(\Delta)$, (6.4.27a, b), diverges and that of $s(\Delta)$, (6.4.26a, b), is wrong in order Δ^2. This is evident from a comparison of the full lines in Figure 6.5 with the numerical data. On the other hand, the direct small-Δ expansion (dashed lines) agrees very well with the numerical experiments.

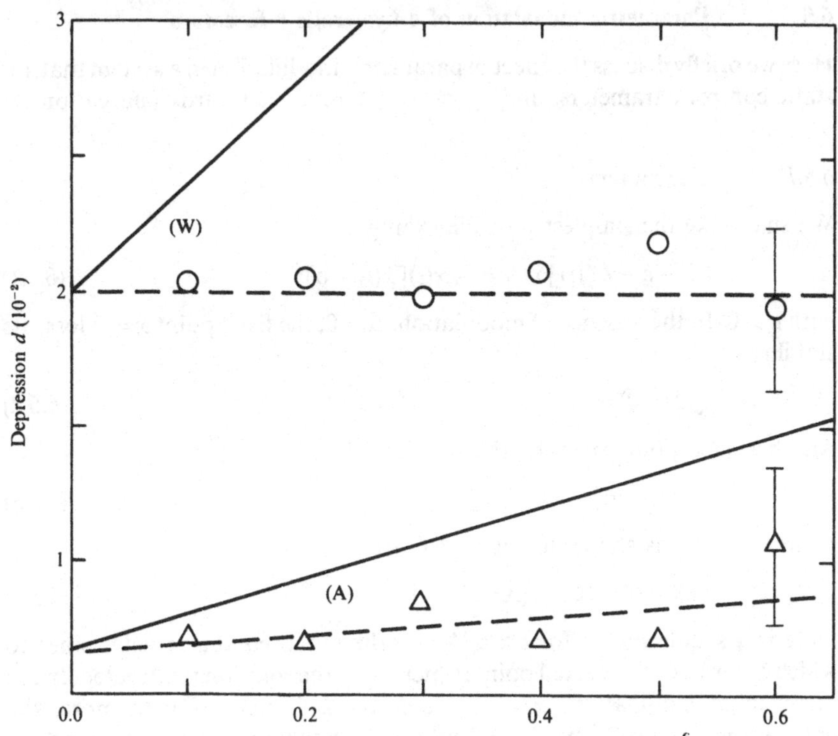

Figure 6.5. Modulation-induced depression of $\langle x^2 \rangle$ as in Figure 6.4. (W) is for white noise ($D_0 = 1$) and (A) is for the spectrum (6.4.44) ($\Gamma = 1$, $\Omega = 2$). The symbols denote numerical results. Full lines represent the lowest order Poincaré–Lindstedt expansion $\langle x^2 \rangle = (\varepsilon - \varepsilon_c)s(\Delta)$ with ε_c (6.3.20), s (6.4.26b) expanded up to Δ^2. Dashed lines represent the direct Δ expansion (6.4.32c). The difference in the slope of these different lines is given in each case by (6.4.39).

The discrepancy when expressed in terms of the difference in slope of the two lines is just given by (6.4.39), $\Delta^2 D(\omega = 0)/(m^2\gamma^3)$, i.e. it decreases with decreasing weight, $D(\omega = 0)$, at zero frequency.

Note that for white-noise modulation the depression is independent of ε in Figure 6.4, i.e. $\langle x^2 \rangle = \varepsilon - \varepsilon_c$, (6.4.40), with $\varepsilon_c = \Delta^2 D_0/(2\gamma m) = 0.02$ for the parameters of Figure 6.4.

We finally discuss the Δ-dependence of the depression d, (6.4.51), at fixed ε. One finds numerically (Lazarov, 1987) that it increases for small Δ proportional to Δ^2 in excellent agreement with the result (6.4.32c) of the direct small-Δ expansion for all types of modulation processes. The secondary small-Δ expansion of the Poincaré–Lindstedt result reproduces this behavior quantitatively only for spectra with vanishing zero-frequency spectral weight. Again the discrepancy of, e.g., the second growth coefficient, d/Δ^2, of the depression increases with growing $D(\omega = 0)$.

6.5 Parametric modulation of a hysteretic bifurcation

Here we briefly discuss the effect of parametric modulation in a system that, for static control parameters, undergoes a hysteretic backwards bifurcation.

6.5.1 *The system*

We shall take the simplest possible example

$$[\partial_t - \varepsilon - \Delta\zeta(t)]x(t) = -x(t)[x(t) - a]^2 \tag{6.5.1}$$

with $a > 0$. In the absence of modulation, $\Delta = 0$, the fixed point $x = 0$ loses its stability at

$$\varepsilon_c(\Delta = 0) = a^2. \tag{6.5.2}$$

Another fixed point given by the solution of

$$\varepsilon = (x - a)^2 \tag{6.5.3}$$

bifurcates at ε_c as shown in Figure 6.6. At

$$\varepsilon_s(\Delta = 0) = 0; \quad x_s(\Delta = 0) = a \tag{6.5.4}$$

there is a saddle node. For $\varepsilon > \varepsilon_s(\Delta = 0)$ the initial condition determines to which of the two stable fixed points (thick lines) the solution is attracted. In the presence of modulation, $x = 0$ remains a fixed point. Furthermore, the bifurcation threshold where $x = 0$ becomes unstable remains unchanged,

$$\varepsilon_c(\Delta) = \varepsilon_c(\Delta = 0) = a^2, \tag{6.5.5}$$

since we are considering for the sake of mathematical convenience in (6.5.1) the overdamped limit case. For small modulation amplitudes, Δ, one expects that the bifurcating solution oscillates around a mean that is close to the fixed point (6.5.3) of the unmodulated system as indicated schematically in Figure 6.6. In the remainder of this section we investigate the bifurcation behavior more quantitatively for a periodic modulation

$$\zeta(t) = \cos \Omega t. \tag{6.5.6}$$

In this case the full lines in Figure 6.6 represent the mean $\langle x \rangle$ of a stable limit cycle and the dashed lines represent the mean $\langle x \rangle$ of an unstable limit cycle that bifurcates transcritically at ε_c from $x = 0$.

6.5.2 *The expansion*

We use the Poincaré–Lindstedt expansion around $x = 0$, $\varepsilon = \varepsilon_c$

$$x = \lambda x_1 + \lambda^2 x_2 + \cdots \tag{6.5.7a}$$

$$\varepsilon - a^2 = \lambda \varepsilon_1 + \lambda^2 \varepsilon_2 + \cdots \tag{6.5.7b}$$

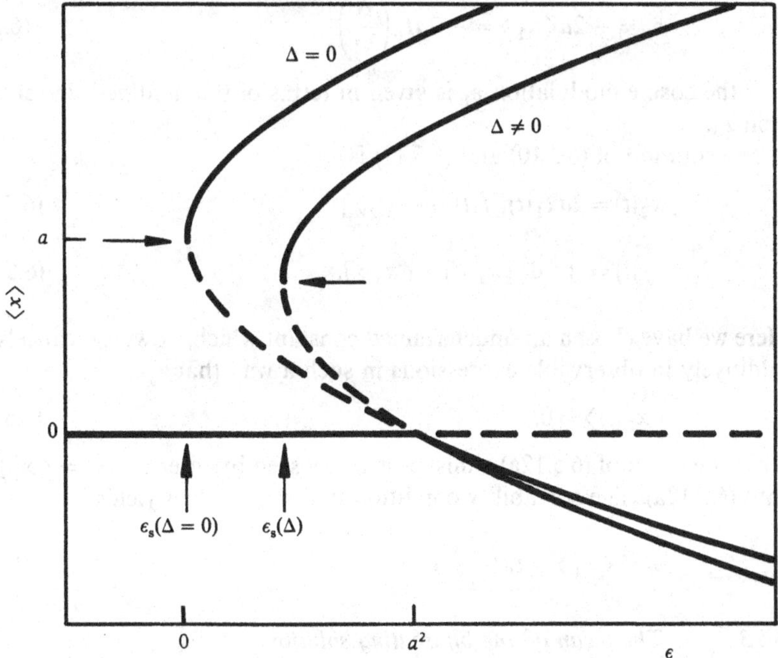

Figure 6.6. Schematic bifurcation diagram of (6.5.1). The fixed point $x = 0$ is stable for $\varepsilon < \varepsilon_c = a^2$. For $\Delta = 0$, full (dashed) lines denote stable (unstable) fixed points. For periodic modulation with small amplitude, $\Delta \neq 0$, the full (dashed) curved line denotes the mean of the periodic stable (unstable) limit cycle solution. Arrows indicate the saddle positions where stable and unstable solution meet.

to approximate these nontrivial solutions. The solution of the equation of motion in order λ

$$(\partial_t - \Delta\zeta)x_1 = 0 \tag{6.5.8a}$$

is

$$x_1(t) = \exp\left[\Delta \int_0^t dt'\zeta(t')\right] = e^{(\Delta/\Omega)\sin\Omega t}. \tag{6.5.8b}$$

Here we have set the undetermined constant $x_1(t = 0) = 1$ since it is anyhow divided out of all observable quantities. The adjoint equation of (6.5.8a) is solved by

$$y_1(t) = 1/x_1(t), \tag{6.5.9}$$

where again a dummy constant was set equal to one. The solvability condition of the equation in order λ^2

$$(\partial_t - \Delta\zeta)x_2 = 2ax_1^2 + \varepsilon_1 x_1 \tag{6.5.10}$$

yields

$$\varepsilon_1 = -2a\langle x_1 \rangle = -2aI_0\left(\frac{\Delta}{\Omega}\right). \tag{6.5.11}$$

For the cosine modulation ε_1 is given in terms of the modified Bessel function I_0.

The solution of (6.5.10) and (6.5.11) is

$$x_2(t) = 2ax_1(t)[f(t) - \langle f(t) \rangle] \tag{6.5.12a}$$

$$f(t) = \int_0^t dt'[x_1(t') - \langle x_1 \rangle]. \tag{6.5.12b}$$

Here we have chosen an undetermined constant which, however, cancels out additively in observable expressions in such a way that

$$\langle x_2(t) \rangle = 0. \tag{6.5.12c}$$

That the mean of (6.5.12a) vanishes is easily seen by inserting $x_1 = \langle x_1 \rangle + \dot{f}$ into (6.5.12a). The solvability condition in order λ^3 then yields

$$\varepsilon_2 = \langle x_1^2 \rangle = I_0\left(\frac{2\Delta}{\Omega}\right). \tag{6.5.13}$$

6.5.3 *The mean of the bifurcating solution*

The easiest way to determine the mean $\langle x \rangle$ of the bifurcating solution x approximated by the Poincaré–Lindstedt expansion is to average (6.5.7a)

$$\langle x \rangle = \lambda \langle x_1 \rangle + O(\lambda^3). \tag{6.5.14a}$$

Thus, since $\langle x_2 \rangle = 0$, one has

$$\lambda = \langle x \rangle / \langle x_1 \rangle + O(\langle x \rangle^3). \tag{6.5.14b}$$

Inserting this into (6.5.7b) and using (6.5.11) and (6.5.13) one finds an equation for ε in terms of $\langle x \rangle$:

$$\varepsilon = a^2 - 2a\langle x \rangle + \frac{\langle x_1^2 \rangle}{\langle x_1 \rangle^2}\langle x \rangle^2 + O(\langle x \rangle^3). \tag{6.5.15}$$

This equation gives a (rotated) bifurcation diagram of ε versus $\langle x \rangle$ in the presence of modulation. Needless to say, (6.5.15) is a good approximation to the true bifurcation diagram in the ε–$\langle x \rangle$ plane as long as the correction terms to (6.5.15) are small.

As a small digression we remark that inverting (6.5.7b) and inserting the resulting series for γ in terms of powers of $\varepsilon - a^2$ into the averaged relation (6.5.7a) yields

$$\langle x \rangle = -\frac{1}{2a}(\varepsilon - a^2) + \frac{\langle x_1^2 \rangle}{8a^3 \langle x_1 \rangle^2}(\varepsilon - a^2)^2 + O(\varepsilon - a^2)^3, \tag{6.5.16}$$

i.e. a power series in the distance $\varepsilon - a^2$ from the bifurcation threshold. However, this cannot reproduce the saddle behavior in Figure 6.6 and thus will no longer be considered here. Given the hysteretic topology of the bifurcation diagram in the absence of modulation, $\varepsilon = (x - a)^2$, the modulated case can also be described more economically by approximating ε as a function of the considered order parameter, e.g. $\langle x \rangle$, instead of approximating $\langle x \rangle$ as a function of ε.

The slope of $\langle x \rangle$ at the bifurcation threshold, $\varepsilon_c = a^2, \langle x \rangle = 0$, is not changed by modulation. The reason is the overdamped, inertia-free dynamics of (6.5.1). However, the curvature

$$\frac{\langle x_1^2 \rangle}{\langle x_1 \rangle^2} = \frac{I_0\left(2\frac{\Delta}{\Omega}\right)}{I_0^2\left(\frac{\Delta}{\Omega}\right)} = 1 + \frac{1}{2}\frac{\Delta^2}{\Omega^2} + O\left(\frac{\Delta^4}{\Omega^4}\right) \tag{6.5.17}$$

is increased by modulation. This gives rise to a shift of the saddle position from $\varepsilon_s(\Delta = 0) = 0$, $x_s(\Delta = 0) = a$ to

$$\langle x_s(t;\Delta) \rangle = a\frac{\langle x_1 \rangle^2}{\langle x_1^2 \rangle}; \quad \varepsilon_s = a^2\left[1 - \frac{\langle x_1 \rangle^2}{\langle x_1^2 \rangle}\right]. \tag{6.5.18}$$

In Figure 6.7 we compare the bifurcation diagram (6.5.15) for a cosine modulation with frequency $\Omega = 1$ and amplitude $\Delta = 0.4$ with the one obtained by a numerical integration of (6.5.1) in the vicinity of the saddle location. The periodic attractor (full circles in Figure 6.7) was determined by integrating forwards in time, and the periodic repeller (open circles) was evaluated by integrating (6.5.1) backwards in time.

In Figure 6.7 we have implied by the full and dashed lines that the periodic attractor and repeller collide at $\varepsilon_s = a^2(1 - \langle x_1 \rangle^2 / \langle x_1^2 \rangle)$. This, in fact, results from a stability analysis (Lazarov, 1987) up to order λ^2 of the periodic solution that is obtained from (6.5.7) up to this order λ^2.

Note that for the value $a = 1$ that was used in Figure 6.7 the saddle position $\langle x_s \rangle \sim 1$, $\varepsilon_s \sim 0$ is 'far away' from the bifurcation threshold, $x = 0, \varepsilon_c = 1$. Nevertheless the Poincaré–Lindstedt result (6.5.15) for the mean $\langle x \rangle$ determined to second order in an expansion around threshold is a surprisingly good approximation, not only up to the saddle position, but even for part of the upper stable branch.

6.5.4 Fluctuations around the mean

Unfortunately the quality (Lazarov, 1987) of the expansion (6.5.7) for the orbit itself deteriorates faster with increasing distance from the bifurcation threshold $\varepsilon_c = a^2$ than for the mean. We demonstrate this here for the root-mean-

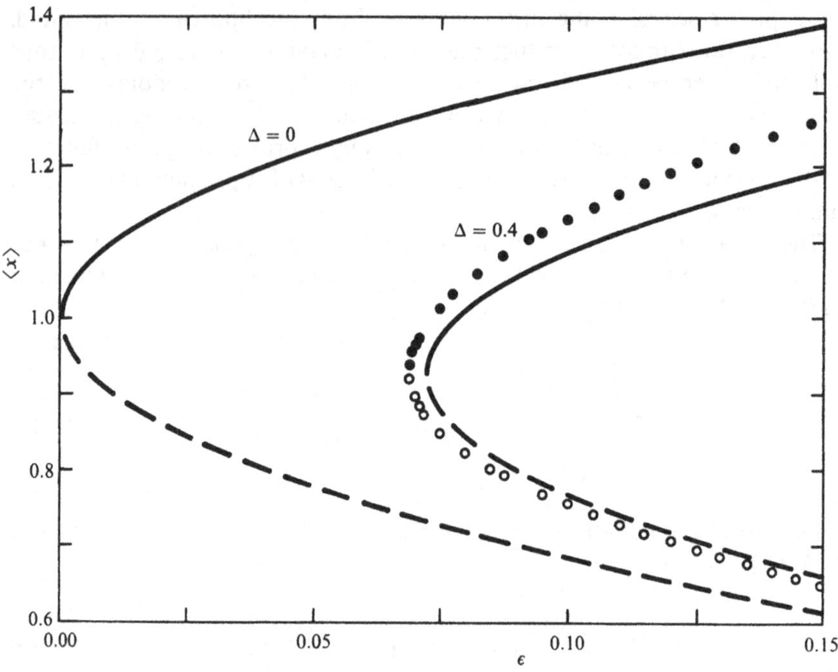

Figure 6.7. Shift of $\langle x \rangle$ induced by modulation $\zeta(t) = 0.4 \cos t$ in the hysteretic system (6.5.1) with $a = 1$. The left-hand parabola shows the fixed points of the unmodulated system in the vicinity of the saddle node. The right-hand parabola represents the mean of the periodic solution obtained from a second order Poincaré–Lindstedt expansion. In each case a full (dashed) line represents an attractor (repeller). Full (open) circles show averages over numerically evaluated stable (unstable) limit cycle solutions of (6.5.1). Integration forwards (backwards) in time was used to follow trajectories onto the periodic attractor (repeller).

square (rms) fluctuation

$$\delta = \langle (x - \langle x \rangle)^2 \rangle^{1/2} \tag{6.5.19}$$

of the bifurcating orbit.

From the expansion (6.5.7a) one finds

$$\delta = \lambda \delta_1 + O(\lambda^3) \tag{6.5.20a}$$

with

$$\delta_1 = (\langle x_1^2 \rangle - \langle x_1 \rangle^2)^{1/2}. \tag{6.5.20b}$$

Here

$$\delta_2 = \frac{\langle x_1 x_2 \rangle - \langle x_1 \rangle \langle x_2 \rangle}{\delta_1^2} = 0, \tag{6.5.20c}$$

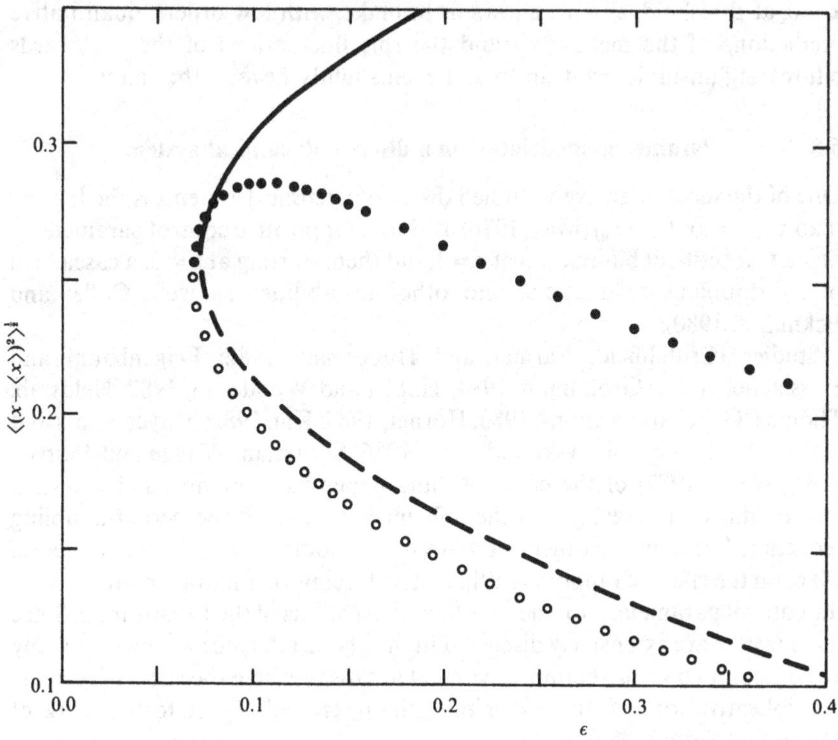

Figure 6.8. Root-mean-square fluctuations of the bifurcated periodic orbit (parameters as in Figure 6.7) in the vicinity of the saddle location. The full (dashed) line represents the result (6.5.21) of a second order Poincaré–Lindstedt expansion for the attractor (repeller). Full (open) circles are numerical results.

since both $\langle x_2 \rangle$ and $\langle x_1 x_2 \rangle$ vanish. Inserting (6.5.20a) into (6.5.7b) one obtains the relation

$$\varepsilon = a^2 - 2a\frac{\langle x_1 \rangle}{\delta_1}\delta + \frac{\langle x_1^2 \rangle}{\delta_1^2}\delta^2 + O(\delta^3) \tag{6.5.21}$$

between the control parameter ε and the rms fluctuation δ up to order δ^2.

In Figure 6.8 we show this quantity in the vicinity of the saddle location by a full (dashed) line for the stable (unstable) periodic orbit in comparison with numerically obtained rms fluctuations. Note that the discrepancy between (6.5.21) and the result of the numerical experiment rapidly grows beyond the saddle on the upper stable branch. That δ decreases on the upper branch for large ε is due to the minimum of the potential associated to (6.5.1) at $\varepsilon = (x - a)^2$ becoming deeper and deeper with growing ε. This depression of δ can not be reproduced by the expansion (6.5.7a, b) to order λ^2. However, it is remarkable that the Poincaré–Lindstedt expansion around the basic solution,

$x = 0$, at threshold, $\varepsilon_c = a^2$, allows us to make, with low orders, quantitative predictions of the mean $\langle x \rangle$ and the rms fluctuations of the backwards bifurcating unstable orbit up to and even slightly beyond the saddle.

6.6 Parametric modulation in a discrete dynamical system

One of the most intensively studied discrete dynamical systems is the logistic map $x_{n+1} = rx_n(1 - x_n)$ (May, 1976). It shows for positive control parameters r first a transcritical bifurcation at $r = 1$ and then, starting at $r = 3$, a cascade of period-doubling bifurcations and other instabilities as well (Collet and Eckmann, 1980).

Studies (Crutchfield, Farmer and Huberman, 1982; Feigenbaum and Hasslacher, 1982; Großmann, 1984; Haken and Wunderlin, 1982; Heldstab, Thomas, Geisel and Radons, 1983; Horner, 1983; Kai, 1982; Mayer-Kress and Haken, 1981; Napiorkowski and Zaus, 1986; Shraiman, Wayne and Martin, 1981; Weiss, 1987) of the effect of 'time'-dependent forcing on this system mostly aimed at investigating the influence of noise on the period-doubling sequence. Here, on the other hand, we review some results (Linz and Lücke, 1986) on the effect of random multiplicative forcing via random modulation of the control parameter r on the very first bifurcations of the logistic map. Since these results are extensively discussed in the above reference we give here only a short summary. Furthermore, we shall not discuss here the effect of *periodic* multiplicative forcing but rather refer the interested reader to the work of Lücke and Saito (1982).

6.6.1 *The system*

In this section we review the statistical properties of the response of the logistic map

$$x_{n+1} = r_n x_n(1 - x_n) \tag{6.6.1}$$

towards stochastic modulation

$$r_n = r(1 + \Delta\xi_n) \tag{6.6.2}$$

in a range of the control parameter r where the unmodulated system, $\Delta = 0$, displays the first two bifurcations (see Figure 6.9): (i) a transcritical bifurcation at the threshold value $r = r_c(\Delta = 0) = 1$ where the fixed point $x^* = 0(x^* = 1 - 1/r)$ being stable (unstable) below threshold becomes unstable (stable) above, and (ii) a pitchfork period-doubling bifurcation at $r = 3$ where the fixed point $x^* = 1 - 1/r$ becomes unstable and a period-two limit cycle is generated.

Averages

We consider here only statistically stationary forces ξ_n with vanishing mean and unit covariance

$$\langle \xi_n \rangle = 0, \quad \langle \xi_n^2 \rangle = 1. \tag{6.6.3}$$

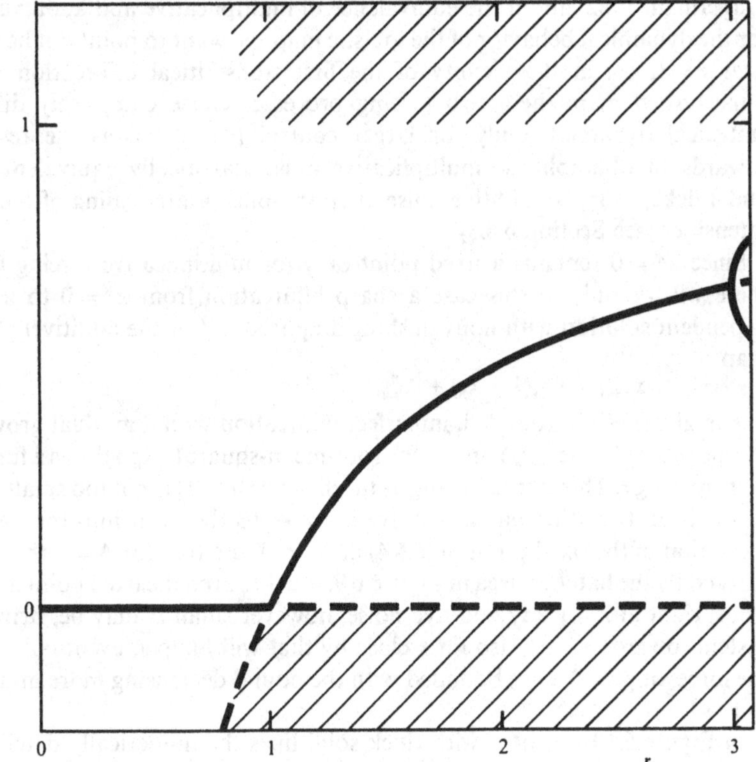

Figure 6.9. Bifurcation diagram of the unmodulated logistic map. The hatched area shows the basin of attraction of the fixed point at $-\infty$.

Averages over functions depending on the forcing process can be thought of as being 'time' averages or ensemble averages over noise realizations. We refer to Linz and Lücke (1986) for a discussion and numerical investigation of the equivalence of these two averages in the context of (6.6.1) and of the dependence of certain averages on the initial value, x_{n_0}, in particular near the pitchfork bifurcation.

Since the forces enter into the equation of motion (6.6.1) only via the combination $\Delta\xi_n$, any average evaluated for system (6.6.1) is an even function of Δ with forces being distributed symmetrically around zero. Furthermore, into a term $\sim\Delta^{2k}$ of a small-Δ expansion of an average there will appear at most $2k$-point correlations of the forces. In particular the statistical properties of the response of (6.6.1) towards small-amplitude modulation (6.6.2) depends up to order Δ^2 only on two-point correlations $\langle\xi_n\xi_m\rangle$ but not on higher order cumulants of the forcing process.

Additive versus multiplicative noise near the first instability

Since the literature contains misleading statements (Crutchfield, Farmer and

Huberman, 1982) about the equivalence of multiplicative and additive noise for the dynamical behavior of the logistic map, we want to point out here that near $r = 1$, i.e. in the vicinity of the first transcritical bifurcation of the unperturbed map, these two forcing processes cause completely different statistical dynamics. Only for larger control parameters is the response towards small-amplitude multiplicative noise statistically equivalent (Linz and Lücke, 1986) to additive noise after an appropriate scaling of the noise intensities (see Section 6.6.3).

Since $x^* = 0$ remains a fixed point only for multiplicative forcing (6.6.1), there will be only in this case a sharp bifurcation from $x^* = 0$ to a time-dependent solution with nonvanishing amplitudes. For the additively forced map

$$x_{n+1} = rx_n(1 - x_n) + \Delta\xi_n \tag{6.6.4}$$

one might expect a rounded, imperfect bifurcation with a gradual growth of the mean amplitude $\langle x_n \rangle$ and of the root-mean-square $(\langle x_n^2 \rangle)^{1/2}$ as a function of increasing r. This is true as long as the distance $|r - 1|$ is not too small. Close to $r = 1$, on the other hand, additive noise kicks the orbit into the basin of attraction of the fixed point of (6.6.4) at $-\infty$. Note that for $\Delta = 0$ this basin, marked by the hatched area in Figure 6.9, touches even the fixed point at $r = 1$. Thus, right at $r = 1$ any additive noise, however small Δ may be, drives the system towards $-\infty$. Also for r close by that will happen eventually unless the force amplitudes are bounded with the bound decreasing more and more as $r \to 1$.

In Figure 6.10 we show with thick solid lines the numerically determined stationary average $\langle x \rangle$ and root-mean-square $\langle x^2 \rangle^{1/2}$ as a function of r for *additive* dichotomous noise. For r-values within the gap of the solid curves the orbits escaped to $-\infty$. There any small-Δ expansion (Weiss, 1987) around the static, $\Delta = 0$, fixed points diverges. Thus additive forcing changes the static, $\Delta = 0$, bifurcation structure near $r = 1$ completely in contrast to multiplicative noise. The latter does not change the bifurcation behavior, e.g., of $\langle x \rangle$ qualitatively as we shall see shortly.

6.6.2 *Bifurcation threshold $r_c(\Delta)$*

The fixed point $x^* = 0$ of the *parametrically* forced map (6.6.1) is linearly stable for $r < r_c(\Delta)$. At

$$r_c(\Delta) = e^{-\langle \ln|1 + \Delta\xi| \rangle} \tag{6.6.5}$$

a time-dependent orbit bifurcates and $x^* = 0$ becomes unstable. The stability threshold (6.6.5) was derived by Linz and Lücke (1986) with the marginality condition that an initially infinitesimal deviation x_0 should neither decay nor grow in the longtime limit.

The above formula (6.6.5) for the bifurcation threshold is exact and holds for arbitrary forcing. With the appropriate interpretation of the average it also

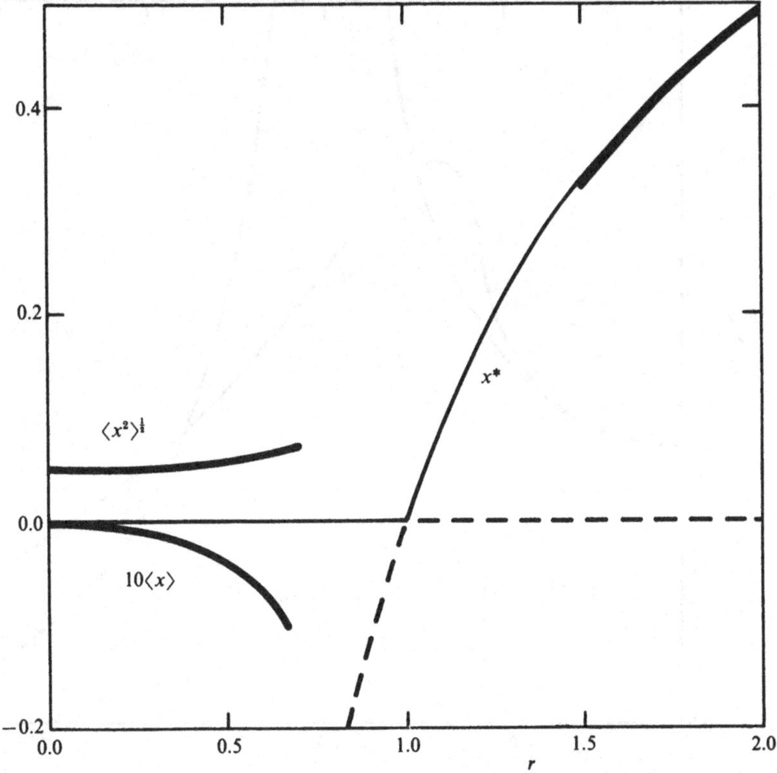

Figure 6.10. Bifurcation diagram of the map with *additive* dichotomous noise of amplitude $\Delta = 0.05$ obtained numerically by Linz and Lücke (1986). Thin solid lines show the bifurcation for $\Delta = 0$. For $0.69 \lesssim r \lesssim 1.45$ the orbits escaped to $-\infty$. For $r > 1.45$ the mean $\langle x\rangle$ and root-mean-square $\langle x^2\rangle^{1/2}$ are identical within a pencil's width.

applies to periodic forcing (Lücke and Saito, 1982). Note that (6.6.5) depends only on the stationary distribution of ξ. Forcing mechanisms with different dynamical correlations but the same stationary distribution cause the same bifurcation threshold $r_c(\Delta)$. It is easy to check this result on a computer by comparing the response of (6.6.1), e.g., against dichotomous noise with that one for deterministic sequences of ± 1.

Also here, like for the Duffing oscillator, *any* small-amplitude parametric modulation whatsoever stabilizes the basic state $x^* = 0$ since

$$r_c(\Delta) = r_c(\Delta = 0) + \tfrac{1}{2}\Delta^2 + O(\Delta^4) \tag{6.6.6}$$

is bigger than the bifurcation threshold in the absence of modulation, $r_c(\Delta = 0) = 1$. Here, however, the threshold shift $\Delta^2/2$ induced by modulation is truly universal. It neither depends on the correlation properties nor on the form of the distribution of the forcing amplitudes.

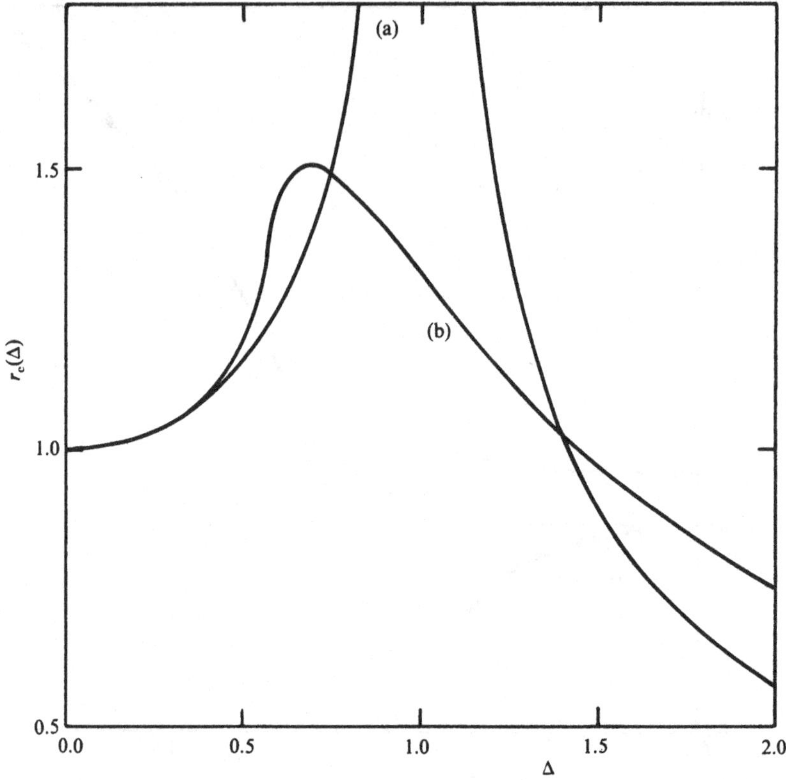

Figure 6.11. Stability boundaries (6.6.8a) and (6.6.8b) of $x^* = 0$ in the presence of *multiplicative* forcing (Linz and Lücke, 1986). Curve (a) is for forces that takes on values ± 1. Curve (b) is for forces that are distributed with equal weight (6.6.7b) over a finite interval. The fixed point $x^* = 0$ is stable for all r, Δ below the respective curves.

Large forcing amplitudes, on the other hand, can cause stabilization or destabilization of the fixed point $x^* = 0$, as for the Duffing oscillator. Take as examples modulation forces that take only values ± 1 with the stationary distribution

$$P(\xi) = \tfrac{1}{2}[\delta(\xi - 1) + \delta(\xi + 1)] \tag{6.6.7a}$$

and forces that are distributed with equal weight over a finite interval

$$P(\xi) = \frac{1}{(12)^{1/2}} \begin{cases} 1 & \text{if } |\xi| < 3^{1/2} \\ 0 & \text{otherwise.} \end{cases} \tag{6.6.7b}$$

The associated bifurcation thresholds

$$r_c(\Delta) = \frac{1}{(|1 - \Delta^2|)^{1/2}} \tag{6.6.8a}$$

Modulated control parameters

$$r_c(\Delta) = \frac{e}{(|1-\tilde{\Delta}^2|)^{1/2}} \left| \frac{1-\tilde{\Delta}}{1+\tilde{\Delta}} \right|^{1/2\tilde{\Delta}}; \quad \tilde{\Delta} = \Delta 3^{1/2} \tag{6.6.8b}$$

shown in Figure 6.11 by curves (a) and (b), respectively, lie partly above and partly below $r_c(\Delta = 0) = 1$. It is easy to understand why the stability threshold (6.6.8a) diverges at $\Delta = 1$. There, $r_n = r(1 + \xi_n)$ will surely become zero if the forces take on only values ± 1 thus driving x for any value of r and also for any initial value towards the fixed point $x^* = 0$. For Δ values nearby, r_n will not vanish but will become very small unless r is very large so that x_{n+1} will be smaller than x_n and so forth. For forces that are distributed according to (6.6.7b), on the other hand, the probability to generate small r_n-values with ξ_n drawn out of a finite interval is too small above a finite critical threshold. Note the structural similarity of the stability boundary curve (a) in Figure 6.11 with the stability boundaries (b) and (c) of Figure 6.3 for the $x = 0$ fixed point of the harmonically modulated Duffing oscillator.

6.6.3 *Response above the threshold $r_c(\Delta)$*

Here we discuss some aspects of the statistical behavior of the map (6.6.1) with a stochastically modulated control parameter above threshold where the fixed point, $x^* = 0$, is no longer stable. The supercritical response against periodic modulation is investigated by Lücke and Saito (1982).

We shall consider here only fluctuating forces with bounded amplitudes $|\xi_n| < \xi_{max}$. Then the restriction to r, Δ combinations such that

$$r(1 + \Delta\xi_{max}) \leqslant 4 \tag{6.6.9}$$

guarantees that x_n remains bounded, while for larger r or Δ there is a good chance that the trajectory is driven to $-\infty$. With the proviso (6.6.9) the system responds above threshold to parametric noise with finite, statistically stationary fluctuations. Here we discuss only the first equal-time moments $\langle x \rangle$ and $\langle x^2 \rangle$ of this response. Further statistical quantities are investigated by Linz and Lücke (1986).

In Figure 6.12 we show the numerically obtained bifurcation diagram of $\langle x \rangle$ and $\langle x^2 \rangle^{1/2}$ in the presence of white parametric noise with the box-shaped stationary distribution (6.6.7b). For other types of small-amplitude noise the bifurcation behavior close to threshold is very similar. Here the parametric noise stabilizes the fixed point up to the threshold value $r_c(\Delta = 0.5) = 1.188$; i.e. the system's statistically stationary response to this noise is trivial, $x = 0$, for all $r < r_c(\Delta)$. Note that both $\langle x \rangle$ and $\langle x^2 \rangle$ increase linearly with the distance $r - r_c$ from threshold so that $\langle x^2 \rangle^{1/2} \sim (r - r_c)^{1/2}$.

Small-Δ perturbation theory

The above described behavior may be reproduced and understood easily with

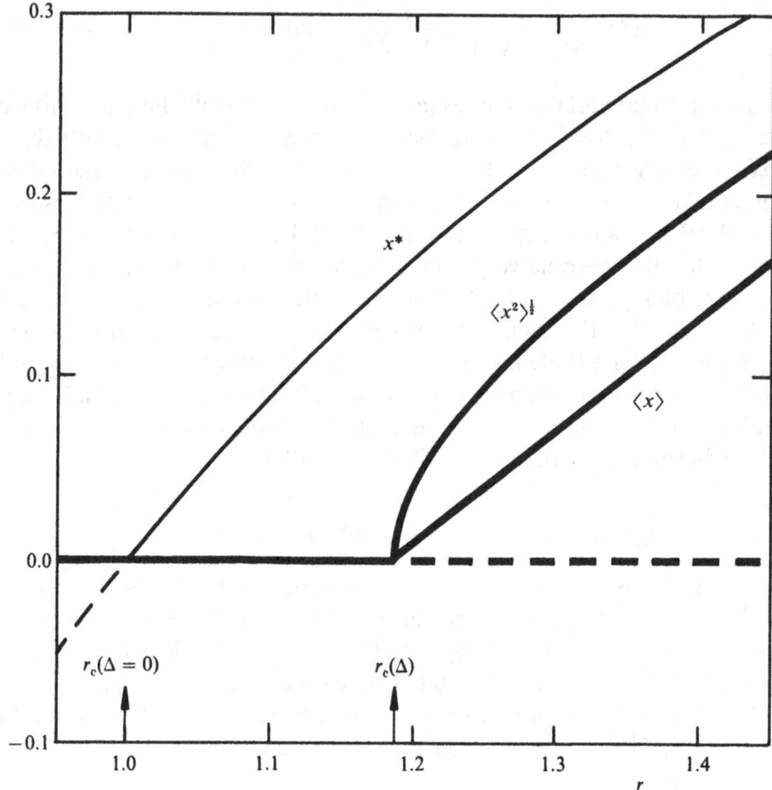

Figure 6.12. Bifurcation diagram obtained numerically by Linz and Lücke (1986). Thick solid lines are for white parametric noise with the box-shaped stationary distribution (6.6.7b) and $\Delta = 0.5$. Thin solid lines denote the unforced system, $\Delta = 0$. The thresholds $r_c(\Delta = 0.5) = 1.188$, (6.6.8b), and $r_c(\Delta = 0) = 1$ are marked by arrows.

a straightforward perturbation expansion of the orbit x_n that is analogous to the one presented in Section 6.4.3 for the modulated Duffing oscillator. The expansion

$$x_n(\Delta) = x_n^{(0)} + \Delta x_n^{(1)} + \Delta^2 x_n^{(2)} + \dots \tag{6.6.10}$$

is done around the bifurcated fixed point

$$x^{(0)} = x^* = 1 - 1/r \tag{6.6.11a}$$

of the unmodulated map. Inserting (6.6.10) into (6.6.1) one obtains inhomogeneous linear difference equations that are solved by

$$x_n^{(1)} = x^* \sum_{i=0}^{n-1} (2 - r)^i \xi_{n-1-i} \tag{6.6.11b}$$

$$x_n^{(2)} = \sum_{i=0}^{n-1} (2-r)^{n-1-i} x_i^{(1)} [(2-r)\xi_i - r x_i^{(1)}]. \tag{6.6.11c}$$

Any statistical quantity of the orbit (6.6.10), e.g. equal-time moments, correlation functions, etc., may be calculated with (6.6.11a–c) and the given statistical properties of the forcing process.

The first two stationary moments of (6.6.10), e.g.,

$$\langle x \rangle = x^* + \Delta^2 \langle x^{(2)} \rangle + O(\Delta^4) \tag{6.6.12a}$$

$$\langle x^2 \rangle = x^{*2} + \Delta^2 \langle (x^{(1)})^2 + 2x^* x^{(2)} \rangle + O(\Delta^4) \tag{6.6.12b}$$

are given by the fixed point, x^*, and in addition two-point correlation functions of the noise which enter into the terms $\sim \Delta^2$ in (6.6.12a, b). These longtime, $n \to \infty$, stationary averages are most easily evaluated if the noise is uncorrelated, $\langle \xi_i \xi_j \rangle = \delta_{ij}$. In that case one finds

$$\langle x \rangle = x^* - \Delta^2 \frac{1}{r(3-r)} + O(\Delta^4) \tag{6.6.13a}$$

$$\langle x^2 \rangle = (x^*)^2 - \Delta^2 \frac{r-1}{r^2(3-r)} + O(\Delta^4). \tag{6.6.13b}$$

Note that the small-Δ expansion breaks down when approaching the period-doubling bifurcation threshold at $r = 3$ (see Linz and Lücke, 1986, and below).

The mean of the fluctuating orbit lies, according to (6.6.13a), below the fixed point of the unforced map (see also Figure 6.12). Both moments $\langle x \rangle$ and $\langle x^2 \rangle$ evaluated to order Δ^2 drop to zero at the bifurcation threshold (6.6.6) obtained from the exact formula (6.6.5) up to order Δ^2. Like for the Duffing oscillator, one can also here reformulate the small-Δ expansion for moments of the bifurcating orbit close to threshold into a power series of the distance, $r - r_c$, from threshold

$$\langle x \rangle = a_1(\Delta)[r - r_c(\Delta)] + a_2(\Delta)[r - r_c(\Delta)]^2 + O(r - r_c)^3 \tag{6.6.14a}$$

$$\langle x^2 \rangle = b_1(\Delta)[r - r_c(\Delta)] + b_2(\Delta)[r - r_c(\Delta)]^2 + O(r - r_c)^3. \tag{6.6.14b}$$

The coefficients are derivatives of the moments with respect to r evaluated to the appropriate power of Δ. For uniformly distributed white noise one obtains

$$a_1 = 1 - \tfrac{3}{4}\Delta^2 + O(\Delta^2); \quad a_2 = -1 + \tfrac{9}{8}\Delta^2 + O(\Delta^4) \tag{6.6.15a}$$

$$b_1 = \tfrac{1}{2}\Delta^2 + O(\Delta^4); \quad b_2 = 1 - \tfrac{9}{4}\Delta^2 + O(\Delta^4). \tag{6.6.15b}$$

Thus according to (6.6.14b) and (6.6.15b) the supercritical growth of the root-mean-square changes from $x^* \sim (r - r_c)$ at $\Delta = 0$ to $\langle x^2 \rangle^{1/2} \sim (r - r_c)^{1/2}$ for finite Δ which is evident in Figure 6.12.

Equivalence of additive and multiplicative noise sufficiently far above $r_c(\Delta)$

If r is sufficiently far above the threshold $r_c(\Delta)$ so that the fixed point at $-\infty$ does not attract the additively perturbed orbits, then small-Δ multiplicative and additive noise cause similar response behavior. This is most easily seen by comparing the Δ expansions

$$\text{(M):}\quad x_n = 1 - \frac{1}{r} + \Delta\left(1 - \frac{1}{r}\right)\sum_{i=0}^{n-1}(2-r)^i \xi_{n-1-i} + O(\Delta^2) \tag{6.6.16a}$$

$$\text{(A):}\quad x_n = 1 - \frac{1}{r} + \Delta \sum_{i=0}^{n-1}(2-r)^i \xi_{n-1-i} + O(\Delta^2) \tag{6.6.16b}$$

of the multiplicatively forced system (M) and of the additively forced system (A) up to linear order in Δ. Up to this order the orbits coincide for the same noise sequence if one scales the amplitudes for the two forcing types according to

$$\Delta_A \leftrightarrow \frac{r-1}{r}\Delta_M . \tag{6.6.17}$$

Since small-Δ averages are strongly dominated by $x_n^{(1)}$ it is not surprising that (6.6.17) turns out to be the appropriate scaling to relate small-amplitude multiplicative and additive noise. Linz and Lücke (1986) checked numerically the validity of (6.6.17) for $r < 3$ as well as for $r \geqslant 3$. They found, e.g., that the stationary distributions of x resulting from multiplicative and additive forces with the same statistics agreed much better with the scaling (6.6.17) than with the one derived by Crutchfield, Farmer and Huberman (1982).

6.6.4 *Effect of noise on the first period-doubling bifurcation*

Here we discuss the effect of small-amplitude noise on the first period-doubling pitchfork bifurcation of the unperturbed map at $r = 3$. The results (Linz and Lücke, 1986) reviewed here apply to multiplicative as well as to additive forcing and may be related to each other quantitatively with the scaling (6.6.17).

In the absence of forcing, $\Delta = 0$, the fixed point $x^* = 1 - 1/r$ of the logistic map loses its stability at $r = 3$ by generating a period-2 limit cycle such that

$$d_n = x_{n+1} - x_n \rightarrow (-1)^n d^* . \tag{6.6.18}$$

The order parameter

$$d^* = \pm\frac{1}{r}[(r-3)(r+1)]^{1/2} \tag{6.6.19}$$

measures how much the time translational invariance is broken. Its magnitude, i.e. the separation of the pitchfork branches, vanishes below threshold, $r = 3$, and grows above. Its phase, ± 1, depends on the initial condition x_0.

Noisy pitchfork bifurcation

In the presence of noise the pitchfork is smeared out as shown in Figure 6.13(a) but its structure is still visible. In fact, for small Δ- and r-values sufficiently far above $r = 3$ the broadened pitchfork branches do not overlap, $|d_n|$ is still roughly of size d^*, and also the sequence of signs $d_n/|d_n|$ is still mostly alternating as for $\Delta = 0$. In this parameter regime the period-2 dynamics of the unforced map dominates the influence of the external noise so that d_n is strongly correlated with its initial value d_0 and averages depend strongly on initial values. On the other hand, very close to threshold the internal period-2 dynamics is disrupted more effectively by the forcing and the smeared pitchfork branches overlap.

An appropriate order parameter describing the time translational symmetry-breaking in the presence of noise is according to Martin (1982)

$$\langle d_n^2 \rangle = \langle (x_{n+1} - x_n)^2 \rangle . \tag{6.6.20}$$

This quantity was evaluated numerically by Linz and Lücke (1986) for various parameters and initial conditions by ensemble averages over up to 30.000 realizations of noise histories and by time averages. They found: (i) the two averaging procedures gave the same result, (ii) $\langle d_n^2 \rangle$ becomes rapidly stationary, and (iii) this stationary value is everywhere independent of initial conditions. (That, however, is not the case for $\langle d_n \rangle$ if r is sufficiently far above three. There ensemble averages either yield $\langle d_{2n} \rangle \simeq |d^*|$ or $\langle d_{2n} \rangle \simeq -|d^*|$ depending on the start value x_0 being close to the lower or upper pitchfork branch where $d_0 > 0$ or $d_0 < 0$, respectively.)

In Figure 6.13(b) we show the rounded bifurcation of $\langle d_n^2 \rangle$ which shows that in the presence of noise time translational symmetry is broken everywhere with $\langle d_n^2 \rangle$ being finite. However, well above threshold noise depresses the order parameter $\langle d_n^2 \rangle$ relative to the unperturbed case, $\Delta = 0$.

It is important to understand the basic difference between the effect of the noise in (6.6.1) on the first transcritical bifurcation close to $r = 1$ and on the first period-doubling bifurcation close to $r = 3$. At the first bifurcation, $r = 1$, the noise couples multiplicatively to the order parameter variable x_n which vanishes for $\Delta = 0$ below threshold $r_c(\Delta = 0) = 1$ and consequently remains zero below threshold $r_c(\Delta)$ with finite forcing. However, the noise in (6.6.1) does not couple multiplicatively to the order parameter variable d_n of the pitchfork bifurcation. Instead, for the dynamics of this variable d_n the noise appears basically as an additive perturbation giving rise to a rounded bifurcation.

Theoretical problems close to $r = 3$

Right at $r = 3$ the first moments of x_n do not vary analytically with Δ for small Δ. For multiplicative, uncorrelated noise with the box-shaped stationary distribution (6.6.7b) one finds (Linz and Lücke, 1986), e.g.

$$\langle x_n \rangle \simeq \tfrac{2}{3} - 0.36|\Delta| \tag{6.6.21a}$$

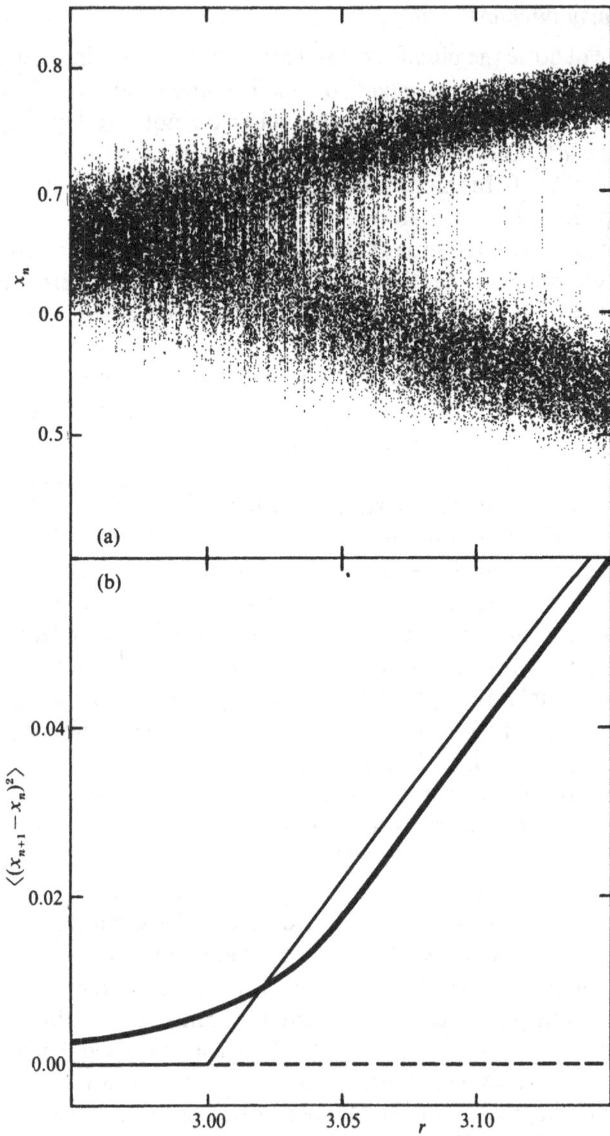

Figure 6.13. The first period-doubling bifurcation with uncorrelated multiplicative noise, $\Delta = 0.015$, and the box-shaped stationary distribution (6.6.7b). (a) shows the positions x_{1000} to x_{1150}. The thick solid line in (b) shows the stationary order parameter $d^2 = \langle (x_{n+1} - x_n)^2 \rangle$. Thin lines show the bifurcation diagram of d^2 in the absence of noise, $\Delta = 0$. The figures are practically identical to Figures 5a and b of Linz and Lücke (1986) showing the same quantities for additive noise with $\Delta = 0.01$. This is explicit support for the equivalence (6.6.17) discussed in Section 6.6.3.

Modulated control parameters

$$\langle (x_n - \langle x_n \rangle)^2 \rangle \simeq 0.24|\Delta|. \tag{6.6.21b}$$

The rounding of the order parameter at $r = 3$ grows as

$$\langle (x_{n+1} - x_n)^2 \rangle \simeq 0.93|\Delta|. \tag{6.6.22}$$

Thus the small-Δ expansion breaks down. Also the small-Δ perturbation theory for additive forcing (Linz and Lücke, 1986; Weiss, 1987) breaks down at $r = 3$. With growing distance $|r - 3|$, however, there is a crossover of the Δ dependence to an analytical behavior.

The breakdown of the perturbation expansion close to the period-doubling pitchfork bifurcation threshold has led to the investigation of approximations that are based on truncating the infinite sequence of equations for moments. The merits and drawbacks of such 'closure' approximations are discussed by Linz and Lücke (1986).

6.7 Conclusion

In summary we have discussed the behavior of three model systems showing simple bifurcations out of a known basic state in the presence of modulation of the control parameter. Stability properties of the basic state were determined and a generalized Poincaré–Lindstedt expansion method was elucidated and used to evaluate the bifurcating nontrivial solution. The power and limitations of this technique were examined in analytical and numerical detail for a Duffing oscillator and an inverted bifurcation as two representative and common examples. We have also reviewed the response of the logistic map under parametric modulation in the vicinity of the first bifurcations.

Acknowledgement

The results reviewed here are largely based on work done together with M. Lazarov, S. J. Linz and F. Schank. Stimulating discussions with them are gratefully acknowledged.

References

Ahlers, G., Hohenberg, P. C. and Lücke, M. 1985a. *Phys. Rev. A* **32**, 3493.

Ahlers, G., Hohenberg, P. C. and Lücke, M. 1985b. *Phys. Rev. A* **32**, 3519.

Andronov, A. A., Vitt, E. A. and Khaiken, S. E. 1966. *Theory of Oscillators.* Oxford: Pergamon Press.

Arnold, L., Papanicolaou, G. and Wihstutz, V. 1986. *SIAM J. Appl. Math.* **46**, 427.

Blanch, G. 1965. In *Handbook of Mathematical Functions* (M. Abramowitz and I. A. Stegun, eds.), New York: Dover Publications.

Collet, P. and Eckmann, J. P. 1980. *Iterated Maps on the Interval as Dynamical Systems.* Basel: Birkhäuser.

Crutchfield, J. P., Farmer, J. D. and Huberman, B. A. 1982. *Phys. Rep.* **92**, 45. See also references cited therein.

Davis, S. H. 1976. *Annu. Rev. Fluid Mech.* **8**, 57.
Davis, S. H. and Rosenblat, S. 1977. *Stud. Appl. Math.* **57**, 59.
Feigenbaum, M. J. and Hasslacher, B. 1982. *Phys. Rev. Lett.* **49**, 605.
Graham, R. and Schenzle, A. 1982. *Phys. Rev. A* **26**, 1676.
Großmann, S. 1984. *Z. Phys. B* **57**, 77.
Haken, H. and Wunderlin, A. 1982. *Z. Phys. B* **46**, 181. See also references cited therein.
Hayashi, C. 1964. *Nonlinear Oscillations in Physical Systems.* New York: McGraw-Hill.
Heldstab, J., Thomas, H., Geisel, T. and Radons, G. 1983. *Z. Phys. B* **50**, 141.
Horner, H. 1983. *Phys. Rev. A* **27**, 1270.
Horsthemke, W. and Kondepudi, D. K., eds. 1984. *Fluctuations and Sensitivity in Nonequilibrium Systems.* Berlin: Springer.
Horsthemke, W. and Lefever, R. 1984. *Noise Induced Transitions: Theory and Applications in Physics, Chemistry, and Biology.* Berlin: Springer.
Joseph, D. D. 1976. *Stability of Fluid Motions* I. Berlin: Springer.
Kai, T. 1982. *J. Stat. Phys.* **29**, 329.
Knobloch, E. and Wiesenfeld, K. A. 1983. *J. Stat. Phys.* **33**, 611.
Kuhlmann, H. 1985. *Phys. Rev. A* **32**, 1703.
Kumar, K., Bhattacharjee, J. K. and Banerjee, K. 1986. *Phys. Rev. A* **34**, 5000. See also work cited therein.
Landau, L. D. and Lifshitz, E. M. 1960. *Course of Theoretical Physics, Vol. I. Mechanics.* Oxford: Pergamon Press.
Lazarov, M. 1987. Diploma Thesis, Universität des Saarlandes, Saarbrücken (unpublished).
Linz, S. J. and Lücke, M. 1986. *Phys. Rev. A* **33**, 2694.
Lücke, M. and Saito, Y. 1982. *Phys. Lett.* **91A**, 205.
Lücke, M. and Schank, F. 1985. *Phys. Rev. Lett.* **54**, 1465.
Martin, P. C. 1982. In *Melting, Localization, and Chaos* (R. K. Kalia and P. Vashista, eds.), p. 179. Amsterdam: North Holland.
May, R. 1976. *Nature* **261**, 459.
Mayer-Kress, G. and Haken, H. 1981. *J. Stat. Phys.* **26**, 149.
Minorsky, N. 1962. *Nonlinear Oscillations.* New York: Van Nostrand.
Napiorkowski, M. and Zaus, U. 1986. *J. Stat. Phys.* **43**, 349.
Nayfeh, A. H. and Mook, D. T. 1979. *Nonlinear Oscillations.* New York: Wiley Interscience.
Risken, H. 1984. *The Fokker–Planck Equation.* Berlin: Springer.
Seshadri, V., West, B. J. and Lindenberg, K. 1981. *Physica* **107A**, 219.
Shraiman, B., Wayne, C. E. and Martin, P. C. 1981. *Phys. Rev. Lett.* **46**, 935.
Venezian, G. 1969. *J. Fluid Mech.* **35**, 243.
Weiss, J. 1987. *Phys. Rev. A* **35**, 879.

7 Period doubling bifurcations: what good are they?

KURT WIESENFELD

7.1 Prologue

The statement that systems near the onset of instabilities are sensitive to fluctuations may be a haggard cliché, but is true nevertheless. The very notion of 'instability' means that a small fluctuation tends to grow with time rather than decay away. This has some distinct disadvantages, as was driven home to me during an experiment performed by Bob Miracky on an electrical circuit. Our goal was to determine the precise parameter value at which a period doubling bifurcation occurred, in order to compare this with analytic results derived from the circuit equation. As an experimental matter, this turned out to be tricky due to some curious interactions of the circuit with unwanted electrical noise.

Figure 7.1 reproduces a typical sequence of power spectra generated by sweeping a control parameter. If one has a good machine, the power spectrum is a very accurate means for determining the onset of period doubling bifurcations: the signature is the birth of a sharp spectral line at one-half the fundamental oscillation frequency ω_0, this line initially growing from zero height. Figure 7.1(a) shows the system well below the bifurcation point; however, in addition to the expected lines at frequencies $\omega = 0$ and $\omega = \omega_0$, there appear two bumps in the broadband noise. Though mysterious, these bumps are not a problem since they are far away from the crucial frequency $\omega_0/2$. Unfortunately, as the bifurcation point is neared these bumps *move together*, eventually colliding at $\omega_0/2$ and obscuring the onset of the sharp line heralding the bifurcation. By the time Figure 7.1(d) is reached, the bifurcation has certainly occurred, but where did it start? What is the origin of the well defined bumps in the first place? What determines their width? Why do they shift – is it coincidence that they converge to precisely the place the experiment needs to measure most accurately?

7.2 Introduction and background

Let me begin by explaining the chapter title: what good *are* period doubling bifurcations, anyway? The short answer is that any physical system, when

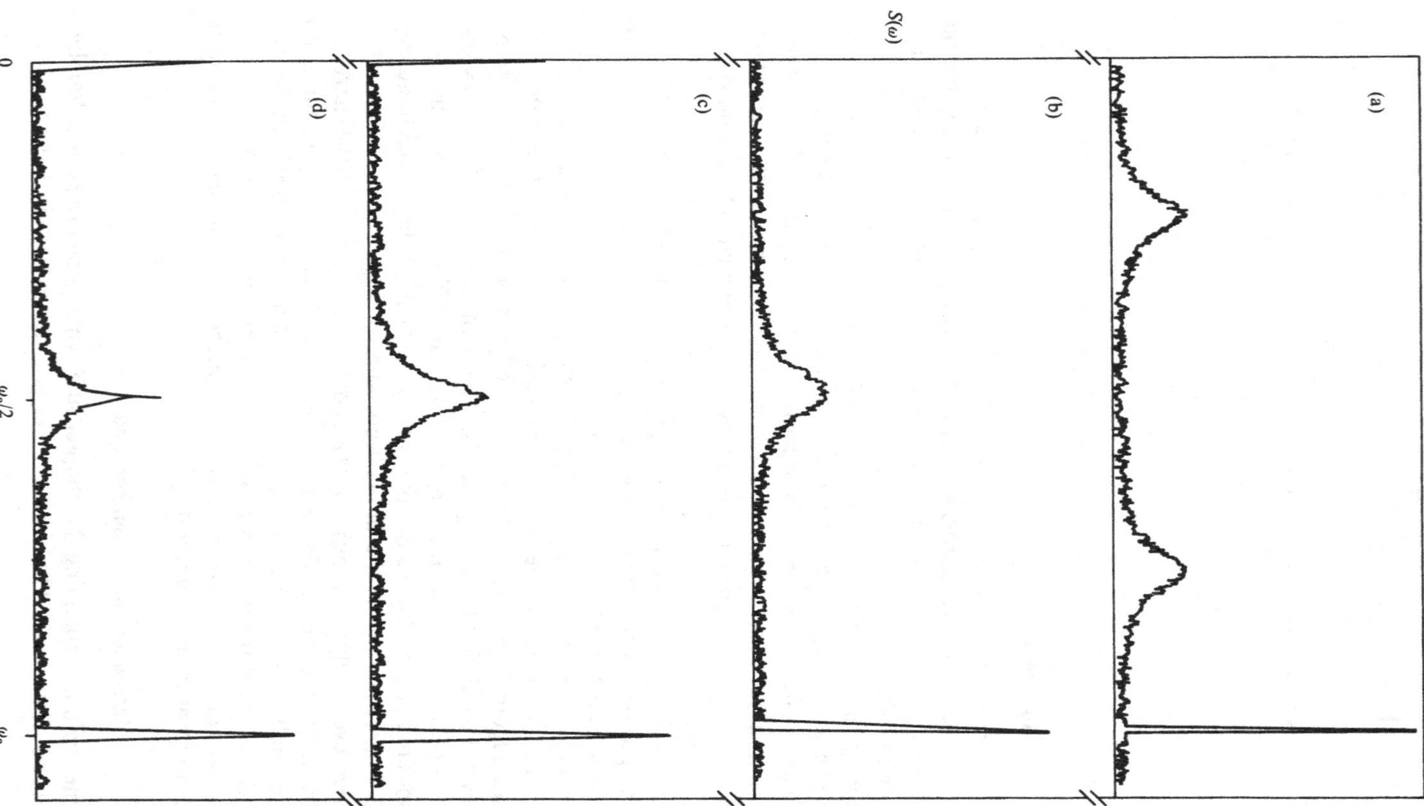

Figure 7.1. Sequence of power spectra observed in an experiment on the Koch–Miracky circuit near the onset of a period doubling bifurcation. The presence of noise obscures the precise parameter value at which the bifurcation occurs.

tuned near the bifurcation point, can act as an amplifier. The underlying reason for this is the inherent sensitivity to fluctuations near points of instability. For the experiment described in the prologue this sensitivity was a negative feature as the amplified noise obscured the measurements. On the other hand, if a coherent perturbation (signal) is purposely coupled in at the 'right' frequency, the same sensitivity can be harnessed to make a parametric amplifier.

Of course, other sorts of bifurcations will do as well: there is nothing special about period doubling in this sense. But it makes for a nice title.

The idea that noise can induce structure in the power spectrum near criticality was appreciated some time ago by McCumber and Haken in their work on the threshold for lasing in optical systems (Haken, 1970; McCumber, 1966). They predicted, and experiments confirmed (Manes and Siegman, 1971), a noise induced bump centered at the incipient lasing frequency. This feature is what I am calling a 'noisy precursor'; it turns out that its existence is not peculiar to laser systems, nor are its major quantitative characteristics.

Even the notion that this sensitivity might be put to practical advantage is not new – not even to the US Government. As early as 1954, Johnny von Neumann filed a patent using period doubling circuits as elements of a digital computer (von Neumann, 1954). The key mechanism involved how coupled elements behave when swept back and forth across the bifurcation point. This idea was developed independently by Goto (1955) and subsequently adapted by Keyes and Landauer as a springboard for discussing the fundamental physical limits to computation (Keyes and Landauer, 1970). So one use of period doubling – though not the use I have in mind – is to build computers.

So, the ideas concerning fluctuation enhancement near the onset of instabilities have been around for years. The new contributions I will discuss are the insights afforded by dynamical systems theory, in particular bifurcation theory. The two central notions that apply are the *classification scheme* for bifurcations and the *center manifold reduction* of phase space (Guckenheimer and Holmes, 1983). By classification scheme, I mean simply that there are a small number of fundamentally distinct bifurcation 'types' that are typically encountered and are easily recognized across even widely different physical systems. It is only the type that matters, even for many quantitative purposes: this is commonly referred to as 'universality', a notion (almost achieving the status of dogma these days) familiar in both the fields of dynamical systems and critical phenomena*. Moreover, the quantitative aspects of universal behavior can be extracted from very simple (i.e. low dimensional) examples; this has its origin in the center manifold reduction, which says that *near a bifurcation point* even high dimensional phase space

* The term 'universality class' seems inherently contradictory – how can there be more than one kind of *universal* behavior? Some two-frequency dynamical studies go so far as to ascribe a different universality class for each irrational number! Jim Yorke has suggested replacing 'universal' with the (only slightly) less sweeping word 'intergalactic'.

dynamics collapse onto a low dimensional center manifold. (This reduction of dimension near an instability will be familiar to laser physicists from Haken's adiabatic elimination procedure: Haken, 1977.)

A simple exercise in this kind of thinking manages to explain the origin in Figure 7.1 of the noise bump at $\omega_0/2$. The argument was first given by James W. Shift, and it considers the situation in phase space illustrated in Figure 7.2 (J. W. Swift, 1983, private communication). Here, the T-periodic (noise-free) solution forms a closed orbit x_0 (Figure 7.2a), and q_0 is an initial condition lying just off x_0. As the orbit through q_0 relaxes to the limit cycle it repeatedly intersects the Poincaré section P at points $q_1, q_2, q_3, \ldots$, asymptotically approaching $q_\infty = x_0 \cap P$. Typically, the $\{q_i\}$ approach q_∞ in some haphazard way; however, near the onset of a period doubling they form a characteristic pattern (Figure 7.2b): successive iterates quickly collapse toward a one-dimensional curve and approach q_∞ in an *alternating fashion.* (This is why a period doubling is sometimes called a 'flip bifurcation' in the theory of discrete mappings.) Consequently, the transient part of the orbit is a damped $2T$-periodic oscillation, contributing a broad peak to the power spectrum at circular frequency $\omega_0/2$ – that is, at half the fundamental frequency of the stable orbit x_0. In a purely deterministic system the trajectory settles down to the attractor x_0 for all times, and the transient contributes only negligibly to the measured power spectrum. The effect of external noise is to continually kick the system off of the stable orbit so that the transient behavior contributes significantly to the observed spectrum.

Notice that this kind of reasoning does not rely on any details about the system generating the dynamics (i.e. 'there is no physics in it'). For example, Figure 7.3 shows data from a chemical system, the Belousov-Zhabotinskii reaction, at parameter values far from (a) and close to (b), a period doubling bifurcation (M. Schumaker, 1984, private communication). Again, the emergence of broadband structure is evident, similar to Figure 7.1.

Swift's explanation really captures the central point in a nutshell – in some sense much of this chapter reports extensions of this picture in one way or another. I will return to Figure 7.2 several times during the chapter; if squeezed hard enough, one can deduce a tremendous amount from this one insight.

The contents of this chapter are organized as follows. The theory of noise amplification together with a discussion of experimental results is reviewed in Sections 7.3 and 7.4. By the end of Section 7.4, the entire sequence of Figures 7.1(a)–(d) is explained. I then turn to the study of coherent perturbations in Sections 7.5 and 7.6, including a thumbnail sketch of several very recent experiments. Aside from the intrinsic theoretical interest, these ideas have practical consequences: in Section 7.7 I discuss an important test bed for these theories, namely the superconducting Josephson junction parametric amplifiers, devices potentially important for observational radioastronomy. The chapter closes with a list of questions that remain unanswered, either by

(a)

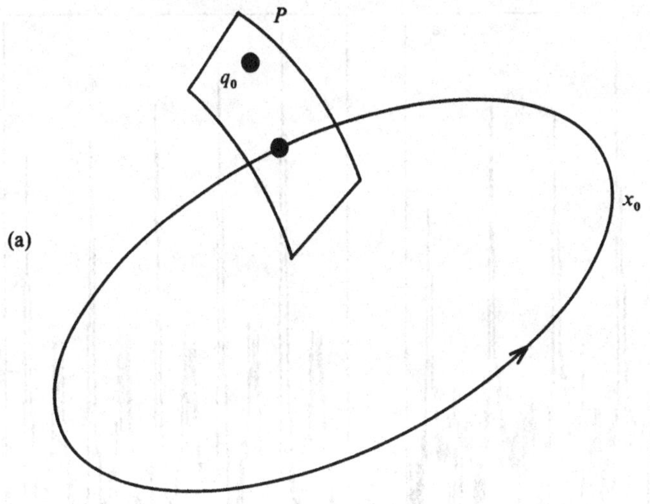

(b)

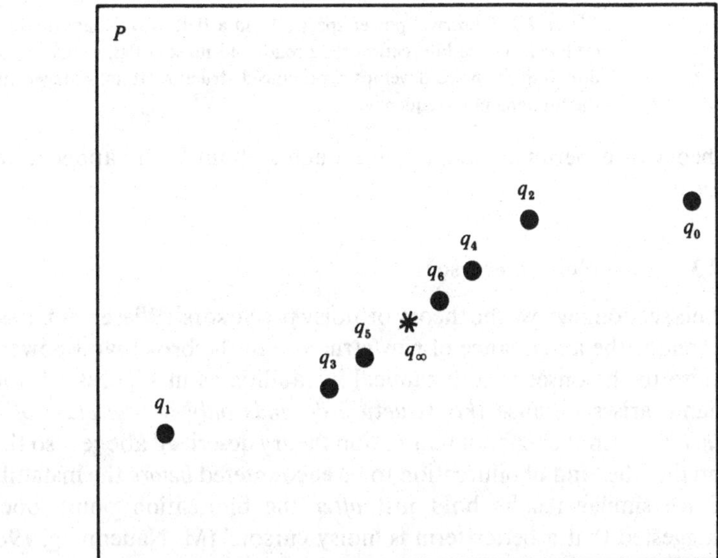

Figure 7.2. (a) Phase space plot of a stable periodic orbit x_0 cut by a transverse Poincaré section P. (b) Near a period doubling, an orbit through q_0 relaxes toward x_0; successive intersections on a one dimensional curve, and approach the fixed point $q_\infty = x_0 \cap P$ in an alternating fashion.

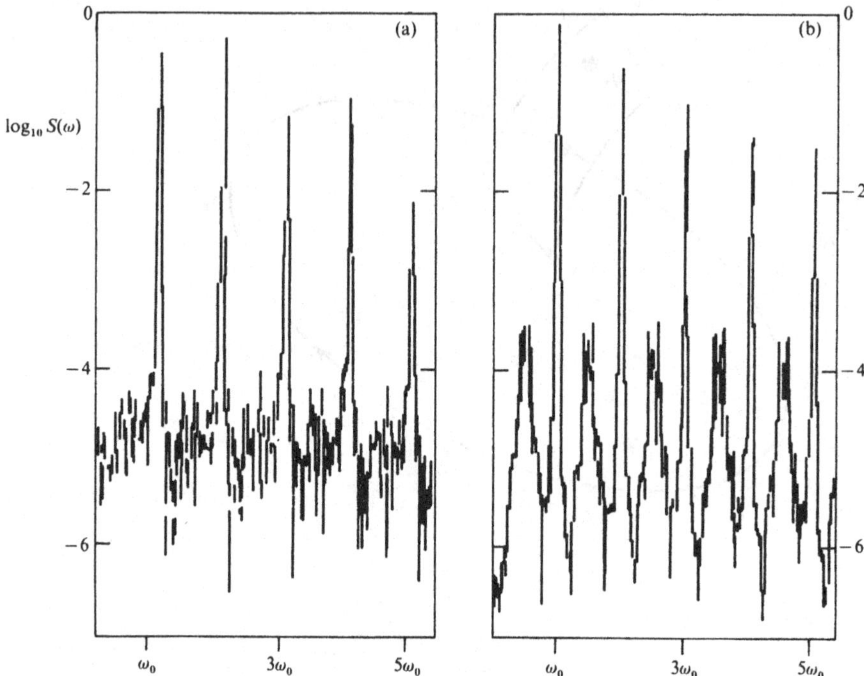

Figure 7.3. Observed power spectra from a Belousov–Zhabotinskii experiment. (a) Far from the bifurcation the broadband noise is flat. (b) Close to the period doubling the noise develops pronounced structure at half-integer multiples of the fundamental frequency.

theory or experiment, and some thoughts about future applications of these ideas.

7.3 Noisy precursors

This section reviews the theory of 'noisy precursors' (Wiesenfeld, 1985a), which explains the appearance of new structure in the broadband power spectrum close to the onset of a dynamical instability, as in Figures 7.1 and 7.2. The name arises because *this structure depends only on the class of instability involved* – in the sense of bifurcation theory described above – so that one can predict the kind of bifurcation to be encountered *before* the instability occurs. Since similar results hold just *after* the bifurcation point, one wag has suggested that a better term is 'noisy cursors' (M. Nauenberg, 1984, private communication). But I shall stick with precursors.

I give here the barest mathematical overview, and quickly move on to a discussion of the results of the theory. Full details are available elsewhere (Hackenbracht and Höck, 1986; Heldstab, Thomas, Geisel and Radons, 1983; Wiesenfeld, 1985a).

The mathematical formulation of the problem begins with the noise-free differential equations governing the system

$$\dot{x} = F(x; \lambda), \; x \in \mathbb{R}^N, \tag{7.3.1}$$

where λ is a control parameter, and F may or may not depend explicitly on time, cases called nonautonomous or autonomous, respectively. (Driven systems such as the damped driven pendulum are nonautonomous, while 'self-oscillators' like the Belusov–Zhabotinskii reaction or the Lorenz equations are autonomous.) Assume that (7.3.1) has a stable time-periodic solution $x_0(t)$ which undergoes a bifurcation at some nearby critical parameter value $\lambda = \lambda_c$. I am interested in what happens just prior to the bifurcation ($\lambda < \lambda_c$), where x_0 is still stable, when an external noise source is present. To model the noise, (7.3.1) is augmented by a random term $\Xi(t)$,

$$\dot{x} = F(x; \lambda) + \Xi(t). \tag{7.3.2}$$

If the noise is weak then the deviation $\eta \equiv x - x_0$ from the deterministic solution will remain small; consequently, the leading behavior is captured by *linearizing* (7.3.2) about x_0,

$$\dot{\eta} = \mathbf{DF} \cdot \eta + \Xi(t), \tag{7.3.3}$$

where $\mathbf{DF}$ is the Jacobian matrix of partial derivatives of F. The plan is to solve this for $\eta(t)$, perform an ensemble average to get the autocorrelation function, and Fourier transform this to find the power spectrum. All of these steps can be carried out in detail (Hackenbracht and Höck, 1986; Wiesenfeld, 1985a). (These references assume the noise is delta-correlated, but this is not crucial: it is sufficient that the noise correlation time is small compared with period of x_0.) The key step is the linearization, for now (7.3.3) is a linear inhomogeneous equation with T-periodic coefficients. Such equations can be solved using results which are known collectively as Floquet Theory (Jordan and Smith, 1977).

The most important quantities in the Floquet analysis are the *Floquet multipliers* μ_k of the orbit x_0. All results of the theory can be stated in terms of these quantities, or alternatively in terms of the *Floquet exponents* ρ_k, which are simply given by

$$\mu_k = e^{\rho_k T}. \tag{7.3.4}$$

As a practical matter the calculations are more easily carried out using the ρ_k, while conceptual discussions are simpler using the μ_k; I will therefore stick with the μ_k here, except to say that the real parts of the ρ_k are also familiarly known as the Liapunov exponents.

The multipliers μ_k are either real or come in complex conjugate pairs, and determine the stability of x_0. They are defined in terms of special solutions $\psi_k(t)$ of the homogeneous counterpart of (7.3.3), with the simple property

$$\psi_k(t + T) = \mu_k \psi_k(t). \tag{7.3.5}$$

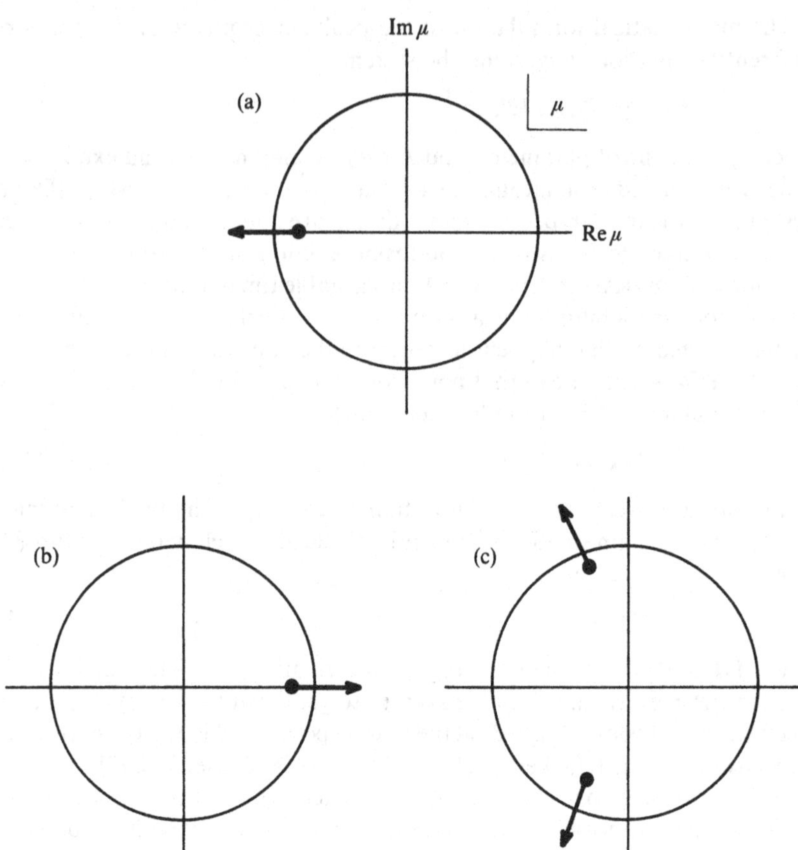

Figure 7.4. Behavior of the critical Floquet multiplier (s) for the three generic types of bifurcation as a single parameter is tuned.

Any solution of the unforced equation ($\Xi = 0$) can be expressed as a linear combination of these functions. Stability of x_0 requires that η remain bounded, which in turn requires $|\mu_k| < 1$ for all k. As a control parameter is varied the μ_k move around the complex plane: an instability occurs when at least one μ_k exits the unit disc. I shall call the exiting multiplier the 'critical multiplier', and denote it by μ_c. If only a single parameter is varied, typically one of three situations will arise (Figure 7.4):

(i) A single μ_c exits at -1. This corresponds to a period doubling, as is easily seen from (7.3.5) since the corresponding ψ_c simply changes sign each time interval T, and so has period $2T$.
(ii) A single μ_c exits at $+1$. This corresponds to a saddle-node bifurcation, which is typically accompanied by a sudden jump in the system's response and hysteresis.
(iii) A pair (μ_c, μ_c^*) exit at $e^{i\theta}$. This signals a 'Hopf' bifurcation, and usually

results in two-frequency dynamics, either quasiperiodic or frequency locked.

(Of course, it is also possible for more complicated situations to arise, e.g. two μ_c exiting at $+1$ simultaneously; however, this generically requires either simultaneous tuning of more than one control parameter or the presence of some sort of symmetry in the problem. While symmetries are common in physical systems, only the simplest case of inversion symmetry (Z_2) has been studied in the context of noisy precursors, and corresponds to a second category of case (ii) above. Incidentally, the influence of symmetry on bifurcation structures is currently a hot field of study which is bound to have important applications in physics: the interested reader can begin with the conference proceedings edited by Golubitski and Guckenheimer, 1986.)

The origin of the 'universality' of noisy precursors is now evident. For example, close to a period doubling bifurcation there is a single long-lived transient function ψ_c which dominates the response of the dynamics to noise *irrespective* of the properties of the other N-1 ψ_k. Moreover, the distinction between different 'universality classes' is simply the distinction between the different situations depicted in Figure 7.4. (Again, the presence of special symmetries or constraints may cause further 'class distinctions', e.g. Figure 7.4(a) also applies to the commonly studied transcritical and pitchfork bifurcations.)

The basic results of the theory are summarized in Figure 7.5. The parameter ε measures the distance of the near-critical multiplier μ_c from the unit circle; expressing μ_c in polar coordinates

$$\mu_c = e^{\varepsilon} e^{i\theta} \simeq (1-\varepsilon)e^{i\theta}. \tag{7.3.6}$$

Thus, the bifurcation occurs at $\varepsilon = 0$. The main point is this: far from the bifurcation point the broadband spectrum is flat and featureless, but near the bifurcation new structure emerges in the form of Lorentzian bumps – these are the noisy precursors. These new 'lines' are much broader than the narrow lines at frequencies $0, \omega_0, 2\omega_0, \ldots$, due to the basic oscillation x_0, but can be quite prominent, growing and sharpening as the bifurcation is approached. Quantitatively, the size and shape of the precursor bumps are determined by ε and the noise strength κ, while their position is determined by θ. One identifies $\theta = 0$ with the frequency ω_0 (and also $2\omega_0, 3\omega_0$, etc.) and $\theta = \pi$ with frequency $\omega_0/2$ (and also $3\omega_0/2, 5\omega_0/2$, etc.). Thus, at a period doubling, precursor bumps are situated halfway between the sharp lines that appear even in the absence of noise; near a Hopf one sees *pairs* of bumps between the sharp lines corresponding to the complex conjugate pair of near-critical multipliers. Though the overall height may differ from bump to bump, each precursor *scales* the same way with ε and κ:

$$\begin{aligned} \text{peak height} &\propto \kappa/\varepsilon^2 \\ \text{peak width} &\propto \varepsilon \\ \text{integrated power} &\propto \kappa/\varepsilon. \end{aligned}$$

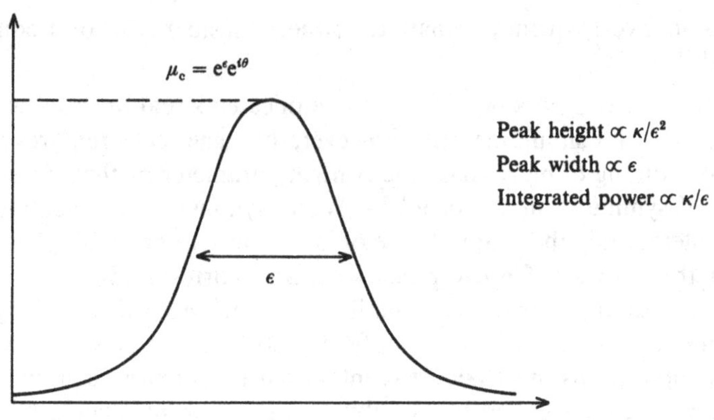

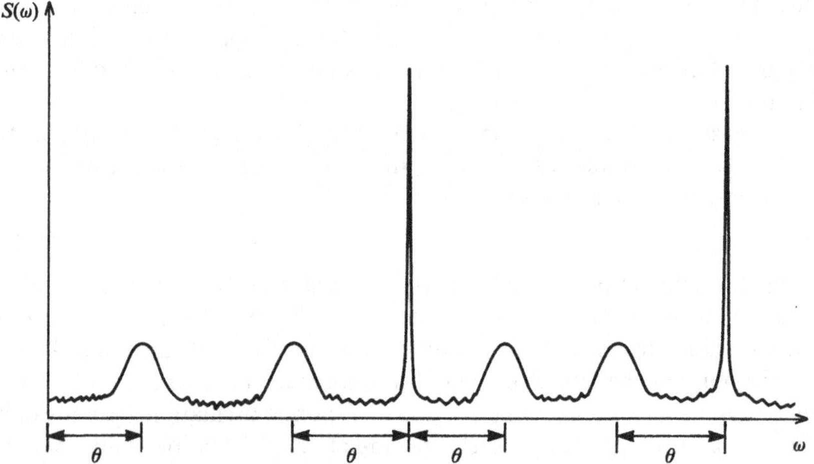

Figure 7.5. Summary of results for the noisy precursor theory.

Qualitatively, the scalings are easy to understand, thinking back to the single-kick picture Figure 7.2. The nearer the bifurcation, the more nearly periodic the transient, so the smaller the bump width; the transient also lasts longer so the greater the overall power (area under the bump). Finally, the height times the width gives the area, so the height must grow as ε diminishes. (The scaling with κ is trivial, simply reflecting the fact that the response is linear.)

There is one more piece needed to complete the theoretical picture, and this concerns the difference between autonomous and nonautonomous

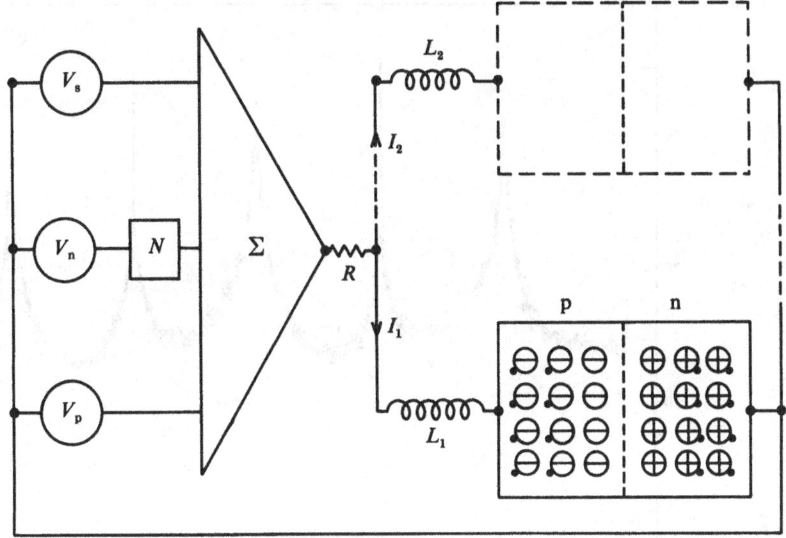

Figure 7.6. Schematic of a nonlinear circuit used to test the noisy precursor theory (Jeffries and Wiesenfeld, 1985).

systems. There is one special fact about autonomous dynamical systems: a periodic orbit *always* has one of its multipliers equal to $+1$. This corresponds to the geometrical fact that the orbit is *neutrally stable* to perturbations along the orbit. As a consequence of this constrained multiplier, the noisy precursors for autonomous systems differ in two ways from their nonautonomous counterparts. First, there are always precursor bumps superimposed on the sharp lines at $\omega = \omega_0$, $2\omega_0$, etc., and these sharp lines themselves acquire a finite width.

To test these predictions, detailed experiments were carried out on an electrical circuit employing p–n junctions as nonlinear elements (Figure 7.6). Essentially, the idea is to take a standard mass-produced chip, hook it up to a voltage generator, and then drive the hell out of it (i.e. well beyond the manufacturer's specifications). With one junction the circuit displays a cascade of period doublings, band mergings, etc., as the driving amplitude is varied, very similar to the behavior of the famous logistic map (Linsay, 1981; Testa, Perez and Jeffries, 1982). With two junctions in series the bifurcation structure is much richer, including Hopf bifurcations to two-frequency dynamics (van Buskirk and Jeffries, 1985). This is an ideal test system because its intrinsic noise level is extremely low. By hooking up a noise generator the input noise intensity κ can be carefully controlled over many decades.

The one-junction circuit was used to test the noisy precursor theory in the vicinity of a period doubling bifurcation, while two junctions in series tested the Hopf precursors (Jeffries and Wiesenfeld, 1985). Figure 7.7 shows the power spectrum near period doubling for three different levels of input noise:

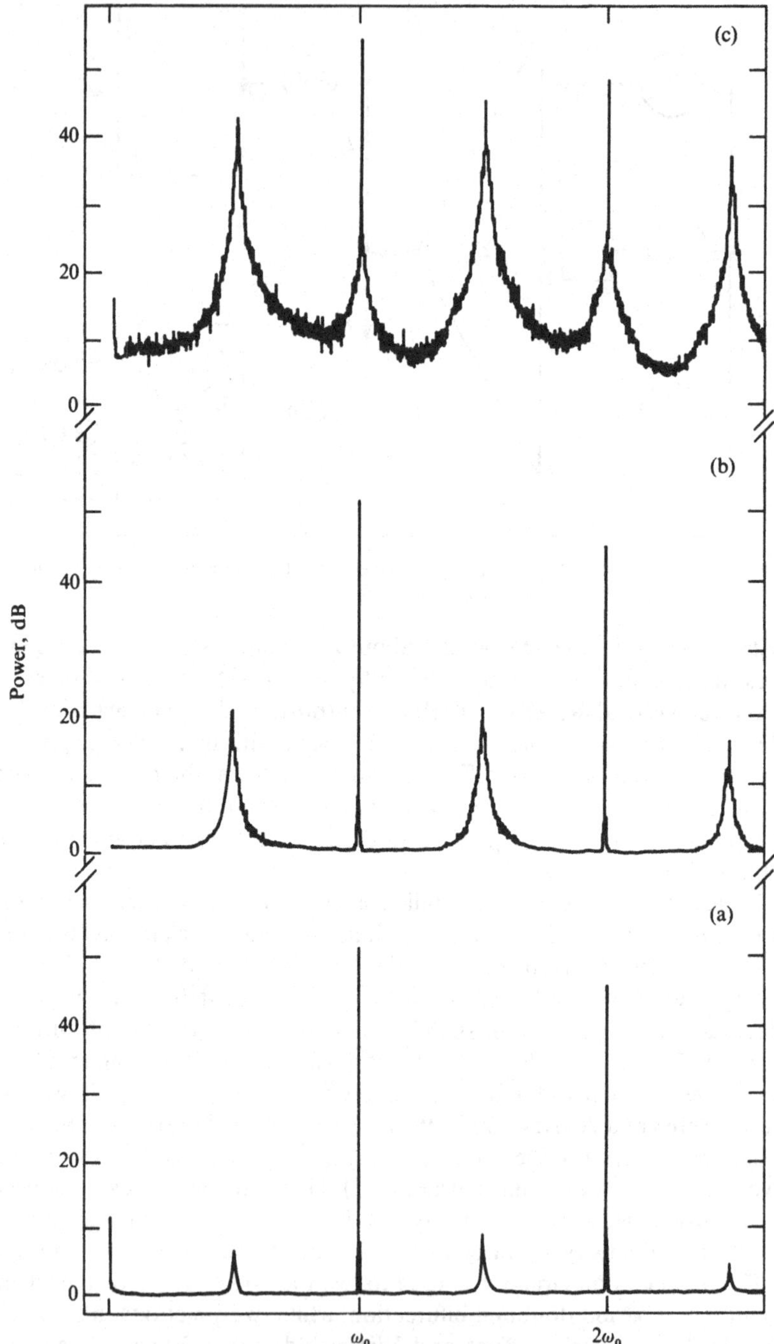

Figure 7.7. Experimental results for a single driven p–n junction near the onset of a period doubling bifurcation, showing power spectra for three different levels of input noise (Jeffries and Wiesenfeld, 1985).

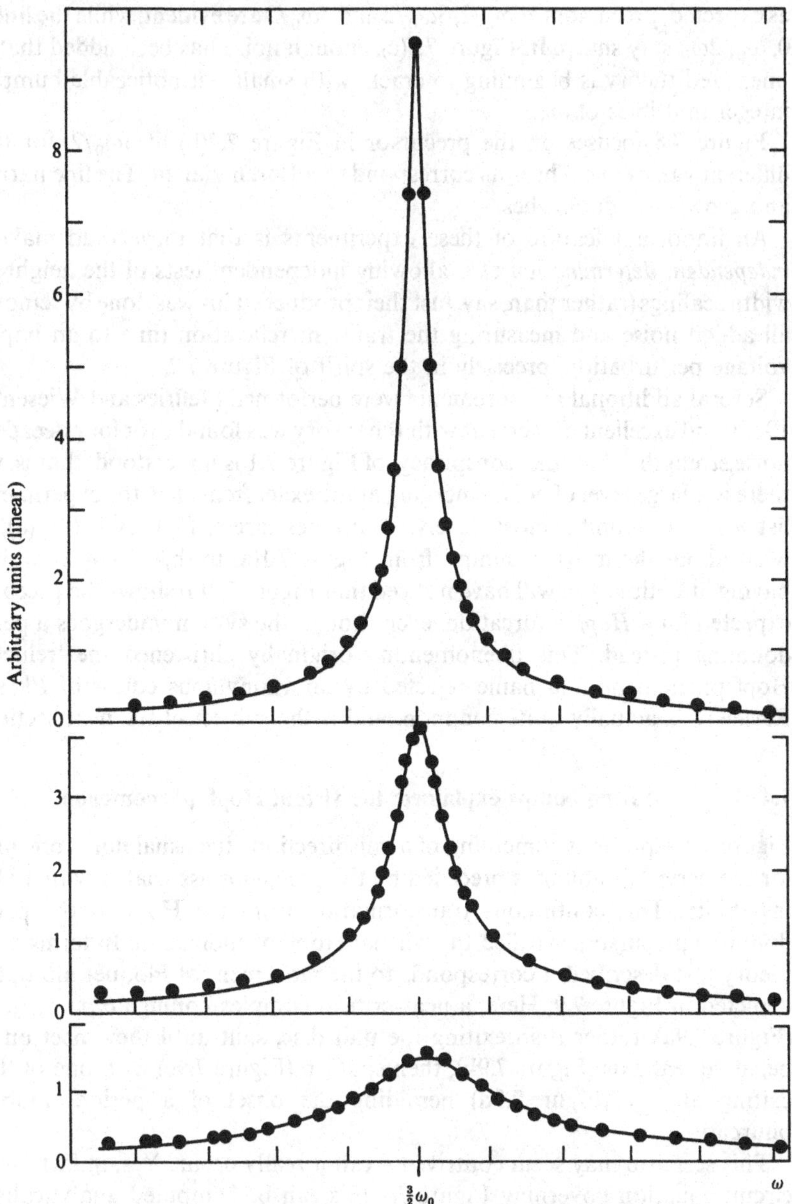

Figure 7.8. Precursor bump at frequency $\frac{3}{2}\omega_0$ from Figure 7.7(b), for three different values of bifurcation parameter. Full line is experiment, filled circles are a fit to a Lorentzian line shape (Jeffries and Wiesenfeld, 1985).

as expected, precursors at $\omega_0/2$, $3\omega_0/2$ and $5\omega_0/2$ are evident, while the lines at 0, ω_0, $2\omega_0$ stay sharp. In Figure 7.7(c), enough noise has been added that the linearized theory is beginning to crack, with small but noticeable bumps at integer multiples of ω_0.

Figure 7.8 focuses on the precursor in Figure 7.7(b) at $3\omega_0/2$, for three different values of ε. The dots correspond to a Lorentzian fit. The line narrows and grows as ε diminishes.

An important feature of these experiments is that they could make an *independent determination* of ε, allowing independent tests of the height and width scalings (rather than, say, just their product). This was done by removing all added noise and measuring the transient relaxation time to an impulse voltage perturbation, precisely in the spirit of Figure 7.2.

Several additional measurements were performed (Jeffries and Wiesenfeld, 1985), and excellent agreement with the theory was found over four decades of noise strength. Thus, the 'conspiracy' of Figure 7.1 is understood: that is, why there is a large level of noise emerging at the exact frequency the experimentalist wishes to monitor most closely. That takes care of Figures 7.1(b)–(d), but what about the moving bumps, from Figure 7.1(a) to (b)? If you have been paying attention, you will have noticed that Figure 7.1(a) shows the precursor expected for a *Hopf* bifurcation, even though the system undergoes a period doubling instead. This phenomenon – originally christened the 'reluctant Hopf phenomenon' (a name rejected by an anonymous editor of *Physical Review*) – is actually quite common, and is the subject of the next section.

7.4 Moving bumps explained: the virtual Hopf phenomenon

Figure 7.1 represents something of a 'misdirection': the usual noisy precursor for the period doubling is preceded by the precursor associated with a Hopf instability. This continuous transformation from the Hopf to the period doubling precursors is called the 'virtual Hopf' phenomenon. In terms of the theory just described it corresponds to the movement of Floquet multipliers depicted in Figure 7.9. Here, a near-critical complex conjugate pair (μ_c, μ_c^*) (Figure 7.9a), rather than exiting the unit disc, shift until they meet on the negative real axis (Figure 7.9b), then split up (Figure 7.9c) with one of them exiting at -1 (Figure 7.9d) heralding the onset of a period doubling bifurcation.

This scenario may seem contrived – can it really occur? Yes, in fact for the circuit equation governing Figure 7.1 this can be computed analytically to good approximation (Wiesenfeld, 1985a), and the multipliers behave just as in Figure 7.9. A similar computation for a modulated semiconductor injection laser likewise reveals a virtual Hopf sequence prior to the first period doubling (Wiesenfeld, 1986).

But there is more: not only *can* it occur, but the virtual Hopf phenomenon is quite common. For example, it occurs in a class of oscillators that includes the

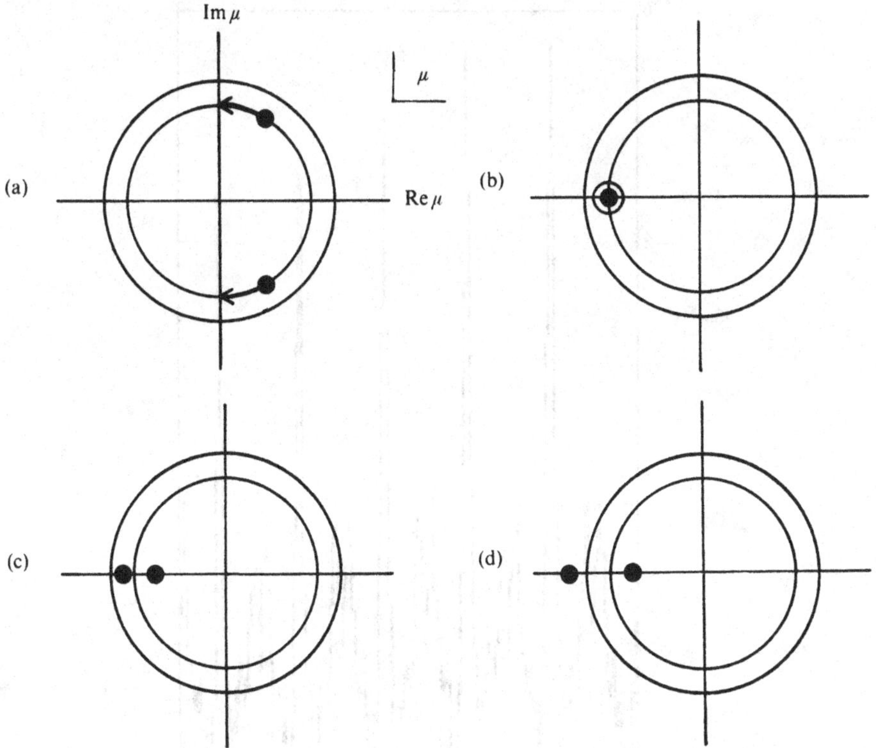

Figure 7.9. Behavior of near-critical Floquet multipliers during the virtual Hopf phenomenon.

driven damped pendulum

$$\ddot{\phi} + \gamma\dot{\phi} + \omega_0^2 \sin\phi = A + B\cos\omega t. \tag{7.4.1}$$

This system can undergo an infinite cascade of period doublings as A is increased (Huberman, Crutchfield and Packard, 1980; D'Humieres, Beasley, Huberman and Libchaber, 1982), and the virtual Hopf scenario is played out *between each successive bifurcation in the infinite cascade.* These facts can be proven analytically; a similar proof exists for a class of autonomous third order systems which includes the Lorenz equations

$$\begin{aligned} \dot{x} &= \sigma(y - x) \\ \dot{y} &= \rho x - y - xz \\ \dot{z} &= -\beta z + xy;\ \sigma, \rho, \beta > 0. \end{aligned} \tag{7.4.2}$$

An important ingredient in these proofs (Crawford and Omohundro, 1984; Wiesenfeld, 1985b) is the low dimensionality of the phase space, and they are hard to extend to higher dimensional systems. Nonetheless, the virtual Hopf

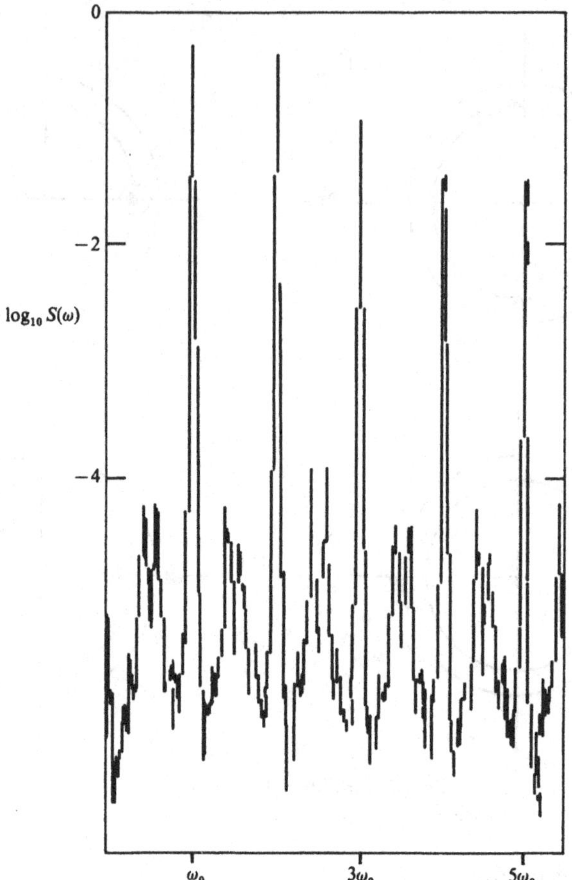

Figure 7.10. Measured power spectrum from the Belousov–Zhabotinskii experiment, for a parameter value in between those of Figure 7.3(a) and (b).

itself is not so restricted. Kazarinoff and Seydel (1986) (and their MAXSYMA program) have shown this for a fourth order autonomous system (originally introduced by Lorenz). And in the Belousov–Zhabotinskii experiment shown earlier (Figure 7.3), there seems to be a hint that the noisy precursor is really double-peaked – is it? I have left out a picture at an intermediate parameter setting and now produce it in Figure 7.10. So here is an example of the virtual Hopf in a very high dimensional system.

To my knowledge, the only previously published experimental data showing the virtual Hopf is for a mechanical 'bouncing ball' system (Pierański and Małecki, 1986).

Is there any period doubling system that does not display the virtual Hopf? Yes: the logistic map is one example. As a one dimensional system, it has only one eigenvalue (which is the discrete map equivalent of the Floquet multiplier)

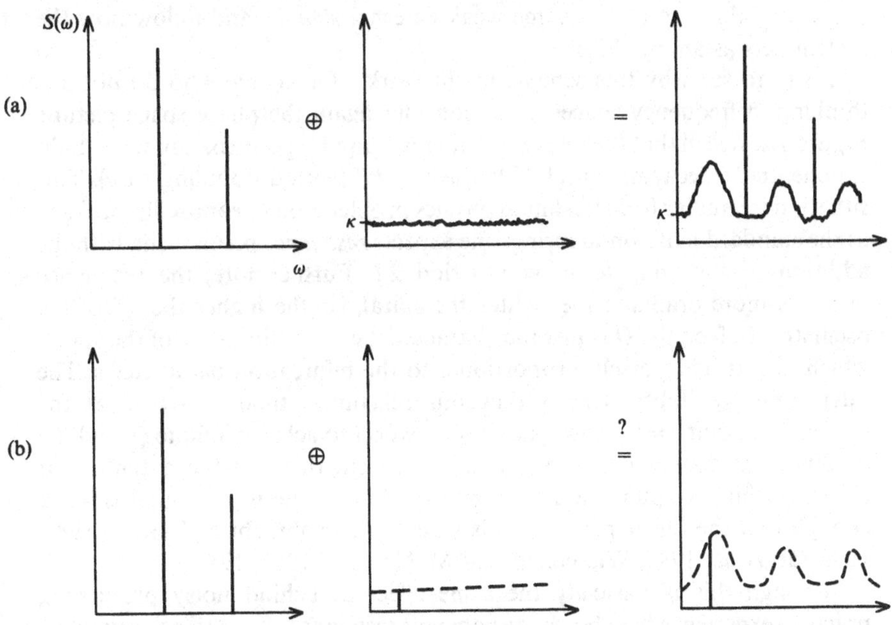

Figure 7.11. (a) Near a bifurcation, the 'sum' of a periodic oscillation and an incoherent (broadband) perturbation results in amplification of certain frequencies. (b) If the input is monochromatic, will amplification of the coherent signal occur?

and so cannot ever have a complex conjugate pair of μ_k.

Aside from being a curiosity, is there anything else the virtual Hopf might be good for? Again the answer is 'yes', provided one leaves the realm of noisy perturbations, and turns to the subject of the next section, that of coherent signal amplification.

7.5 Amplification of coherent signals

The lesson so far is summarized in Figure 7.11. Think about it in terms of combining the power spectrum of the deterministic system – a series of delta functions at $\omega = \omega_0, 2\omega_0, \ldots$ – with that of an external noise source, i.e. a flat, featureless spectrum. Ordinarily, the total output is the 'incoherent sum' of these pictures, but near a bifurcation point things are different: it is as if certain perturbation frequencies are amplified.

So if it works with noise, why not with signal?

The hope is that an input 'noise' with a narrowband spectrum will get similarly amplified, without generating all sorts of interfering structure at all other frequencies (Figure 7.11b). It turns out that things are not quite so simple, but basically this hope is realized. Thus, where before the concern was the effects of unwanted external noise, the game now is to consider purposely 'perturbing' a system – think of this as some weak external *signal* – and follow how the system acts as an amplifier.

A way to see why this scheme might work – for people who do not like thinking in frequency space – is to consider again the phase space picture, Figure 7.2. Recall that the relaxation after an impulse perturbation was a tight exponentially decaying spiral of frequency $2T$ (period doubling case). This situation is familiar from freshman physics problems on linear oscillators such as the standard mass-on-a-spring: one expects *resonance* phenomena with the addition of a driving term with period $2T$. Furthermore, the resonance becomes more dramatic the tighter the spiral, i.e. the higher the 'Q' of the oscillator. In fact, the Q is just the characteristic relaxation time of the spiral, which in turn is inversely proportional to the bifurcation parameter ε. The interesting possibility here is that this relaxation time is *infinite* at the bifurcation point $\varepsilon = 0$. Now, one cannot expect to achieve infinite Q, and the nonlinear effects enter at some stage to prevent this, but the potential for extremely high amplification is there, especially if the input signal is weak enough that the linear picture holds water (Hackenbracht and Höck, 1986; Heldstab *et al.*, 1983; Wiesenfeld and McNamara, 1985, 1986).

Although this is basically the same effect as behind noisy precursors, practical experience has shown the coherent resonance effects to be particularly easy to observe in the laboratory. Of course, the effects are more dramatic the closer the bifurcation point and the closer the perturbing signal is to one of the resonant frequencies.

The mathematical formulation of this problem is similar to the noisy precursor problem; in fact the coherent theory is somewhat more general (allowing either additive or multiplicative perturbations), the calculations are a bit easier (no ensemble averaging is needed), and certain 'technical difficulties' are avoided (the autonomous case does not lead to 'nonperturbative' corrections as in the noisy case). Beginning with the general equation

$$\dot{x} = F(x; \lambda), \quad x \in \mathscr{R}^N, \tag{7.5.1}$$

the parameter λ is modulated a small amount about its mean value λ_0,

$$\lambda(t) = \lambda_0 + \lambda_1 \cos \omega_s t \tag{7.5.2}$$

which causes a deviation $\eta \equiv x - x_0$ from the unperturbed T-periodic solution x_0. Linearizing in both η and λ_1 yields

$$\dot{\eta} = \mathbf{DF} \cdot \eta + \partial_\lambda F, \tag{7.5.3}$$

where both the matrix $\mathbf{DF}$ and the vector $\partial_\lambda F$ are evaluated at $x = x_0$, $\lambda = \lambda_0$. Again this is a linear inhomogeneous equation with periodic coefficients, and the machinery of Floquet theory can be pressed into service.

The basic results of the theory are summarized in Figure 7.12. Again, the near-critical Floquet multiplier is $\mu_c = e^{\varepsilon} e^{i\theta} \approx (1 - \varepsilon) e^{i\theta}$. In addition to the bifurcation parameter ε, an important quantity is the *frequency detuning*

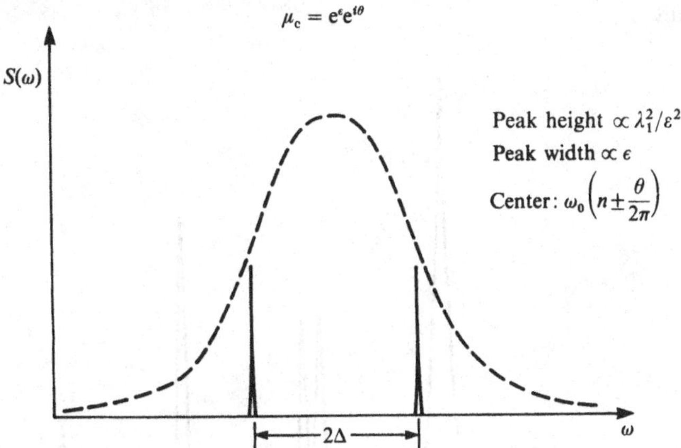

Figure 7.12. Summary of results of small signal amplification near the onset of simple bifurcations.

parameter Δ which measures the difference between the signal frequency ω_s and the nearest resonance frequency $\theta\omega_0/(2\pi)$. (For example, near a period doubling $\theta = \pi$, so that

$$\Delta = \frac{\omega_s - \frac{1}{2}\omega_0}{\omega_0}). \tag{7.5.4}$$

The signal introduces a new set of pairs of lines in the power spectrum at frequencies $n\omega_0 \pm \omega_s$, each of which scales as

$$S(\omega) \propto \frac{\lambda_1^2}{\varepsilon^2 + \Delta^2} \tag{7.5.5}$$

for nonautonomous systems; for autonomous systems there are corrections to this Lorentzian law.

The first tests of the theory were performed on an analog circuit for a driven Duffing equation

$$\ddot{x} + \gamma\dot{x} + ax + bx^3 = A + B\cos\omega_p t + \lambda_1\cos(\omega_s t + \phi) \tag{7.5.6}$$

which is Newton's second law for a damped particle moving in a single-well ($a > 0$, $b > 0$) quartic potential. For zero perturbation, this nonautonomous system displays period doubling (Novak and Frehlich, 1982) as B is increased (and a variety of other bifurcations depending on the values of γ, a, b, ω_p, and A) – say, at $B = B_*$ the oscillation at frequency ω_p period doubles. The circuit was tuned to B just less than B_*, and a small signal ($\lambda_1 > 0$) was added, yielding the power spectrum of Figure 7.13: this is a low resolution plot in order to get a wide range of frequencies on one graph. The signal frequency was chosen to be close to $3\omega_p/2$ (similar effects occur with $\omega_s \simeq \omega_p/2$, or $5\omega_p/2$, etc.) and the anticipated pairs of lines are evident. Note that each pair consists of two nearly

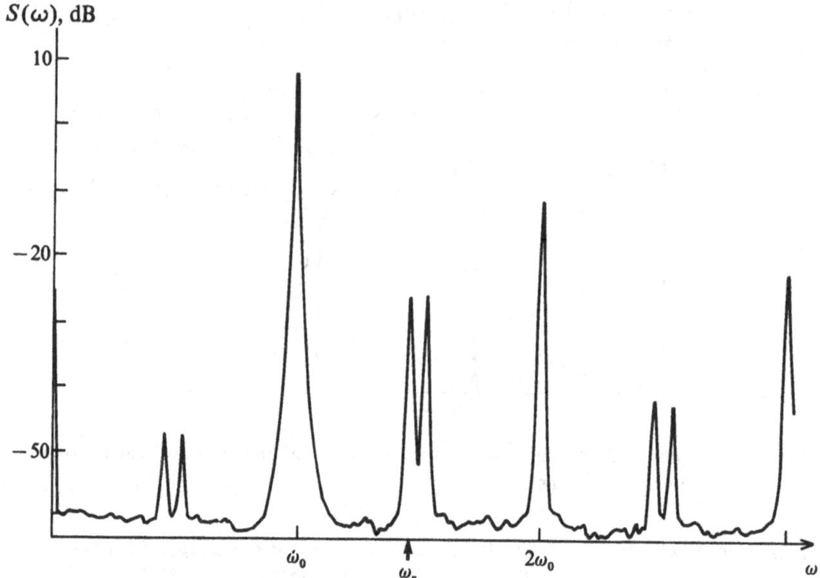

Figure 7.13. Output from an analog simulation of a driven Duffing equation subject to a small near-resonant perturbation at frequency ω_s (Wiesenfeld and McNamara, 1985).

equal lines, as predicted. The input power level at ω_s is about -60 dB on this graph, so there is a relatively modest power gain of about 25 dB, which is a factor of seventeen or so. Much bigger factors are obtained by reducing the detuning frequency, and tuning B closer to B_*. Figure 7.14 focuses on the output at ω_s, showing the *amplitude* ($\propto (\text{power})^{1/2}$) gain as a function of Δ for three different values of $B < B_*$. The resonance curves sharpen and grow as the bifurcation point is approached, and the data follow nicely the predicted square-root Lorentzian law. The highest gain shown is about a factor of 500 in amplitude, and thus $\sim 250\,000$ in power gain.

In effect, this theory represents a new perspective on the old subject of parametric amplification: the emphasis is on quantitative scaling results that apply regardless of the details of any specific physical system. Of course, there must be *some* results that cannot be captured by such a general analysis. Specifically, what cannot be predicted are overall amplification factors – predictions like (7.5.5) involve scaling, but do not fix the overall proportionality constants. Thus, in Figure 7.13 the theory predicts pairs of lines with equal heights, and how these heights change with the parameters λ_1, ε and Δ; but the fact that the first pair of lines is 20 dB lower than the second pair requires a detailed solution of the full differential equation, which typically requires numerical integration.

With this theory in hand, one can stop and look for experimental consequences either by going forward or backward in time. By 'going

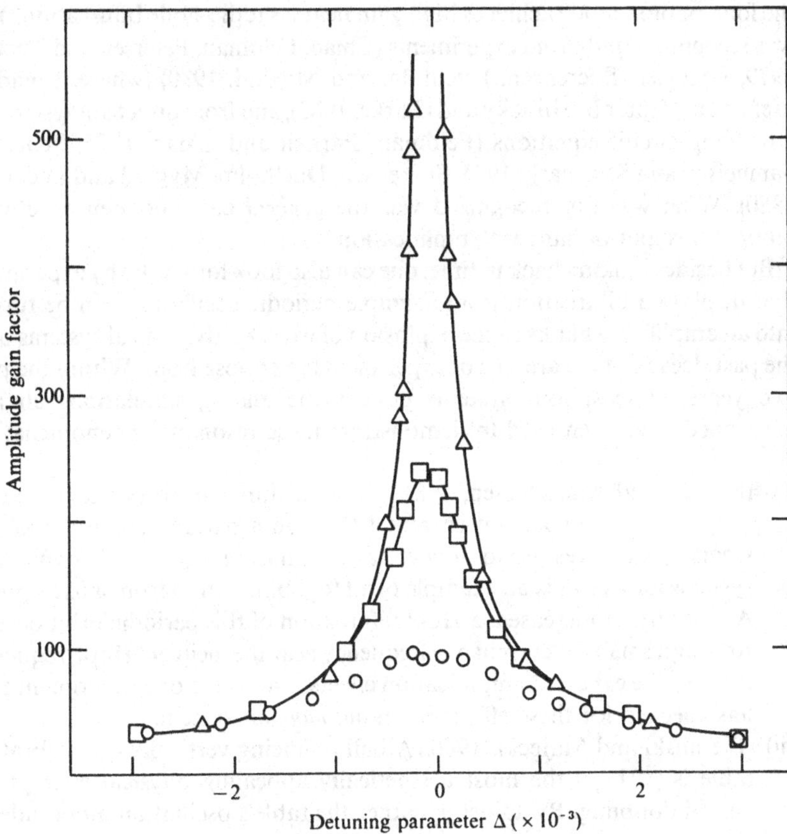

Figure 7.14. Output amplitude at the perturbation frequency as a function of detuning Δ, for three different values of bifurcation parameter ε. Open symbols are measured from analog circuit; solid lines are fits to theoretical line shape (Wiesenfeld and McNamara, 1985).

backward', I mean looking for old experiments that report parametric amplification and seeing if these systems were operating near bifurcation points. I have done this explicitly for two examples. First, an experiment on modulated semiconductor injection lasers (Grothe, Harth and Russer, 1976) demonstrated 'parametric sideband amplification', an effect that had been predicted theoretically just one year earlier (Russer, Hillbrand and Harth, 1975). Sure enough, their experimental parameters were very close to a period doubling bifurcation point (Wiesenfeld, 1986). A second example involves devices known as Josephson junction parametric amplifiers, a subject I will return to in detail in Section 7.7. Widely studied since about 1973, these devices operate in one of two 'modes'. In both cases investigators recognized that the high gain limit corresponds to dynamical instabilities – in the language of dynamical systems, the three-photon mode operates best close to a period doubling, while the four-photon mode achieves high gain near a saddle-node bifurcation. This was recognized both from experiments (Chiao, Feldman, Petersen and Tucker, 1979; Pedersen, Soerensen, Duelholm and Mygind, 1980) (where it made a useful rule of thumb: Miracky and Clarke, 1983), and from direct analysis of the governing circuit equations (Feldman, Parrish and Chiao, 1975; Pedersen, Samuelsen and Saermark, 1973; Soerensen, Duelholm, Mygind and Pedersen, 1980). What was not recognized was the *general* correspondence between bifurcations and parametric amplification.

But besides looking back in time, one can also look forward: any experiment that displays a bifurcation from a simple periodic oscillation can be turned into an amplifier. Thanks to the explosion of work on dynamical systems over the past decade, there are lots of experiments to choose from. Within the past two years, at least four systems besides the analog simulations already mentioned have been used to demonstrate these resonance phenomena:

(i) (Martin and Martienssen, 1986.) Held within the proper temperature range, passing a constant current I through a barium–sodium–niobate single crystal gives rise to slow voltage oscillations ($\lesssim 1\,\mathrm{Hz}$), in dynamical systems terms this is an example of a Hopf bifurcation from a fixed point. As I is further increased, a Hopf bifurcation of this periodic orbit occurs. Adding a small a.c. current at a frequency near the incipient Hopf frequency leads to the expected amplification effects. This is the one experiment that has checked for these effects in an *autonomous* system.

(ii) (Pierański and Małecki, 1986.) A ball bouncing vertically on a vibrating table is perhaps the most aesthetically appealing physical example of period doubling. By adjusting either the table's oscillation amplitude or frequency the ball's motion can change from all equal-height bounces (period one) to alternately high and low bounces (period two). (The set up I witnessed at Bryn Mawr College is particularly elegant – one can actually *hear* the bifurcation! – and can be reconstructed readily for the classroom: Tufillaro and Albano, 1986.) The 'signal' here is incorporated by driving the table at a second frequency, and resonance effects are readily observed. A big advantage of this example is that a thorough analytical study of the nonlinear dynamics can be carried out, including a center manifold reduction (Wiesenfeld and Tufillaro, 1987), which is especially important for elucidating the theory of *nonlinear* effects discussed in Section 7.6.

(iii) (H. Savage and C. Adler, 1986, private communication.) A thin ribbon, when stood on end, can support its own weight without buckling only if its length is less than some critical value. In this experiment the ribbon is *magnetostrictive*, so that this critical length changes substantially when placed in an external magnetic field. Thus, for a fixed length of ribbon, increasing the field B causes buckling at a critical value B_*; if the field has an a.c. component, the dynamics can show period doubling. (Typical dimensions are $10\,\mathrm{cm} \times 2\,\mathrm{mm} \times 18\,\mu\mathrm{m}$, $B_* \sim 0.3\,\mathrm{G}$, a.c. frequency

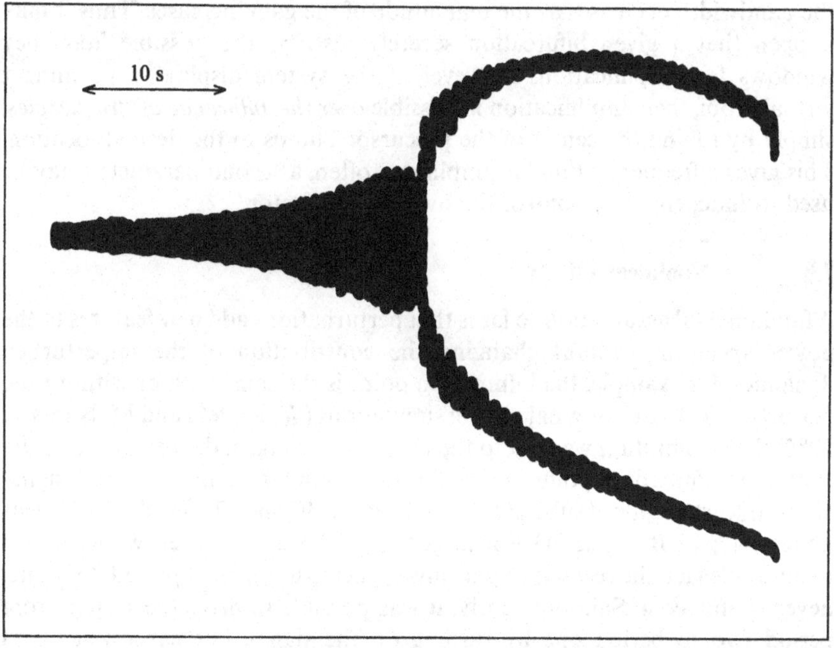

Figure 7.15. Experimentally determined bifurcation diagram from an NMR laser, showing unexpected oscillations as the control parameter is slowly swept through the period doubling bifurcation point (Derighetti *et al.*, 1985).

~ 20 kHz.) Superimposing a second a.c. field – the signal – produces the resonance effects described in this and the next section.

(iv) (Derighetti *et al.*, 1985.) This experiment shows how the amplification can appear in unexpected ways. The system employs a nuclear magnetic resonance experiment in such a way as to achieve stimulated emission during the relaxation of the nuclear spins. Figure 7.15 shows a standard kind of bifurcation diagram obtained by sweeping the control parameter slowly past the period doubling bifurcation point, and has the characteristic 'pitchfork' obtained by sampling the output once per drive period. However, there are also prominent wiggles to the pitchfork; these are particularly large near the bifurcation point. Peter Meier explained that, since Figure 7.15 represents a ~ 40s sweep, these wiggles correspond to a *temporal* modulation of 1.35 Hz, while the incipient period doubling frequency was 51.35 Hz. Thus the wiggles were clearly due to amplification of stray signal. (Americans will not understand this until told that European power lines run at 50 Hz, as opposed to 60 Hz in the States.) Careful shielding of stray fields eliminated the mysterious wiggles.

Finally, I return to the possible utility of the virtual Hopf phenomenon. The amplification just described occurs over a relatively narrow bandwidth, in fact the bandwidth decreases as the magnitude of the gain increases. Thus it may happen that a given bifurcation severely restricts the possible frequency windows for amplification. However, if the system displays a prominent virtual Hopf, then amplification is possible *over the full range of frequencies*, simply by tuning the center of the precursor bumps to the desired location. This gives a frequency-tunable amplifier – often, a second parameter may be used to independently control the overall gain factor.

7.6 Nonlinear effects

A fundamental assumption so far is that perturbations add new features to the power spectrum without changing the contribution of the unperturbed dynamics: for example, the bifurcation point is the same with or without the perturbation. In our original analog simulations (Wiesenfeld and McNamara, 1985) this assumption was put to the test; to our chagrin the perturbation *did* shift the bifurcation point. In order to 'eliminate' this effect the signal amplitude was reduced substantially to generate Figures 7.13 and 7.14. (λ_1 was about 0.1% of B in (7.5.6).) For larger signal levels, however, we noticed a remarkable fact: the resonant signal always acted to *suppress* period doubling, never to induce it. Said differently, it was possible to drive the system from period two to period one by turning on the signal, but never vice versa (Figure 7.16).

Besides the shifted bifurcation point, we observed a host of new phenomena, which I will describe later. These all result from 'feedback' of the nonlinear dynamics on the basic oscillation. It was Paul Bryant who, after hearing a description of these effects, developed a theory to explain them by taking fuller advantage of the center manifold reduction. The theory described in Section 7.5 considers the solution of the full (though linearized) high dimensional problem, and then notices that this decomposes into pieces, with the dominant contribution coming from the piece associated with the 'softest' phase space direction, i.e. the center manifold. Bryant's new step was to discover a dynamical equation valid on the (low dimensional) center manifold which includes nonlinear terms and the effective perturbation. (Again, this style of thinking should be familiar to laser physicists versed in Haken's adiabatic elimination method.)

This idea has been studied in exquisite detail for the case of period doubling (Bryant and Wiesenfeld, 1986). The centerpiece of the theory is an evolution equation for a coordinate z along the center manifold

$$\dot{z} = \mu z - z^3 + \lambda \cos \Delta t; z \in \mathbb{R}, \tag{7.6.1}$$

where μ is the control parameter ($\mu = 0$ at the bifurcation point of the unperturbed system), λ is proportional to the signal amplitude, and Δ measures the frequency detuning. Here, z is a *scalar* variable measuring the displacement along the one dimensional curve on the Poincaré section P in Figure 7.2(b).

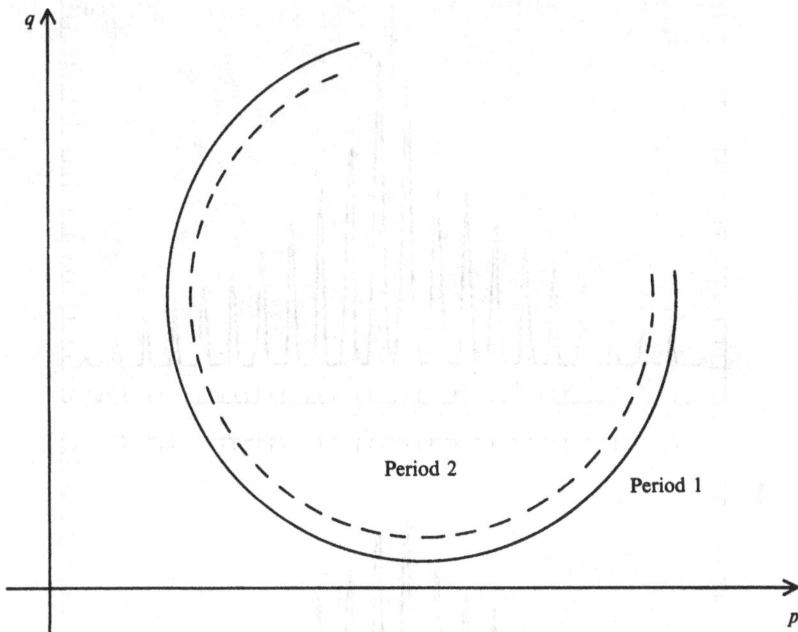

Figure 7.16. Schematic representation of parameter space, illustrating the observation that coherent perturbations always serve to suppress (supercritical) period doubling bifurcations. Key: ——— no signal; --------- finite signal.

Three remarks are in order. First, since successive iterates flip back and forth near a period doubling, (7.6.1) is really a continuous time approximation to the second iterate of the return map. Said differently, the relationship between z and the full blown dynamical variable $X \in \mathbb{R}^N$ takes the form

$$X = X_0(t) + z(t)X_1(t), \tag{7.6.2}$$

where $X_0(t+T) = X_0(t), X_1(t+T) = -X_1(t)$ (the 'flipping' behavior), and T is the unperturbed period. Second, although this equation correctly describes the leading nonlinear effects for nonautonomous systems like the driven Duffing equation (7.5.6) and the bouncing ball experiment, it cannot capture frequency locking behavior and so must be an incomplete description of moderately perturbed autonomous systems. Finally, similar reduced equations hold for symmetry breaking and cusp bifurcations (the latter is a codimension two bifurcation), though the link (7.6.2) between z and X involves somewhat different properties for X_0 and X_1 (Bryant, Wiesenfeld and McNamara, 1987).

Remarkably, even the seemingly innocuous (7.6.1) cannot be solved in detail without resorting to numerical simulations. (Is it any wonder that the explosion of advances in nonlinear dynamics coincides with the advent of powerful yet inexpensive computers?) One prediction that can be derived

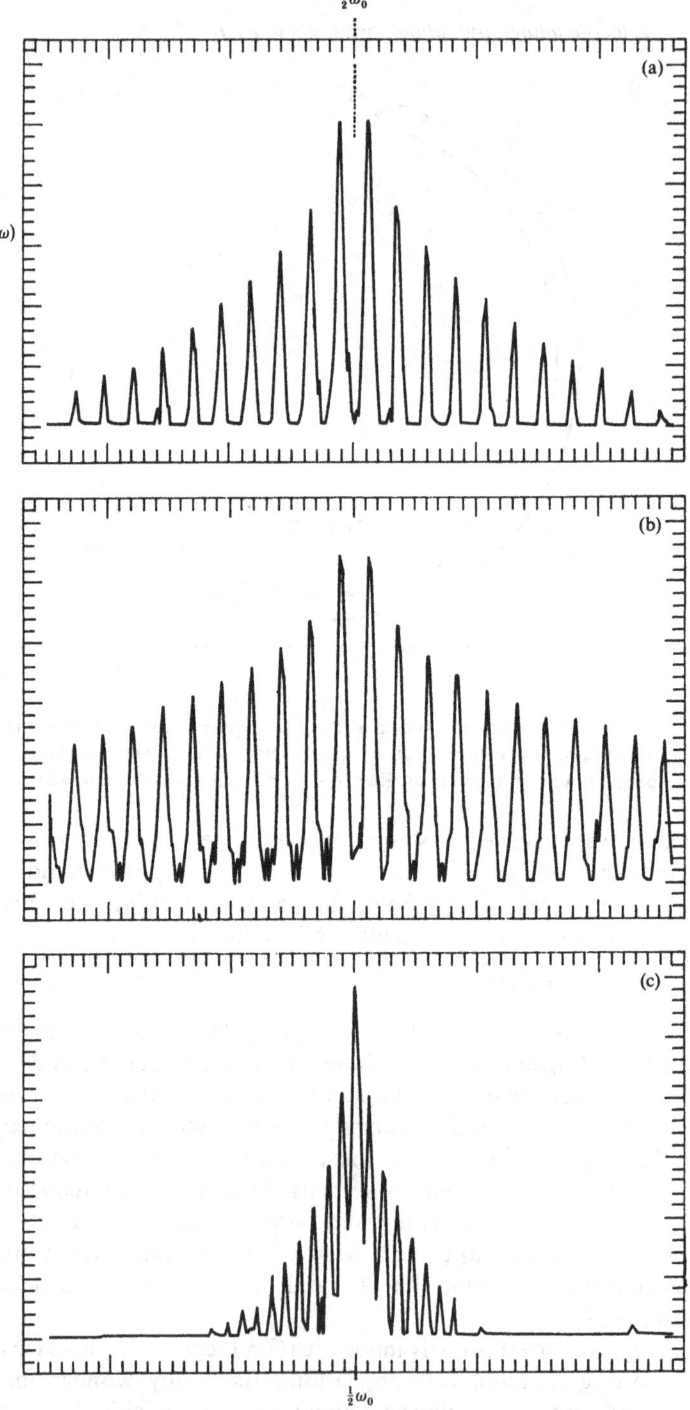

Figure 7.17. Output from analog simulations on the driven Duffing oscillator showing the rise of the 'menorah' in the power spectrum very close to the shifted period doubling bifurcation point (Bryant and Wiesenfeld, 1986).

analytically gives the shift of the bifurcation point from $\mu = 0$ to $\mu = \mu_B$ as a function of signal amplitude,

$$\mu_B \propto \lambda^{2/3}. \tag{7.6.3}$$

This unusual power law is in excellent agreement with numerical and digital simulations of the driven Duffing equation (7.5.6), and in reasonable agreement with experiments on the bouncing ball (P. Pierański, manuscript in preparation; Wiesenfeld and Tufillaro, 1987).

A number of other effects also follow from (7.6.1), and are seen in simulations as well. Figure 7.17 shows the rise (and fall) of the 'menorah', i.e. the emergence of a host of closely spaced spectral lines flanking the usual pair of lines from the linear theory (compare Figure 7.13). These extra lines are prominent for $0 < \mu < \mu_B$, growing toward a characteristic sloping envelope reminiscent of candles in a menorah. Another result is the correction to the physically impossible 'infinite gain' prediction of the linearized theory, i.e. the $\varepsilon \to 0$, $\Delta \to 0$ limit of (7.5.5). The nonlinear terms saturate the gain, making it an asymmetric function about the bifurcation point $\mu = \mu_B$, with an *abrupt* drop just beyond the bifurcation point. Without too much imagination, this drop can be seen in the NMR laser data, Figure 7.15. Still other nonlinear effects also appear (Bryant and Wiesenfeld, 1986).

Very recently some work in Denmark (Svensmark, Hansen and Pedersen, 1987) examined suppression of period doubling in Josephson circuits, which is a problem of some practical importance (see Section 7.7). At one bifurcation point suppression is observed, but at another the signal seemed to *enhance* subharmonic oscillations. Does this contradict the above theory? It does not, but it brings up an important and somewhat subtle point. The discussion in this section has implicitly assumed that the period doubling bifurcation is *supercritical*, in which a stable period one orbit loses stability and gives birth to a stable period two orbit (Figure 7.18a). But there is a second kind of period doubling called *subcritical* in which a stable period one orbit and a coexisting unstable period two orbit collide leaving behind an unstable period one orbit (Figure 7.18b). The supercritical case is the one most commonly reported in experiments; indeed, the subcritical case can only be deduced from indirect evidence (its hallmarks are the sudden appearance of finite amplitude subharmonic oscillations, and associated hysteresis). From a theoretical viewpoint, the subcritical case involves only a sign change of the cubic term in (7.6.3) (Bryant and Wiesenfeld, 1986; Thompson and Stewart, 1986). This leads to enhancement rather than suppression of subharmonic oscillations, though a proper treatment must also include still higher order terms to stabilize the center manifold dynamics (Wiesenfeld and Pedersen, 1987).

7.7 Origin of noise rise in SUPARAMPS

About a dozen years ago, the first Josephson junction parametric amplifiers were built and studied (Feldman, Parrish and Chiao, 1975; Mygind, Pedersen

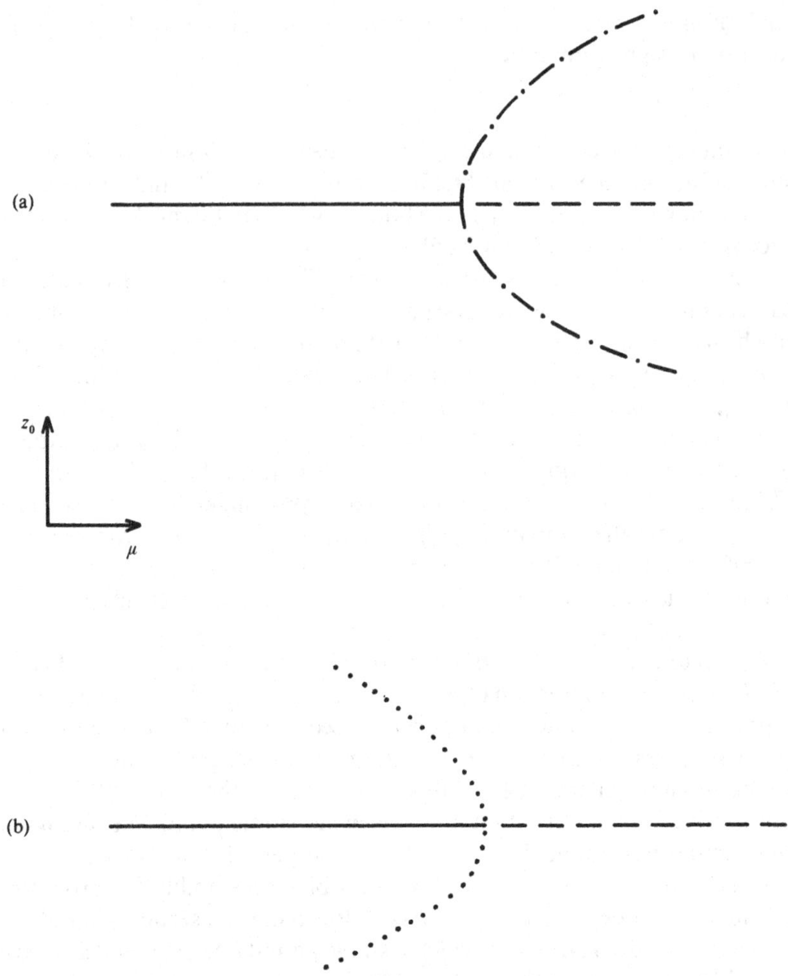

Figure 7.18. Bifurcation diagrams illustrating the two types of period doubling bifurcations: (a) supercritical, (b) subcritical. Key: ——— stable period one; ----- unstable period one; –·–·–·– stable period two; ..... unstable period two.

and Soerensen, 1978; Mygind *et al.*, 1979; Soerensen, Mygind and Pedersen, 1978; Taur and Richards, 1977; Wahlstein, Rudner and Claeson, 1978). These devices achieved good gain of electromagnetic radiation in the frequency range ~ 1–100 GHz; building good detectors in this range continues to pose a challenge for observational radioastronomers. Unfortunately, the Josephson junction amplifiers suffer from a mysterious noise problem – the so-called noise rise – previously unseen in other kinds of parametric amplifiers. One ordinarily expects the ratio of signal amplification G_S to broadband noise amplification G_N to be a constant independent of parameter settings: this ratio is proportional to the noise temperature T characterizing a given device.

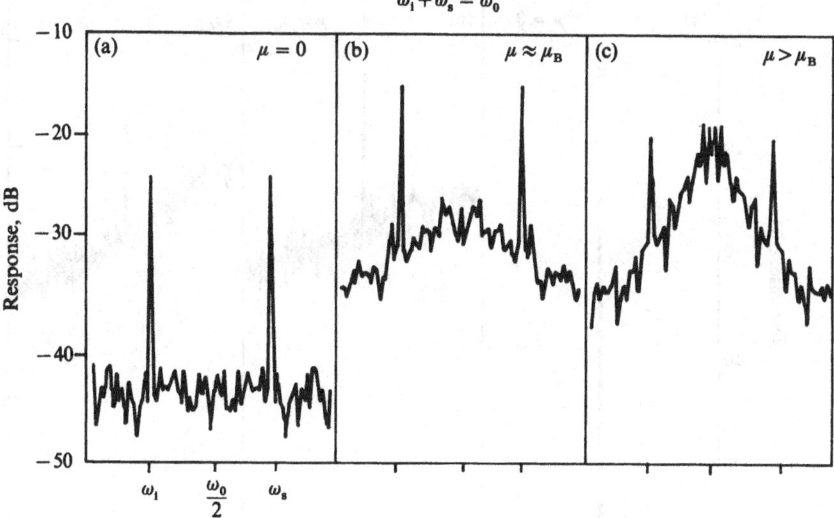

Figure 7.19. Analog simulation of the three-photon Josephson junction parametric amplifier, illustrating the noise rise. (a) Modest signal gain achieved at a signal frequency close to $\omega_0/2$. Roughly equal power is generated at the idler frequency. (b) Though the signal gain has increased, the signal-to-noise ratio decreased. (c) As the bifurcation parameter is further increased, the signal gain eventually decreases (Bryant, Wiesenfeld and McNamara, 1987).

Instead, the Josephson devices display a noise temperature which increases with increasing G_S – that is, the greater the signal gain the worse the signal-to-noise ratio until the noise output overwhelms the signal. This effect is illustrated in Figure 7.19, which shows the results of analog simulations of the Josephson junction parametric amplifier equations, including an external noise source (Bryant, Wiesenfeld and McNamara, 1987).

Josephson junction parametric amplifiers have been built using a variety of arrangements, employing either a single junction (microbridge, point contact, or tunnel junction) or an array of many junctions in series. The noise rise has been observed in all of these designs. What is its origin? Over the years a variety of explanations have been forwarded, but the issue is still unresolved. A nice, concise review of some theories and their shortcomings has been written by Feldman and Levinsen (1981). One interesting suggestion that the noise rise was the result of deterministic chaos was proposed by Huberman, Crutchfield and Packard (1980); however, the chaotic parameter regime does not coincide with the conditions for large G_S. Instead, the high gain limit coincides with the simplest instabilities, either a period doubling bifurcation or a saddle-node bifurcation, corresponding to the so-called three-photon and four-photon modes, respectively. (The latter mode lacks an external bias voltage, and led to the colorful acronym SUPARAMP for **s**uperconducting **u**nbiased **par**ametric **amp**lifier.)

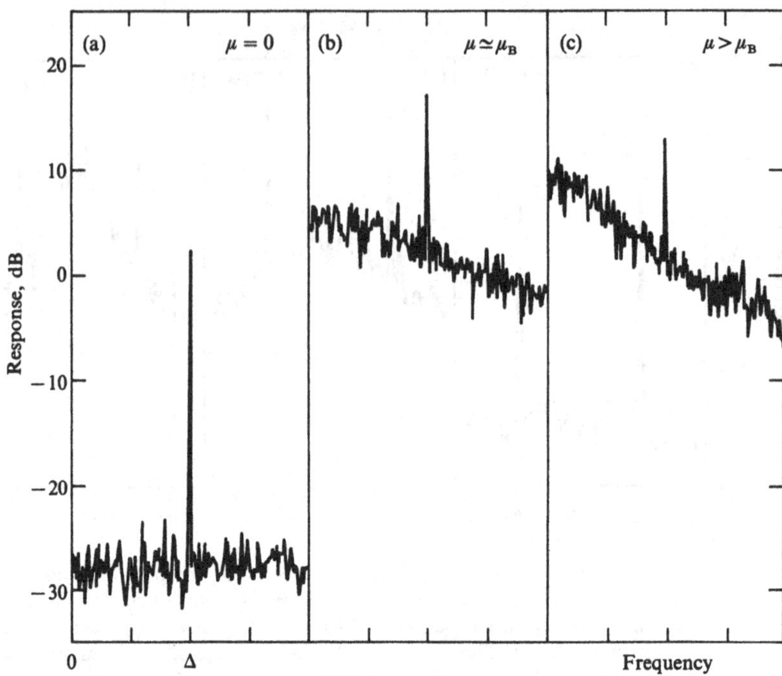

Figure 7.20. Results from the reduced equation (7.7.1) governing the center manifold dynamics, reproducing the noise rise observed in the full equation. Zero frequency here corresponds to $\omega_0/2$ in Figure 7.19, so only one-half of the signal-idler pair is shown (Bryant, Wiesenfeld and McNamara, 1987).

This last fact suggests a theory of the simultaneous effects of signal and noise at the onset of simple bifurcations might explain the noise rise. Without the center manifold reduction this would be a terribly difficult problem, but instead the theory is quite simple (Bryant, Wiesenfeld and McNamara, 1987). Along the one dimensional center manifold the essential dynamics reduces to (for the three-photon case)

$$\dot{z} = \mu z - z^3 + \lambda \cos \Delta t + \xi(t); z \in \mathbb{R} \tag{7.7.1}$$

which is just (7.6.2) with a random term $\xi(t)$ added. Again, the deep results behind a center manifold reduction lead us to a disarmingly simple equation. And again, this is still too complicated to analyze in detail without heavy reliance on numerical simulations.

What do we find? First of all, digital simulations of (7.7.1) reproduce the noise rise, as shown in Figure 7.20. In terms of phase space dynamics, the external noise causes random jumps between two closely spaced branches of the noise-free attractor. This is very similar to an earlier suggestion that the noise rise corresponds to random hopping between coexisting attractors (Miracky and Clarke, 1983).

Besides the qualitative reproduction of the noise rise, the theory based on (7.7.1) makes a number of quantitative predictions (Bryant, Wiesenfeld and McNamara, 1987) for future experiments to test. (It is the kind of theory experimentalists should like; the predictions are crisp enough that there is the potential to *disprove* the theory.) These results are, for the most part, based on numerical studies of (7.7.1), and the corresponding equation for the four-photon mode.

One final point. The theory leading to (7.7.1) is quite general: so why has the noise rise only been reported in the Josephson devices and not in other kinds of parametric amplifiers? The theory provides an answer, as follows. In order to see a substantial noise rise, very small frequency detunings Δ are required, and the high frequencies of the Josephson devices enable these small Δ to be achieved (see (7.5.4)). It follows that modulated semiconductor lasers, which also operate in the gigahertz regime, make excellent alternative candidate systems to explore the noise rise and associated phenomena. But, in the end, such high frequency systems are not required to see the noise rise, as evidenced by existing analog simulations of the Josephson devices (Bryant, Wiesenfeld and McNamara, 1987; Miracky and Clarke, 1983).

7.8 Future vistas

Finally, here are some unresolved issues that require further work.

Although several experiments have tested aspects of the existing theories, careful quantitative studies are lacking for autonomous systems: recall that for both noisy and coherent perturbations the results are a bit more complicated in this instance than for nonautonomous systems. Besides checking existing predictions, there is the opportunity for experiments to drive further theory because no nonlinear theory analogous to that of Section 7.6 has been developed for autonomous systems. This problem is especially interesting since there is the additional possibility of frequency locking between the signal and the unperturbed oscillator to complicate matters.

The treatment of simultaneous coherent and noisy perturbations also awaits careful experiments, especially the quantitative behavior of the noise rise (Section 7.7). An important qualitative prediction is that effects similar to those observed in Josephson junctions should be observed elsewhere, with modulated semiconductor lasers being an obvious place to look.

An interesting question is whether random noise can suppress the onset of period doubling. I think it should: after all, since coherent modulations at *any* detuning frequency act to stabilize supercritical period doubling, should not a superposition of modulations – and by extension broadband noise – also act to stabilize? There is a major obstacle confronting attempts to answer this, namely how to define the bifurcation point in an operationally meaningful way. When broadband noise is present there is always power at all frequencies, so the signature cannot be taken as the onset of power at half the fundamental frequency. The Swift criterion offers one alternative, defining the onset by extrapolating backward from where the sharp line at $\omega_0/2$ is clearly discernable (Wiesenfeld, 1985a). A second criterion has been suggested by Frank Moss, identifying the bifurcation point with the point of most rapid growth of power at $\omega_0/2$. Both alternatives give the correct answer in the noise-free limit, which is reassuring. It is even possible that different definitions will give rise to different answers to this question.

Another issue that remains a mystery: why does the virtual Hopf phenomenon appear to be so common? It is understandable in a low dimensional system, but it also shows up in the undeniably high dimensional Belousov–Zhabotinskii reaction (Figures 7.3 and 7.10); and why should that be? Perhaps global topological arguments akin to the discussion by Crawford and Omohundro (1984) can be extended to answer this question.

Finally, if these ideas correctly describe the origin of the noise rise in Josephson junction parametric amplifiers, then the same theory can be studied to see how to minimize this undesirable effect, perhaps leading to performances competitive with the so-called superconductor–insulator–superconductor mixers (Smith, Sandell, Burch and Silver, 1985).

I think this chapter adequately answers the question posed by its title. A number of people have already asked me the next question: So what good is chaos? The answer may not be far off.

Acknowledgements

A number of people's contributions are not sufficiently represented by the references cited in this chapter. Bob Miracky first suggested that these ideas might be relevant to the noise rise problem in superconducting parametric amplifiers; likewise Herb Winful suggested their relevance to sideband amplification in semiconductor injection lasers. The references to the early work of von Neumann and Goto were pilfered straight from papers by Rolf Landauer (see his Chapter 1 in Volume 1), which lends a fascinating historical perspective. I thank a number of people for keeping me abreast of their own work on this subject: Charlie Adler, Theo Geisel, D. Hackenbracht, Sam Martin, Peter Meier, Frank Moss, Piotr Pierański, Neils Pedersen, Howard Savage, Mark Schumaker and Carl Weiss. I owe a great debt to my collaborators, Paul Bryant, Carson Jeffries, Bruce McNamara and Nick Tufillaro. Also important has been the encouragement by Neal Abraham, John Clarke, John Guckenheimer, Carson Jeffries, Edgar Knobloch and Mike Nauenberg. Finally, the central idea of this work – fusing the elements of bifurcation theory and stochastic differential equations – is due to Edgar Knobloch.

This chapter is dedicated to the memory of David Wiesenfeld, December 11, 1916–March 5, 1987.

This chapter has been written with support by the Division of Material

Sciences, US Department of Energy, under contract number DE-AC02-76CH00016.

References

Bryant, P. and Wiesenfeld, K. 1986. *Phys. Rev. A* **33**, 2525.
Bryant, P., Wiesenfeld, K. and McNamara, B. 1987. *Phys. Rev. B* **36**, 752.
Chiao, R. Y., Feldman, M. J., Petersen, D. W. and Tucker, B. A. 1979. *AIP Conf. Proc.* **44**, 259.
Crawford, J. D. and Omohundro, S. 1984. *Physica* **13D**, 161.
Derighetti, B., Ravani, M., Stoop, R., Meier, P., Brun, E. and Badii, R. 1985. *Phys. Rev. Lett.* **55**, 1746.
D'Humieres, D., Beasley, M. R., Huberman, B. A. and Libchaber, A. 1982. *Phys. Rev. A* **26**, 3438.
Feldman, M. J. and Levinsen, M. T. 1981. *IEEE Trans. Magn.* **MAG-17**, 834.
Feldman, M. J., Parrish, P. T. and Chiao, R. Y. 1975. *J. Appl. Phys.* **46**, 4031.
Golubitski, M. and Guckenheimer, J. 1986. *Multiparameter Bifurcation Theory.* Contemporary Mathematics **56**. Providence, RI: American Mathematical Society.
Goto, E. 1955. *J. Elec. Commun. Engrs. Japan* **38**, 770.
Grothe, H., Harth, W. and Russer, P. 1976. *Electron. Lett.* **12**, 522.
Guckenheimer, J. and Holmes, P. 1983. *Nonlinear Oscillations, Dynamical Systems and Bifurcations of Vector Fields*, Springer Series in Applied Mathematical Sciences **42**. New York: Springer.
Hackenbracht, D. and Höck, K.-H. 1986. *J. Phys. C* **19**, 4095.
Haken, H. 1970. *Handbuch der Physik* (*Encyclopedia of Physics*), vol. XXV/2C (L. Genzel, ed.). Heidelberg: Springer.
Haken, H. 1977. *Synergetics: An Inroduction.* New York: Springer.
Heldstab, J., Thomas, H., Geisel, T. and Radons, G. 1983. *Z. Phys. B* **50**, 141.
Huberman, B. A., Crutchfield, J. P. and Packard, N. H. 1980. *Appl. Phys. Lett.* **37**, 750.
Jeffries, C. and Wiesenfeld, K. 1985. *Phys. Rev. A* **31**, 1077.
Jordan, D. W. and Smith, P. 1977. *Nonlinear Ordinary Differential Equations.* Oxford University Press.
Kazarinoff, N. D. and Seydel, R. 1986. *Phys. Rev. A* **34**, 3387.
Keyes, R. W. and Landauer, R. 1970. *IBM J. Res. Dev.* **14**, 152.
Linsay, P. S. 1981. *Phys. Rev. Lett.* **47**, 1349.
McCumber, D. 1966. *Phys. Rev.* **141**, 306.
Manes, K. and Siegman, A. 1971. *Phys. Rev. A* **4**, 373.
Martin, S. and Martienssen, W. 1986. *Phys. Rev. A* **34**, 4523.
Miracky, R. F. and Clarke, J. 1983. *Appl. Phys. Lett.* **43**, 508.
Mygind, J., Pedersen, N. F. and Soerensen, O. H. 1978. *AIP Conf. Proc.* **44**, 246.
Mygind, J., Pedersen, N. F., Soerensen, O. H., Duelholm, B. and Levinsen, M. T. 1979. *Appl. Phys. Lett.* **35**, 91.
Novak, S. and Frehlich, R. G. 1982. *Phys. Rev. A* **26**, 3660.
Pedersen, N. F., Samuelsen, M. R. and Saermark, K. 1973. *J. Appl. Phys.* **44**, 5120.
Pedersen, N. F., Soerensen, O. H., Duelholm, B. and Mygind, J. 1980. *J. Low Temp. Phys.* **38**, 1.

Pierański, P. and Małecki, J. 1986. *Phys. Rev. A* **34**, 582.
Russer, P., Hillbrand, H. and Harth, W. 1975. *Electron. Lett.* **11**, 87.
Smith, A. D., Sandell, R. D., Burch, J. P. and Silver, A. H. 1985. *IEEE Trans. Magn.* **MAG-21**, 1022.
Soerensen, O. H., Duelholm, B., Mygind, J. and Pedersen, N. F. 1980. *J. Appl. Phys.* **51**, 5483.
Soerensen, O. H., Mygind, J. and Pedersen, N. F. 1978. *AIP Conf. Proc.* **44**, 246.
Svensmark, H., Hansen, J. B. and Pedersen, N. F. 1987. *Phys. Rev. A* **35**, 1457.
Taur, Y. and Richards, P. L. 1977. *J. Appl. Phys.* **48**, 1321.
Testa, J., Perez, J. and Jeffries, C. 1982. *Phys. Rev. Lett.* **48**, 714.
Thompson, J. M. T. and Stewart, H. B. 1986. *Nonlinear Dynamics and Chaos.* Chichester: Wiley and Sons.
Tufillaro, N. B. and Albano, A. 1986. *Am. J. Phys.* **54**, 939.
van Buskirk, R. and Jeffries, C. 1985. *Phys. Rev. A* **31**, 3332.
von Neumann, J. 1954. *US Patent 2,815,488* (filed April 28, 1954, issued December 3, 1957).
Wahlstein, S., Rudner, S. and Claeson, T. 1978. *J. Appl. Phys.* **49**, 4248.
Wiesenfeld, K. 1985a. *J. Stat. Phys.* **38**, 1071.
Wiesenfeld, K. 1985b. *Phys. Rev. A* **32**, 1744.
Wiesenfeld, K. 1986. *Phys. Rev. A* **33**, 4026.
Wiesenfeld, K. and McNamara, B. S. 1985. *Phys. Rev. Lett.* **55**, 13; and erratum in *Phys. Rev. Lett.* **56**, 539 (E), (1986).
Wiesenfeld, K. and McNamara, B. S. 1986. *Phys. Rev. A* **33**, 629.
Wiesenfeld, K. and Tufillaro, N. B. 1987. *Physica D* **26**, 321.
Wiesenfeld, K. and Pedersen, N. F. 1987. *Preprint.*

8 Noise-induced transitions

WERNER HORSTHEMKE and RENÉ LEFEVER

8.1 Introduction

All macroscopic systems are subject to irregular perturbations that cannot be modelled in a deterministic description. Following common usage, we shall call these perturbations 'noise'. These may be 'internal fluctuations', due to the complex interaction of parts of a composite system, or they may be 'external noise', due to an irregular influence imposed on the system from the environment to which it is coupled. In this chapter we shall be interested particularly in the latter type of fluctuations and their effects on the macroscopic behavior of open nonlinear systems. It is well known now that nonlinear systems, such as the laser (Haken, 1978), the Couette system (Di Prima and Swinney, 1981), and the Belousov–Zhabotinskii reaction (Field and Burger, 1985), can display a rich variety of transition phenomena leading to temporal (and spatial) structuring: bistability, limit cycle oscillations, period doubling behavior, aperiodic behavior, etc. These transitions are due to the nonequilibrium constraints imposed by the environment on the system. It is under nonequilibrium conditions that the intrinsic nonlinearities of the system come fully into play; at and near equilibrium the behavior of these macroscopic systems can be well described by linear laws.

In light of this fact, it is particularly important to take into account that environments of open systems, and thus the external constraints imposed on the system, fluctuate in general more or less strongly. Rapid random fluctuations are always present in natural systems and their amplitude is not necessarily small, contrary to laboratory systems. In the latter systems the experimenter will of course try to minimize the effect of random perturbations, though it is impossible to eliminate noise completely. Clearly, random noise is ubiquitous in the external constraints of open systems. Since these constraints are at the origin of the transition phenomena, the question naturally arises if external noise has any effect on the structuring of nonlinear systems. Is noise a mere nuisance we have to live with or is there any interesting physics to be found? The intuitive, and wrong, answer would be that noise is indeed a nuisance and has no positive aspects: the system averages out rapid fluctuations and the only trace of the external noise would be a certain fuzziness in the state of the system. Of course, if the state of the system becomes unstable as the (average) value of the external constraints is changed, the fluctuations initiate the departure from the unstable state. Then the dynamics of the system takes over and the system evolves to a new stable state. In this chapter we will summarize the results, obtained over the last decade, which show that besides these trivial effects external noise can give rise to unexpected and interesting phenomena.

Random external fluctuations, in addition to the expected disorganizing effect, also tend to impose a structure that is not present when the fluctuations are absent. Indeed, noise can change the stability properties of a system, namely stabilize or destabilize a steady state. This effect was first theoretically predicted in studies on oscillations in radio circuits (Kuznetsov, Stratonovic and Thikhonov, 1965), on the survival of populations (May, 1973), and on oscillating enzyme systems (Hahn, Nitzan, Ortoleva and Ross, 1974). External noise can also create new states which never exist under deterministic environmental conditions (Horsthemke and Lefever, 1984). Clearly, there is interesting physics to be found in the study of noise-induced transitions.

The organization of this chapter is as follows. In Section 8.2.1 we will discuss the modelling of nonlinear systems coupled to a randomly fluctuating environment, and in particular we discuss the limit of Gaussian white noise. Then we will apply the results of the white noise analysis to the unfolding of the pitchfork bifurcation in Section 8.2.2 and to the so-called genetic model, which displays a pure noise-induced transition, in Section 8.2.3. This analysis of steady state behavior will be followed by a study of the critical dynamics of noise-induced transitions in Section 8.3, as well as a study of the mean first passage times between noise-induced states in the large noise limit. We will discuss methods to deal with colored noise in Section 8.4. Section 8.5 is devoted to a discussion of Poisson white noise, which is the appropriate white noise model in those situations where the fluctuating external parameter has to be positive for physical reasons. Section 8.6 shows that monochromatic periodic modulations do not induce the rich variety of transition behavior seen for broadband random perturbations. The current experimental evidence for noise-induced transitions is reviewed in Section 8.7.

8.2 Transition phenomena in fluctuating environments

8.2.1 Nonlinear systems and external noise

For the sake of clarity we will discuss the effect of external noise for as simple a situation as possible. We will therefore assume that the system has the following three properties: (i) It is spatially homogeneous. This corresponds to the limit of fast transport. (ii) The system is macroscopic and can be described by intensive variables. This corresponds to the thermodynamic limit and implies that any finite-size effects, such as internal fluctuations whose amplitude scales with an inverse of the system size, can be neglected. (iii) The state of the system can be described by one variable. This is only a point of mathematical convenience; explicit analytical results can be obtained for one variable systems.

These three properties imply that the evolution of the system can be modelled by the following kinetic equation:

$$\dot{x}(t) = f(x, \lambda) = h(x) + \lambda g(x). \tag{8.2.1}$$

Here x denotes the state of the system at time t, for instance the concentration of a chemical species. The parameter λ describes the effect of the environment on the system, the coupling between the system and the environment. It is the bifurcation parameter for the system and it is this external parameter which we will consider below to be a fluctuating quantity, modelling the environmental fluctuations. All other external parameters are for simplicity assumed to have fluctuations of negligible amplitude and are not explicitly written in (8.2.1). In most applications f is nonlinear in the state variable x but linear in the bifurcation parameter, the external parameter λ. The case of nonlinear bifurcation parameters can be treated as is shown in Horsthemke and Lefever (1984) and Horsthemke, Doering, Lefever and Chi (1985). Associated with (8.2.1) is a typical time scale, t_{sys}, which is the characteristic time scale for the macroscopic evolution of the system.

To model the fluctuations of the environment, we start with the experimental fact that the environment is noisy and can be modelled by a random process. The exact origin of the irregular variations in the external constraints need not concern us here. The fluctuations of the environment manifest themselves on the level of the phenomenological description (8.2.1) as fluctuations in the bifurcation parameter λ. Thus to take into account the influence of external noise we replace the parameter λ in (8.2.1) by a random process λ_t. In other words, (8.2.1) describes the system for a deterministic environment and represents the limit of vanishing noise amplitude. The fact that the external parameter is a stochastic process for a randomly fluctuating environment implies obviously that the state of the system at a given instant of time is a random variable. The state is no longer characterized by a simple number but by a probability distribution. To model the external noise, we assume that the environment has the following three properties:

(i) It is stationary, i.e. in particular

$$\langle \lambda_t \rangle = \lambda, \quad \langle (\lambda_t - \lambda)(\lambda_{t+\tau} - \lambda) \rangle = C(\tau). \tag{8.2.2}$$

Here $\langle\ \rangle$ denotes the mathematical expectation or average. Stationarity is assumed in order to assess the effects of noise separately from any effects due to a systematic evolution of the environment. Furthermore, this assumption is fulfilled in most applications to a good degree of approximation. We write the external noise as

$$\lambda_t = \lambda + Z_t, \tag{8.2.3}$$

where Z_t is now a process with mean value zero.

(ii*a*) Unless a different type of noise is deliberately applied to the system, it is reasonable to assume that Z_t is Gaussian distributed. The basis for this assumption is the Central Limit Theorem. The external parameter often represents the cumulative effect of a large number of small, additive contributions, which are only weakly coupled.

(ii*b*) In experiments it is of course possible to impose any desired probability distribution on the external parameter. Besides the Gaussian distribution, the so-called dichotomous noise is frequently chosen. In this case, Z_t takes on only two values,

$$Z_t \in \{\Delta_-, \Delta_+\}. \tag{8.2.4}$$

Such a noise is easy to generate electronically.

(iii) Since we are interested in macroscopic systems, we will observe the system usually only on macroscopic time scales. It is then reasonable to assume that Z_t is a Markov process. Indeed, it is found in applications that the environment fulfils the Markov property to a satisfactory degree of approximation. Furthermore, we expect that Markovian noise qualitatively exhausts all the physics of the real-noise case: (i) The system is non-Markovian and is correlated with its environment. (ii) The time evolution of the system is smooth, i.e. it has differentiable realizations, though the environment may display sharp variations and even jumps.

Properties (i)–(iii) above uniquely specify the noise process. In the case of (ii*a*) we find, in light of Doob's theorem (Doob, 1942), that Z_t is given by a stationary Ornstein–Uhlenbeck process, i.e. it obeys the following Langevin equation:

$$\dot{Z}_t = -\gamma Z_t + \sigma \xi_t. \tag{8.2.5}$$

This equation was originally introduced (Uhlenbeck and Ornstein, 1930) to model the velocity of a Brownian particle. The random force ξ_t is Gaussian white noise,

$$\langle \xi_t \rangle = 0, \quad \langle \xi_{t+\tau} \xi_t \rangle = \delta(\tau). \tag{8.2.6}$$

Gaussian white noise is a very irregular random process with no memory at all; the correlation function is a Dirac delta function. This implies that Gaussian white noise is a generalized stochastic process (Gelfand and Wilenkin, 1964). Fortunately the use of generalized functions can be avoided in the description of the effect of external noise on nonlinear systems. We can work entirely within the framework of ordinary stochastic processes by exploiting the fact that

$$\int_0^t \xi_s \, ds = W_t. \tag{8.2.7}$$

Here W_t is the Wiener process, which describes the position of a Brownian particle (Chandrasekhar, 1943), and which is characterized by

$$\langle W_t \rangle = 0, \quad \langle W_t W_s \rangle = \min(t, s), \tag{8.2.8}$$

and

$$p(w, t | u, s) = [2\pi(t-s)]^{-1/2} \exp\{-(w-u)^2/2(t-s)\}. \tag{8.2.9}$$

Here $p(w, t|u, s)$ is the probability density to find a value w for W_t given that $W_s = u$.

Multiplying (8.2.5) formally by dt and using (8.2.7), we obtain the stochastic differential equation (SDE)

$$dZ_t = -\gamma Z_t \, dt + \sigma \, dW_t. \tag{8.2.10}$$

The probability density of Z_t obeys the following evolution equation, a Fokker–Planck equation,

$$\partial_t p(z, t | z_0, 0) = -\partial_z(-\gamma z) p(z, t | z_0, 0) + \tfrac{1}{2} \partial_{zz} \sigma^2 p(z, t | z_0, 0). \tag{8.2.11}$$

The stationary solution of (8.2.11) is given by

$$\bar{p}(z) = [\pi(\sigma^2/\gamma)]^{-1/2} \exp\{-z^2/(\sigma^2/\gamma)\}. \tag{8.2.12}$$

If the Ornstein–Uhlenbeck process is started with this probability density, then it is a stationary process and

$$\langle Z_t \rangle = 0, \quad \langle Z_t Z_{t+\tau} \rangle = (\sigma^2/2\gamma) e^{-\gamma|\tau|} = C(\tau). \tag{8.2.13}$$

The correlation time, which characterizes the memory of the environment, is given by

$$t_{\mathrm{cor}} = \gamma^{-1}. \tag{8.2.14}$$

The power spectrum of Z_t is

$$S(\omega) = (2\pi)^{-1} \int e^{-i\omega\tau} C(\tau) \, d\tau = (\sigma^2/2\pi)(\omega^2 + \gamma^2)^{-1}. \tag{8.2.15}$$

Thus in case (ii*a*), i.e. the environment is described by an Ornstein–Uhlenbeck process, the evolution of the system is governed by the SDE

$$\dot{X}_t = h(X_t) + (\lambda + Z_t) g(X_t). \tag{8.2.16}$$

If we multiply (8.2.16) by dt we obtain

$$dX_t = [h(X_t) + \lambda g(X_t)] \, dt + Z_t g(X_t) \, dt, \tag{8.2.17}$$

$$dZ_t = -Z_t \, dt + \sigma \, dW_t. \tag{8.2.18}$$

Here time has been rescaled, so that $t_{\mathrm{cor}} = \gamma^{-1} = 1$.

The joint probability density $p(x,z,t)$ for the system and its environment obeys the following Fokker–Planck equation:

$$\partial_t p(x,z,t) = -\partial_x[h(x)+\lambda g(x)+zg(x)]p(x,z,t) \\ -\partial_z(-z)p(x,z,t)+(\sigma^2/2)\partial_{zz}p(x,z,t). \qquad (8.2.19)$$

After the system has been coupled to the environment for a sufficiently long time, it will, in general, settle down to a steady state. To describe this state, we have to find the stationary solution of (8.2.19), i.e. $\partial_t p = 0$. It turns out that this is, in most cases, an intractable problem. However, the stationary solution can be obtained in a physically relevant limiting case. Often the external noise is very rapid on the macroscopic time scale, i.e. $t_{\text{cor}} \ll t_{\text{sys}}$. This fact suggests that most of the physics of the problem is captured if we pass to the idealization of $t_{\text{cor}} = 0$, the so-called white noise limit. In other words, we replace the real system and environment by an approximate one in which the external noise has no memory. The white noise idealization has to be approached, however, with some circumspection, so as indeed not to lose any of the essential features of the real system. The most convenient way to take the white noise limit is to 'speed up' the noise, i.e. replace the time scale of the Ornstein–Uhlenbeck process by t/ε^2, $Z_t \to Z_{t/\varepsilon^2}$, where $\varepsilon \to 0$, and scale up the amplitude of the noise by a factor $1/\varepsilon$ (Blankenship and Papanicolaou, 1978). In this limit the power spectrum $S^\varepsilon(\omega)$ of the external noise $\varepsilon^{-1}Z_{t/\varepsilon^2}$ converges to a flat spectrum:

$$S^\varepsilon(\omega) = (\sigma^2/2\pi)(\varepsilon^4\omega^2+1)^{-1} \to \sigma^2/2\pi. \qquad (8.2.20)$$

Thus the limit $\varepsilon \to 0$ does indeed correspond to the white noise limit. The above scaling procedure amounts to replacing (8.2.17) and (8.2.18) by the following set of SDEs:

$$\mathrm{d}X_t = [h(X_t)+\lambda g(X_t)]\,\mathrm{d}t + \varepsilon^{-1}Z_t g(X_t)\,\mathrm{d}t, \qquad (8.2.21)$$

$$\mathrm{d}Z_t = -\varepsilon^{-2}Z_t\,\mathrm{d}t + \varepsilon^{-1}\sigma\,\mathrm{d}W_t. \qquad (8.2.22)$$

The Fokker–Planck equation (8.2.19) reads now:

$$\partial_t p^\varepsilon(x,z,t) = -\partial_x\,[h(x)+\lambda g(x)]p^\varepsilon(x,z,t) - \varepsilon^{-1}\partial_x z g(x)p^\varepsilon(x,z,t) \\ +\varepsilon^{-2}[\partial_z z + (\sigma^2/2)\partial_{zz}]p^\varepsilon(x,z,t). \qquad (8.2.23)$$

The form of (8.2.23) suggests the following perturbation expansion for $p^\varepsilon(x,z,t)$:

$$p^\varepsilon(x,z,t) = p_0(x,z,t) + \varepsilon p_1(x,z,t) + \varepsilon^2 p_2(x,z,t) + \cdots. \qquad (8.2.24)$$

This perturbation expansion is known as the wide-band perturbation expansion (Horsthemke and Lefever, 1980; Horsthemke and Lefever, 1984). We will skip the technical details here and just present the main results. It turns out that to lowest order the joint probability density factorizes,

$$p_0(x,z,t) = p(x,t)\bar{p}(z), \qquad (8.2.25)$$

as it should. The system and environment are independent at some instant of time in the white noise case. $\bar{p}(z)$ is given by (8.2.12), and $p(x,t)$ obeys the Fokker–Planck equation

$$\partial_t p(x,t) = -\partial_x[h(x) + \lambda g(x) + (\sigma^2/2)g'(x)g(x)]p(x,t) + (\sigma^2/2)\partial_{xx}g^2(x)p(x,t). \tag{8.2.26}$$

(Prime denotes derivative with respect to x.) The following Stratonovic SDE corresponds to this Fokker–Planck equation (Arnold, 1974):

$$\mathrm{d}X_t = [h(X_t) + \lambda g(X_t)]\,\mathrm{d}t + \sigma g(X_t)\circ \mathrm{d}W_t. \tag{8.2.27}$$

Here $\circ$ denotes the Stratonovic integral (Arnold, 1974; Schuss, 1980).

This shows that in the white noise limit the system is described by a random process, which is the solution of a Stratonovic SDE and thus a Markovian diffusion process. The form of this Stratonovic SDE is the one which would be obtained, if the external parameter λ in the phenomenological rate equation (8.2.1) were simply replaced by the white noise process $\lambda + \sigma\xi_t$. The steady state probability density of the system coupled to a white noise environment, i.e. the stationary solution of (8.2.27), can be written down exactly:

$$\bar{p}(x) = Ng^{-1}(x)\exp\{(2/\sigma^2)\int^x \mathrm{d}y[h(y) + \lambda g(y)]/g^2(y)\}. \tag{8.2.28}$$

Here N is a normalization constant and $\bar{p}(x)$ exists if N is finite.

8.2.2 Noise-induced transitions

The phenomenon of nonequilibrium transitions for nonfluctuating external constraints is by now a familiar one and well understood (Haken, 1978; Nicolis and Prigogine, 1977; Swinney and Gollub, 1981). A transition corresponds to a qualitative change in the state of the system. For instance, a steady state may lose stability for a certain value of the external parameter, e.g. the conducting state in the Rayleigh–Bénard system at a critical value of the temperature difference, and new branches of stable states bifurcate, e.g. the convecting-rolls state for the Rayleigh–Bénard system. In direct analogy we say that a nonequilibrium transition occurs in a system with noise, if the state of the system changes qualitatively. As discussed above, the state of the system is a random variable and is thus characterized by a probability distribution. In order to determine when a transition occurs, we need to monitor $\bar{p}(x)$ for qualitative changes. To do so, we need to find a suitable indicator. As we deal with a probability distribution, moments come to mind. Consider, however, the simple Landau equation

$$\dot{x} = \lambda x - x^3 = f(x,\lambda). \tag{8.2.29}$$

Here $x\in(-\infty,\infty)$ and $\lambda\in(-\infty,+\infty)$. This equation is frequently encountered in modelling equilibrium and nonequilibrium critical phenomena (Haken, 1978; Lifshitz and Pitaevskii, 1980). We will come back to this particular equation in the next section. The Landau equation is the normal form for the pitchfork bifurcation. In the deterministic case the steady states $\bar{x}$ are given by

$$\bar{x}=0, \text{ and } \bar{x}= \pm\lambda^{1/2} \text{ if } \lambda>0. \tag{8.2.30}$$

A stability analysis shows that $\bar{x}=0$ is stable for $\lambda<0$ and loses its stability at a critical value $\lambda_c=0$, where two new stable branches $\bar{x}=\pm\lambda^{1/2}$ bifurcate. Let us now take into account other rapid degrees of freedom in the system by adding a white noise term

$$\dot{X}_t=\lambda X_t-X_t^3+\sigma\xi_t. \tag{8.2.31}$$

We have

$$\bar{p}(x)=N\exp\{(2/\sigma^2)(\lambda x^2/2-x^4/4)\}. \tag{8.2.32}$$

It is easily verified that for $\lambda<\lambda_c=0$, $\bar{p}(x)$ consists of a single peak centered on $x=0$, i.e. the stable deterministic steady state. For $\lambda>0$, the stationary probability density consists of two peaks centered on $+\lambda^{1/2}$ and $-\lambda^{1/2}$, i.e. again on the stable steady states. Clearly, the system described by (8.2.31) undergoes at $\lambda_c=0$ a transition completely similar to the deterministic system (8.2.29). However, this transition is *not* reflected in the moments of the stationary probability. The mean value and all higher odd moments are zero for all values of the bifurcation parameter λ, and the even moments are infinitely often differentiable with respect to the bifurcation parameter. Clearly, the transition of the system cannot be detected by monitoring the moments. Moments are *not* a reliable indicator of nonequilibrium transitions; the integration $\int dx\, x^n\bar{p}(x)$ washes out a lot of information about the state of the system. Furthermore, moments are also a bad choice on general theoretical grounds, since moments often do not uniquely characterize a probability distribution (Prohorov and Rozanov, 1969).

It is obvious from the above example that the appropriate indicators of a transition are the extrema of the probability density: (i) They reflect the qualitative features of $\bar{p}(x)$, for instance if $\bar{p}(x)$ is single-humped or multi-humped. (ii) It can be shown that the extrema x_m of the stationary probability density converge towards the deterministic steady states $\bar{x}$ as the noise is turned off; see below. (iii) The maxima are the most probable values and are preferentially observed in an experiment; they correspond, so to speak, to the 'phases' of the system. It should be stressed that it is not a difficult task to measure the stationary probability density of a system experimentally; see (Griswold and Tough, 1987; Smythe, Moss and McClintock, 1983). The experimental observation and detection of transitions in noisy systems is not more difficult than in 'noise-free' systems. In the white noise limit the extrema of the steady state probability density of the system are the zeroes of the following equation:

$$[h(x_m)+\lambda g(x_m)]-(\sigma^2/2)g'(x_m)g(x_m)=0. \tag{8.2.33}$$

For additive noise, i.e. $g(x) = \text{const.}$, (8.2.33) reduces to the equation for the deterministic steady states and the extrema coincide always with the deterministic steady states. This is an additional argument to consider the extrema of the stationary probability density and not its moments. In the case of state dependent or multiplicative noise, i.e. $g(x) \neq \text{const.}$, we have the following situation. For σ^2 very small, the second term in (8.2.33) is negligible and the extrema of the stationary probability density of the system coincide approximately with the steady states of the deterministic system, $x_m \simeq \bar{x}$. However, as the intensity σ^2 of the white noise is increased, the term due to the noise will become more and more important. If $g'g$ is more nonlinear than h or g, then the extrema x_m of the stationary probability density can be very different, in number and location, from the deterministic steady states. In other words, a qualitative change in $\bar{p}(x)$, i.e. a transition in the noisy system, can occur as the noise intensity is varied. Since this change in the steady state behavior of the system arises without any changes in the average values of the external constraints, we call this transition phenomenon a noise-induced transition. In the following two subsections we present two typical examples of noise-induced transitions.

8.2.3 *The imperfect pitchfork bifurcation with noise*

As already mentioned above, the Landau equation or the pitchfork bifurcation is frequently encountered in nonlinear systems. The reason is that the pitchfork bifurcation is a bifurcation of low codimension. The codimension measures the complexity of a bifurcation, so to speak. It corresponds to the number of secondary parameters that have to be adjusted to a particular value, in order for the bifurcation to occur, as a distinguished external parameter, the bifurcation parameter, is varied (Golubitsky and Schaeffer, 1985). A bifurcation has codimension zero if a singular point, i.e. a steady state at which the Jacobian of the system becomes singular, will typically be encountered as the distinguished parameter is varied, without the need to adjust any secondary parameters. The only codimension zero bifurcation of steady states is the limit point and is described by the normal form:

$$\lambda - x^2 = 0. \tag{8.2.34}$$

A bifurcation has codimension one if it is necessary to adjust a single secondary parameter to a particular value in order to encounter the singularity as λ is varied. High codimension bifurcations require that a large number of secondary parameters take on particular values, and for this reason they are not very likely to be observed in experiments. The pitchfork bifurcation is a codimension two bifurcation, and it is the only codimension two bifurcation that mediates a continuous transition between steady states. Many different mathematical models contain a given bifurcation. It can be shown that all of these models are in some sense equivalent to certain polynomials, called normal forms. A normal form for the pitchfork bifurcation is the Landau equation:

$$\lambda x - x^3 = 0. \tag{8.2.35}$$

The variable x characterizes the state of the system and denotes generally the deviation from some reference state. It can therefore be positive or negative. As already mentioned, the pitchfork bifurcation has codimension two. So two secondary parameters must in general be adjusted to achieve the singularity at $x = 0, \lambda = 0$. Small perturbations will generally destroy this singularity. Close to the singularity, it can be shown that all analytic perturbations are equivalent to those generated by two terms with coefficients α_0 and α_2 in the 'universal unfolding' of the normal form (8.2.35):

$$f(x, \lambda) = \alpha_0 + \lambda x + \alpha_2 x^2 - x^3. \tag{8.2.36}$$

It is interesting to note that the unfolding of the pitchfork bifurcation (8.2.36) contains the transcritical and hysteresis bifurcations, which are the only codimension one bifurcations giving rise to continuous transitions between steady states (Golubitsky and Schaeffer, 1985). Thus let us consider the following model system:

$$\dot{x} = \alpha_0 + \lambda x + \alpha_2 x^2 - x^3. \tag{8.2.37}$$

This model system displays all the codimension one and two bifurcations giving rise to continuous transitions between steady states. The perfect pitchfork bifurcation occurs at

$$\lambda_c = 0, \quad x_c = 0, \quad \alpha_0 = \alpha_2 = 0, \quad \text{pitchfork, codim 1.} \tag{8.2.38}$$

If α_0 and α_2 are small, but nonzero, we will say that the system (8.2.37) displays an imperfect pitchfork transition. The locus of transcritical bifurcations of the system is the axis $\alpha_0 = 0$:

$$\lambda_c = 0, \quad x_c = 0, \quad \alpha_0 = 0, \quad \text{transcritical bifurcations, codim 1.} \tag{8.2.39}$$

The curve $\alpha_0 = \alpha_2^3/27$ is the locus of hysteresis bifurcations:

$$\lambda_c = -\alpha_2^2/3, \quad x_c = \alpha_2/3, \quad \alpha_0 = \alpha_2^3/27, \quad \text{hysteresis bifurcations, codim 1.} \tag{8.2.40}$$

Let us now study the effect of noise on the unfolded Landau model (8.2.37). We will assume that the parameters λ, α_0 and α_2 display independent white noise fluctuations:

$$\mathrm{d}X_t = [\alpha_0 + \lambda X_t + \alpha_2 X - X_t^3]\mathrm{d}t + \sigma_0 \,\mathrm{d}W_t^{(0)} + \sigma_1 X_t \circ \mathrm{d}W_t^{(1)} + \sigma_2 X_t^2 \circ \mathrm{d}W_t^{(2)}. \tag{8.2.41}$$

We will further assume that the intensity σ_i of the parameter fluctuations is small, so that the probability to be outside the range of validity of the unfolding of the pitchfork bifurcation is small. If the intensity of the white noise fluctuations in the parameter α_2 becomes too large, infinity will cease to be a natural boundary and explosion can occur. However, this reflects only the fact that in the derivation of a normal form higher order saturating terms, like $-x^4$ or $-x^5$, are neglected since they are small near the transition point $x=0$; only the lowest order nontrivial nonlinear terms are retained. If fluctuations cause the system to take on large values with non-negligible probability, then an appropriate number of higher terms have to be retained in the model.

According to (8.2.33) the extrema of the stationary probability density of (8.2.40) are given by:

$$[\alpha_0 + \lambda x_m + \alpha_2 x_m^2 - x^3] - \tfrac{1}{2}\sigma_1^2 x_m - \sigma_2^2 x_m^3 = 0, \tag{8.2.42}$$

or

$$\alpha_0 + (\lambda - \sigma_1^2/2)x_m + \alpha_2 x_m^2 - (1+\sigma_2^2)x_m^3 = 0. \tag{8.2.43}$$

Dividing (8.2.43) by $1+\sigma_2^2$, we obtain

$$\tilde{\alpha}_0 + \tilde{\lambda} x_m + \tilde{\alpha}_2 x_m^2 - x_m^3 = 0, \tag{8.2.44}$$

with

$$\tilde{\alpha}_0 = \alpha_0/(1+\sigma_2^2), \quad \tilde{\lambda} = (\lambda - \sigma_1^2/2)/(1+\sigma_2^2),$$
$$\tilde{\alpha}_2 = \alpha_2/(1+\sigma_2^2). \tag{8.2.45}$$

Thus, in the presence of noise the extrema of the unfolded Landau model are still given by the universal unfolding of the pitchfork bifurcation; however, the coefficients in the unfolding are 'renormalized' by the noise. From (8.2.38) we obtain that (8.2.44) shows a perfect pitchfork bifurcation at (Figure 8.1)

$$\tilde{\lambda}_c = 0, \quad x_m = 0, \quad \tilde{\alpha}_0 = \tilde{\alpha}_2 = 0,$$

i.e.

$$\lambda = \sigma_1^2/2, \quad x_m = 0, \quad \alpha_0 = \alpha_2 = 0. \tag{8.2.46}$$

In other words, the pitchfork bifurcation is postponed. The locus of transcritical bifurcations in the noise-driven system is $\tilde{\alpha}_0 = 0$:

$$\tilde{\lambda}_c = 0, \quad x_m = 0, \quad \tilde{\alpha}_0 = 0,$$

i.e.

$$\lambda = \sigma_1^2/2, \quad x_m = 0, \quad \alpha_0 = 0. \tag{8.2.47}$$

Thus the locus of the transcritical bifurcation also is unchanged, $\alpha_0 = 0$, but the bifurcation itself is again postponed by $\sigma_1^2/2$. In fact, the notion of noise-induced transitions was first introduced in studies of transcritical bifurcations (Horsthemke and Lefever, 1977; Horsthemke and Malek-Mansour, 1976). While the pitchfork bifurcation and the transcritical bifurcations are only affected by the noise in the bifurcation parameter λ, the hysteresis bifurcations are influenced by fluctuations both in λ and α_2. As expected, fluctuations in the coefficient α_0 have no effect on the various transition phenomena, since they

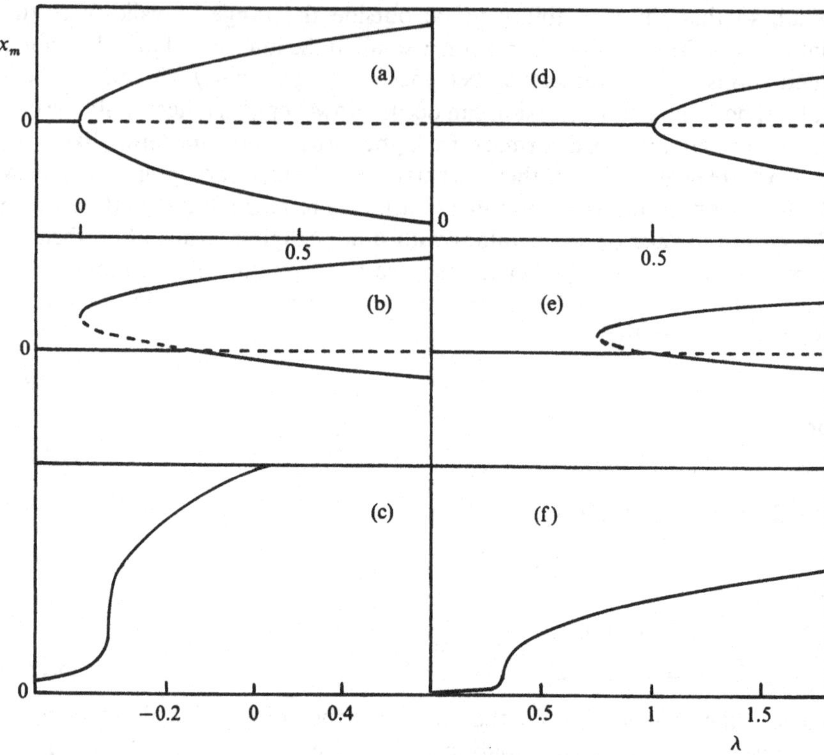

Figure 8.1. Comparison of bifurcation diagrams in the absence and presence of external noise. In (a), (b) and (c), the deterministic steady states obtained by putting $\sigma_1^2 = \sigma_2^2 = 0$ in (8.2.42) and corresponding to the pitchfork bifurcation, transverse bifurcation and hysteresis (see (8.2.38), (8.2.39) and (8.2.40)) have been plotted as a function of λ. In (b) and (c), $\alpha_2 = 1$. In (d), (e) and (f), the extrema of the probability density have been plotted for $\sigma_1^2 = \sigma_2^2 = 1$, and $\alpha_2 = 1$. ($\alpha_0 = 1/54$ in f). The broken lines correspond to the unstable deterministic steady states or to the minima of the probability density.

give rise to additive noise. The locus of the hysteresis bifurcation is $\tilde{\alpha}_0 = \tilde{\alpha}_2^3/27$:

$$\tilde{\lambda}_c = -\tilde{\alpha}_2^2/3, \quad x_m = \tilde{\alpha}_2/3, \quad \tilde{\alpha}_0 = \tilde{\alpha}_2^3/27,$$

i.e.

$$\left.\begin{aligned} &\lambda = -\tfrac{1}{3}\alpha_2^2/(1+\sigma_2^2)^2 + \tfrac{1}{2}\sigma_1^2, x_m = \tfrac{1}{3}\,\alpha_2/(1+\sigma_2^2), \\ &\alpha_0 = \tfrac{1}{27}\alpha_2^3/(1+\sigma_2^2)^2. \end{aligned}\right\} \tag{8.2.48}$$

The locus of the hysteresis bifurcation is changed by fluctuations in the coefficient α_2, the critical value of x is decreased by these fluctuations and they, together with the fluctuations in the bifurcation parameter, delay the hysteresis bifurcation.

The Landau model, (8.2.37), is an example for those systems where the noise-induced term $g'g$ in (8.2.33) for the extrema does not give rise to higher nonlinearities than already contained in the equation for the deterministic steady states. In such cases, the effect of parametric noise is a simple 'renormalization' of the coefficients and the bifurcation parameter of the system. This leads to either postponement, as in the example above, or advancement of the various transition phenomena of the deterministic system. In the following subsection we will present a model system where external noise induces new transition phenomena.

8.2.4 *Noise-induced bistability in the genetic model*

We consider a constant population of haploid individuals and assume that two alleles, say A and a, are competing for a particular gene locus. Let x and $1-x$ be the frequency of the alleles in the population. We assume that there are two mechanisms that change the frequencies of the alleles, namely mutation between the two alleles and natural selection. Then the kinetic equation for x reads (Kimura and Ohta, 1971), assuming equal mutation rates:

$$\dot{x} = 0.5 - x + \lambda x(1-x). \qquad (8.2.49)$$

Here $x \in [0, 1]$ and $\lambda \in (-\infty, +\infty)$. The parameter λ is the selection coefficient. The following chemical model reaction scheme can be associated with (8.2.49):

$$A + X + Y \rightleftarrows 2Y + A^*,$$

$$B + X + Y \rightleftarrows 2X + B^*,$$

where, A, B, A* and B* are assumed to be in large excess. The total concentration of X and Y remains constant in the above reaction scheme. After some scaling (Horsthemke and Lefever, 1984), we find that the fraction of X in the system obeys (8.2.49). The steady states of (8.2.49) are

$$\bar{x} = (2\lambda)^{-1}[\lambda - 1 + (1 + \lambda^2)^{1/2}]. \qquad (8.2.50)$$

In other words, for each value of the extremal parameter λ there exists a unique steady state, which is globally stable as is easily verified. Thus the genetic model (8.2.49) does not display any transition for deterministic external constraints. Since the fitness of an allele depends on how well it is adapted to a particular environment, the selection coefficient becomes a random quantity if the environment of the haploid population fluctuates. Similarly in the context of the chemical model, if the concentration of A and B fluctuates, the parameter λ, which is essentially the difference of these concentrations, has to be replaced by a random process. In the Gaussian white noise idealization, we obtain, using (8.2.33), the following equation for the extrema of the stationary probability density of the genetic model:

$$0.5 - x_m + \lambda x_m(1 - x_m) - (\sigma^2/2)x_m(1 - x_m)(1 - 2x_m) = 0. \qquad (8.2.51)$$

Note that this model belongs to the class of systems where the noise-induced term $g'g$ of (8.2.33) produces a higher nonlinearity than the one encountered in

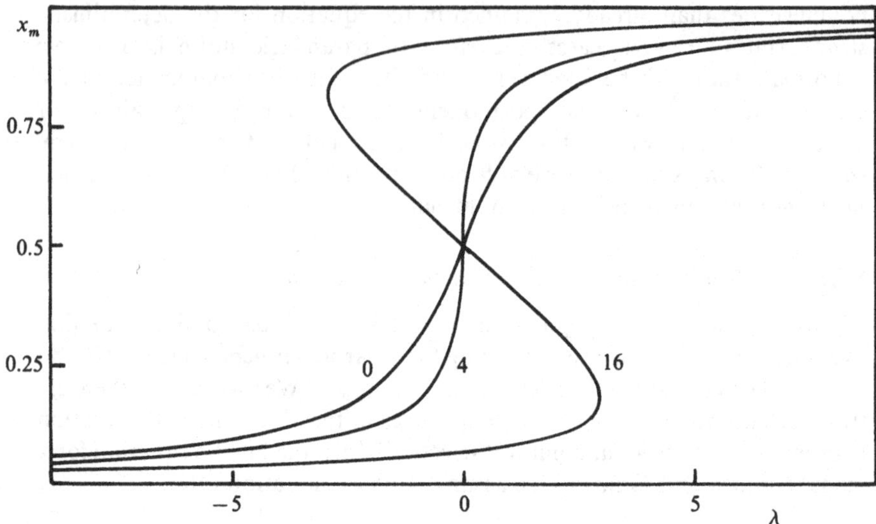

Figure 8.2. Noise-induced transition in the genetic model. The extrema x_m of the probability density given by (8.2.51) are plotted as a function of λ for increasing values of σ^2. For $\sigma^2 > 4$ one has appearance of the hysteresis loop.

the equation for the deterministic steady states. The term $g'g$ changes the equation from a second order polynomial to a third order polynomial. Consider the case that on the average neither allele confers an advantage to the individual, i.e. $\lambda = 0$. Then in the deterministic case, the steady state is of course: $\bar{x} = 0.5$. From (8.2.51) we find for the system with external noise that

$$x_{m,0} = 0.5, \quad \text{and} \quad x_{x,\pm} = [1 \pm (1 - (4/\sigma^2))^{1/2}]/2 \qquad \text{for } \sigma^2 \geqslant \sigma_c^2 = 4. \tag{8.2.52}$$

Thus the system with external white noise has a noise-induced critical point at $\lambda_c = 0$, $\sigma_c^2 = 4$, $x_c = 0.5$ (Arnold, Horsthemke and Lefever, 1978). For $\sigma^2 < 4$, the stationary probability density has a single peak centered on $x_{m,0} = 0.5$. At $\sigma^2 = 4$, this maximum becomes a double maximum and for $\sigma^2 > 4$ it splits in two; the probability density becomes double-humped. The external noise has induced bistable behavior! For deterministic external constraints no bistability is possible. Like equilibrium critical behavior, the noise-induced critical point is characterized by critical exponents. According to our discussion in Section 8.2.2, the order parameter is given by

$$m = x_{m,+} - 0.5. \tag{8.2.53}$$

We have

$$m \sim (\sigma^2 - \sigma_c^2)^{1/2} \quad \text{for } \lambda = \lambda_c = 0, \quad \text{i.e. } \beta = \tfrac{1}{2}, \tag{8.2.54}$$

$$m \sim \lambda^{1/3} \quad \text{for } \sigma^2 = \sigma_c^2 = 4, \quad \text{i.e. } \delta = 3, \tag{8.2.55}$$

and

$$(\partial m/\partial \lambda)|_{\lambda=0} \sim |\sigma^2 - \sigma_c^2|^{-1}, \quad \text{i.e.} \quad \gamma = \gamma' = 1. \tag{8.2.56}$$

We find the classical values for the critical exponents, which is expected, since our model excludes any spatial inhomogeneities. In the next section we will study the dynamics of noise-driven systems, in particular the dynamics near the noise-induced critical point.

8.3 Dynamics of noise-driven systems

The dynamics of a system near an equilibrium critical is characterized by so-called critical slowing down. The closer the critical point is approached, the more sluggish the system becomes. Consider, for example, the Landau model (8.2.29). Linearizing around the equilibrium state $\bar{x} = 0$, we find that for $\lambda < 0$ small perturbations δx decay as

$$\dot{\delta x} = \lambda\, \delta x \tag{8.3.1}$$

or

$$\delta x(t) = \delta x(0) \exp(\lambda t). \tag{8.3.2}$$

In other words, the exponential decay of perturbations becomes slower and slower as the critical point is approached, i.e. $\lambda \to 0$. At the critical point itself, the linear term vanishes and the temporal evolution of small perturbations is governed by the nonlinear term $-x^3$, resulting in algebraic decay of deviations from equilibrium. This is the phenomenon of critical slowing down. Let us, however, now take into account the neglected degrees of freedom of the system and study the frequently used stochastic Landau equation, already introduced above,

$$\dot{X}_t = \lambda X_t - X_t^3 + \sigma \xi_t. \tag{8.3.3}$$

Here the amplitude σ of the (internal) noise is kept fixed. As discussed above, the critical point is not affected by the additive noise. The temporal evolution of the probability density of the system is of course governed by the following Fokker–Planck equation:

$$\partial_t p(x,t) = -\partial_x(\lambda x - x^3) p(x,t) + (\sigma^2/2) \partial_{xx} p(x,t). \tag{8.3.4}$$

The dynamics of the system can be studied by finding the lowlying eigenvalues of the Fokker–Planck operator. The lowest eigenvalue is $\mu_0 = 0$ and corresponds to the steady state. The next eigenvalue μ_1 characterizes the long-term approach to equilibrium. In general, it is an intractable problem to calculate the eigenvalues exactly, but variational principles and Weinstein's intermediate theorem can be used to obtain upper and lower bounds for the eigenvalues (Brand, Schenzle and Schröder, 1982; Weinstein, 1934). Using the supersymmetry of the Fokker–Planck equation (Bernstein and Brown, 1984), Doering (1986) has recently derived the following result for the eigenvalue μ_1 of (8.3.4) at the critical point, $\lambda = 0$,

$$0.56\,\sigma \leqslant \mu_1 \leqslant 0.98\,\sigma. \tag{8.3.5}$$

Thus, as Doering stresses, there is *no* critical slowing down in terms of the spectrum in the Landau model. The spectrum displays no anomolous behavior near the transition. The usual statement that the Landau model exhibits critical slowing is based upon the study of either a linearized or a deterministic system. If fluctuations are included in the model, it is only in the *linearized* problem that an eigenvalue tends to zero as the critical point is approached. It is worthwhile to repeat that the full, nonlinear model shows no anomalous behavior near the equilibrium critical point.

Noise-induced transitions are an *intrinsically nonlinear* phenomenon. As is clear from the form of (8.2.33), in particular the noise-induced term $g'g$, noise-induced transitions result from an interplay of the nonlinear kinetics of the system and the external noise. Thus, if we want to study the dynamics near noise-induced transitions, we cannot analyze a linearized problem. There is no corresponding deterministic state with respect to which a linearization could be carried out. Our task is thus to study on the full, nonlinear problem the relaxation of the probability density $p(x,t)$ from one form to another as the noise-induced critical point is approached. As already mentioned, it is generally an intractable problem to find the time-dependent solution of a Fokker–Planck equation. We will therefore only consider the following class of models, of which the genetic model is a member: (i) The boundaries of the state space are entrance boundaries (Horsthemke and Lever, 1984; Karlin and Taylor, 1981). (ii) The state space $[b_1, b_2]$ is a finite interval, or the real line. (iii) $f(x) = h(x) + \lambda g(x)$ has the following property: $f(z+u) = -f(z-u)$, where z is the midpoint of the interval. (iv) $g(x)$ has the following property: $g(z+u) = g(z-u)$.

The eigenvalue prcblem for a Fokker–Planck equation reads

$$-\mu\psi(x) = -\partial_x f(x)\psi(x) + (\sigma^2/2)\partial_{xx} g^2(x)\psi(x), \tag{8.3.6}$$

with no flux boundary conditions. Since the solution of a Fokker–Planck equation is a probability density, it must in particular be integrable. Thus the eigenvalue problem (8.3.6) has to be solved on the space of integrable functions, $L_1(b_1, b_2)$. The eigenvalue problem for the Fokker–Planck equation can be cast into a Sturm–Liouville problem (Wong, 1964), a problem well studied in the literature (Hille, 1969), by the following transformation

$$\psi(x) = \phi(x)\bar{p}(x). \tag{8.3.7}$$

Here $\bar{p}(x)$ is the stationary probability density. Inserting (8.3.7) into (8.3.6) we obtain the following Sturm–Liouville problem:

$$(\sigma^2/2)[g^2(x)\bar{p}(x)\phi'(x)]' + \mu\bar{p}(x)\phi(x) = 0, \tag{8.3.8}$$

with the boundary conditions

$$g^2(x)\bar{p}(x)\phi'(x)|_{x=b_1,b_2} = 0. \tag{8.3.9}$$

Note that there is one crucial difference between the eigenvalue problem for the Fokker–Planck equation and the Sturm–Liouville problem. The Sturm–Liouville problem is usually studied on an L_2 space, a space of square integrable functions (Hille, 1969), whereas the Fokker–Planck equation has to be solved on an L_1 space. Probability densities have to be normalizable, i.e. integrable, but not necessarily square integrable. This gap between the Sturm–Liouville problem and the eigenvalue problem of the Fokker–Planck equation has been bridged by Elliott (1955). Under certain conditions, which our class of models defined above fulfils, the results of the Sturm–Liouville problem can be applied directly to the Fokker–Planck problem. We have then the following facts for our class of systems:

(i) The spectrum is purely discrete.
(ii) The eigenvalues are real and non-negative and

$$0 = \mu_0 < \mu_1 < \cdots < \mu_n < \cdots, \quad \lim \mu_n = +\infty.$$

(iii) The eigenfunctions $\phi_n(x)$ form a complete system in $L_1(b_1, b_2)$.
(iv) The eigenfunction $\phi_n(x)$ has exactly n simple zeroes in (b_1, b_2).

Let us now consider the following situation: in the genetic model both alleles are equally fit on the average, i.e. $\lambda = 0$, and the system is initially prepared to be in the state $x = 0.5$, $p(x,0) = \delta(x)$. Let the intensity of the environmental noise σ^2 be larger than the critical noise intensity, $\sigma_c^2 = 4$. Then, as time goes to infinity, the system will relax from the initially single-humped distribution to the double-humped stationary probability density. Using the facts about the eigenvalue problem, we can write the time-dependent probability density in the following way (Horsthemke and Lefever, 1984),

$$p(x,t|z) = \bar{p}(x) \sum_0^\infty \phi_{2n}(z)\phi_{2n}(x) \exp(-\mu_{2n}t), \quad z = 0.5. \tag{8.3.10}$$

The symmetry properties of f and g imply that the symmetry of the initial probability density $p(x,0) = \delta(x)$ will not be destroyed. In other words, $p(x,t|z)$ will be symmetric with respect to the midpoint z for all times. Only the eigenfunctions with an even index have this symmetry property in light of the fact of their number of zeroes. For the long-term behavior we have

$$p(x,t|z) = \bar{p}(x)[1 + \phi_2(z)\phi_2(x)\exp(-\mu_2 t)]. \tag{8.3.11}$$

We define the critical time t_c to be the time when the initially single-humped probability density develops a double maximum and then becomes double-humped. We find that

$$\begin{aligned} t_c = -\mu_2^{-1} \ln\{ &-k(z)/\phi_2(z)[k(z)\phi_2(z) \\ &+ (\sigma^2/2)g^2(z)\tilde{\phi}_2'(z)]\}, \end{aligned} \tag{8.3.12}$$

where

$$(x - z)k(x) \equiv f(x) - (\sigma^2/2)g'(x)g(x),$$

$$(x - z)\tilde{\phi}_2'(x) \equiv \phi_2'(x).$$

It can be shown that $\phi_2(z)g^2(z)\tilde{\phi}'_2(z)$ does not vanish as σ_c^2 is approached from above. On the other hand, z is a triple root of (8.2.33) at the critical point, which implies that $k(z)$ vanishes for $\sigma^2 \downarrow \sigma_c^2$. Thus we find that t_c diverges logarithmically as the noise-induced critical point is approached from above.

The lifetime of noise-induced states, i.e. their stability, has recently been analyzed by Doering (1986), particularly for systems deep in the bistable regime, i.e. in the limit of large noise. He finds that the lifetime depends essentially on the answers to two questions: (i) Is the stochastic differential equation interpreted as an Ito equation or as Stratonovich equation? (ii) Is the state space finite or infinite? If the SDE

$$\dot{X}_t = f(X_t) + \sigma g(X_t)\xi_t \tag{8.3.13}$$

is a continuous-time version of a certain discrete-time problem, then it should be interpreted as an Ito equation (Arnold, 1974). If it is the white noise limit of a colored-noise problem, then it should be interpreted as a Stratonovic equation according to Section 8.1; see also Arnold (1974). The Fokker–Planck equation associated with the SDE in these two interpretations can be written as

$$\begin{aligned}\partial_t p(x,t) = & -\partial_x[f(x) + (2-v)(\sigma^2/2)g'(x)g(x)]p(x,t) \\ & + (\sigma^2/2)\partial_{xx}g^2(x)p(x,t),\end{aligned} \tag{8.3.14}$$

where $v = 2$ corresponds to the Ito interpretation and $v = 1$ to the Stratonovic interpretation. The difference between the two interpretations is the noise-induced drift $(\sigma^2/2)g'(x)g(x)$. Clearly, this noise-induced drift will become more important for large noise intensities. In fact, Doering finds that, as expected, identical systems will evolve faster in the Stratonovic interpretation. For systems with symmetric double-humped stationary probability densities, like the genetic model, we can use the mean first passage time $\langle T \rangle$ between a maximum and the minimum to characterize the lifetime of the most probable states. For the following discussion it is convenient to rescale the state variable of the genetic model, $x = 0.5(1 + u)$, $u \in [-1, +1]$, and to display explicitly in the equation the deterministic time scale, $\omega = t_{\text{sys}}^{-1} = 1$,

$$\dot{U}_t = -\omega U_t + \sigma(1 - U_t^2)\xi_t, \quad \text{for} \quad \lambda = 0. \tag{8.3.15}$$

In the Ito interpretation of (8.3.15) the noise-induced critical point occurs at $\sigma_c^2 = \omega/2$, and in the Stratonovic interpretation at $\sigma_c^2 = \omega$. As the intensity of the external noise increases, the stationary probability density will be more and more concentrated at the boundaries of the state space. Doering has shown that in the Ito interpretation the mean first passage time $\langle T \rangle$ approaches t_{sys} in the limit of infinite noise intensity; the lifetime of the noise-induced states is thus nonzero even for extremely large noise. The switching between the noise-induced states occurs on the macroscopic time scale of the system. Further, it turns out that the (random) first passage time T is exponentially distributed. This implies that the Ito genetic model reduces to a two-level Markov process, the so-called dichotomous Markov noise or random telegraph signal (see next section), in the infinite noise limit. On the other hand, the mean first passage time $\langle T \rangle$ in the Stratonovic genetic model is on the order of t_{sys} near the noise-induced critical point (Brand, Schenzle and Schröder, 1982) and goes to zero as the noise intensity goes to infinity. The system degenerates into a two-level white noise.

Doering shows further that the nature of the state space affects the results on the lifetime of noise-induced states and that the behavior of systems on an unbounded state space is different. He considers as an example Hongler's model (Hongler, 1979),

$$\dot{U}_t = -\omega \tanh(U_t) + \sigma\xi_t/\cosh(U_t), \quad U \in (-\infty, +\infty), \tag{8.3.16}$$

A noise-induced transition to bistability occurs at $\sigma_c^2 = \omega$ (Ito) and at $\sigma_c^2 = 2\omega$ (Stratonovic). The peaks of the bimodal probability density are uniformly exponentially localized and move away from each other proportionally to $\ln(\sigma)$. For the Stratonovic interpretation of Hongler's model the mean first passage time $\langle T \rangle$ approaches the value $(1.11\,\omega)^{-1}$ as the noise intensity goes to infinity; i.e. the noise-induced states in this model have a nonzero lifetime even in the Stratonovic interpretation for extremely large noise. In the Ito interpretation of Hongler's model the mean first passage time goes to infinity as the noise intensity goes to infinity. The noise-induced states become *absolutely* stable. Doering finds that the second eigenvalue μ_2 of the Fokker–Planck operator, however, remains finite in this limit. Thus the system does not relax infinitely fast to one of the noise-induced states; it takes the system a finite amount of time to diffuse into one of the two noise-induced states.

8.4 The effect of colored noise

In the preceding sections we have used the Gaussian white noise idealization to discuss the behavior of systems coupled to a rapidly fluctuating environment. The white noise idealization was derived via the wide-band perturbation expansion; it represents the lowest order. Continuing this expansion to higher orders affords a first means to attack the difficult problem of analyzing the macroscopic behavior of nonlinear systems driven by colored noise. This procedure shows in particular that the white noise results are robust. The noise-induced transitions which occur in systems coupled to white noise occur also in identical systems coupled to environments whose fluctuations have a small but nonzero correlation time (Horsthemke and Lefever, 1980, 1984).

In the following we will not consider a general type of colored noise, but we will study the case (ii*b*) mentioned in Section 8.2, i.e. the two-level noise case. This noise is also known as the random telegraph signal and is described by the following master equation:

$$\frac{\mathrm{d}}{\mathrm{d}t}\begin{pmatrix} p(\Delta_-, t) \\ p(\Delta_+, t) \end{pmatrix} = \begin{pmatrix} -\beta & \alpha \\ \beta & -\alpha \end{pmatrix} \begin{pmatrix} p(\Delta_-, t) \\ p(\Delta_+, t) \end{pmatrix}. \tag{8.4.1}$$

Here α is the average frequency of transitions from Δ_+ to Δ_- and β is the average frequency of transitions from Δ_- to Δ_+. The stationary solution of (8.4.1) is given by

$$\bar{P}(\Delta_+) = \beta/\gamma, \quad \bar{P}(\Delta_-) = \alpha/\gamma, \tag{8.4.2}$$

where

$$\gamma = \alpha + \beta. \tag{8.4.3}$$

If the dichotomous noise is started with probability (8.4.2), then it is stationary and

$$\langle Z_t \rangle = (\alpha\Delta_- + \beta\Delta_+)/\gamma, \tag{8.4.4}$$

$$C(\tau) = (\alpha\beta/\gamma^2)(\Delta_+ - \Delta_-)^2 \exp(-\gamma|\tau|). \tag{8.4.5}$$

Since we have to impose $\langle Z_t \rangle = 0$, according to (8.2.3), then

$$\alpha\Delta_- = -\beta\Delta_+ \tag{8.4.6}$$

must hold. The evolution equation for the probability density of a system driven by a dichotomous Markov noise is (Horsthemke and Lefever, 1984)

$$\begin{aligned}\partial_t p(x,t) = &-\partial_x[h(x) + \lambda g(x)]p(x,t) + (\alpha\beta/\gamma^2)(\Delta_- - \Delta_+)^2 \partial_x g(x) \\ &\times \int_{-\infty}^{t} dt' \exp\{-[\gamma + \partial_x f(x,\lambda) \\ &+ \gamma^{-1}(\beta\Delta_- + \alpha\Delta_+)\partial_x g(x)](t-t')\}\partial_x g(x) p(x,t'),\end{aligned} \tag{8.4.7}$$

where ∂_x acts on everything to its right up to, and including, $p(x,t)$. The memory kernel in (8.4.7) is a clear indication that the system is described by a non-Markovian process X_t. This agrees with the well known fact that the system is Markovian if and only if it is driven by a white noise process. Contrary to the case of Ornstein–Uhlenbeck-noise, the stationary solution of (8.4.7) can be found in an explicit way:

$$\begin{aligned}\bar{p}(x) = Ng(x)[F_+(x)F_-(x)]^{-1} \exp\Big\{-\gamma \int^x [h(y) + \lambda g(y)] \\ \times [F_+(y)F_-(y)]^{-1} dy\Big\},\end{aligned} \tag{8.4.8}$$

for $x \in U$, $U = [\bar{x}(\lambda + \Delta_-), \bar{x}(\lambda + \Delta_+)]$ and $\bar{p}(x) \equiv 0$ for $x \notin U$. In (8.4.8),

$$F_\pm = [h(x) + (\lambda + \Delta_\pm)g(x)]. \tag{8.4.9}$$

Let us now consider the symmetric dichotomous Markov noise, $-\Delta_- = \Delta_+ = \Delta$, $\alpha = \beta = \gamma/2$. The white noise limit can be discussed directly in this case. The speeding up of the noise corresponds to letting γ go to infinity and the scaling up of the amplitude to letting Δ go to infinity, such that $\Delta^2/\gamma = \text{const.} = \sigma^2/2$. For symmetric dichotomous Markov noise the extrema of the stationary probability density (8.4.8) obey the following equation:

$$\begin{aligned}&[h(x_m)+\lambda g(x_m)]-(\Delta^2/\gamma)g'(x_m)g(x_m)\\&\quad+(2/\gamma)[h(x_m)+\lambda g(x_m)][h'(x_m)\\&\quad+\lambda g'(x_m)]-(1/\gamma)[h(x_m)+\lambda g(x_m)]^2 g'(x_m)/g(x_m)=0.\end{aligned} \tag{8.4.10}$$

This equation has a very interesting structure: The first term equal to zero yields the deterministic steady states. In the white noise limit only the first two terms survive and (8.4.10) reduces indeed to (8.2.33). The last two terms are corrections due to the nonvanishing correlation time of the noise. Thus (8.4.10) illustrates the general result, already mentioned above, that the effects induced by Gaussian white noise are also observed if the system is driven by a nonwhite noise with a sufficiently short correlation time. However, there can be additional modifications of the macroscopic behavior of the system if the correlation increases, i.e. γ decreases, as reflected by the last two terms of (8.4.10).

The advantage of the dichotomous Markov noise is the fact that the steady state behavior of the noise can be determined explicitly for arbitrary noise amplitude and for arbitrary correlation time of the noise. This allows us to construct a 'phase diagram' of the behavior of the system in the Δ–γ plane. Such phase diagrams have been constructed for the genetic model and other model systems (Horsthemke *et al.*, 1985; Horsthemke and Lefever, 1984; Kitahara, Horsthemke and Lefever, 1979; Kitahara, Horsthemke, Lefever and Inaba, 1980).

8.5 Poisson white noise

If we consider asymmetric dichotomous Markov noise and take the white noise limit in the following way,

$$\Delta_+ \to \infty, \quad \alpha \to \infty \text{ such that } \Delta_+/\alpha = \text{const.} = \omega, \tag{8.5.1}$$

Poisson white noise is obtained (van den Broeck, 1983) instead of Gaussian white noise. Note that the requirement (8.4.6) implies that $\Delta_- = -\beta\omega$. Further, a stochastic process on a discrete state space is Markovian, if and only if its waiting time T in an arbitrary state is exponentially distributed (Karlin and Taylor, 1975). For the dichotomous Markov process, the random waiting time in the upper state is exponentially distributed with $\langle T \rangle = 1/\alpha$. Thus the area of a Δ_+-pulse, $T\Delta_+$, is exponentially distributed with a mean of ω. This leads to the following picture of the white noise limit: the noise spends essentially all the time in the state $\Delta_- = -\beta\omega$, i.e. on the baseline. At random times, which form a Poisson process and which occur with an average frequency β, this state is interrupted by a Dirac delta spike with a weight that is a random variable. The weights are exponentially distributed with mean value ω. The Poisson white noise limit of (8.4.7) is

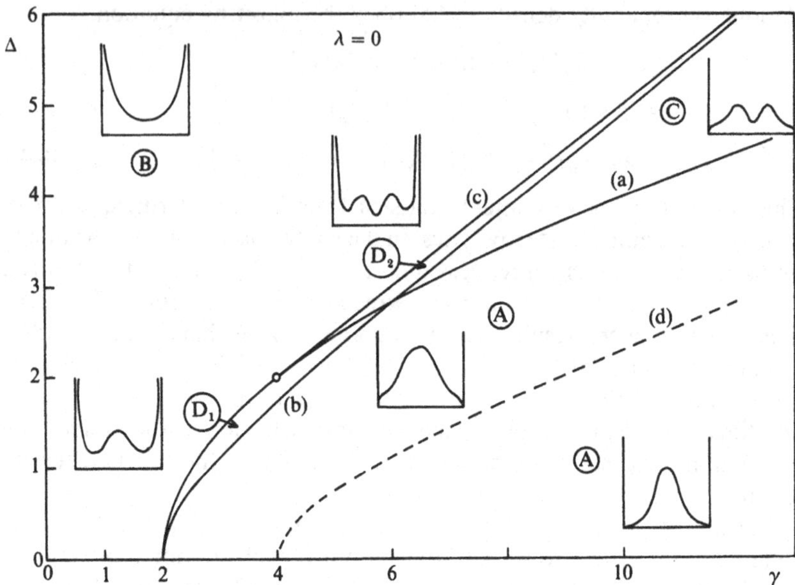

Figure 8.3. Phase diagram in the (Δ–γ)-plane of the genetic model submitted to dichotomous noise. In doman Ⓐ, the probability has a unique extremum (maximum) and a zero tangent at the boundaries of its support. When the broken line (d) is crossed, the behavior of $\bar{p}(x)$ near the boundaries changes, but the maximum remains unique. This peak splits into two on the line (a). This yields the double-peaked densities observed in domains Ⓒ and (D_2). The transition line (b) marks the appearance of the divergence of $\bar{p}(x)$ at the boundaries of its support while (c) corresponds to the disappearance of the maxima.

$$\partial_t p(x,t) = -\partial_x[h(x) + \lambda g(x)]p(x,t) + \beta\omega^2 \partial_x g(x)[1 + \omega\partial_x g(x)]^{-1}\partial_x g(x)p(x,t), \tag{8.5.2}$$

whose stationary solution is given by

$$\bar{p}(x) = N[h(x) + (\lambda - \beta\omega)g(x)]^{-1}\exp\left\{-\omega^{-1}\int^x [h(y) + \lambda g(y)] \times g^{-1}(y)[h(y) + (\lambda - \beta\omega)g(y)]^{-1}\,dy\right\}, \tag{8.5.3}$$

for $x \in U$. The boundaries of the support U are given by

$$g(x) = 0, \quad \text{and} \quad h(x) + (\lambda - \beta\omega)g(x) = 0. \tag{8.5.4}$$

The equation for the extrema of (8.5.3) reads

$$[h(x_m) + \lambda g(x_m)] - \beta\omega^2 g'(x_m)g(x_m) + \omega g(x_m)[h'(x_m) + \lambda g'(x_m)] = 0. \tag{8.5.5}$$

Since Poisson white noise is bounded from below by $-\beta\omega$, contrary to Gaussian white noise which is unbounded, it is particularly useful in modelling situations where the parameter should remain positive for physical reasons. This can be achieved by imposing that $\lambda > \beta\omega$.

To illustrate the application of Poisson white noise, consider the simple photochemical model (Horsthemke and McCarty, 1986):

$$\mathrm{A} + \mathrm{X} \rightleftarrows 2\mathrm{X}, \quad \mathrm{X} + h\nu \rightarrow \mathrm{C}. \tag{8.5.6}$$

We assume that the species A is fed into the reactor such that it is in excess, so that the concentration of A is constant. The species C is immediately removed from the reactor. For a well-mixed system, the concentration of X has no spatial dependence. We can therefore integrate the light absorption over the entire volume of the reactor. With appropriate scaling we obtain

$$\dot{x} = ax - x^2 - kI(1 - \mathrm{e}^{-\varepsilon x}). \tag{8.5.7}$$

Here a is proportional to the concentration of A. The last term in (8.5.7) represents the light absorbed in the reactor and which is involved in the third reaction step of (8.5.6). The light intensity incident upon the reactor is I and ε is the absorption coefficient of X times the thickness of the reacting volume. This simple model shows the following steady state behavior under deterministic conditions: $\bar{x} = 0$ is a steady state for all values of I. For sufficiently large values of the incident light intensity I, this state is stable. This is the expected result, since X is photochemically degraded. As I is decreased, $\bar{x} = 0$ becomes unstable at $I_0^{\mathrm{d}} = a/k\varepsilon$. At this point a nonequilibrium transition occurs and a branch of nontrivial steady states bifurcates from the branch $\bar{x} = 0$. It is interesting to note that the character of this bifurcation depends on the concentration of A, i.e. the strength of autocatalysis in the system. The bifurcation is subcritical for $a > a_{\mathrm{c}}^{\mathrm{d}}$ and supercritical for $a < a_{\mathrm{c}}^{\mathrm{d}}$, where $a_{\mathrm{c}}^{\mathrm{d}} = 2/\varepsilon$. If the bifurcation is subcritical, the new branch exists for values of I bigger than I_0^{d}. These states are unstable till the branch turns around at some I_{u} and becomes stable. The system thus displays bistability for $I_0^{\mathrm{d}} < I < I_{\mathrm{u}}$; a nonzero state coexists with $\bar{x} = 0$. If the bifurcation is supercritical, the new branch exists for values of I smaller than I_0^{d} and is stable.

Let us now study the effect of light intensity fluctuations on the transition phenomena of the photochemical system. The fluctuations can be intrinsic to the source, for instance in the case of thermal sources, or can result from scattering in a turbulent medium, a situation realized in the experiments in (De Kepper and Horsthemke, 1978; Micheau, Horsthemke and Lefever, 1984). Obviously, the incident light intensity should always be a positive quantity. Thus, Poisson white noise is a more adequate way to model rapid fluctuations in the light intensity than Gaussian white noise. Let I now denote the mean value of the incident light intensity. To guarantee positivity, we must have

$$\beta\omega < I. \tag{8.5.8}$$

Applying (8.5.5) to the photochemical model, we find that $x_m=0$ is an extremum of the stationary probability density for all values of I. It is a maximum for $I>I_0^p$ and a minimum for $I<I_0^p$, where

$$I_0^p=(a/k\varepsilon)-\beta\omega^2 k\varepsilon/(1-\omega k\varepsilon), \tag{8.5.9}$$

which yields for small noise intensities

$$I_0^p=I_0^d-\beta\omega^2 k\varepsilon+O(\omega^3). \tag{8.5.10}$$

It is at this point that a branch of nonzero extrema bifurcates. This bifurcation changes its character from supercritial to subcritical at the point

$$a_c^p=2(1-3\omega k\varepsilon-\beta\omega^2k^2\varepsilon^2+2\omega^2k^2\varepsilon^3)\varepsilon^{-1}(1-\omega k\varepsilon)^{-1} \times(1-2\omega k\varepsilon)^{-1}, \tag{8.5.11}$$

which yields for small, but frequent, spikes

$$a_c^p\simeq a_c^d-2\beta\omega^2k^2\varepsilon. \tag{8.5.12}$$

Though rapid fluctuations of parameters which have to remain non-negative for physical reasons are better modelled by Poisson white noise, Gaussian white noise nevertheless has often been used in the past because of its practical convenience. It has been argued qualitatively that the negative portions of the realizations will have a negligible effect on the behavior of the system as long as the noise intensity is not too large. The photochemical model offers a means to test this argument. If we model the light intensity fluctuations by Gaussian white noise, we obtain for the bifurcation point and the critical point, i.e. the point of change from supercritical to subcritical behavior, the following values:

$$I_0^g=I_0^d-\sigma^2k\varepsilon/2,\quad a_c^g=a_c^d-\sigma^2k^2\varepsilon. \tag{8.5.13}$$

These results look very similar to (8.5.10) and (8.5.12). In fact, Poisson white noise goes over into Gaussian white noise in the following limit (van den Broeck, 1983). Make the frequency of the spikes larger and larger, $\beta\to\infty$, but make the average weight of each spike smaller and smaller, $\omega\to 0$, such that $\beta\omega^2$ is held constant, $\beta\omega^2=\sigma^2/2$. Thus, comparing (8.5.13) and (8.5.10) and (8.5.12), we find that for not too large noise intensities the 'unphysical' Gaussian white noise yields the same results for the transition points as the Poisson white noise. It turns out that the extrema of the stationary probability density in the Gaussian white noise case are also qualitatively similar to the Poisson white noise case.

8.6 The effect of periodic dichotomous fluctuations

The preceding sections illustrate that random external fluctuations can qualitatively change the macroscopic behavior of nonlinear systems. It is natural to ask if the observed noise-induced effects depend strongly on the random nature of the environment. Do the same phenomena appear when the external constraints vary in a similar, but regular way? To study this question, we need a random process with a regular analog and a criterion for the comparison of a random system with a 'quasi-random' system (only the initial conditions are random when we consider periodic fluctuations). Although studies and comparison of the moments of some systems subjected to random and periodic fluctuations have been made (Lindenberg, Seshadri and West, 1980), we adopt, for reasons expounded above, the stationary probability density as the criterion for comparison of systems under the influence of random and regular modulations. In this section we will consider only a particular type of fluctuations, namely (symmetric) dichotomous noise. This is a natural choice since an unambiguous regular analog exists for this random process. The periodic modulation is one which switches between the same two values as the noise process at a fixed frequency, equal to the average switching frequency of the noise.

So we consider the case where the external parameter $\lambda(t)$ switches between the values $\lambda \pm \Delta$, and the switching occurs at definite time intervals of length $T/2$. This periodic variation will mimic the dichotomous Markov process with correlation time γ^{-1}, if we choose

$$2/T = \gamma/2. \tag{8.6.1}$$

As was shown in Doering (1984) and Doering and Horsthemke (1985), the stationary probability density for the system described by the evolution equation

$$\dot{x} = F(x, \lambda(t)), \tag{8.6.2}$$

where $\lambda(t)$ is the above defined regular periodic modulation, is given by

$$\bar{q}(x) = T^{-1}\{F(x, \lambda + \Delta)^{-1} - F(x, \lambda - \Delta)^{-1}\} \tag{8.6.3}$$

for $x \in V = [y_-, y_+]$ and vanishes identically outside of V. Equation (8.6.3) holds if F fulfils some technical conditions (Doering and Horsthemke, 1985), namely that the system has a unique stable steady state for each value of the external parameter, i.e. $\lambda - \Delta$ and $\lambda + \Delta$, and has a unique stable steady state for the average force $\bar{F} \equiv \frac{1}{2}\{F(x, \lambda + \Delta) + F(x, \lambda - \Delta)\}$. The two values $y_\pm$ are the solutions to the simultaneous equations

$$T/2 = \int_{y_-}^{y_+} \mathrm{d}x\, F(x, y + \Delta)^{-1}, \tag{8.6.4}$$

$$T/2 = \int_{y_+}^{y_-} \mathrm{d}x\, F(x, \lambda - \Delta)^{-1}. \tag{8.6.5}$$

The results of the studies reported in Doering and Horsthemke (1985) are the following: (i) For very small or very large fluctuation rates γ and *fixed* amplitudes, systems behave similarly under the influence of the random or periodic modulations. (ii) If the time scale of the noise is of the order of the characteristic time of the system, various fluctuation-induced transitions can occur, but the effects of the two types of noise are quite distinct. This is not surprising. When a system is fast enough to respond significantly to variations in its external constraints, it is highly correlated with its environment. However, in general there is more transition structure when the fluctuations are random. The appearance of the external parameter must become increasingly more complex (multiplicative, nonlinear) for the various transitions to occur with the periodic variations, whereas even more phenomena can occur in the simplest systems under the influence of the random fluctuations. (iii) In the white noise limit ($\gamma \to \infty, \Delta \to \infty$ such that $\Delta^2/\gamma = \sigma^2/2 = \text{constant}$) the effect of random dichotomous noise and of periodic dichotomous noise is drastically different. Whereas there can be noise-induced transitions for very rapid random fluctuations, as shown above, periodic modulations have no effect in the white noise limit. In fact, the support V of $\bar{q}(x)$ collapses to the steady state of the average force $\bar{F}$ in the limit of rapid noise. (The support U of $\bar{p}(x)$ for symmetric dichotomous noise becomes the state space of the system.) This difference in the behavior of the support of the stationary probability density gives us some insight into the nature of the mechanism responsible for the difference of the influence of the types of noise. This difference can be further elucidated by comparing the spectra of the two noises. The spectral density of the random dichotomous Markov process is the Lorentzian

$$S_{\text{ran}}(\omega) = \gamma\Delta^2/\pi(\omega^2 + \gamma^2), \tag{8.6.6}$$

while that of the periodic dichotomous fluctuations is

$$S_{\text{per}}(\omega) = \sum_{n=\pm 1, \pm 3, \ldots} [\Delta^2\gamma^2/\omega^2]\delta(\omega - 0.5\, n\pi\gamma). \tag{8.6.7}$$

When $\gamma \ll t_{\text{sys}}^{-1}$, the δ-functions in (8.6.7) are very closely spaced and may be averaged from the viewpoint of the system. In this case, the spectrum (8.6.7) closely resembles the Lorentzian (8.6.6). However, when the fluctuations are fast, i.e. $\gamma \gg t_{\text{sys}}^{-1}$, we are no longer justified in smoothing out the δ-functions in (8.6.7). There are large voids in the spectral density of the periodic modulations, most significantly at the low frequencies. The presence of these low-frequency components in $S_{\text{ran}}(\omega)$, to which the system can respond, is the agent preventing the collapse of the support of the stationary probability density in the random noise case. In the white noise limit, where $S_{\text{ran}}(\omega) \to \sigma^2/2\pi$, there is an important low-frequency contribution to the power spectrum that is missing in the spectral density of the periodic fluctuation. A common characteristic of all genuinely random processes is the vanishing of the correlation function for long times and hence the continuity of the spectral density. No matter how complex the regular variations, any periodicity in the fluctuations will result in a discrete spectrum and the absence of key low-frequency components.

8.7 Applications to nonlinear systems and experimental evidence of noise-induced transitions

The concept of noise-induced transitions has found numerous applications in theoretical studies of the effect of external random fluctuations on nonlinear systems. Some examples, without any claim to completeness, are studies on optical bistability (Brambilla, Lugiato, Strini and Narducci, 1986; Bulsara, Schieve and Gragg, 1978; Lugiato, Colombo, Broggi and Horowicz, 1986), on dye lasers (Dixit and Sahni, 1983; Graham, Hohnerbach and Schenzle, 1982; Lett Gage and Chyba, 1987), on liquid crystals (Behn and Muller, 1985; Brand and Schenzle, 1980; Horsthemke *et al.*, 1985; Sagues and San Miguel, 1985; San Miguel, 1985, San Miguel and Sancho, 1981), on superfluid turbulence (Moss and Welland, 1982), on plasma physics (Hastings and Auerbach, 1985; Shaing, 1984), on nuclear reactors (Karmeshu, 1981), on nonlinear oscillators (Wiesenfeld and Knobloch, 1982), on catalytic and enzymatic reactions (de la Rubia, Garcia-Sanz and Velarde, 1984; de la Rubia and Velarde, 1978; Hahn *et al.*, 1974), on limit cycles in chemical systems (Lefever and Turner, 1986), and on wave-front propagation in chemical systems (Engel, 1985).

The number of experimental studies is unfortunately smaller, though experimental evidence for noise-induced transitions has been accumulating at a steady rate since 1977. The first experimental observation of a noise-induced transition was presented at the 'Statphys 13' in 1977 by Kabashima (1978). More detailed accounts of the experiments are found in Kabashima and Kawakubo (1979) and Kabashima, Kogure, Kawakubo and Okada (1979). Kabashima and coworkers studied an electrical parametric oscillator. Their system consists of a primary and secondary *RC* circuit loop coupled by two ferrite toroidals. On the primary side a d.c. current is applied in an additional bias circuit to shift the operating point of the parametric oscillator into a domain where second order nonlinearities come into play in the relation between the magnetic flux of the coils and the currents. A 50 kHz a.c. current is supplied to the primary circuit. This current acts as a pump and excites subharmonic oscillations in the secondary circuit, if its amplitude surpasses some threshold value. In addition to the pumping current a random, quasi-white current (spectrum is flat between 0.01 and 100 kHz) can be supplied to the primary circuit. Kabashima *et al.* show that the amplitude of the secondary current is an order parameter for their system and that it obeys the Landau equation (8.2.29). Their experimental results show good agreement with the theoretical predictions of Section 8.2.3, in particular (8.2.46). The threshold for the excitation of subharmonic oscillations in the secondary circuit is increased by $\sigma^2/2$.

These experiments were followed by experimental studies of the effect of external fluctuations on an oscillating chemical reaction, namely the influence of light intensity fluctuations on the Briggs–Rauscher reaction (De Kepper and Horsthemke, 1978) and on nematic liquid crystals, namely the effect of noise in the applied voltage on transitions to the Williams domain and the dynamic scattering mode (Brand, Kai and Wakabayashi, 1985; Kai, Kai and Hirakawa, 1979; Kawakubo, Yanagita and Kabashima, 1981). Experimental results on the behavior of dye lasers were understood in terms of fluctuations in the pump due to turbulence in the active medium (Kaminishi, Roy, Short and Mandel, 1981; Roy, Yu and Zhu, 1985). Photochemical systems, for which the reaction mechanism is well understood, were shown to be good candidates for a quantitative comparison of theoretical predictions and experimental results (Micheau, Horsthemke and Lefever, 1984). Moss and McClintock and their coworkers pioneered the use of analog electrical systems to explore the effect of noise on nonlinear model systems (see Chapter 9, Volume 3), and were the first to measure the critical exponents of a noise-induced critical point (Smythe, Moss and McClintock, 1983).

The most exciting development in the experimental study of noise-induced effects is the recent work by Griswold and Tough on the TI/TII transition in superfluid helium (Griswold and Tough, 1986, 1987); see also Chapter 1 by Tough in Volume 3. Tough and Griswold have studied the effect of external noise in the heat current on superfluid turbulence in thermal counterflow. They have observed for the first time a *pure noise-induced transition to bistability in a physical system.* They find that for noise with a relative variance of about 15%, or greater, the turbulent state TI can coexist with the turbulent state TII. Such coexistence does not occur under deterministic conditions or under conditions with a small amount of noise. Superfluid turbulence promises to be an exciting system for the experimental study of noise-induced transitions.

References

Arnold, L. 1974. *Stochastic Differential Equations: Theory and Applications.* New York: Wiley.
Arnold, L., Horsthemke, W. and Lefever, R. 1978. *Z. Phys. B* **29**, 367.
Behn, U. and Muller, R. 1985. *Phys. Lett. A* **113**, 85.
Bernstein, M. and Brown, L. S. 1984. *Phys. Rev. Lett.* **52**, 1933.
Blankenship, G. and Papanicolaou, G. C. 1978. *SIAM J. Appl. Math.* **34**, 437.
Brambilla, M., Lugiato, L. A., Strini, G. and Narducci, L. M. 1986. *Phys. Rev. A* **34**, 1237.
Brand, H. R., Kai, S. and Wakabayashi, S. 1985. *Phys. Rev. Lett.* **54**, 555.
Brand, H. and Schenzle, A. 1980. *J. Phys. Soc. Jpn.* **48**, 1382.
Brand, H., Schenzle, A. and Schröder, G. 1982. *Phys. Rev. A* **25**, 2324.
Bulsara, A. R., Schieve, W. C. and Gragg, R. F. 1978. *Phys. Lett. A* **68**, 294.
Chandrasekhar, S. 1943. *Rev. Mod. Phys.* **15**, 1.
De Kepper, P. and Horsthemke, W. 1978. *C. R. Acad. Sci. Paris C* **287**, 251.
de la Rubia, F. J., Garcia-Sanz, L. and Velarde, M. G. 1984. *Surface Sci.* **143**, 1.
de la Rubia, J. and Velarde, M. G. 1978. *Phys. Lett. A* **69**, 304.
Di Prima, R. C. and Swinney, H. L. 1981. In *Hydrodynamic Instabilities and the*

Transition to Turbulence (H. L. Swinney and J. P. Gollub, eds.) p. 139. Berlin: Springer.
Dixit, S. N. and Sahni, P. S. 1983. *Phys. Rev. Lett.* **50**, 1273.
Doering, C. R. 1984. *Springer Proc. Phys.* **1**, 253.
Doering, C. R. 1986. *Phys. Rev. A* **34**, 2564.
Doering, C. R. and Horsthemke, W. 1985. *J. Stat. Phys.* **38**, 763.
Doob, J. L. 1942. *Ann. Math.* **43**, 351.
Elliott, J. 1955. *Trans. Am. Math. Soc.* **78**, 406.
Engel, A. 1985. *Phys. Lett. A* **113**, 139.
Field, R. J. and Burger, M., eds. 1985. *Oscillations and Travelling Waves in Chemical Systems.* New York: Wiley.
Gelfand, I. N. and Wilenkin, N. J. 1964. *Verallgemeinerte Funktionen IV.* Berlin: VEB Deutscher Verlag der Wissenschaften.
Golubitsky, M. and Schaeffer, D. G. 1985. *Singularities and Groups in Bifurcation Theory.* New York: Springer.
Graham, R., Hohnerbach, M. and Schenzle, A. 1982. *Phys. Rev. Lett.* **48**, 1396.
Griswold, D. and Tough, J. T. 1986. *Bull. Am. Phys. Soc.* **31**, 1739.
Griswold, D. and Tough, J. T. 1987. *Phys. Rev. A* **36**, 1360.
Hahn, H. S., Nitzan, A., Ortoleva, P. and Ross, J. 1974. *Proc. Natl. Acad. Sci. USA* **71**, 4067.
Haken, H. 1978. *Synergetics*, 2nd edn. Berlin: Springer.
Hastings, D. E. and Auerbach, S. P. 1985. *Phys. Fluids* **28**, 2219.
Hille, E. 1969. *Lectures on Ordinary Differential Equations.* Reading, MA: Addison-Wesley.
Hongler, M. O. 1979. *Helv. Phys. Acta* **52**, 280.
Horsthemke, W., Doering, C. R., Lefever, R. and Chi, A. S. 1985. *Phys. Rev. A* **31**, 1123.
Horsthemke, W. and Lefever, R. 1977. *Phys. Lett. A* **64**, 19.
Horsthemke, W. and Lefever, R. 1980. *Z. Phys. B* **40**, 241.
Horsthemke, W. and Lefever, R. 1984. *Noise-Induced Transitions.* Berlin: Springer.
Horsthemke, W. and McCarty, P. 1986. *Phys. Lett. A* **117**, 10.
Horsthemke, W. and Malek-Mansour, M. 1976. *Z. Phys. B* **24**, 307.
Kabashima, S. 1978. *Ann. Israel Phys. Soc.* **2**, 710.
Kabashima, S. and Kawakubo, T. 1979. *Phys. Lett. A* **70**, 375.
Kabashima, S., Kogure, S., Kawakubo, T. and Okada, T. 1979. *J. Appl. Phys.* **50**, 6296.
Kai, S., Kai, T. and Hirakawa, K. 1979. *J. Phys. Soc. Jpn.* **47**, 1379.
Kaminishi, K., Roy, R., Short, R. and Mandel, L. 1981. *Phys. Rev. A* **24**, 370.
Karlin, S. and Taylor, H. M. 1975. *A First Course in Stochastic Processes.* New York: Academic Press.
Karlin, S. and Taylor, H. M. 1981. *A Second Course in Stochastic Processes.* New York: Academic Press.
Karmeshu, 1981. *Ann. Nucl. Energy* **8**, 41.
Kawakubo, T., Yanagita, A. and Kabashima, S. 1981. *J. Phys. Soc. Jpn.* **50**, 1451.
Kimura, M. and Ohta, T. 1971. *Theoretical Aspects of Population Genetics.* Princeton University Press.
Kitahara, K., Horsthemke, W. and Lefever, R. 1979. *Phys. Lett. A* **70**, 377.

Kitahara, K., Horsthemke, W., Lefever, R. and Inaba, Y. 1980. *Prog. Theor. Phys.* **64**, 1233.

Kuznetsov, P. I., Stratonovic, R. L. and Thikhonov, V. I. 1965. In *Nonlinear Transformations of Stochastic Processes* (P. I. Kuznetsov, R. L. Stratonovic and V. I. Thikhonov, eds.), p. 223. Oxford: Pergamon.

Lefever, R. and Turner, J. W. 1986. *Phys. Rev. Lett.* **56**, 1631.

Lett, P., Gage, E. C. and Chyba, T. H. 1987. *Phys. Rev. A* **35**, 746.

Lifshitz, E. M. and Pitaevskii, L. P. 1980. *Statistical Physics*, 3rd edn., part 1. Oxford: Pergamon.

Lindenberg, K., Seshadri, V. and West, B. J. 1980. *Phys. Rev. A* **22**, 2171.

Lugiato, L. A., Colombo, A., Broggi, G. and Horowicz, R. J. 1986. *Phys. Rev. A* **33**, 4469.

May, R. M. 1973. *Stability and Complexity in Model Ecosystems.* Princeton University Press.

Micheau, J. C., Horsthemke, W. and Lefever, R. 1984. *J. Chem. Phys.* **81**, 2450.

Moss, F. and Welland, G. V. 1982. *Phys. Rev. A* **25**, 3389.

Nicolis, G. and Prigogine, I. 1977. *Self-Organization in Nonequilibrium Systems.* New York: Wiley.

Prohorov, Y. V. and Rozanov, Y. A. 1969. *Probability Theory.* Berlin: Springer.

Roy, R., Yu, A. W. and Zhu, S. 1985. *Phys. Rev. Lett.* **55**, 2794.

Sagues, F. and San Miguel, M. 1985. *Phys. Rev. A* **32**, 1843.

San Miguel, M. 1985. *Phys. Rev. A* **32**, 3811.

San Miguel, M. and Sancho, J. M. 1981. *Z. Phys. B* **43**, 361.

Schuss, Z. 1980. *Theory and Applications of Stochastic Differential Equations.* New York: Wiley.

Shaing, K. C. 1984. *Phys. Fluids* **27**, 1924.

Smythe, J., Moss, F. and McClintock, P. V. E. 1983. *Phys. Rev. Lett.* **51**, 1062.

Swinney, H. L. and Gollub, J. P., eds. 1981. *Hydrodynamic Instabilities and the Transition to Turbulence.* Berlin: Springer.

Uhlenbeck, G. E. and Ornstein, L. S. 1930. *Phys. Rev.* **36**, 823.

van den Broeck, C. 1983. *J. Stat. Phys.* **31**, 467.

Weinstein, D. H. 1934. *Proc. Natl. Acad. Sci. USA* **20**, 529.

Wiesenfeld, K. A. and Knobloch, E. 1982. *Phys. Rev. A* **26**, 2946.

Wong, E. 1964. *Proc. Symp. Appl. Math.* **16**, 264.

9 Mechanisms for noise-induced transitions in chemical systems

RAYMOND KAPRAL and EDWARD CELARIER

9.1 Introduction

The state of chemical equilibrium is endowed with stability: perturbations away from equilibrium always decay to this state. This is a fundamental feature of the statistical mechanics of relaxation processes near equilibrium. The situation is not so simple in the far-from-equilibrium regime. Bifurcations can lead to the appearance of more exotic system states, ranging from simple stationary states akin to the equilibrium state, to periodic, quasiperiodic or even chaotic states. In addition, the asymptotic state may not be unique; given different initial conditions, different final states may be reached. While these phenomena are widespread in nature, chemically reacting fluids, driven to the nonequilibrium regime by external constraints, constitute a ubiquitous and important class of systems of this type (Nicolis and Prigogine, 1977).

The presence of noise can have significant effects on nonlinear systems. It can modify the nature of existing states, create new states, or induce transitions between coexisting states (Horsthemke and Lefever, 1984). Noise may enter the description through internal fluctuations that have their origin in the molecular nature of the system, or through external means due to the inability to control system constraints. In fact, external noise may be intentionally applied to the system in order to achieve a given transition process.

The principal focus of this article is on transitions between bistable states, induced by external noise. The use of external noise sources to promote noise-induced transitions has several convenient features. In contrast to internal fluctuations, the statistical properties of the noise are often known, due to the nature of the experimental conditions leading to the noisy constraints or, in the case of intentionally applied noise, they may be specified by the investigator; also the manner in which noise is applied to the system can be varied. Noise may be applied to some or all of the control parameters, or to the dynamical variables. This flexibility provides a powerful means to probe the system's dynamics, since the rate of the transition, or even its character, is a strong function of the type of noise and its statistical properties. Some of these features are explored in this chapter.

One of the main experimental tools used to study nonequilibrium chemical systems is the continuously stirred tank reactor (CSTR) (Field and Burger, 1985). In experiments using this apparatus, reagents are fed into the reaction vessel at a fixed flow rate in order to maintain the system in a nonequilibrium state; the reacting mixture is stirred rapidly to effect nearly homogeneous conditions, and portions of the reacting mixture are removed to maintain a constant volume. By controlling parameters like the flow rate, chemical composition of the feed, or the vessel temperature, the system state may be altered, and such parameter variations have revealed a rich bifurcation structure that rivals or surpasses that in hydrodynamics (Field and Burger, 1985; Roux, 1983; Swinney, 1983; Vidal and Hanusse, 1986). Parametric noise processes can be implemented by making some or all of the control parameters stochastic variables, and other types of noise may be introduced by chemical or other perturbations.

While the examples chosen for study in this chapter originate in chemistry, the phenomena have a much wider domain. The mechanisms we discuss apply equally well to nonlinear models that arise in other fields, and are common features of far-from-equilibrium nonlinear stochastic systems. Other chapters in these volumes provide a wealth of examples.

In Section 9.2 we give some examples of the types of chemical systems under investigation, and discuss how external noise might arise in, or be applied to, these systems. For bistable deterministic systems, the phase space can be partitioned into regions, called *basins of attraction*, whose points, regarded as initial conditions, evolve to one of the two asymptotic system states. The structure of these basins and their boundaries play an important part in the study of noise-induced transitions, and Section 9.3 is devoted to an examination of some of their properties. In some situations the dynamics generated by the set of nonlinear differential equations used to model the reaction may be reduced to that of a one-dimensional map. An example of such a reduction of the dynamical description is given in Section 9.4. Stochastic versions of the discrete map may then be used to explore noise-induced transitions, and these can suggest mechanisms for analogous processes in the stochastic differential equation models.

Bistability involving steady states occurs commonly in far-from-equilibrium chemical systems; in fact the origin of relaxation oscillations in chemical systems has been linked to steady-state bistability nearby in parameter space (DeKepper and Boissonade, 1985). Oscillating states are also common and important system states, and examples abound in biology (Winfree, 1980) and chemistry (Field and Burger, 1985). The aim of this chapter is the study of noise-induced transitions involving oscillating states. The transition mechanisms are more varied and less well studied than those for steady states, although some mechanisms operate for both steady and oscillatory states. Section 9.5 provides casebook studies of a number of mechanisms that are expected to operate commonly in systems possessing oscillating states. Two situations are examined: the noise-induced transitions that may occur between coexisting limit cycles, and those that may occur between a limit cycle and a fixed point.

9.2 Chemical reactions and external noise

There is an inexhaustible supply of chemical reaction schemes that arise in modeling chemical systems in the nonlinear domain. Provided spatial degrees of freedom play no role, the chemical rate equations can be written in the general form,

$$\frac{\mathrm{d}\mathbf{c}(t)}{\mathrm{d}t} = \mathbf{R}[\mathbf{c}(t);\alpha], \tag{9.2.1}$$

where $\mathbf{c}(t)$ is a vector of concentration variables, $\mathbf{R}$ is a nonlinear vector-valued function of the composition and the set of control parameters, α. $\mathbf{R}$ can take many different forms, depending on the specific application. We shall employ two models to illustrate the phenomena. Our emphasis will be on those features of the noise-induced transition process that are of a general nature and robust enough to be observed in a wide variety of circumstances; hence the results obtained here should apply to other reaction schemes.

As a first example we consider an irreversible exothermic reaction $A \rightarrow B$ taking place in a CSTR thermostated by a heat bath at temperature T_c (Mankin and Hudson, 1984; Uppal, Ray and Poore, 1974). The relevant dynamical variables describing the system are the chemical concentration of A, c_A, and the temperature of the reacting mixture, T_r, whose evolution equations follow from considerations of mass and energy balance. Assuming that reactants with concentration c_f at temperature T_f are fed into the reactor at a flow rate F, we have

$$\left.\begin{aligned} V\frac{\mathrm{d}c_A}{\mathrm{d}t} &= F(c_f - c_A) - Vk(T_r)c_A, \\ V\rho C_p\frac{\mathrm{d}T_r}{\mathrm{d}t} &= F\rho C_p(T_f - T_r) \\ &\quad + V(-\Delta H)k(T_r)c_A - hA(T_r - T_c). \end{aligned}\right\} \tag{9.2.2}$$

Here V and A are the volume and surface area of the vessel, ρ and C_p are the density and heat capacity of the reacting mixture, ΔH is the molar enthalpy of the reaction, and h is heat transfer rate per unit area to the heat bath. The rate coefficient of the reaction, $k(T_r)$, is taken to have an Arrhenius form, $k(T_r) = k_0 \exp(E_a/RT_r)$, where E_a is the activation energy. Following Golubitsky and Keyfitz (1980), these equations may be written in dimensionless form as

$$\left.\begin{aligned} \frac{\mathrm{d}c}{\mathrm{d}t} &= -\varepsilon c + D(1-c)\mathrm{e}^{\gamma T/(1+T)}, \\ \frac{\mathrm{d}T}{\mathrm{d}t} &= -(1+\varepsilon)T + BD(1-c)\mathrm{e}^{\gamma T/(1+T)} + \eta. \end{aligned}\right\} \tag{9.2.3}$$

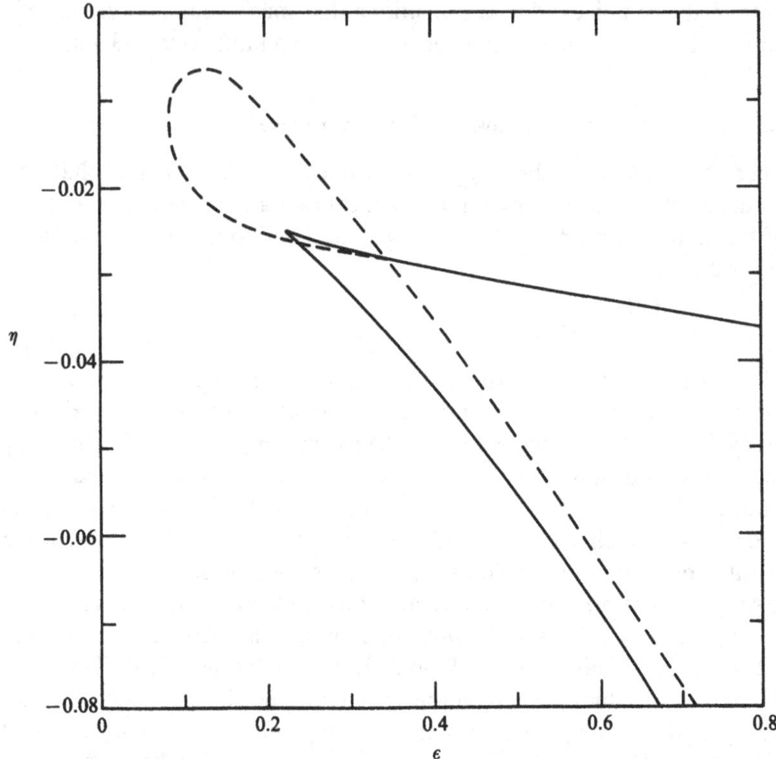

Figure 9.1. Phase diagram for the GK model. Fixed system parameters are $B = 0.22$, $D = 0.25$ and $\gamma = 100$. The solid curves denote the locus of saddle-node bifurcations where the number of fixed points changes, while the dashed curve represents a Hopf bifurcation that gives rise to a limit cycle.

The dimensionless concentration and temperature are $c = (c_f - c_A)/c_f$, and $T = (T_r - T_f)/T_f$, while $\eta = (T_c - T_f)/T_f$ is the scaled bath temperature. Notice that c is restricted to the interval $0 \leqslant c \leqslant 1$. The remaining parameters are: ε, the dimensionless flow rate; B, a constant proportional to the heat of reaction; D, the Damköhler number characterizing the heat gain and loss terms; and γ, the activation energy in units of RT_f. In the sequel we shall fix $B = 0.22, D = 0.25$ and $\gamma = 100.0$ and study the bifurcation structure in the $\varepsilon\eta$-plane. Equation (9.2.3) will be referred to as the GK model.

The phase diagram (Figure 9.1) for this deterministic set of equations shows the nature of the solutions for different values of the parameters ε and η. The solid curves define a cusp-shaped region where the system possesses three fixed points, at least one of which is stable; outside this region there is a single stable fixed point. The dashed line in the figure is a Hopf bifurcation boundary. The solution structure (values of c for the stable and unstable fixed points and the limit cycle versus η) along the line $\varepsilon = 0.35$ is presented in Figure 9.2; bistability

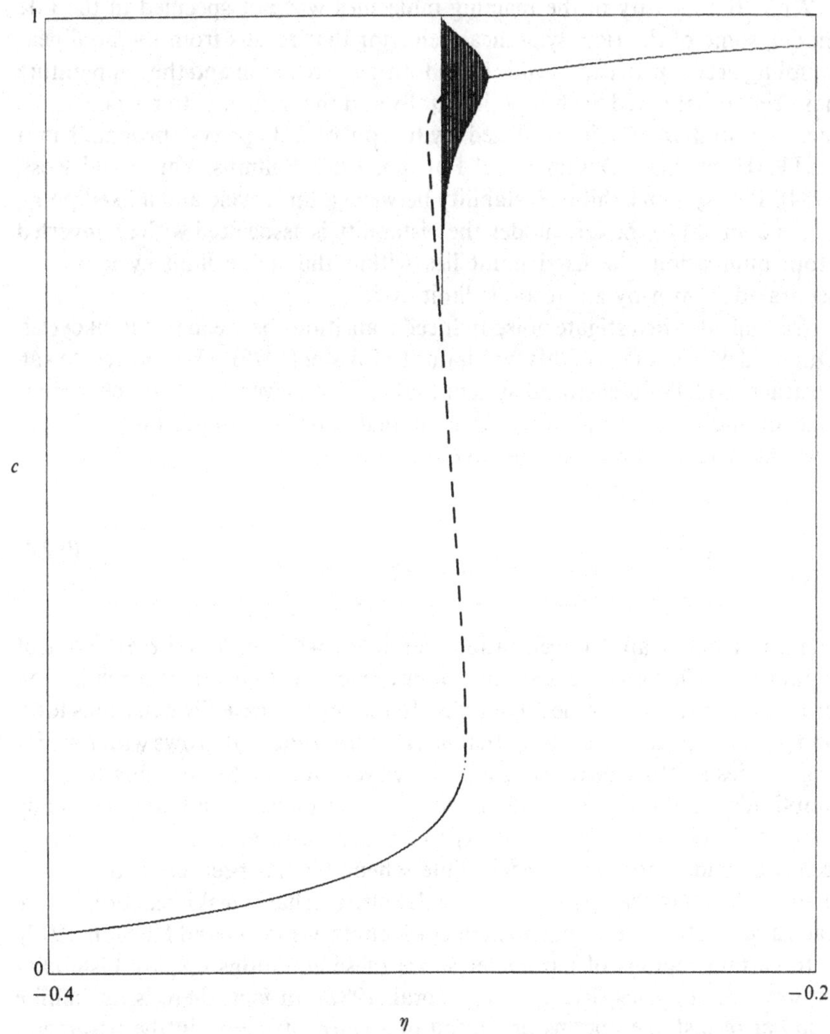

Figure 9.2. Solution structure (c versus η) along a line $\varepsilon = 0.35$ through the phase diagram. The lower branch corresponds to a steady state, while on the upper branch the fixed point that exists at large η disappears, and a stable limit cycle is born by a supercritical Hopf bifurcation as η decreases.

and hysteresis clearly exist. Notice that, since the fixed point lying on the upper branch bifurcates to a limit cycle when the Hopf bifurcation boundary is crossed, the system exhibits bistability between a limit cycle and a fixed point, in addition to bistability between two fixed points, depending on η. The nature of noise-induced transitions involving an oscillatory state and a steady state will be studied in the context of this model.

While the identity of the reacting molecules was not specified in the GK model, some of the rich dynamical behavior that results from the nonlinear coupling between an exothermic or endothermic reaction and the temperature has been investigated both experimentally and theoretically. In particular, we mention studies of acid-catalyzed hydration of 2, 3-epoxy-1-propanol in a CSTR (Heemskirk, Dammas and Fortuin, 1980; Rehmus, Vance and Ross, 1984). This system exhibits bistability between a limit cycle and a fixed point, but in contrast to the GK model, the bistability is associated with an inverted Hopf bifurcation; the fixed point lies within the stable limit cycle, and is separated from it by an unstable limit cycle.

We shall also investigate noise-induced transitions between two limit cycles. Our model for systems of this type is due to Rössler (1979), who constructed an abstract model of a chemical system that exhibits several features characteristic of the phase space flow seen in real far-from-equilibrium chemical systems. The kinetic equations take the form

$$\begin{aligned} \dot{x} &= -y - z, \\ \dot{y} &= x + ay, \\ \dot{z} &= bx - cz + zx. \end{aligned} \tag{9.2.4}$$

Here, x, y and z are 'concentration' variables, while a, b and c are control parameters. This set of equations mimics reactions that utilize a reinjection process to produce complex dynamics. In the present case the equations for x and y constitute a suboscillator that has an amplitude that grows with time for trajectories starting near the origin. However, due to the coupling to z, the amplitude does not grow indefinitely, and the phase points are eventually reinjected near the origin. This type of process can give rise to oscillatory behavior and chaos. The model, while schematic, has been used to describe certain aspects of real systems, like the Belousov–Zhabotinskii reaction, where such a reinjection mechanism operates (Schmitz, Graziani and Hudson, 1977).

In certain regions of parameter space these equations display bistability between limit cycles (Fraser and Kapral, 1982). In fact, there is an infinite number of bistable regions organized in a spiral structure in the parameter plane about the end point of a line of homoclinic systems (Gaspard, Kapral and Nicolis, 1984). This full structure is not of interest to us here, but any of the bistable regions will display the phenomena we consider. We shall study the most prominent of these bistable regions, that supporting bistable three-loop limit cycles, an example of which is shown in Figure 9.3. Bistable limit cycle oscillations have been observed experimentally in the Belousov–Zhabotinskii reaction (Roux, 1983) as well as the chlorite–bromate–iodide system (Alamgir and Epstein, 1983), although the underlying description of the bistability is different from that in Rössler's model.

Bistable limit cycle oscillations, or even more exotic types of bistability involving quasiperiodic or chaotic attractors, are not rare. Such behavior is

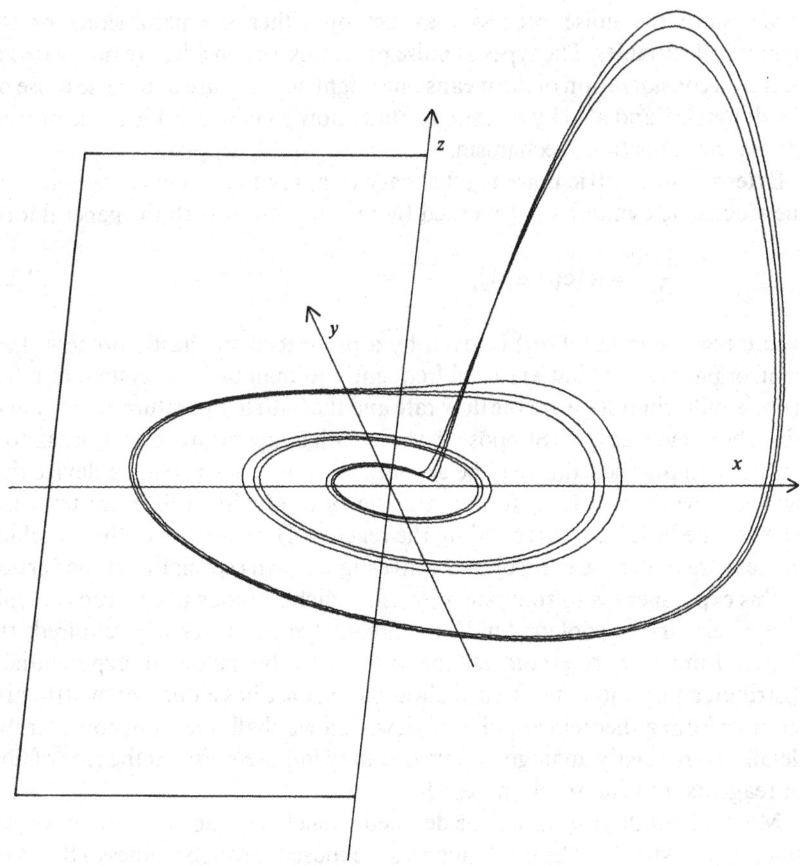

Figure 9.3. Bistable limit cycles for Rössler's model. The outer and inner cycles in this figure are stable and the limit cycle lying between them is unstable. A trajectory on the cycle makes three loops around the origin before it closes itself; hence our terminology, three-loop limit cycle. Also shown is a portion of a typical Poincaré surface (the xz-plane, $x \leqslant 0$) that may be used to construct the return map.

observed easily in forced systems (see, for instance, Huberman and Crutchfield, 1979; Kapral, Schell and Fraser, 1982; Rehmus, Vance and Ross, 1984). A model of an enzymatic reaction has also displayed bistable limit cycle oscillations (Decroly and Goldbeter, 1982).

These two models do not exhaust the possible types of bistabilities involving oscillating states. However, their study will allow us to identify four different mechanisms for noise-induced transition processes that should operate in a variety of circumstances.

The investigation of noise-induced transitions entails the study of stochastic versions of the chemical rate equations, which can take a number of different forms since the noise process may act on either the parameters or the dynamical variables. The types of noise processes we consider are motivated in part by a consideration of the means one might actually use to impose noise on the dynamics, and also by a desire to study how a change in the noise process affects the transition mechanism.

External parametric noise is often easy to apply to a chemical system, and the stochastic dynamics is governed by rate equations with the general form

$$\frac{\mathrm{d}\mathbf{c}(t)}{\mathrm{d}t} = \mathbf{R}[\mathbf{c}(t); \alpha(t)], \tag{9.2.5}$$

where the dynamics of $\alpha(t)$ is given by a prescribed stochastic process. Two control parameters that are used frequently to maintain the system in a far-from-equilibrium state are the flow rate and the bath temperature, if the system absorbs or evolves heat. Suppose the heat bath temperature is made a random variable. In practice, this may be accomplished by constructing a device that switches the coolant feeds from a number of reservoirs at different temperatures to the heat bath surrounding the reactor. The statistics of the switching process are under the control of the investigator. An especially simple version of this experiment is to simulate a Poisson–dichotomous noise process. Only two reservoirs of coolant fluid at different temperatures are required; the switch from one reservoir to the other can be made at exponentially distributed time intervals. Also, dichotomous noise has a number of attractive features from a theoretical point of view, and we shall treat it in considerable detail. In an exactly analogous way, one may impose noise on the rate of flow of reagents into the reaction vessel.

Much of our discussion will be devoted to such parametric noise processes, since many aspects of the mechanisms are exposed through studies of this type. Specifically, we shall consider the GK model with noise on ε and η, and the Rössler model with noise on the parameter a. In Section 9.6 we shall refer briefly to additive and velocity-additive noise processes, which can be used to probe different aspects of the system's dynamics.

Even within the class of parametric noise processes different kinds of parametric noise can be studied. Noise statistics can have dramatic effects on the rates and mechanisms of transitions. We shall consider both bounded and unbounded noise sources, but our emphasis will be on bounded noise, since precise criteria for the onset of noise-induced transitions can be established.

9.3 Basins and their boundaries

If the system possesses bistable states, phase space can be partitioned into regions whose phase points evolve to the same attractor. The structure of these basins of attraction and their boundaries are very important factors determining the mechanisms of noise-induced transitions. For the two-dimensional GK model the basins are simple and the basin boundaries may be constructed

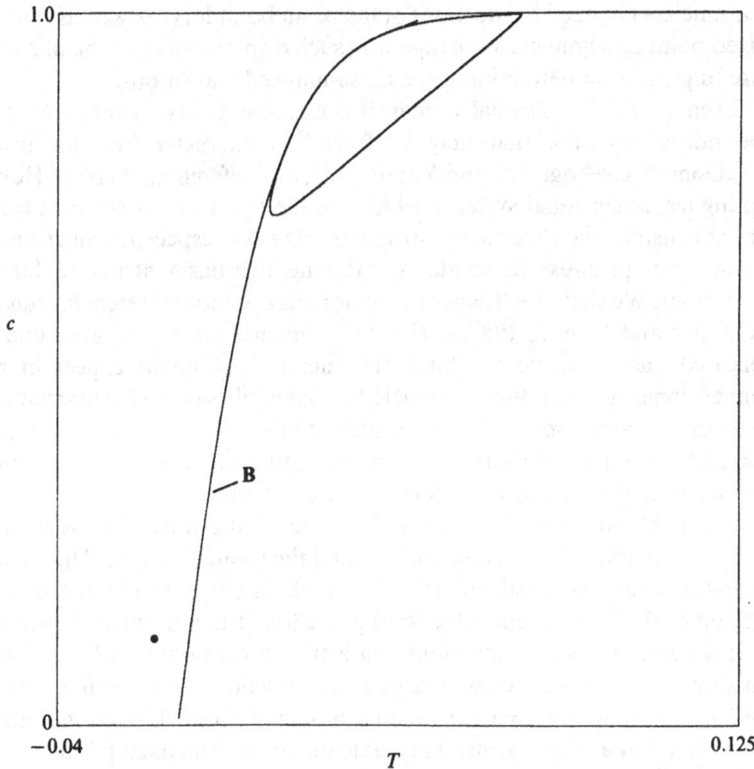

Figure 9.4. Plot of the coexisting limit cycle and fixed point in the cT-plane, along with the boundary **B** separating their basins of attraction. System parameters: ($\varepsilon = 0.35$, $\eta = -0.0312$).

easily. For the three-dimensional Rössler model the boundaries are not simple, but have a well-organized structure. We briefly discuss the boundaries for these two models since our discussion of noise-induced transitions relies heavily on their nature.

In the bistable region of the GK model there are three fixed points: the lower branch fixed point, that at low concentration, is stable within this region; the central fixed point is always unstable; and the upper fixed point is stable for part of the parameter range, but undergoes a Hopf bifurcation to a stable limit cycle, as described earlier. A linear stability analysis shows that the central fixed point is hyperbolic, and stable and unstable manifolds of this point can be found. The stable manifold constitutes the basin boundary separating the two basins of attraction. A phase plane (c, T) plot of the coexisting limit cycle and fixed point, along with the boundary, **B**, separating their basins of attraction, is shown in Figure 9.4. In this case a simple curve (the stable manifold) separates the two basins; all phase points to the left of this curve are attracted to the fixed point, while all points to the right are attracted to the limit cycle. As the system parameters change, the locations of the basin boundary, as well as that of the fixed point and limit cycle, change, and such displacements in the phase plane are important in determining the noise-induced transitions.

Even in low-dimensional systems it is possible to have a very complicated boundary structure that may be fractal in character (see, for instance, McDonald, Grebogi, Ott and Yorke, 1985, and references therein). However, in higher-dimensional systems, which are the rule rather than the exception in chemistry, the boundary structure may be especially intricate, and some attempt must be made to examine the basin structure for these situations. We shall use Rössler's equations as our model system for this study (Celarier and Kapral, 1987a). The three-dimensional basins are quite complicated and illustrate the kind of structure one might expect in multi-dimensional chemical kinetic models. Similar studies of the basin structure for differential equation models have been carried out by other investigators (Arecchi, Badii and Politi, 1985; Grebogi, Ott and Yorke, 1986; Gwinn and Westervelt, 1985; Levi, 1981; Moon and Li, 1985).

The stable and unstable manifolds of the fixed points of Rössler's model serve to organize the phase space flow and the basin structure. This system of equations has two fixed points, both of which are unstable in the bistable region of the phase plane. One fixed point lies at the origin, and possesses a one-dimensional stable manifold and a two-dimensional unstable manifold arising from a complex-conjugate pair of eigenvalues; the other fixed point lies at $(ak, -k, k)$, where $k = (c/a) - b$, and possesses a one-dimensional unstable manifold and a two-dimensional stable manifold. The fixed points and their stable and unstable manifolds are shown in Figure 9.5. The three-dimensional basin structure can be constructed numerically in the following way. First, a reference Poincaré plane containing the z-axis is chosen to lie in the third quadrant of the xy-plane, and is used to identify the stable fixed points. Next, in a series of planes containing the z-axis and oriented at an angle θ to the positive x-axis, an ensemble of uniformly and randomly distributed initial phase points is selected. Each phase point is evolved in time until it lies sufficiently close to one of the fixed points in the reference plane. In this way the initial phase point can be classified as belonging to one or the other of the coexisting period-three limit cycles. The results of this calculation are shown in Figure 9.6 for $\theta = 180°$ and $\theta = 310°$. The basin structure is highly fragmented and develops a spiral structure in some regions of the phase plane.

The spiral structure arises from the fact that the phase space flow is governed by the manifolds of the fixed points. Consider a phase point in the vicinity of the origin and with positive z. Due to the strongly contracting character of the flow, arising from the large negative eigenvalue of the stable manifold which lies nearly along z, this phase point will be injected rapidly onto the xy-plane near the origin. Once there, it will spiral out from the origin and evolve to one of the stable limit cycles. Since points in the vicinity of the origin can evolve to either attractor, we expect a spiral basin structure near the origin in the

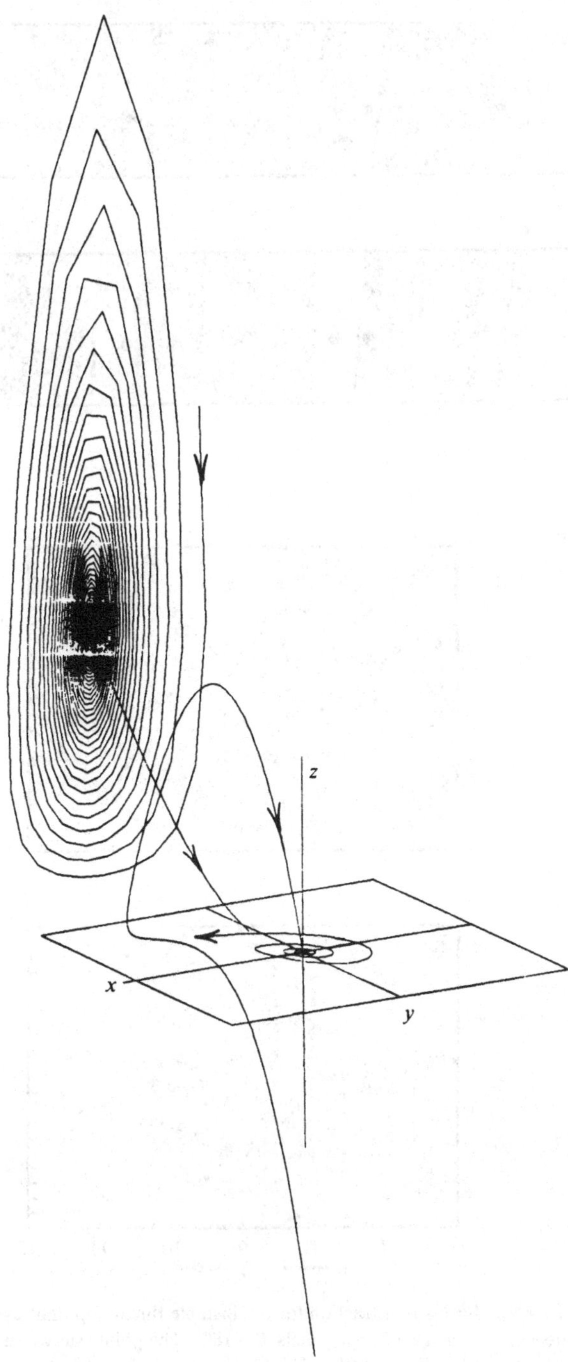

Figure 9.5. Unstable fixed points for Rössler's model and their stable and unstable manifolds. System parameters: ($a = 0.3425$, $b = 0.4$, $c = 5.8$).

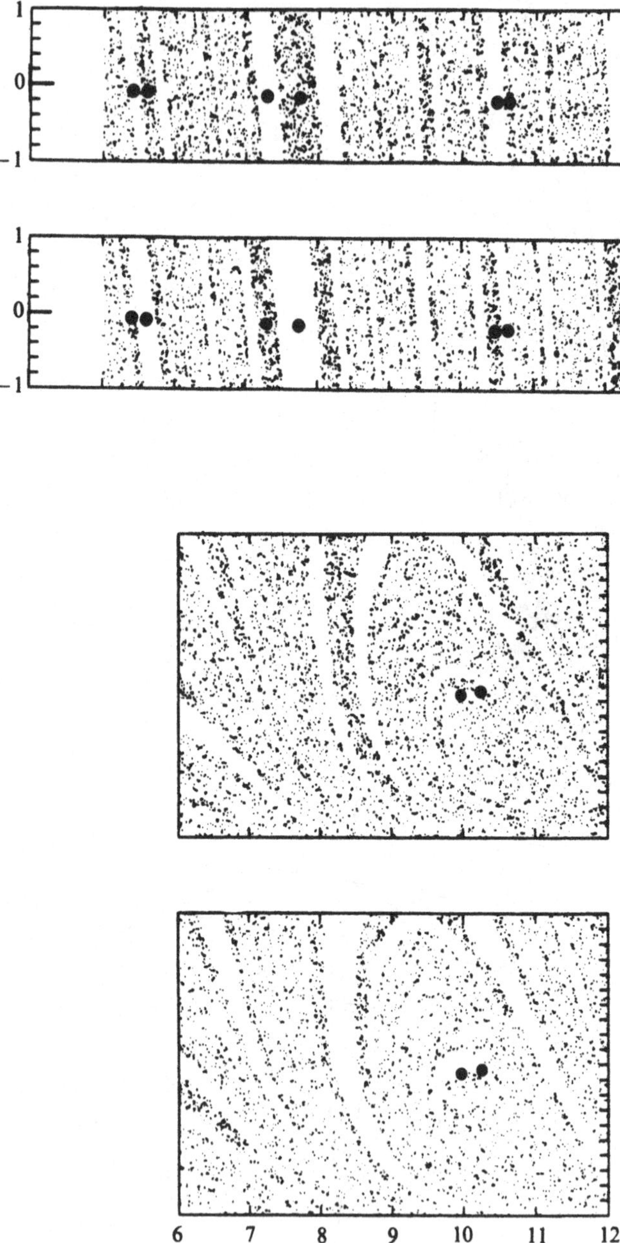

Figure 9.6. Basins of attraction for the bistable three-loop limit cycles of Rössler's model. Two upper panels, $\theta = 180°$: The points shown in each panel evolve to different limit cycles. The two lower panels correspond to $\theta = 310°$. The dark points indicate where the stable limit cycles pierce the section plane.

xy-plane. However, all preimages of these points must also lie in the basin of attraction. So one may imagine propagating the phase points backward in time to generate additional features of the basin structure. Phase points near the origin that are evolved backward in time will closely follow the stable manifold which loops through phase space as shown in Figure 9.5. Therefore, the spiral basin structure near the origin occurs in other regions of the phase space as it follows the stable manifold. This gives rise to the rather complicated basin structure observed in the simulations. The following terminology is useful: In each section plane of Figure 9.6 there is a continuous region surrounding each attractor, whose phase points belong to the basin of attraction; this is the *primary basin*. Other basin regions are preimages of these primary basins.

We shall refer to this basin structure in the sequel, but it is convenient to point out some of its features here. The character of the basins is not the same everywhere along the limit cycles. In particular, when the limit cycles lie very close to the xy-plane the local basin boundaries are roughly parallel to the z-axis. In the fourth quadrant, where the reinjection mechanism operates, the limit cycles lie at varying angles to the spiral structure in this plane created by the passage of the stable manifold. Hence, one might expect that different parts of the cycle will have different susceptibility to noise. This feature will be important when noise-induced transitions between these cycles are considered.

Basin boundaries may be determined experimentally using techniques analogous to those described above; perturbations in the chemical composition can serve to change the initial condition, and the initial phase points can be labeled according to the attractor to which they evolve. Experiments of this type have been carried out on a two-variable chemical system possessing bistable steady states, and the stable manifold which constitutes the basin boundary has a configuration similar to that of the GK model (Pifer, Ganapathisubramanian and Showalter, 1985). Perturbation experiments of this type have been performed on more complex chemical reactions (Bar-Eli and Geiseler, 1983; DeKepper, Epstein and Kustin, 1981; Orban and Epstein, 1982).

9.4 Discrete-time models

Two of the mechanisms are discussed most conveniently in terms of the one-dimensional map underlying the phase space flow. Therefore, we shall first show how certain aspects of the bistable limit cycle regime in Rössler's model can be understood in terms of maps, and then, in Sections 9.5.3 and 9.5.4, we consider stochastic versions of these maps to study noise-induced transitions.

The three-loop limit cycles in Rössler's model arise from a chaotic state by a saddle-node bifurcation. The nature of the chaotic state changes from spiral-type chaos in the vicinity of C_1 to screw-type chaos around C_2 as one moves in

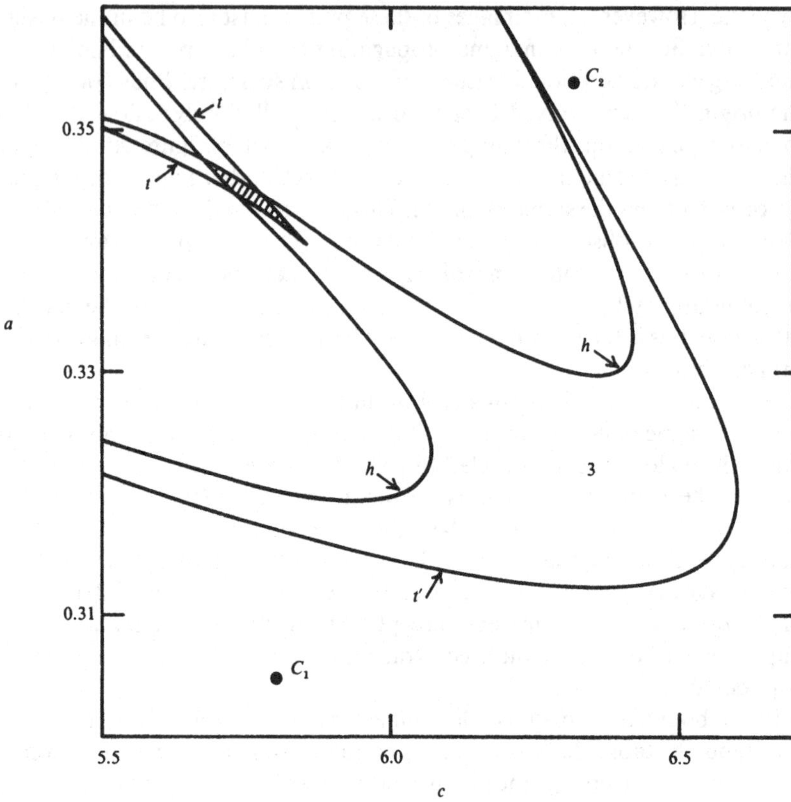

Figure 9.7. A portion of the phase diagram for Rössler's model. The type of solution is shown as a function of the a and c parameters for $b = 0.4$. The cusp-shaped lines labelled t enclose a region (shaded) where bistable three-loop limit cycles exist. Singly stable three-loop limit cycles exist within the area bordered by t' boundaries, which signal the transition from periodic to chaotic dynamics by an intermittency mechanism, and h boundaries, which determine when the limit cycle doubles its period by a subharmonic bifurcation.

the parameter plane around the region of stable three-loop limit cycles (see Figure 9.7).

A one-dimensional map model of the dynamics can be constructed in the chaotic regions of the parameter plane using the following procedure. Select a Poincaré surface (planes containing the z-axis and lying in the second and third quadrants are convenient), and plot the value of one coordinate, say $-x(t)$, at the nth intersection of the plane, x_n, versus its value at the $n+1$st intersection, x_{n+1}. A return map corresponding to screw chaos is shown in Figure 9.8. Notice that the map possesses two extrema. The spiral chaotic attractor has a simpler structure and its return map has a single extremum; the second extremum appears as an additional fold in the flow develops, and it is

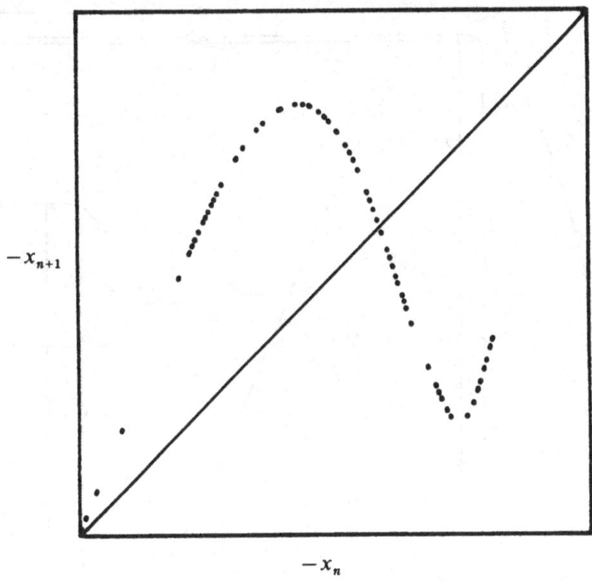

Figure 9.8. A return map for screw chaos near C_2 (cf. Figure 9.7), constructed from successive crossing of the Poincaré surface.

this feature that distinguishes screw from spiral chaos (Rössler, 1979).

While our interest in this chapter is not in chaotic dynamics, the existence of the second extremum in the map also has important consequences for the periodic orbit structure. Bistable three-loop limit cycles are very closely related to this feature of the map. In fact, neither the existence of bistability nor the appearance of periodic orbits from chaos depends on the details of the map. Hence, it suffices to study a two-parameter cubic polynomial,

$$x_{n+1} = ax_n^3 + (1-a)x_n + b(1-x_n^2) \equiv C(x_n; a, b), \tag{9.4.1}$$

which displays a bifurcation structure similar to that of Rössler's model (cf. Figure 9.7) (Celarier, Fraser and Kapral, 1983; Fraser and Kapral, 1982; Kapral and Fraser, 1984). The period-three orbits of the map correspond to three-loop limit cycles in Rössler's model, since such periodic orbits pierce the Poincaré section in three points, which are visited cyclically (see Figure 9.3). The period-three orbit can be determined from the fixed points of the third-iterate map:

$$x_i = C(C(C(x_i))) = C^{(3)}(x_i; a, b), \quad (i = 1, 3), \tag{9.4.2}$$

and its stability depends on the product of the slopes along the orbit,

$$s = \prod_{i=1}^{3} C'(x_i), \tag{9.4.3}$$

with stability for $|s| < 1$ (Collet and Eckmann, 1980). Stable period-three orbits

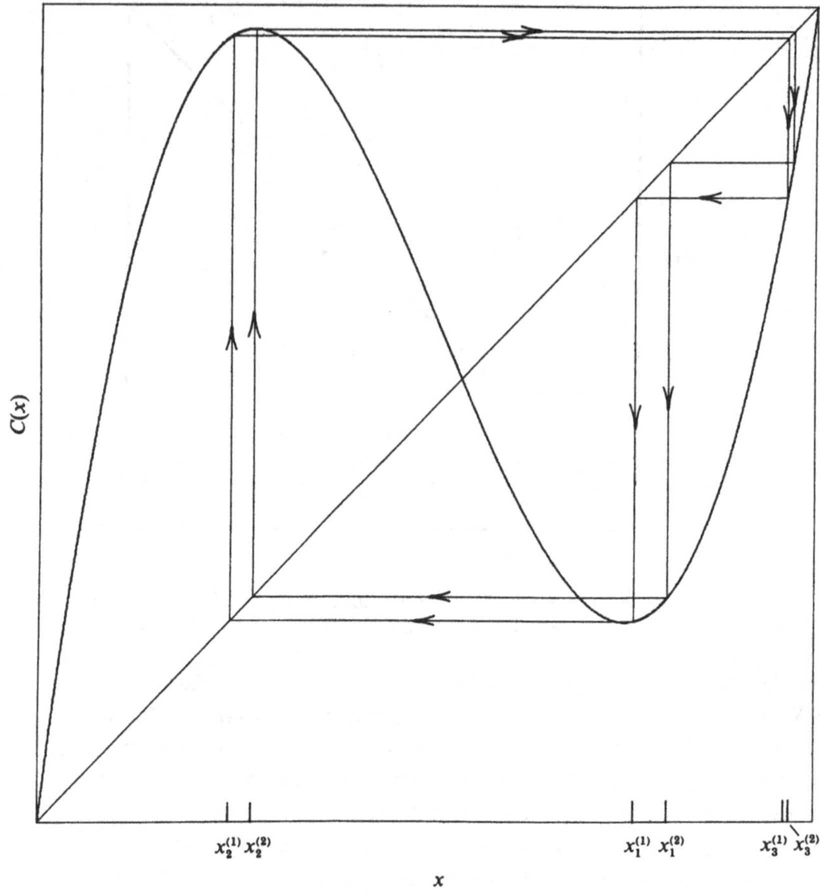

Figure 9.9. Cubic map and bistable period-three orbits. The superscripts on the fixed points label the orbit to which they belong.

appear by a tangent bifurcation process where $s = +1$ at the bifurcation point; a second stable period-three orbit is born by a bifurcation where $s = +1$ also. Examples of bistable period-three orbits for the cubic map are shown in Figure 9.9. Since $C'(x) = 0$ at the maximum x_M and minimum x_m of the map, these points and their vicinities provide sources of stability for the periodic orbits. Because there are two extrema it is possible to simultaneously stabilize two periodic orbits.

9.5 Mechanisms for noise-induced transitions

The following subsections present studies of four mechanisms for noise-induced transitions that have been identified in the model systems described above. While these examples do not exhaust the list of possible mechanisms, they do provide some insight into how noise-induced transitions take place in nonequilibrium chemical systems. Furthermore, they illustrate some of the peculiar features of transitions involving limit cycles. Even for the simple case of irreversible escape from a limit cycle to a fixed point two mechanisms operate; these are described in Subsections 9.5.1 and 9.5.2. Two additional mechanisms that have been identified for transitions between coexisting limit cycles in Rössler's model are discussed in Subsections 9.5.3 and 9.5.4.

The mechanisms take their simplest forms for parametric dichotomous noise processes. Since only two parameters are involved, the operation of the mechanisms entails the study of the dynamics of two system states. This leads to a great simplification of the description. For Poisson–dichotomous noise the switch between parameter values occurs at exponentially distributed times characterized by the noise correlation time, τ_n. More specifically, if a parameter $\alpha(t)$ takes on the two values α_1 and α_2, the probability of finding the system in $\alpha(t) = \alpha_{1,2}$, $p(\alpha_{1,2}, t)$ is given by the solution of the pair of equations,

$$\frac{\mathrm{d}}{\mathrm{d}t}\begin{pmatrix} p(\alpha_1, t) \\ p(\alpha_2, t) \end{pmatrix} = -\tau_n^{-1}\begin{pmatrix} 1 & -1 \\ -1 & 1 \end{pmatrix}\begin{pmatrix} p(\alpha_1, t) \\ p(\alpha_2, t) \end{pmatrix}. \tag{9.5.1}$$

Once the operation of a mechanism for this type of noise process is understood, generalization to other situations is made easily. We shall take that up briefly in Section 9.6.

9.5.1 Broken-limit-cycle mechanism

Consider the GK model and imagine that the system is in an oscillating state characterized by a flow rate ε and a cooling temperature $\eta(t)$, whose time dependence is governed by a Poisson–dichotomous noise process. Thus $\eta(t)$ takes on two values η_1 and η_2 for time intervals selected from an exponential distribution with correlation time τ_n. In some circumstances the imposition of dichotomous noise can induce transitions to the stable fixed point. Consider again the solution structure in Figure 9.2. As η decreases the amplitude of the limit cycle grows until, at a critical value of $\eta = \eta_c$, it touches the locus of unstable fixed points (dashed line). When this occurs the limit cycle is 'broken' at a point; it disappears and the phase point will evolve to the stable fixed point. This process is akin to the crises described by Grebogi, Ott and Yorke (1983) for the destruction of chaotic attractors. If the parametric noise process is such that $\eta(t)$ takes on a value that exceeds η_c, transitions to the fixed point may take place. This rather straightforward mechanism possesses a number of interesting features and will operate only if certain conditions are met.

We present the results of simulations of this stochastic version of the GK model in order to explore some of these features. We suppose that $\varepsilon = 0.35$ and the dichotomous noise process is such that one value of η is fixed at $\eta_1 = -0.031$, which corresponds to a stable limit cycle coexisting with a fixed point.

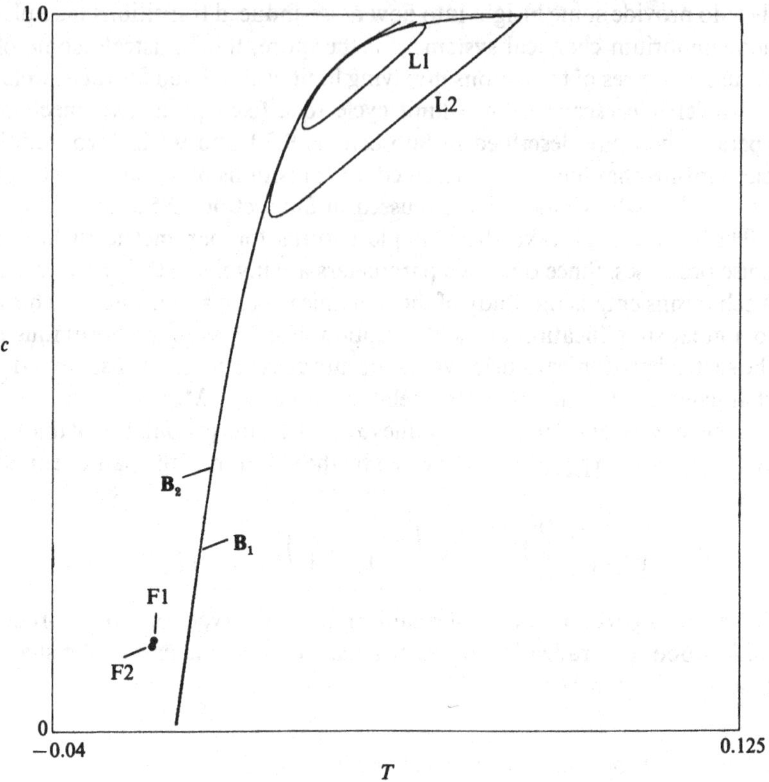

Figure 9.10. Phase plane plot of coexisting limit cycles and fixed points, along with basin boundaries, for two values of the parameter η for $\varepsilon = 0.35$. Notation: $\eta_1 = -0.031$: L1, F1, $\mathbf{B}_1$; $\eta_2 = -0.0312$: L2, F2, $\mathbf{B}_2$, where L, F and **B** label the limit cycles, fixed points and basin boundaries, respectively.

The stochastic dynamics may then be examined for a series of cases where η_2 varies until it moves beyond η_c.

As η_2 decreases the amplitude of the limit cycle increases, as does its period. The limit cycles corresponding to η_1 and one value of η_2, $\eta_2 = -0.0312$, along with the basin boundaries, are shown in Figure 9.10. Since no basin boundary is crossed when η changes, no noise-induced transition to the fixed point will occur; only a noisy limit cycle is produced. However, since the limit cycle grows as η_2 decreases, eventually it will collide with the unstable fixed point and lose its stability at η_c. Now a noise event that changes η to a value $\eta_2 < \eta_c$ can induce a transition to the stable fixed point.

An examination of this transition mechanism shows that direct escape occurs from a particular region of phase space where the destruction of the limit cycle took place. The dynamics of the transition process depends on the nature of the phase space flow for parameters corresponding to the broken limit cycle. This is revealed by the following simulation. Suppose the system is

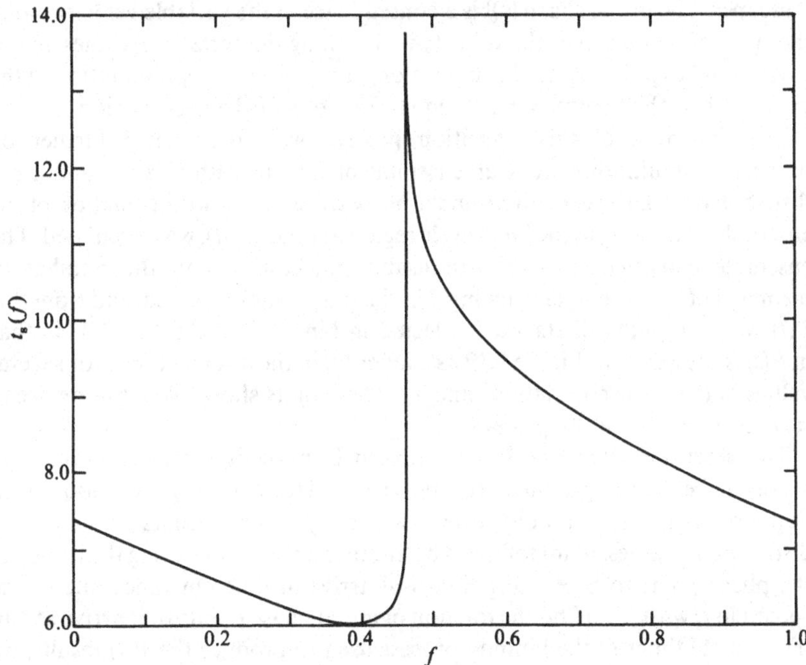

Figure 9.11. Time taken to reach S starting from a point f on the limit cycle L1, $t_s(f)$, versus f.

initially in a limit cycle with $\eta = \eta_1$(L1), and an arbitrary marker point on this cycle, taken here as the point of minimum c, is picked. The system is evolved under the deterministic dynamics with $\eta = \eta_1$ for a time $t_f = f\tau_1$, where $0 \leqslant f \leqslant 1$ and τ_1 is the period of L1, at which point the parameter η is set to η_2. Evolution under the deterministic dynamics with $\eta = \eta_2$ is then carried out until the system reaches a small neighborhood of the stable fixed point. Setting the origin of time at the point when η is set to η_2, the time interval to reach the fixed point can be determined as a function of f. Actually, it is sufficient to follow the trajectories until evolution to the fixed point is certain, regardless of the value of η. Clearly, any phase point that lies in the basin of attraction of F1 will satisfy the condition, since return to L1 will be impossible. Operationally, it is easier to follow trajectories until they reach a surface **S** defined by $c = 0.4$ and $T \leqslant T_B$, where T_B is determined from the solution of $\mathbf{B}_1(c = 0.4, T_B) = 0$, the equation for the basin boundary. Phase points crossing **S** will evolve to the fixed point region regardless of the value of η. There is a great deal of freedom in the selection of **S**, but this choice suffices for our purposes. Let $t_s(f)$ be the time required to reach **S** starting from point f on the cycle L1. Figure 9.11 shows $t_s(f)$ versus f. The sudden jump in t_s at $f \approx 0.5$ from near 6 to more than 13 arises from the fact that beyond a certain value of f the phase point must execute a loop around the unstable fixed point that gave rise to the cycle before it escapes. The initial sharp fall is a consequence of the variable velocity along the cycle. Note that since the velocity varies along the cycle the f values do not correspond to points spaced equally along a phase space representation of the cycle; rather, they correspond to equal fractions of the cycle period.

The dynamics of this transition process was investigated further by numerical simulations where an ensemble of systems with f values uniformly distributed on L1 was evolved under noisy dynamics, and the fraction of the ensemble remaining in the limit-cycle region at time t, $P(t)$, was calculated. The ensemble consisted of several-thousand trajectories, and the number of members of the ensemble reaching **S** in the time t was measured, and from this $P(t)$ was computed. Data was collected in bins $i-1/2 \leqslant t \leqslant i+1/2$ so that $\ln P(t)$ is presented in Figure 9.12 as a function of the discrete time t for several values of the noise correlation time, τ_n. These plots show two separate decay regions at short and long times.

The short time dynamics follows directly from the information in the $t_s(f)$ versus f behavior, provided the noise correlation time is not too short. Suppose that the first noise event, which by design occurs at uniformly distributed f values, is not followed by another noise event during the transit of the phase point to **S**. Phase points will arrive at different times due to the spread in t_s with f. To find the fraction of initial phase points that arrive at **S** in the discrete times of the binning process used to produce the $P(t)$ results, we may calculate the fraction of the limit cycle, Δf_t, yielding $t_s(f)$ values in the range $i-1/2 \leqslant t \leqslant i+1/2$. These results are shown in the last column of Table 9.1. So, provided no noise event acts before the phase point reaches **S**, we may determine from these values the fractions of phase points reaching **S** as a function of time. In the absence of any subsequent noise events in the stochastic evolution, the decay of the initial ensemble could be computed directly from Δf_t. Of course, noise may act on the system during its evolution toward **S**, and we must take this into account. Actually, from the remarks made above, noise will be effective during only part of the evolution toward **S**, that lying within the basin of L1. Therefore, we must weight the Δf_t values by the probability that no noise event acts during the time the system is within the basin of L1. If we let this time interval be $t_B(f)$ for a phase point starting at f, then, for exponentially distributed noise events, the probability that no noise event acts in t_B is $\exp[-t_B(f)/\tau_n]$. We may estimate t_B crudely from t_s. Most phase points starting from L1 cross $\mathbf{B}_1$ within a small interval of the basin boundary. Numerical simulations show that it takes approximately five time units for a phase point move from this interval of $\mathbf{B}_1$ to **S**, and we may estimate $t_B(f)$ as $t_B(f) = t_s(f) - 5$. Hence, the fraction of phase points reaching **S** (and, thus, the fixed point) is

$$P_s(t) = \Delta f_t \, e^{-t_B/\tau_n}. \tag{9.5.2}$$

The results are presented in Table 9.1 and compared with the simulations for various values of τ_n. Note in particular the maximum at $t = 7$, which is a

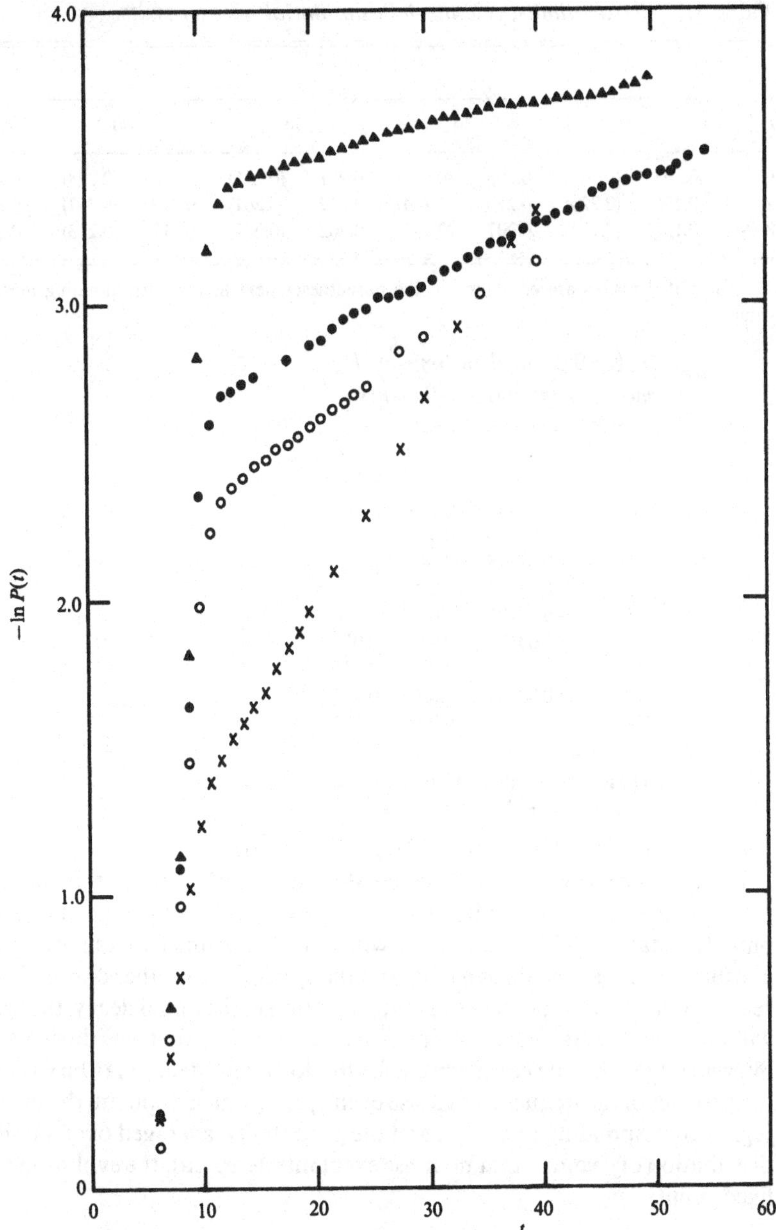

Figure 9.12. Plot of the natural logarithm of the fraction of phase points that did not reach S in the (discrete) time t, $\ln P(t)$, versus t for several values of τ_n and $\eta_2 = -0.03125$. The simulations were carried out by evolving an ensemble of 4000 systems in L1 with uniformly and randomly distributed f values, and recording the time at which each member of the ensemble reached S. Key: ▲, $\tau_n = 100$; ●, $\tau_n = 50$; ○, $\tau_n = 30$; ×, $\tau_n = 10$.

Table 9.1. *Broken-limit-cycle mechanism: initial decay results.*

	τ_n								
t	10		30		50		100		Δf_t
6	0.193	(0.208)	0.205	(0.222)	0.221	(0.225)	0.226	(0.228)	0.230
7	0.206	(0.223)	0.250	(0.254)	0.242	(0.261)	0.249	(0.267)	0.272
8	0.155	(0.163)	0.207	(0.199)	0.202	(0.207)	0.212	(0.213)	0.220

The calculated results are contained in the parentheses next to the corresponding measured values.

Table 9.2. *Broken-limit-cycle mechanism: rate coefficients.*

τ_n	η_2	k	k^*
1	−0.03135	0.124	
1	−0.03130	0.088	
1	−0.03125	0.048	
1	−0.03123	0.028	
1	−0.03121	0.0025	
10	−0.03125	0.070	0.077
30	−0.03125	0.029	0.031
50	−0.03125	0.018	0.019
100	−0.03125	0.010	0.010

The measured rate coefficient is k, while that for the stochastic model is k^*.

consequence of the large value of Δf_t at this time.

The long-time decay may be calculated in a similar manner. If the initial phase point does not reach $\mathbf{B}_1$ before a noise event acts, it will be kicked back onto the stable cycle. The system will remain confined to this cycle until another noise event changes η to η_2 and the system has another opportunity to reach the stable fixed point. As in the calculation of the initial decay, the system will evolve to the stable fixed point provided no noise event acts during a time interval $t_B(f)$. The rate coefficient, k, for this long-time decay may be written as the product of the frequency of noise events, τ_n^{-1}, which populate the unstable region corresponding to $\eta = \eta_2$, and the probability, averaged over a uniform distribution of f values, that no noise event interferes with the evolution to the fixed point,

$$\mu = \int_0^1 \mathrm{d}f \exp[-t_B(f)/\tau_n]. \tag{9.5.3}$$

Thus, $k = \mu/\tau_n$. The numerical results for k are compared with this stochastic model in Table 9.2 for different values of τ_n for $\eta_1 = -0.031$ and $\eta_2 = -0.03125$. Notice that in the limit of $\tau_n \gg \tau_1$ the rate coefficient is simply given by τ_n^{-1}, since the rate-limiting step is clearly the supply of phase points to the unstable region. Once there, they will surely escape if the noise correlation time is long enough. This produces the obvious result that as the noise correlation time decreases the transition rate increases.

When $\tau_n < \tau_1$ new effects come into play and the transition rate does not decrease as rapidly as predicted by the simple model. When the correlation time is short the noise acts many times during the course of a cycle and transient dynamics are important. Now it is not only the supply of phase points to the unstable region that matters, but also the fact that phase points may be kicked back and forth in the stable and unstable regions before escaping.

The transition rate increases as the penetration into the broken-limit-cycle region increases. Table 9.2 presents results for $\tau_n = 1$ for a number of different η_2 values in the broken-limit-cycle region. Now a smaller fraction of the phase points loop around the unstable fixed point that gave rise to the limit cycle, leading to a larger transition probability.

9.5.2 *Moving boundary mechanism*

The mechanism described above involved parameter excursions into a region where the limit cycle no longer existed. However, it is possible to induce transitions to the fixed point when both parameter values in the dichotomous noise process correspond to stable limit cycles on the upper branch. The mechanism can best be described with reference to Figure 9.13, where the two stable limit cycles and their associated basin boundaries are shown. We denote the inner cycle by L1 and the outer cycle by L2. Coexisting with these two limit cycles we have stable fixed points F1 and F2. Note that the basin boundary corresponding to L1 crosses a portion of L2. We shall refer to this part of L2 as the vulnerable phase region for the following reason: if a noise event acts while the system is in this part of the L2 cycle the phase point will suddenly find itself in the basin of the fixed point F1. A transition will have been induced by the sudden shift of the basin boundary.

This mechanism has a number of interesting features. As noted above, only certain phase regions on L2 are vulnerable to noise-induced transitions. This will have important consequences for the dynamics of the transition process. Furthermore, the transition to the fixed point can only occur for the noise process that takes the parameters from L2 $\rightarrow$ L1, and not the reverse.

The vulnerable phase region can be determined by a procedure similar to that used to determine t_s. Suppose the system is initially in the stable cycle L2. As earlier, a marker on this cycle, the point of minimum c, is picked as the origin. At a given fraction f of the limit cycle period, τ_2, measured from the marker, the parameters are changed to those appropriate to L1, and the time it takes the phase point to reach S, t_s, is determined. Of course, if the phase point never reaches S, $t_s = \infty$, while in the vulnerable region t_s will be finite. Thus, the

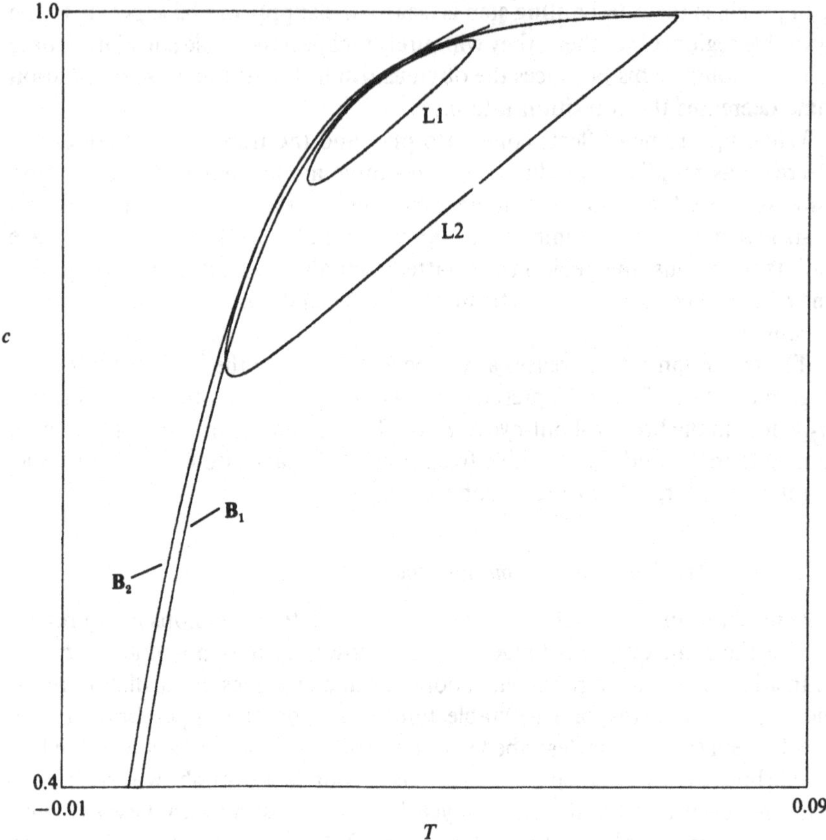

Figure 9.13. Configuration of limit cycles and basin boundaries involved in the moving boundary mechanism. The inner limit cycle L1 corresponds to ($\varepsilon_1 = 0.375$, $\eta_1 = -0.034$), while the outer limit cycle L2 has parameters ($\varepsilon_2 = 0.35$, $\eta_2 = -0.0312$). The associated basin boundaries are $\mathbf{B}_1$ and $\mathbf{B}_2$.

$t_s(f)$ versus f plots have a form different from that in the broken-limit-cycle mechanism.

All of our simulations are carried out with the parameters for L1 fixed at ($\varepsilon_1 = 0.375$, $\eta_1 = -0.034$), and those of L2 are ($\varepsilon_2 = 0.35$, η_2), with η_2 variable. The dichotomous noise process now acts simultaneously on ε and η. While this type of noise process requires switching the flow rate and bath temperature simultaneously in the context of the GK model and is more difficult to realize experimentally, this need not be the case in all circumstances. This feature is a function of the region of parameter space under study as well as the details of the model, and other parameter regions or reaction schemes may very well exhibit this mechanism for variations in a single parameter.

Figure 9.14 presents the results of such calculations for two values of the system parameters. In the figure one can see the change in size of the vulnerable

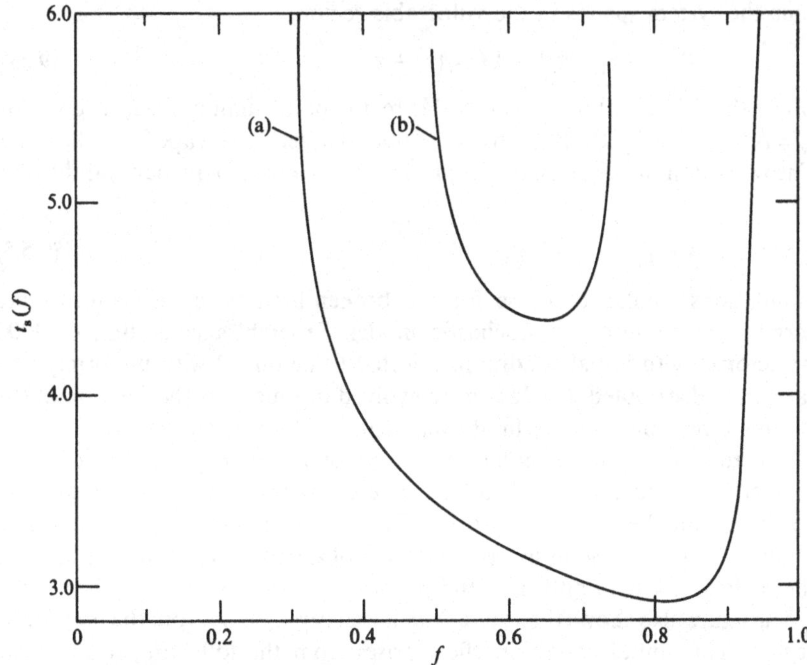

Figure 9.14. Plot of $t_s(f)$ versus f for parameter changes (a) L2 ($\varepsilon_2 = 0.35$, $\eta_2 = -0.0312$) $\rightarrow$ L1 ($\varepsilon_1 = 0.375$, $\eta_1 = -0.034$), and (b) L2 ($\varepsilon_2 = 0.35$, $\eta_2 = -0.0305$) $\rightarrow$ L1 ($\varepsilon_1 = 0.375$, $\eta_1 = -0.034$), for the moving boundary mechanism.

region as the parameters are tuned. We can imagine tuning the η parameter of the L2 cycle until it lies completely within the basin boundary of L1: no noise-induced transition will occur for any value of f for η beyond this critical value.

The parameter variations considered here lead to one-way transitions out of the limit-cycle state. The configuration of the basin boundaries in the fixed-point region is like that in Figure 9.10. A change in parameter values as a result of a noise event will not cause the system to be found in the basin of the limit cycle. The system will hop from F1 to F2 giving rise to a noisy fixed point; hence the transition from the limit cycle is irreversible. Two-way transitions occur in other parameter regions, but we shall not study these here.

A very crude stochastic model accounts for most aspects of the results for the transition rate coefficient. Provided the noise correlation time is long compared to the periods of the limit cycles, the dynamics will be confined largely to the regions of the cycles. If noise-induced transitions are possible due to the existence of a range of vulnerable phases, we can imagine the system in the limit-cycle region as composed of a single noisy limit-cycle state with a 'hole' whose size is a measure of the range of vulnerable phases. Naturally, such a simplified description does not take into account the internal structure of the limit cycles. The measure of the 'hole', β, can be estimated as the fraction of time the system spends in the vulnerable region:

$$\beta = t_v/(\tau_1 + \tau_2) = \Delta f \tau_2/(\tau_1 + \tau_2), \tag{9.5.4}$$

where the time spent in the vulnerable region of the limit cycle, t_v, is given by $t_v = \Delta f \tau_2$, with Δf the fraction of f values that lead to escape from the cycle. The transition rate is given by the product of the noise frequency and the hole size:

$$k = \beta/\tau_n. \tag{9.5.5}$$

Simulations similar to those for the broken-limit-cycle mechanism were carried out to test this stochastic model. Ensembles consisting of 4000 trajectories with initial conditions selected to lie on L2 with uniformly and randomly distributed f values were evolved in time, and the fraction of the ensemble remaining in the limit-cycle region, $P(t)$, was computed.

The results of these simulations are presented in Figure 9.15, in which $\ln P(t)$ is plotted versus t. The long-time exponential decay yields the rate coefficient for the transition process. Table 9.3 compares the measured and predicted rate coefficients for several values of τ_n and η_2. These results confirm the scaling of k with both τ_n^{-1} and β.

The figure also shows that there is an initial rapid decay from the limit-cycle region. This initial condition effect arises from the following source. The system was prepared in L2 with a uniform distribution of f values. Phase points in the vulnerable region will decay directly to the fixed point provided no noise acts before the system reaches the point of no return. The considerations parallel exactly those for the broken-limit-cycle mechanism. The initial decay, for large τ_n, is given by the fraction of phase points in the vulnerable regions which reach **S** at the measured discrete time intervals.

Also shown in the figure is the result of one simulation where the system was prepared in L1 with uniformly and randomly distributed f values. Now the system must hop to L2 before escape can occur, and one would expect the initial decay region to be absent, but the long-time decay should be the same as that when the system is prepared in L2. This is indeed the case, confirming this aspect of the mechanism.

9.5.3 Tangent mechanism

We turn now to study transitions between two limit cycles using Rössler's model. The first mechanism we describe has some features in common with the intermittency route to chaos (Pomeau and Manneville, 1980) and, like that process, is discussed conveniently in terms of one-dimensional maps.

We argued in Section 9.4 that many aspects of the appearance of three-loop limit cycles in Rössler's model and the existence of bistability can be understood in terms of a cubic one-dimensional map. First we consider a stochastic version of this map and show how noise can induce transitions between the

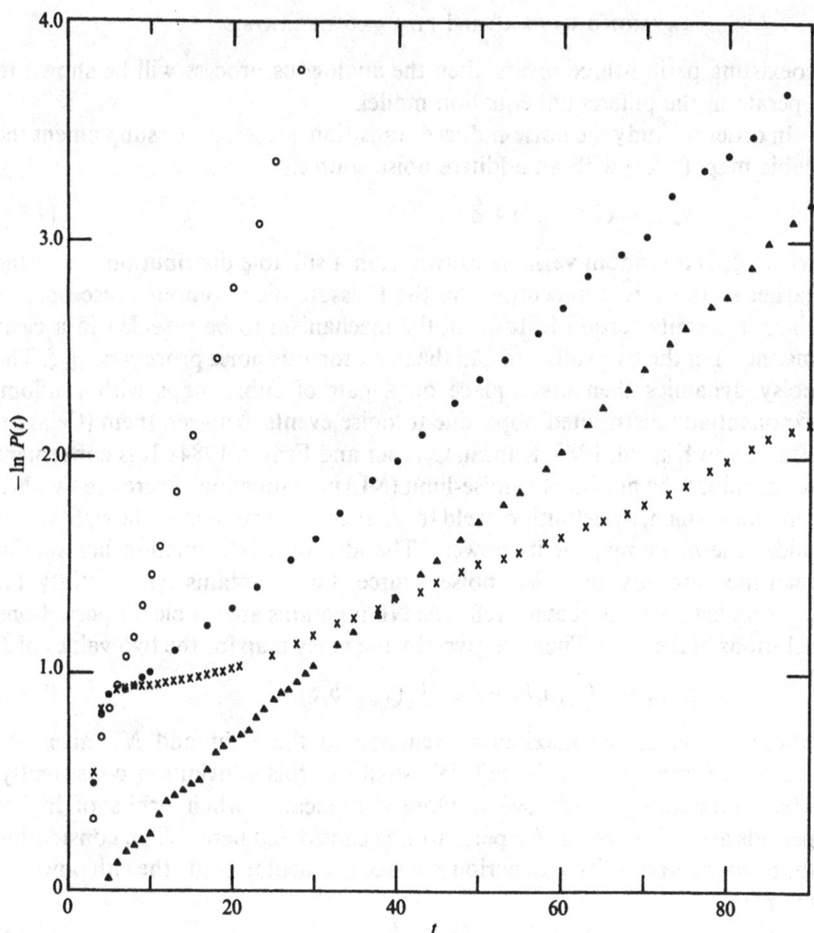

Figure 9.15. Plot of ln $P(t)$ versus t for several values of τ_n, starting from an initial state on L2. The parameter values in the Poisson–dichotomous noise process are: ($\varepsilon_1 = 0.35$, $\eta_1 = -0.0312$) and ($\varepsilon_2 = 0.375$, $\eta_2 = -0.034$). Symbols: ○, $\tau_n = 1$; ●, $\tau_n = 10$; ×, $\tau_n = 20$. Also shown as triangles is the result of a simulation with $\tau_n = 10$ starting on L1.

Table 9.3. *Moving boundary mechanism: rate coefficients.*

τ_n	η_2	k	k^*
1	−0.03120	0.130	
10	−0.03120	0.035	0.040
20	−0.03120	0.016	0.020
10	−0.03115	0.031	0.031
10	−0.03050	0.010	0.012

The measured rate coefficient is k, while that for the stochastic model is k^*.

coexisting period-three orbits, then the analogous process will be shown to operate in the differential equation model.

In order to study the noise-induced transition processes, we supplement the cubic map, (9.4.1) with an additive noise source:

$$x_{n+1} = C(x_n; a, b) + \xi_n, \tag{9.5.6}$$

where ξ_n is a random variable drawn from a suitable distribution. As in the earlier sections, we concentrate on the Poisson–dichotomous noise process since it permits certain features of the mechanism to be revealed in a clear manner. Let the two values of ξ in the dichotomous noise process be $\pm\xi$. The noisy dynamics then takes place on a pair of cubic maps with random, exponentially distributed hops, due to noise events, between them (Celarier, Fraser and Kapral, 1983; Kapral, Celarier and Fraser, 1984). It is convenient to introduce the notion of a noise-limit (NL) map function. There are two NL functions which, by definition, yield the maximum excursion to the right or left under the noisy map or its powers. The idea of a NL function has special relevance for any bounded noise source, but it retains some utility for unbounded noise sources as well. The NL functions are simple for period-one solutions of the map. They are given by the noisy map for the two values of ξ,

$$x_{n+1} = C(x_n; a, b) \pm \xi \equiv N_{\pm}(x_n; a, b; \xi), \tag{9.5.7}$$

where N_+ gives the maximum excursion to the right and N_- gives the maximum excursion to the left. We shall use this convention consistently. These functions are somewhat more complicated when orbits of higher periods are of interest, as for period-three considered here. When considering the properties of orbits with period m it is convenient to study the mth power of the map,

$$C^{(m)}(x; a, b) = C(C \ldots C(x; a, b) \ldots; a, b), \tag{9.5.8}$$

since each of the m points on the orbit is a fixed point of the mth power of the map. Likewise, for the noisy dynamics one may ask whether these fixed points simply become fuzzy under the influence of noise or if loss of phase coherence (the cyclic property is lost under the noisy dynamics) and noise-induced transitions take place. The complication with the NL functions arises from the fact that in different regions of the interval on which the map dynamics occurs, different values of ξ will be responsible for maximum excursions to the right or left. Consequently, we must define the NL functions corresponding to powers of the map by:

$$N_{\pm}^{(n)}(x; \xi) = \begin{cases} \max\limits_{\xi_1, \xi_2, \ldots, \xi_{n-1}} C(C(\ldots C(x; \xi_1); \ldots; \xi_{n-1}); +\xi) \\ \min\limits_{\xi_1, \xi_2, \ldots, \xi_{n-1}} C(C(\ldots C(x; \xi_1); \ldots; \xi_{n-1}); -\xi), \end{cases} \tag{9.5.9}$$

where the intermediate values ξ_i can be either $\pm\xi$. For this dichotomous noise process the NL functions are continuous functions of x with piecewise

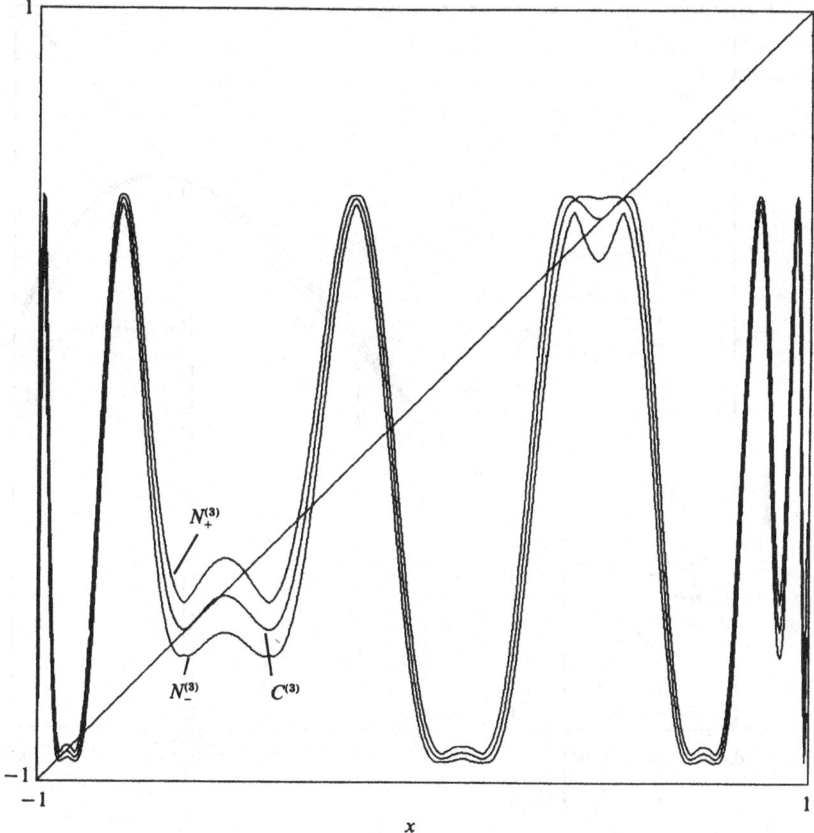

Figure 9.16. Noise-limit cubic map function for a dichotomous noise process. The noisy dynamics is confined to these two curves. For comparison the cubic map with $\xi = 0$ is also shown, although this function is not involved in the noisy dynamics. Note the change in character of the NL maps in some regions of the interval.

continuous derivatives. The NL function for the third power cubic map, $N_\pm^{(3)}$, is sketched in Figure 9.16; the switch in the character of the NL function in certain regions of the interval is evident. The curve lying between the two NL functions is $C^{(3)}$, and parameters were chosen so that bistable period-three cycles exist. Notice that the dynamics of a map with Poisson–dichotomous noise is due entirely to the NL functions, and not at all to the underlying deterministic map.

The stochastic dynamics of the cubic map in the period-three region may be described in terms of the NL functions for different values of the noise amplitude. An enlargement of the map functions near one of the bistable pairs of fixed points is shown in Figure 9.17(a). Since the noise has a finite amplitude it is possible to produce orbits that are noisy, yet periodic. Consider the

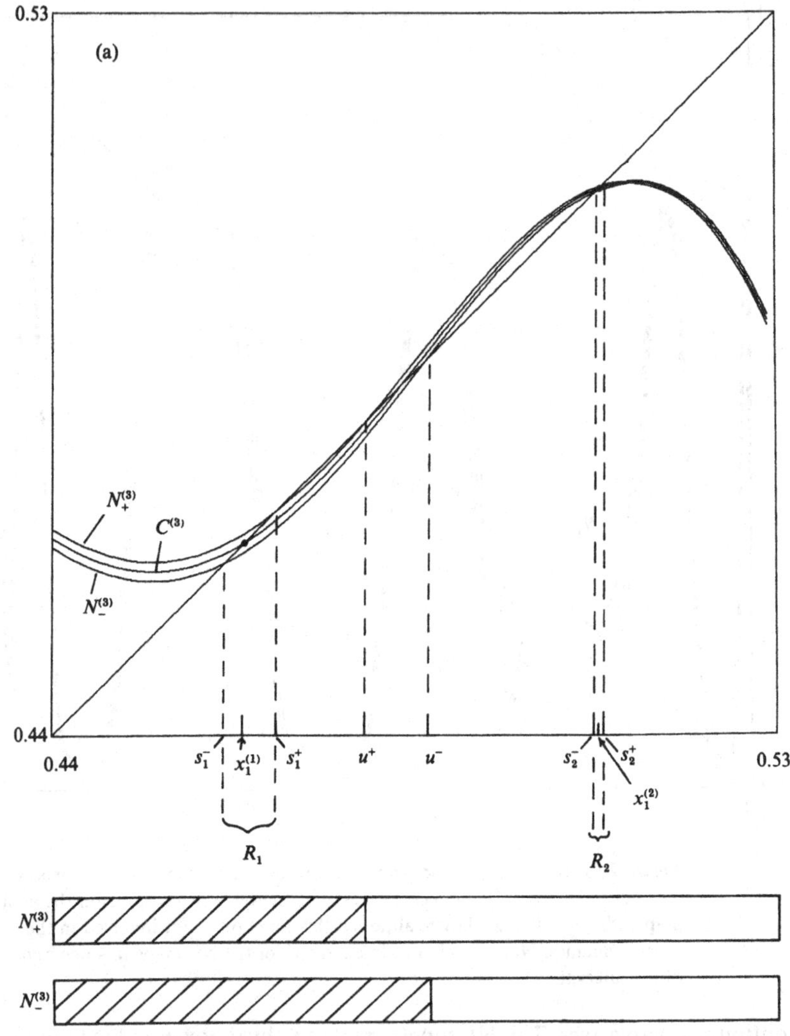

Figure 9.17. Noise-limit cubic map functions near one of the bistable pairs of fixed points. (a) No noise-induced transitions occur, only a blurring of the fixed points results upon application of noise. Lower strips show portions of the primary basins for the $N_+^{(3)}$ and $N_-^{(3)}$ NL functions. The primary basins for s_1^+ and s_1^- are shaded, while those for s_2^+ and s_2^- are white. (b) Noise-limit functions pass through tangency and noise-induced transitions take place. Note the basin structure in the lower strips; now s_1^+ and its basin have disappeared.

dynamics in the vicinity of the fixed point $x_1^{(1)}$. In the configuration shown $N_+^{(3)}$ intersects the bisectrix at the stable fixed point s_1^+, while $N_-^{(3)}$ intersects the bisectrix at the stable fixed point s_1^-. Since these fixed points are locally attracting, it is impossible for a map iterate to escape from the interval

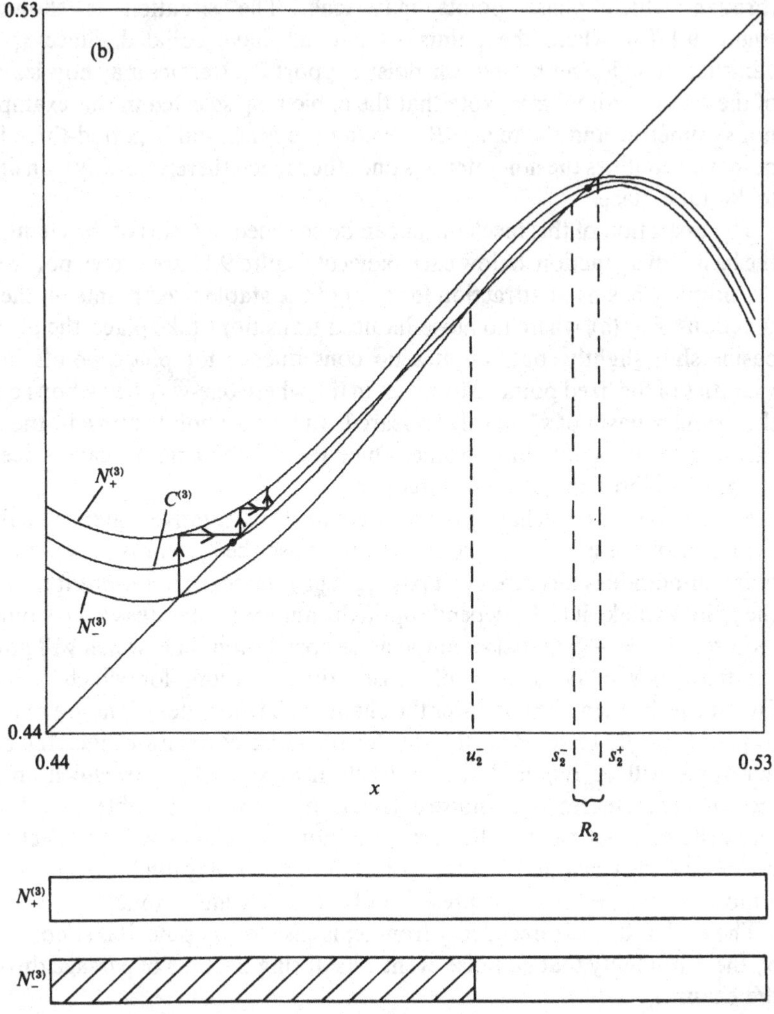

$R_1 = [s_1^-, s_1^+]$ once it has arrived there. A similar construction can be carried out for the bistable mate of this fixed point, as well as for the analogous regions near the two other pairs of fixed points. Other points on the entire interval $(-1, +1)$ will be attracted to one of these regions and remain there forever. Thus, for this noise amplitude the map supports noisy bistable period-three orbits: no noise-induced transitions occur.

As the magnitude of ξ increases, the splitting between the NL functions and the deterministic map increases, and, as a result, the stable and unstable fixed points of the NL functions move. Eventually, at some value, ξ_c, a stable–unstable pair of points collide. For $\xi > \xi_c$ these points move into the complex plane, leaving a small channel between the NL functions and the bisectrix through which phase points may leak. The situation is shown in Figure 9.17(b), where the points s_1^+ and u^+ have collided. Since s_1^+ has vanished, it no longer bounds the noisy support R_1. Iterates may now leak out of the region around $x_1^{(1)}$. Note that the cubic map selected in this example is not symmetric, and the region R_2 around the other stable period-three fixed point still confines the noisy iterates once they reach there: one-way transitions to R_2 take place.

The operation of this mechanism can be couched in terms of the changes in the basins of attraction. Below each panel of Figure 9.17 are shown portions of the primary basins of attraction for each of the stable fixed points on the NL functions. For (a), where no noise-induced transitions take place, the primary basins shift slightly, but this has no consequence for phase points in the vicinities of the fixed points. However, in (b), where one-way transitions occur, the primary basin of s_1^+ has disappeared, and phase points move to the only existing attractor for this ξ value. Once there, they cannot leave since the channel is closed in the other direction.

For the tangent mechanism, we may make quantitative statements about the onset of noise-induced transitions. Suppose the system is in R_1, and the noise amplitude is increased just past ξ_c. The number of iterations it takes for the point to make it to R_2 depends upon the number of iterates which would be required if $\xi_i = +\xi$, and also on the noise correlation time, which will govern the frequency of occurrence of consecutive iterations for which $\xi_i = +\xi$. Qualitatively, then, the narrower the channel, the more iterations are required to pass from R_1 to R_2. And the shorter the noise correlation time, the more iterations will be required, as the fraction of started walks which end up incomplete increases. If the noise correlation time is comparable to, or longer than, the channel passage time, τ_{ch}, the noisy dynamics will be effective in promoting the transition. Shorter correlation times may inhibit the transition almost entirely, even in the presence of a rather wide channel.

The probability of direct decay from R_1 is easy to compute. It is simply given by the probability that no noise event acts during the time of passage through the channel,

$$p_{ch} = \exp[-\tau_{ch}/\tau_n]. \tag{9.5.10}$$

The channel passage time is clearly an important quantity in determining the transition rate, and methods similar to those used in the treatment of intermittency may be applied to quantitatively estimate it. First, the threshold for the tangent mechanism, ξ_c, may be calculated from the condition that the NL function is tangent to the bisectrix:

$$x_\pm^* = N_\pm^{(3)}(x_\pm^*; a, b; \xi_c), \quad \left(\frac{\mathrm{d}N_\pm^{(3)}(x; a, b; \xi_c)}{\mathrm{d}x}\right)_{x=x_\pm^*} = +1. \tag{9.5.11}$$

Next, we may estimate the time necessary to traverse a channel by a deterministic walk on one of the NL functions. The NL functions are locally parabolic in the vicinities of the tangent points $x^*_\pm$, so we may expand $N^{(3)}_\pm$ in a Taylor series in x about $x^*_\pm$ and in ξ about ξ_c:

$$N^{(3)}_\pm(x; a, b; \xi) \approx x + A(x^*_\pm; a, b, \xi_c)(x - x^*_\pm)^2 + B(x^*_\pm; a, b; \xi_c)(\xi - \xi_c), \tag{9.5.12}$$

where A and B are the first and second derivatives of the NL function with respect to x and ξ, respectively, evaluated at the tangent point. Provided the channel width is small, the dynamics in the channel may be approximated by the continuous-time equation,

$$\frac{\mathrm{d}x(t)}{\mathrm{d}t} = A(x - x^*_\pm)^2 + B(\xi - \xi_c). \tag{9.5.13}$$

The time, τ_{ch}, it takes a phase point that enters the channel at x_s and leaves at x_e to traverse the channel can be found from (9.5.13):

$$\tau_{ch} = (AB\delta\xi)^{-1/2}\left[\tan^{-1} y_e\left(\frac{A}{B\delta\xi}\right)^{1/2} - \tan^{-1} y_s\left(\frac{A}{B\delta\xi}\right)^{1/2}\right], \tag{9.5.14}$$

where $y = x - x^*_\pm$ and $\delta\xi = \xi - \xi_c$. If $\delta\xi$ is small, this expression reduces to

$$\tau_{ch} = \pi(AB\delta\xi)^{-1/2}. \tag{9.5.15}$$

Thus, the channel passage time scales as $\delta\xi^{-1/2}$. The channel passage time is a measure of the number of map iterates necessary to move from the vicinity of R_1 to that of R_2. In the example of Figure 9.17 this is an irreversible process, but in the case where both channels are open transitions of the type $R_1 \leftrightarrows R_2$ occur. Under noisy dynamics the map iterates will not remain on one of the NL functions but will hop randomly between them. The rate of the transition process will then depend on the relative values of the noise correlation time, τ_n, and $n\tau_{ch}$, where n is the period of the cycle: If the noise correlation time is long compared to $n\tau_{ch}$, the deterministic dynamics is likely to take the system to the other attractor before a noise event causes the system to hop to the other NL function and evolve back to the original state.

The relation between the stochastic discrete-time map and a stochastic differential equation is not obvious. We described earlier how the Rössler dynamics in the three-loop limit cycle region could be reduced to that of a two-extremum, one-dimensional map by viewing the dynamics in a suitably chosen Poincaré surface. Consider the stochastic dynamics of Rössler's equations when the parameter a is regarded as a random variable governed by a Poisson–dichotomous noise process. We may view the noisy dynamics in a Poincaré plane. Note one important difference between this process and the stochastic map: the noise may now act at any point along the cycle, not just at the surface of section as is the case for the map. Nonetheless, provided the noise correlation time is not too short, the system dynamics will occur largely on the deterministic cycles, and conditions are favorable for the operation of the tangent mechanism.

Simulations of Rössler's equations can be carried out to investigate this point. We select parameters ($a_0 = 0.34018, b = 0.4, c = 5.86$) so that the system is within the bistable region, and take the dichotomous noise process to be such that $a(t) = a_0 + \xi(t)$, where $\xi(t) = \pm\xi$ with time intervals selected from an exponential distribution with correlation time τ_n. The amplitude of the noise, ξ, is selected so that $a(t)$ lies outside the bistable region, hence, inducing a transition process from one attractor to the other. A noisy third-iterate return map may be constructed in the usual way by recording the value of $-x(t)$ at the nth intersection of the Poincaré plane against its value at the $n+$ third intersection. The Poincaré plane is chosen to be the xz-plane with $x \leqslant 0$. The results of a simulation are shown in Figure 9.18 for $\tau_n = 100$. This is long compared to the cycle period ($\tau_c \sim 18$). The noisy return map demonstrates the existence of well-defined NL functions underlying the dynamical behavior. The probability density can serve as an indicator of the operation of the tangent mechanism since it possesses several distinctive features. In the limit where the noise correlation time is long compared to the channel passage time, the noisy dynamics will be confined largely to the NL functions, as discussed above. Hence, the phase point will tend to be attracted to one of the stable fixed points of the NL functions. A channel walk will ensue when a noise event occurs (the ξ parameter changes), and the evolution will be deterministic until the next noise event. Thus, the probability density will contain a series of sharp spikes corresponding to the deterministic steps in the channel. The locations of the spikes are easily determined by evolving the fixed point of one NL function under the dynamics of the other NL function. The lower panel of Figure 9.18 shows the probability density for the noisy dynamics of Rössler's model for parameter values corresponding to the return map in the upper panel. Note the two large peaks at the ends of the bounded interval that contains the probability density: these correspond to the stable fixed points of the NL functions, and are the leaky, but still identifiable, three-loop limit cycles. Between these large peaks is a series of smaller spikes arising from the channel walks the phase points must take in order to reach the other limit-cycle state. The locations of the peaks correspond exactly to the mapping of the large spikes under the NL functions. This distinctive structure of the probability density can be used to identify the operation of the tangent mechanism in the absence of other information.

The stochastic dynamics is especially simple when $\tau_n \gg \tau_c$, since the motion will be confined largely to the stable cycles. However, if there are several noise events during one period, the system will not have time to relax and the behavior off the stable cycles must be taken into account. This implies that in the crossings of the Poincaré plane the phase points will not necessarily lie on the NL functions, since they correspond to deterministic dynamics with one or

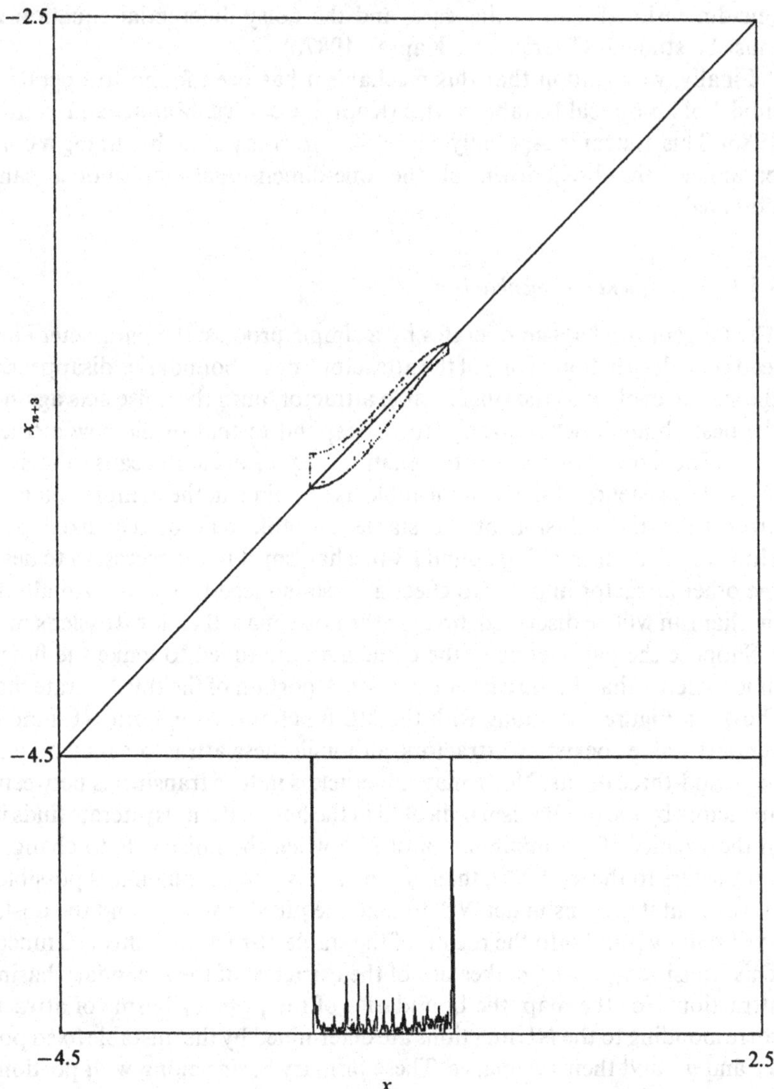

Figure 9.18. Third-iterate return map for Rössler's model with Poisson-dichotomous parametric noise (upper panel) and probability density (lower panel). System parameters are ($a_0 = 0.34018$, $b = 0.4$, $c = 5.86$), $\tau_n = 100$, $\xi = 0.0004$.

the other of the dichotomous noise parameters. Iterates of the noisy return map now fill the region between the two deterministic curves. The system is no longer adequately characterized as random hops between two deterministic attractors; the transient evolution toward these attractors occurs for a major portion of the time. The stochastic dynamics cannot be described by a noisy one-dimensional map in this case, and the noisy differential equation itself must be studied (Celarier and Kapral, 1987b).

Finally, we mention that this mechanism has been found to operate in a model of an optical bistable device (Kapral, Celarier, Mandel and Nardone, 1986). This system is especially convenient to study since by tuning a control parameter the breakdown of the one-dimensional-map model can be observed.

9.5.4 Backdoor mechanism

The tangent mechanism operates by a simple process: the parameter change leads to a destruction of one of the attractors, basin boundaries disappear, and the system evolves to the single stable attractor, until the noise acts again and the basin boundaries rearrange to correspond to that of the new attracting state. The slow evolution in the channel region occurs because the system 'feels' the existence of the now unstable fixed points in the complex plane that arose from the collision of the stable–unstable pair of real fixed points. However, as in the moving boundary mechanism, it is not necessary to destroy the other attractor in order to effect a noise-induced transition. An alternate mechanism will be discussed, first for the noisy map, then for Rössler's model.

Suppose the parameters in the cubic map are tuned to make the function much steeper than in the earlier example. A portion of the third-iterate map is shown in Figure 9.19 along with the NL functions. Now both NL functions support stable coexisting attractors, although these attractors may no longer be period-three orbits. Noise may nonetheless induce transitions between the attractors by the mechanism indicated in the figure. If a noisy iterate finds itself in the vicinity of the minimum m of $N_{-}^{(3)}$ when the noise acts to change the parameters to those of $N_{+}^{(3)}$, then, if the map is steep enough, it is possible for subsequent iterations under $N_{+}^{(3)}$ to take the phase point beyond the unstable fixed point u_{+} and into the region of the stable attractor on this NL function. This transition process makes use of the existence of the secondary basins of attraction. For the map, the boundaries of the primary basins of attraction corresponding to the NL functions are determined by the unstable fixed points u_{+} and u_{-} and their preimages. These primary basins along with portions of the secondary basins are shown below the graph of the map function. The trajectory in the figure demonstrates that a phase point in the primary basin of the stable attractor near m on $N_{-}^{(3)}$ may find itself, as a result of a noise event that changes the parameters of the system, in the secondary basin of the stable attractor near M on $N_{+}^{(3)}$. In this sense the mechanism resembles the moving boundary mechanism. Both induce the transition by moving a basin boundary past a phase point.

It is not difficult to construct criteria for the onset of transitions by this process. Maximum excursions to the left occur from the minimum of $N_{-}^{(3)}$, and thus map onto the highest point on $N_{+}^{(3)}$. If the image of this point maps to the

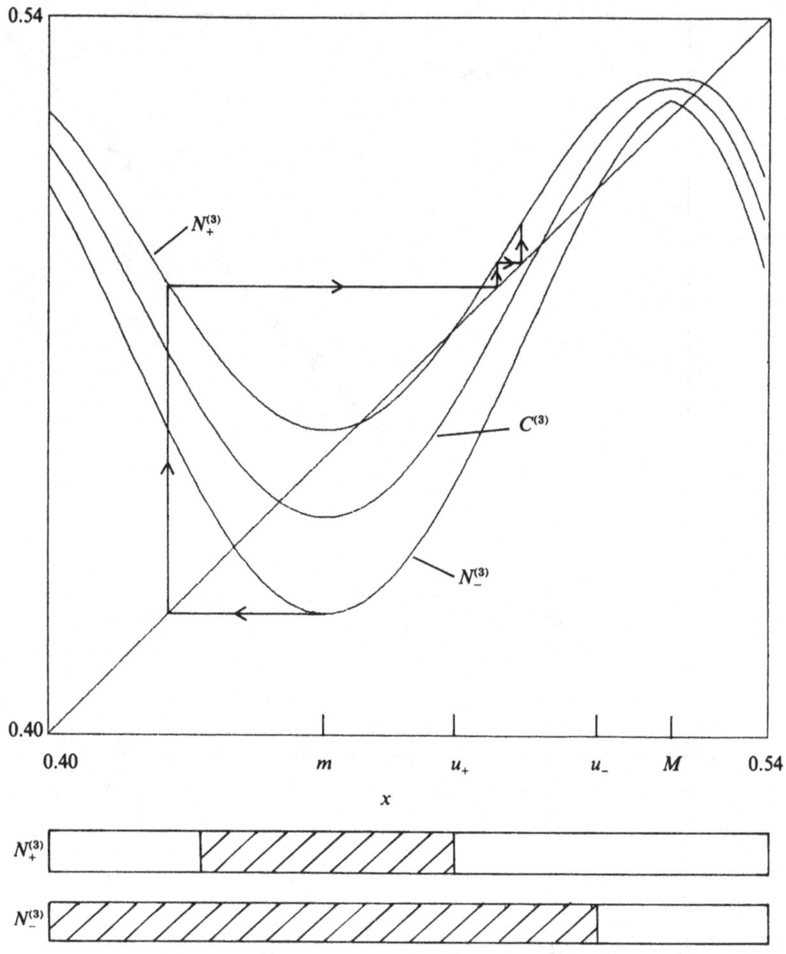

Figure 9.19. Noise-limit map functions, $N_{\pm}^{(3)}$, and $C^{(3)}$ for the backdoor mechanism. The lower strips show the basins for the attractors on the NL functions. The basin for the attractor in the vicinity of m on each NL function is shown as the shaded region, while the basins for the attractors near M are the white regions.

right of u_+ subsequent evolution under $N_+^{(3)}$ will carry the phase point to the other attractor, while if it maps to the left no transition will occur. The critical condition for a transition is given by

$$N_+^{(3)}(N_-^{(3)}(m)) = u_+, \tag{9.5.16}$$

or

$$N_+^{(3)}(N_+^{(3)}(N_-^{(3)}(m))) = N_+^{(3)}(N_-^{(3)}(m)). \tag{9.5.17}$$

The solution of (9.5.16) or (9.5.17) yields a critical value of the noise amplitude, ξ_c, beyond which noise-induced transitions may take place.

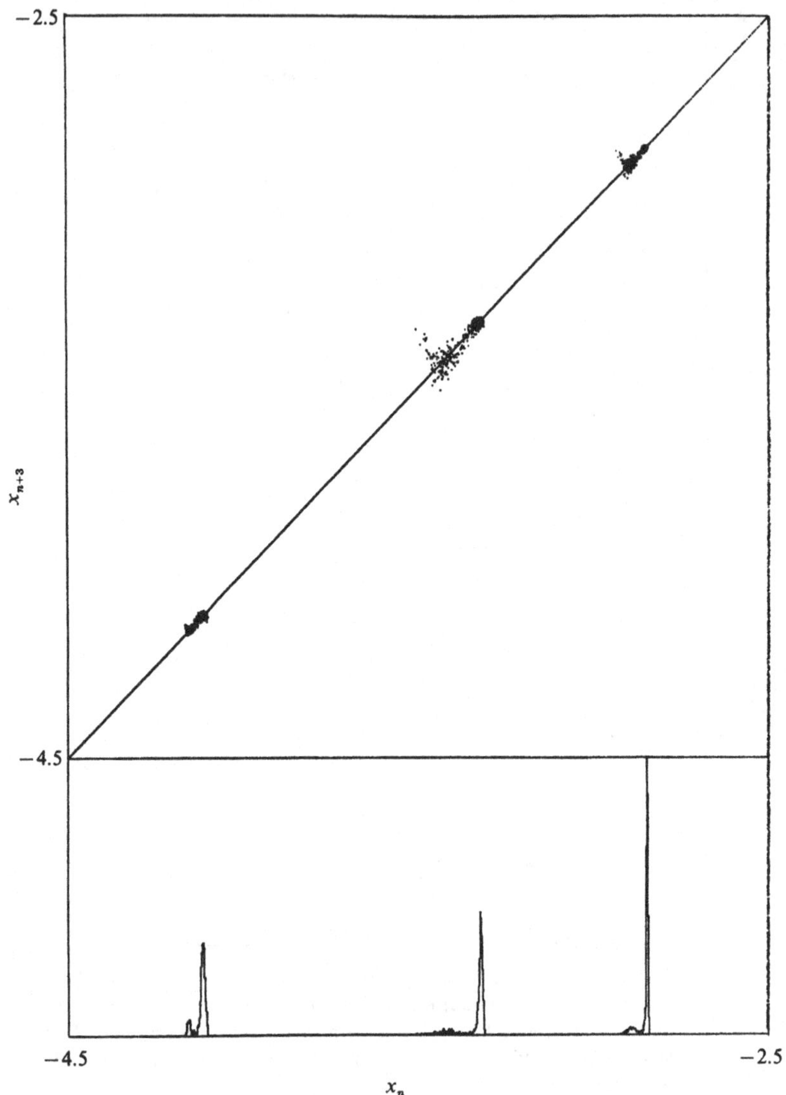

Figure 9.20. Third-iterate return map (upper panel) and probability density (lower panel) for Rössler's model. The figure shows the operation of the backdoor mechanism. System parameters are given in the text.

There is one final feature of this process worth noting here. Since both NL functions support bistable attractors, the noisy attractors may be quite complex since they will originate from hops between the two right hand or the two left hand attractors among themselves. In addition, if the region near the minimum maps very close to the unstable fixed point, phase density will tend to accumulate there since evolution away from this point is slow. In an observation this high density region will appear to be a new system state. In this way the noise process may be said to create states where none existed before.

We have identified this mechanism in Rössler's model. Figure 9.20 shows the third-iterate return map and the probability density for ($a_0 = 0.3425, b = 0.4, c = 5.8$) and $\xi = 0.0045$, a parameter setting within the cusp region. The noise correlation time in this simulation is $\tau_n = 0.1$; the noise acts approximately 180 times per cycle. Examination of the central fixed point region indicates that iterates starting at the extreme left of this region are mapped beyond the unstable fixed point, indicating the operation of the backdoor mechanism. Similar comments apply to the other two regions as well. Well-defined NL functions do not exist in this case, due to the short correlation time of the noise, leading to transient dynamics in the flow that does not take place on these functions. Nonetheless, the noisy return map provides evidence for the operation of the mechanism, as predicted by the one-dimensional map analysis.

The effect of the noise correlation time on the transition rate is somewhat more involved for this mechanism. Once an iterate is injected in the region past the unstable fixed point, its passage to the other stable region will be favored by a long noise correlation time, as can be seen from an examination of the graphical iteration process depicted in Figure 9.19. If the injection takes place very close to the fixed point, this may require very long correlation times indeed. However, only a small fraction of phase points in one of the 'stable' regions may actually be injected into the region beyond the unstable fixed point: those that lie in an interval near the minimum m of $N_-^{(3)}$. To populate this region requires a more complex series of noise events, and is favored by a short noise correlation time.

9.6 Discussion

The consideration of Poisson–dichotomous noise processes allows one to easily visualize and discuss the mechanisms for noise-induced transitions, and permits the construction of very simple analytical models of the statistical properties of the noisy dynamical system. However, many of the results are quite easily generalized to noisy processes with different statistical properties.

Consider, for example, a Poisson noise process with a uniform distribution on a bounded interval; one may still define noise-limit functions, as in (9.5.9), but the domain from which the ξ_i are chosen is now the interval $[-\xi, \xi]$, instead of the set $\{-\xi, \xi\}$. It is clear that the realizations of the noisy map lie between the NL functions, rather than on them. The NL functions still provide criteria for the onset of noise-induced transitions, though the rates of transitions, particularly near their onset, will be very different. For example, in the case of the tangent mechanism the dynamics in the channel will be rather different for a uniform noise process. Even though the NL function may correspond to the opening of a channel, there will be a distribution of map functions; many of these will correspond to a closed channel, or a channel with a much greater walk time than that of the NL functions. Similarly, for the moving boundary mechanism, the transition rate is determined by the fraction of time the system spends in the vulnerable region of phase space. If the noisy parameter is allowed to take any value from a continuous range, the rate of transitions may either increase or decrease relative to that of a Poisson–dichotomous noise process with the same correlation time.

Many aspects of the effects of noise statistics on the rate have been explored for noisy map models (Celarier, Fraser and Kapral, 1983; Kapral, Celarier and Fraser, 1984). The problem also may be approached through the use of integral (Chapman–Kolmogorov) equation representations of the noisy discrete-time dynamics (see, for instance, Fraser, Celarier and Kapral, 1983; Haken and Wunderlin, 1982; Talkner, Hänggi, Freidkin and Trautman, 1988).

As a further generalization, one may consider noise processes that affect the dynamical variables directly (additive noise) or, in the case of continuous-time systems, that affect the time derivatives directly (velocity-additive noise). Additive noise could arise from the direct infusion of chemical species into a reaction vessel; here, apart from random perturbations of the chemical composition, the evolution of the system is deterministic. Velocity-additive noise arises in the GK model from the presence of fluctuations in the cooling bath temperature.

In the presence of an additive noise process, the basin boundaries of the system do not move, since they are functions of the parameters. Transitions arise when the phase point is pushed across a boundary by the fluctuations. The analogs of vulnerable phase regions exist for this noise process. For the GK model additive perturbations will be ineffective while the phase point is not near the basin boundary; even near the basin boundary they will be largely ineffective if they do not push the phase point toward lower temperature, hence across the boundary. This sensitivity of the noise-induced transition process to the phase space direction of the applied noise is a general feature of additive noise processes.

Rössler's model also exhibits vulnerable phases. In some portions of the cycle the basin boundaries lie parallel to the cycle, and noise events transverse to the cycle are effective in inducing transitions; in other portions of the cycle the basin structure is more intricate, as was discussed in Section 9.3, and the system is much more susceptible to any type of noise event. This additive noise process has been studied for Rössler's model (Celarier and Kapral, 1987b). Finally, we point out that additive stochastic chemical perturbations of this type have been studied experimentally for an enzymatic reaction (Boiteux, Goldbetter and Hess, 1975).

The analysis of velocity-additive noise processes is considerably less straightforward, since there is no simple relation between the fixed point and basin structure of the deterministic equation and that of the stochastic model with the appended noise term. Nonetheless the underlying deterministic structure is important for the study of the effects of noise on the system dynamics (Huberman and Crutchfield, 1979; Kapral, Schell and Fraser, 1982). Even for this type of noise process the operation of the tangent and backdoor mechanisms have been observed.

Further investigations will certainly turn up additional mechanisms for noise-induced transitions, particularly as the study of more complicated chemical systems is undertaken. We feel confident that there will be a sufficiently rich repertoire of variations to insure that their continued investigation will prove rewarding.

Acknowledgements

This research was supported in part by a grant from the Natural Sciences and Engineering Research Council of Canada.

References

Alamgir, M. and Epstein, I. 1983. *J. Am. Chem. Soc.* **105**, 2500–2.

Arecchi, F. T., Badii, R. and Politi, A. 1985. *Phys. Rev. A* **32**, 402–8.

Bar-Eli, K. and Geiseler, W. 1983. *J. Phys. Chem.* **87**, 1352–7.

Boiteux, A., Goldbetter, A. and Hess, B. 1975. *Proc. Natl. Acad. Sci. USA* **72**, 3829–33.

Celarier, E., Fraser, S. and Kapral, R. 1983. *Phys. Lett.* **94A**, 247–50.

Celarier, E. and Kapral, R. 1987a. *J. Chem. Phys.* **86**, 3357–65.

Celarier, E. and Kapral, R. 1987b. *J. Chem. Phys.* **86**, 3366–72.

Collet, P. and Eckmann, J.-P. 1980. *Iterated Maps on the Interval as Dynamical Systems.* Boston: Birkhäuser.

Decroly, O. and Goldbeter, A. 1982. *Proc. Natl. Acad. Sci. USA* **79**, 6917–21.

DeKepper, P. and Boissonade, J. 1985. In *Oscillations and Travelling Waves in Chemical Systems* (R. J. Field and M. Burger, eds.), pp. 223–56. New York: Wiley.

DeKepper, P., Epstein, I. and Kustin, K. 1981. *J. Am. Chem. Soc.* **103**, 6121–7.

Field, R. J. and Burger, M. 1985. *Oscillations and Travelling Waves in Chemical Systems.* New York: Wiley.

Fraser, S., Celarier, E. and Kapral, R. 1983. *J. Stat. Phys.* **33**, 341–70.

Fraser, S. and Kapral, R. 1982. *Phys. Rev. A* **25**, 3223–33.

Gaspard, P., Kapral, R. and Nicolis, G. 1984. *J. Stat. Phys.* **35**, 697–727.

Golubitsky, M. and Keyfitz B. L. 1980. *SIAM J. Math. Anal.* **11**, 316–39.

Grebogi, C., Ott, E. and Yorke, J. A. 1983. *Physica* **7D**, 181–200.

Grebogi, C., Ott, E. and Yorke, J. A. 1986. *Phys. Rev. Lett.* **56**, 1011–14.

Gwinn, E. G. and Westervelt, R. M. 1985. *Phys. Rev. Lett.* **54**, 1613–16.

Haken, H. and Wunderlin, A. 1982. *Z. Phys. B* **47**, 181–7.

Heemskirk, A. H., Dammas, W. R. and Fortuin, J. M. H. 1980. *Proceedings of the International Symposium on Chemical Reactions of Engineering 6 Nice.* New York: Pergamon.

Horsthemke, W. and Lefever, R. 1984. *Noise Induced Transitions.* Berlin: Springer.

Huberman, B. A. and Crutchfield, J. P. 1979. *Phys. Rev. Lett.* **43**, 1743–7.
Kapral, R., Celarier, E. and Fraser, S. 1984. In *Fluctuations and Sensitivity in Nonlinear Systems* (W. Horsthemke and D. Kondepudi, eds.), pp. 179–86. New York: Springer.
Kapral, R., Celarier, E., Mandel, P. and Nardone, P. 1986. *SPIE Optical Chaos* **667**, 175–82.
Kapral, R. and Fraser, S. 1984. *J. Phys. Chem.* **88**, 4845–52.
Kapral, R., Schell, M. and Fraser, S. 1982. *J. Phys. Chem.* **86**, 2205–17.
Levi, M. 1981. *Mem. Am. Math. Soc.* **244**, 1–147.
McDonald, S. W., Grebogi, C., Ott, E. and Yorke, J. A. 1985, *Physica* **17D**, 125–53.
Mankin, J. C. and Hudson, J. L. 1984. *Chem. Eng. Sci.* **41**, 2651–61.
Moon, F. C. and Li, G.-X. 1985. *Phys. Rev. Lett.* **55**, 1439–42.
Nicolis, G. and Prigogine, I. 1977. *Self-Organization in Nonequilibrium Systems.* New York: Wiley.
Orban, M. and Epstein, I. 1982. *J. Am. Chem. Soc.* **104**, 5918–22.
Pifer, T., Ganapathisubramanian, N. and Showalter, K. 1985. *J. Chem. Phys.* **83**, 1101–10.
Pomeau, Y. and Manneville, P. 1980. *Commun. Math. Phys.* **74**, 189–97.
Rehmus, P., Vance, W. and Ross, J. 1984. *J. Chem. Phys.* **80**, 3373–80.
Rössler, O. 1979. *Ann. NY Acad. Sci.* **316**, 376–92.
Roux, J.-C. 1983. *Physica* **7D**, 57–68.
Schmitz, R. A., Graziani, K. R. and Hudson, J. L. 1977. *J. Chem. Phys.* **67**, 3040–4.
Swinney, H. L. 1983. *Physica* **7D**, 3–15.
Talkner, P., Hänggi, P., Freidkin, E. and Trautman, D. 1988. *J. Stat. Phys.* (in press).
Uppal, A., Ray, W. H. and Poore, A. B. 1974. *Chem. Eng. Sci.* **29**, 967–85.
Vidal, C. and Hanusse, P. 1986. *Int. Rev. Phys. Chem.* **5**, 1–55.
Winfree, A. T. 1980. *The Geometry of Biological Time.* Berlin: Springer.

10 State selection dynamics in symmetry-breaking transitions

DILIP K. KONDEPUDI

10.1 Introduction

In nature, states of broken symmetry are found at all levels, from the microscopic to the macroscopic. The *state* of a physical system is determined by – depending on the level of description – atomic and molecular *interactions* or by irreversible thermodynamic *processes*. We speak of a 'broken symmetry' if the state does not have the symmetries of the interactions or the processes that determine it; transitions to such states are the *symmetry-breaking transitions*. From a thermodynamic viewpoint, symmetry-breaking transitions can occur in systems in equilibrium as well as systems far from equilibrium. In equilibrium systems, magnetic transitions in crystalline solids are the best known examples (Briss, 1964; Cracknell, 1975). In nonequilibrium systems, such transitions are known to occur in hydrodynamic, chemical and laser systems (Haken, 1977; Nicolis and Prigogine, 1977). A most striking example of a state of broken symmetry is the 'state of life'. The chemistry of life has a definite handedness: all proteins are made of L-amino acids and DNA and RNA contain exclusively D-sugars (Miller and Orgel, 1974). Biochemical function depends very delicately on this molecular chirality. D-amino acids are a rarity and they occur only as polypeptides (Bentley, 1969). Though there are many conjectures, the processes that resulted in a transition to such a state of broken chiral symmetry have not yet been identified.

A transition to a state of broken symmetry is always a transition to one of the many possible states. In the absence of a strong bias towards any of these possible states, which of the states will be realized is a matter of chance; generally they are all equally probable. This chapter is about the dynamics of transition to a symmetry-breaking state in the presence of an extremely small bias – smaller in magnitude than the r.m.s. value of the fluctuations. Considering situations in which the system is made to pass through the transition point, a theory is developed to calculate the probability of transition to any of the symmetry-breaking states. This theory, whose range of validity and limitations are verified through numerical and electronic simulation, reveals an interesting enhancement of sensitivity to extremely small biases as a result of passage through the transition point. Enhanced sensitivity of this type has interesting implications in switching processes and in the considerations of extremely small influences, such as weak interaction parity violation, in nonequilibrium chemical systems; and it is generally applicable to all symmetry-breaking transitions. (For other recent studies of dynamical aspects of equilibrium and nonequilibrium transitions see Erneux and Mandel, 1984a, b; Erneux and Mandel, 1986; Joshua, Goldburg and Onuki, 1985; Joshua, Maher and Goldburg, 1983; Kapral and Mandel, 1985.)

The theory developed is based on general considerations of symmetry. In Section 10.2, a general formulation of the equations for symmetry-breaking transitions is presented. In Section 10.3, the process of state selection in a one-dimensional system with two symmetry-breaking states is discussed and the main ideas and results are presented. In Section 10.4, higher-dimensional transitions with four, six and eight symmetry-breaking states are considered. We conclude this chapter with a few remarks on the implications of this phenomenon of sensitivity due to slow passage through the transition point.

10.2 Symmetry-breaking transitions

By considering the symmetries of the irreversible processes or the atomic/molecular interactions one can formulate a general theory of symmetry-breaking transitions. To every symmetry group and its representation one can associate an equation that describes the dynamics of the transition which breaks the symmetries of that group. An overview of this theory is presented in this section.

Let $X(\bar{r},t) \equiv [X_1(\bar{r},t), X_2(\bar{r},t), \ldots, X_N(\bar{r},t)]$ be the state of an N-variable system, which is a function of position $\bar{r}$ and time t. The time evolution of such a state is obtained from a Hamiltonian or from phenomenological rate laws. Let this evolution be denoted by:

$$\frac{\partial X}{\partial t} = \mathrm{D}(X;\lambda), \tag{10.2.1}$$

in which D, in general, is a partial differential operator (with appropriate boundary conditions). λ is a parameter (for simplicity we consider only one such parameter), such as temperature or flow rate of a chemical component; we assume that if λ exceeds a critical value, λ_c, a symmetry-breaking transition will occur. If the phenomenological rate laws or the Hamiltonian is invariant under a symmetry group G then (10.2.1) will have the following covariance property:

$$\frac{\partial gX}{\partial t} = gD(X;\lambda) = \mathrm{D}(gX;\lambda) \tag{10.2.2}$$

for all $g \in G$ (we have assumed here that λ is invariant under G). The state X has broken the symmetry g if $gX \neq X$. From (10.2.2) it immediately follows that if

X is a solution of (10.2.1) then gX is also a solution; thus breaking of symmetry is associated with multiple states all of which can be realized.

Transition to a symmetry-breaking state generally occurs when the value of a parameter, such as λ, exceeds a critical value λ_c. We assume that when λ is below this critical value the system is in a stable symmetric state X_0, i.e. $gX_0 = X_0$ for all $g \in G$. The state X of the system can be written as:

$$X = X_0 + \sum_k \beta_k(t)\psi_k, \tag{10.2.3}$$

in which ψ_k are an appropriate set of modes and β_k are the corresponding mode amplitudes. When λ is less than λ_c, $\beta_k \to 0$ as $t \to \infty$ for all k. In the vicinity of λ_c, some of the modes experience 'critical slowing', i.e. their mode amplitudes evolve on a time scale that is much slower than that of the rest. Consequently, the evolution of the slow modes may be separated from the fast modes by adiabatic elimination of the fast modes (Haken, 1977; Van Kampen, 1985). Let there be s slow modes and let α_k be their amplitudes. Then by the above procedure one can arrive at a reduced equation of the form:

$$\frac{d\alpha_k}{dt} = F_k(\alpha_i; \lambda), \quad k = 1, 2, \ldots, s. \tag{10.2.4}$$

On a slow time scale, the state of the system is now of the form:

$$X \simeq X_0 + \sum_{k=1}^{s} \alpha_k(t)\psi_k. \tag{10.2.5}$$

In the vicinity of λ_c the behavior of α_k is such that, when λ is less than λ_c, $\alpha_k \to 0$ as $t \to \infty$, but when λ is greater than λ_c, $\alpha_k(t) \to \alpha_{k0} \neq 0$ as $t \to \infty$. Equations of the type (10.2.4) are similar to those derived under the well-known mean field approximation.

The form of (10.2.4) depends on the symmetries in the group G that are being broken (Briss, 1964; Sattinger, 1979). It can be shown that (Ruelle, 1973; Sattinger, 1977) covariance of (10.2.1) under G as expressed in (10.2.2) implies that (10.2.4) is also covariant under G. Sattinger (1979) has elaborated the general procedure through which one can obtain the form of $F_k(\alpha_i, \lambda)$ from a knowledge of the representation of G on the α_k-space. In the following sections we shall consider the functions F_k derived for some simple groups.

For a discussion of selection probabilities of the modes corresponding to the α_k's, we must add fluctuations to (10.2.4) along with small biasing terms, which we shall write as $C_k g_k (k = 1, 2, \ldots, s)$, for the s slow modes. We would then have Langevin equations of the form:

$$\frac{d\alpha_k}{dt} = F_k(\alpha_i; \lambda) + C_k g_k + \varepsilon_k^{1/2} f_k(t) \tag{10.2.6}$$

in which we included *additive* fluctuations $\varepsilon_k^{1/2} f_k(t)$ of r.m.s. value $\varepsilon_k^{1/2}$. As we shall see below, for the considered processes of state selection *multiplicative* fluctuations, which enter the equations because some of the parameters in F_k might be fluctuating, are not very significant.

Taking (10.2.5) to be an adequate description of the system, we consider the following process. We begin with a value of λ well below the transition point λ_c. Here the average value of α_k will be nearly zero, only perturbed from zero by the very small biasing terms $C_k g_k$. In all our considerations we shall adopt units of time which are appropriate for the relaxation time scales of the system (a precise definition will be given below). In these units of time, the magnitude of $C_k g_k$ will be assumed to be smaller than $\varepsilon_k^{1/2}$. With these values, we let λ sweep through the transition point, λ_c, at a rate γ. Our objective is to obtain the probability of transition to each of the possible states as a function of $C_k g_k$, ε_k and γ.

10.3 The process of state selection

If the symmetry group G consists of only two elements (as in the case of parity) it can be shown, in general, that in the vicinity of the transition point λ_c the amplitude α of the symmetry-breaking solution obeys a Langevin equation of the form:

$$\frac{d\alpha}{dt} = -A\alpha^3 + B(\lambda - \lambda_c)\alpha + \varepsilon^{1/2} f(t) \qquad (10.3.1)$$

in which A and B are positive constants, λ is the critical parameter and we assume that the noise $\varepsilon^{1/2} f(t)$ is a Gaussian white noise with $\langle f(t) f(t') \rangle = \delta(t - t')$. Under the symmetry operation, α changes to $-\alpha$; hence, in order to have covariance, the quadratic term in α is absent. For $\lambda < \lambda_c$, the only stable steady state is $\alpha = 0$; but when $\lambda > \lambda_c$, this state becomes unstable and the system can evolve to one of the two new steady states $\alpha_{\pm} = \pm(B(\lambda - \lambda_c)/A)^{1/2}$. The symmetry of the system is reflected in the fact that α_+ and α_- are on equal footing: as λ sweeps from a value below λ_c to a value above λ_c, either state will be reached with equal probability.

Now, if we introduce an interaction g that destroys the perfect symmetry, it can be easily shown that (10.3.1) will be altered by the addition of a term containing g:

$$\frac{d\alpha}{dt} = -A\alpha^3 + B(\lambda - \lambda_c)\alpha + Cg + \varepsilon^{1/2} f(t), \qquad (10.3.2)$$

where C is a constant characterizing the coupling between g and the system. If we take (10.3.2) to be a mean field description of a magnetic transition, for example, g would be proportional to a weak external field; in the case of chiral symmetry-breaking in chemical systems, g can be an interaction that makes the reaction rates of left- and right-handed molecules unequal. In many instances, g can be expressed as a dimensionless parameter $g = \Delta E/kT$, ΔE being an interaction energy (k is the Boltzmann constant and T is the temperature). To make our discussion general, it is most appropriate to choose units of time that are characteristic of the system's relaxation time scale. Accordingly, we define

$$\tau \equiv (B\lambda_c)t \tag{10.3.3}$$

and scale all the terms in (10.3.2) accordingly. (Note that $\varepsilon^{1/2}$ scales to $(\varepsilon/B\lambda_c)^{1/2}$.) We now have the equation:

$$\frac{d\alpha}{d\tau} = -\bar{A}\alpha^3 + \bar{\lambda}\alpha + \bar{C}g + (\bar{\varepsilon})^{1/2} f(\tau). \tag{10.3.4}$$

Here $\bar{A} = A/(B\lambda_c)$; $\bar{C} = C/(B\lambda_c)$; $\bar{\varepsilon} = \varepsilon/(B\lambda_c)$ and $\bar{\lambda} = (\lambda/\lambda_c - 1)$. In most of our considerations $\bar{\lambda}$ will vary from -1 to $+1$, and $\bar{C}g \ll 1$. In the vicinity of the critical point ($\bar{\lambda} \simeq 0$) it is easy to see that the macroscopic steady states are separated by a distance $S \sim (\bar{C}g/\bar{A})^{1/3}$ (see Figure 10.1a). The effect of the interaction $g = \Delta E/kT$ appearing with a fractional exponent has important consequences for the sensitive behavior of the system (Kondepudi and Prigogine, 1981). In all our considerations we shall assume that the r.m.s. value of the fluctuations, $(\bar{\varepsilon})^{1/2}$, is very small compared to the distance of $\alpha_\pm$ from $\alpha = 0$. This means that the relaxation time for the probability distribution to reach its stationary value is very long compared to the time scale of the sweep rate $\bar{\lambda}$ – an assumption easily met in most physical systems. (In our numerical simulations we use the values $(\bar{\varepsilon})^{1/2} = 10^{-4}$, $\bar{C}g = 10^{-5}$ and $\bar{A} = 1$.) Because of the fluctuations, we can associate a probability distribution $P(\alpha, \tau)$ with (10.3.4). $P(\alpha, \tau)$ obeys the well-known Fokker–Planck equation:

$$\frac{\partial P}{\partial \tau} = -\frac{\partial}{\partial \alpha}\left(-\bar{A}\alpha^3 + \bar{\lambda}\alpha + \bar{C}g - \frac{\bar{\varepsilon}}{2}\frac{\partial}{\partial \alpha}\right)P(\alpha, \tau). \tag{10.3.5}$$

When $\bar{\lambda} \simeq -1$, the macroscopic steady state $\alpha_0 \simeq \bar{C}g \ll 1$. In fact, we will be considering $\bar{C}g$ an order ot magnitude smaller than $(\bar{\varepsilon})^{1/2}$. The steady state probability distribution, $P_0(\alpha)$, when $\bar{\lambda} \simeq -1$ can be easily obtained from (10.3.5). It is essentially a Gaussian centered at $\bar{C}g$:

$$P_0(\alpha) = N \exp\left[\left(-\frac{\bar{A}\alpha^4}{4} - \frac{\alpha^2}{2} + \bar{C}g\alpha\right)\Big/(\varepsilon/2)\right]$$
$$\simeq N' \exp[-(\alpha - \bar{C}g)^2/(\varepsilon/2)], \tag{10.3.6}$$

where N and N' are appropriate normalization constants. Since the width of the Gaussian is $(\bar{\varepsilon})^{1/2}$, if $\bar{C}g \ll (\bar{\varepsilon})^{1/2}$, then this Gaussian is hardly distinguishable from a Gaussian centered at $\alpha \simeq 0$. In other words, in this state, the influence of the interaction g is insignificant. The same is true for the stationary states of $P(\alpha, \tau)$ when $\bar{\lambda} \simeq 1$. Now if $\bar{\lambda}$ sweeps from its initial value -1 to a final value $\bar{\lambda} = +1$, this probability distribution will spread and split into two parts, one concentrating around the state $\alpha_+ = +(\bar{\lambda}/\bar{A})^{1/2}$ and the other at $\alpha_- = -(\bar{\lambda}/\bar{A})^{1/2}$. (Note that this bimodal distribution is *not* the stationary

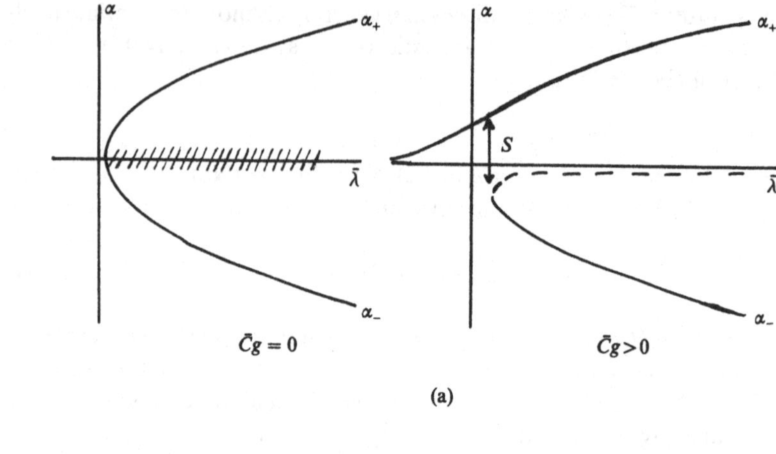

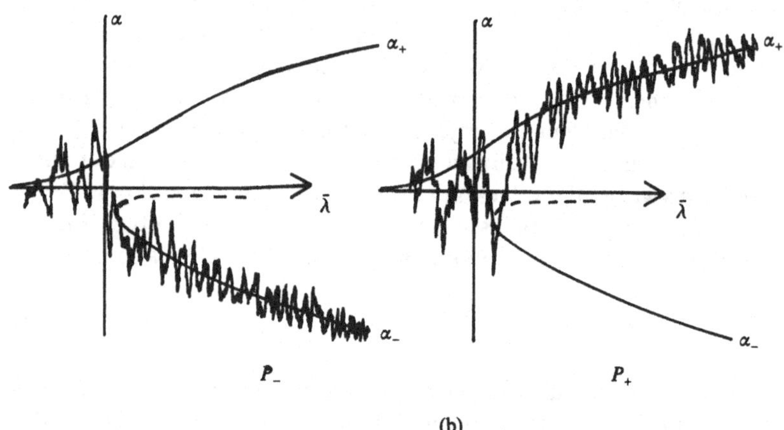

Figure 10.1. State selection in a symmetry-breaking transition. (a) Steady states of (10.3.1) and (10.3.2) with no noise term. The separation S of the branches is proportional to $(\bar{C}g)^{1/3}$ in the vicinity of $\bar{\lambda} \simeq 0$. (b) Trajectories showing state selection in the presence of noise, as $\bar{\lambda}$ sweeps through the transition point. The system evolves to the branches α_+ and α_- randomly with probability P_+ and P_-, respectively. Here $\bar{C}g > 0$. (c) Evolution of probability density $P(\alpha, \tau)$ obtained from an ensemble of 2000 trajectories, such as those shown in (b), through electronic analogue simulation (courtesy F. Moss and P. V. E. McClintock).

distribution; it will slowly evolve to the stationary distribution on the time scale of the first passage time – which may be astronomical!) What we require are the fractions P_+ and $P_- = (1 - P_+)$, of $P(\alpha)$ that go to either state in this process – which, of course, will be the probabilities for the system to reach the states $\alpha_\pm$ (see Figure 10.1). It comes as a surprise to us when we realize that the $\bar{C}g$, that seemed so insignificant, can have a profound influence on

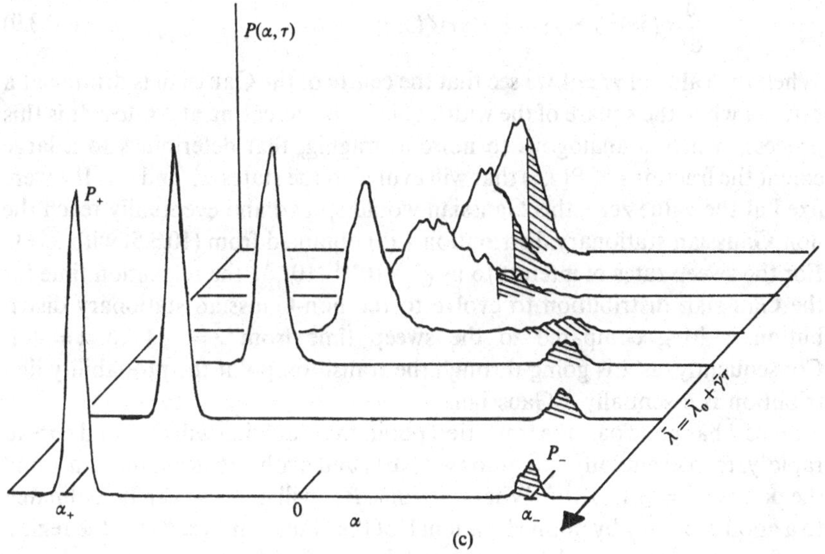

(c)

the probabilities of state selection P_+ and P_-, depending on the speed with which $\bar{\lambda}$ sweeps from -1 to $+1$. This can be seen as follows.

Let us assume that $\bar{\lambda}$ sweeps from -1 to $+1$ at a rate $\bar{\gamma}$, i.e. $\bar{\lambda} = -1 + \bar{\gamma}t$, until it reaches the value $+1$, where it stops. We shall be considering $\bar{\gamma}$ in the range 10^{-1}–10^{-5}, which means the sweep time is at most five orders of magnitude larger than the linear relaxation time (we mean the relaxation time in the region where $|\bar{\lambda}| \simeq 1$). Initially $\bar{\lambda} = -1$, and the probability distribution is a Gaussian as in (10.3.6). As $\bar{\lambda}$ sweeps through the critical point $\bar{\lambda}$, the essential process of state selection, which is the process that determines the fractions of $P(\alpha)$ going to the two states $\alpha_\pm$, occurs in the vicinity of $\bar{\lambda} \simeq 0$; and it is in this region that the approximations that lead to reduced equations of the type (10.3.1) are valid. This observation greatly simplifies the computation of $P_\pm$. First, we see that the probability distribution (10.3.2) is quite sharply peaked, essentially near $\alpha \simeq 0$. Returning to the Langevin equation (10.3.4), this implies that the α^3 term in it has an insignificant role. Hence we only need to consider the approximate equation

$$\frac{d\alpha}{d\tau} \simeq (-1 + \bar{\gamma}\tau)\alpha + \bar{C}g + (\bar{\varepsilon})^{1/2} f(t). \tag{10.3.7}$$

It is then straightforward to show (using, for example, lemma I in chapter 2 of Chandrasekhar's 1943 classic review) that the probability $P(\alpha, \tau)$ associated with (10.3.7) is an evolving Gaussian. The time evolutions of the mean $\langle\alpha\rangle$ and the deviation from the mean $\langle(\delta\alpha)^2\rangle = \langle(\alpha - \langle\alpha\rangle)^2\rangle$ are given by:

$$\frac{d}{d\tau}\langle\alpha\rangle = (-1 + \bar{\gamma}\tau)\langle\alpha\rangle + \bar{C}g \tag{10.3.8}$$

$$\frac{d}{d\tau}\langle(\delta\alpha)^2\rangle = 2(-1+\bar{\gamma}\tau)\langle(\delta\alpha)^2\rangle + \bar{\varepsilon}. \tag{10.3.9}$$

When the value of $\bar{\gamma}\tau \simeq 1$ we see that the centre of the Gaussian is drifting at a rate Cg while the square of the width $\langle(\delta\alpha)^2\rangle$ is increasing at a rate $\bar{\varepsilon}$. It is this process, which is analogous to noise averaging, that determines to a large extent the fractions of $P(\alpha, \tau)$ that will evolve to the states α_+ and α_-. If $\bar{\lambda}$ were fixed at the value zero, this Gaussian would spread and eventually reach the non-Gaussian stationary distribution $P(\alpha)$ obtained from (10.3.5) with $\bar{\lambda}=0$. For the sweep rates of interest to us ($\bar{\gamma} \simeq 10^{-1}$–$10^{-5}$), the relaxation time for the Gaussian distribution to evolve to the non-Gaussian stationary distribution is long compared to the sweep time from $\bar{\lambda}=-1$ to $\lambda=+1$. Consequently, as $\bar{\lambda}$ is going through the transition point the probability distribution is essentially a Gaussian.

Once $\bar{\lambda}$ has gone past the transition point, the Gaussian will drift and spread rapidly, then eventually split into two parts and evolve to accumulate around the points $\alpha_\pm = \pm(\bar{\lambda}/\bar{A})^{1/2}$. What fractions, $P_\pm$, will go to $\alpha_\pm$ can be estimated to a good accuracy by noting how much of the Gaussian $P(\alpha, \tau)$ is in the region $\alpha > 0$ and how much in the region $\alpha < 0$; i.e. $\alpha > 0$ and $\alpha < 0$ are assumed to be *domains of attraction* of α_+ and α_-, respectively (the corrections due to Cg being small). Here we are assuming that the process of selection occurs almost completely in the region where $P(\alpha, \tau)$ is a Gaussian; the consequent evolution of $P(\alpha, \tau)$ to a bimodal distribution does not alter the fractions P_+ and P_-. The validity of this assumption can be checked through numerical and electronic simulation of (10.3.4).

With these ideas, P_+ and P_- can be calculated by defining

$$N = \operatorname*{Lt}_{\tau\to\infty} \frac{\langle\alpha\rangle_\tau}{(\langle(\delta\alpha)^2\rangle_\tau)^{1/2}} \tag{10.3.10}$$

where $\langle\alpha\rangle_\tau$ and $\langle(\delta\alpha)^2\rangle_\tau$ are the solutions of (10.3.8) and (10.3.9), respectively. It can easily be shown that (Kondepudi, Moss and McClintock, 1986a)

$$N = \frac{\bar{C}g}{(\bar{\varepsilon}/2)^{1/2}}(\pi/\bar{\gamma})^{1/4}. \tag{10.3.11}$$

This tells us how many standard deviations the centre of the Gaussian is from the $\alpha = 0$ line (which is the demarcation line between the domains of attraction of α_+ and α_-). It then follows, according to the above ansatz:

$$P_+ = \frac{1}{(2\pi)^{1/2}}\int_{-\infty}^{N} e^{-x^2/2}\,dx \tag{10.3.12}$$

and, of course,

$$P_- = (1 - P_+). \tag{10.3.13}$$

A comparison of the selection probabilities calculated from (10.3.11) and (10.3.12) and those obtained from direct numerical simulation of (10.3.4) (in

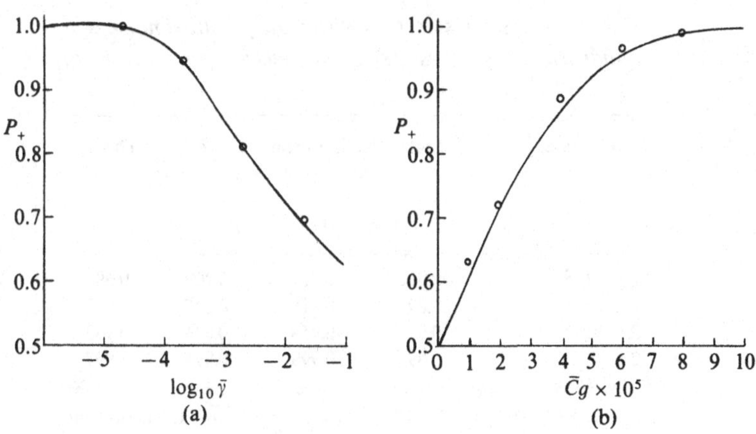

Figure 10.2. A comparison of the theory, (10.3.11) and (10.3.12), and results of numerical simulation for the value of P_+ as a function of $\bar{\gamma}$ and $\bar{C}g$. (a) $\varepsilon^{1/2} = 10^{-4}$, $\bar{C}g = 10^{-5}$ and $\bar{A} = 1$. (b) $\varepsilon^{1/2} = 3.3 \times 10^{-4}$, $\bar{\gamma} = 2 \times 10^{-3}$ and $\bar{A} = 1$. The solid curves are from theory. Simulation results are from 1000 trajectories.

which $f(t)$ was assumed to be Gaussian white noise) is shown in Figure 10.2. The results of electronic simulation of this process can be found in Kondepudi, Moss and McClintock (1986a).

Effects of multiplicative noise and quadratic terms

The validity of $P_\pm$ calculated using the above ansatz is virtually unaffected by inclusion of multiplicative noise into the system (Kondepudi, Prigogine and Nelson, 1985) – unless, of course, the magnitude of the multiplicative noise is very large. The same is true if a quadratic term, $\bar{D}\alpha^2$, is added to (10.3.2). This can be verified by considering a Langevin equation of the type:

$$\frac{d\alpha}{d\tau} = -\bar{A}\alpha^3 + \bar{D}\alpha^2 + (\bar{\lambda} + \Lambda^{1/2} f_\lambda(t))\alpha + \bar{C}g + (\bar{\varepsilon})^{1/2} f(\tau), \quad (10.3.14)$$

in which $f_\lambda(t)$ is assumed to be Gaussian white noise. The values of P_+ obtained through numerical simulation of (10.3.14) are compared with theoretical values calculated from (10.3.11) and (10.3.12) in Table 10.1. When $\bar{\Lambda}^{1/2} \simeq 10^{-2}$ and $(\bar{\varepsilon})^{1/2} \simeq 10^{-4}$, the effects of multiplicative fluctuations are hardly noticeable. Table 10.2 shows the effect of increasing $\bar{\Lambda}^{1/2}$; only when $\bar{\Lambda}^{1/2} \simeq 0.3$, i.e. when the fluctuations of $\bar{\lambda}$ are almost 30% of the maximum value of $\bar{\lambda}$ do we see significant effects. Also it is clear that inclusion of a quadratic term $\bar{D}\alpha^2$ makes no difference.

The reason for the system's insensitivity to multiplicative fluctuations can easily be understood. As described above, the process of selection occurs essentially in the vicinity of $\bar{\lambda} \simeq 0$, where α is very small. In this region the evolution of α is mainly due to the last two terms, $\bar{C}g$ and $(\bar{\varepsilon})^{1/2} f(\tau)$, in (10.3.14);

Table 10.1 *State selection with multiplicative noise and quadratic term: results of numerical simulation of (10.3.14).*

Sweep rate, $\bar{\gamma}$	State selection probability P_+ (1000 trajectories)			Theory P_+
	$\bar{D}=0$	$\bar{D}=1$	$\bar{D}=-1$	
2×10^{-2}	0.732	0.717	0.709	0.691
2×10^{-3}	0.841	0.818	0.847	0.813
2×10^{-4}	0.953	0.959	0.941	0.943
2×10^{-5}	0.995	0.999	0.991	0.997

$\bar{\Lambda}^{1/2}=5.47\times10^{-2}$, $\varepsilon^{1/2}=10^{-4}$, $\bar{C}g=10^{-5}$ and $\bar{A}=1$. Theoretical P_+ from (10.3.11) and (10.3.12). (From Kondepudi, Prigogine and Nelson, 1985.)

Table 10.2. *State selection with increasing strength in the multiplicative noise.*

Multiplicative noise strength $\bar{\Lambda}$	State selection probability P_+ (1000 trajectories)		
	$\bar{D}=0$	$\bar{D}=1$	$\bar{D}=-1$
0	0.802	0.833	0.807
3.0×10^{-3}	0.841	0.818	0.847
1.0×10^{-2}	0.835	0.817	0.840
5.0×10^{-2}	0.771	0.754	0.748
1.0×10^{-1}	0.682	0.692	0.720

$\varepsilon^{1/2}=10^{-4}$, $\bar{C}g=10^{-5}$, $\bar{\gamma}=2\times10^{-3}$ and $\bar{A}=1$. (From Kondepudi, Prigogine and Nelson, 1985.)

hence addition of terms of the order α or higher do not effect the selection process significantly.

Effects of coloured noise

We have also studied (Kondepudi, Moss and McClintock, 1986b) the effects of exponential time correlation in the noise term in (10.3.4). The Langevin equation for this system may be written as:

$$\frac{d\alpha}{d\tau}=-\bar{A}\alpha^3+\bar{\lambda}\alpha+\bar{C}g+\xi(\tau) \tag{10.3.15}$$

$$\frac{d\xi}{d\tau}=-\xi/\tau_N+W(\tau), \tag{10.3.16}$$

in which τ_N is the noise correlation time and $W(\tau)$ is a Gaussian white noise term for which:

$$\langle W(\tau) W(\tau')\rangle = \bar{\sigma}\delta(\tau - \tau'). \tag{10.3.17}$$

Equations (10.3.16) and (10.3.17) imply that

$$\langle \xi(\tau)\xi(\tau')\rangle = (\tau_N\bar{\sigma}/2)\exp(-|\tau - \tau'|/\tau_N). \tag{10.3.18}$$

By defining $\varepsilon \equiv \tau_N^2\bar{\sigma}$ we can write:

$$\langle \xi(\tau)\xi(\tau')\rangle = (\varepsilon/2\tau_N)\exp(-|\tau - \tau'|/\tau_N). \tag{10.3.19}$$

The limit $\tau_N \to 0$, $\bar{\sigma} \to \infty$ such that ε is finite gives us the white noise limit for which $\langle \xi(\tau)\xi(\tau')\rangle = \varepsilon\delta(\tau - \tau')$.

From (10.3.15) and (10.3.16) it can be shown that (see Kondepudi, Moss and McClintock, 1986b for details), in the vicinity of $\bar{\lambda} = 0$ the probability distribution $P(\alpha, \tau)$ is essentially a Gaussian but whose width increases at a slower rate than the white noise width. Hence we may expect the probability P_+ to increase with the noise correlation time. Under the assumption $\tau_N < 1$ one can show that (Kondepudi, Moss and McClintock, 1986b) if $\bar{\lambda} = \bar{\lambda}_0 + \bar{\gamma}\tau$ then:

$$P_+ = \frac{1}{(2\pi)^{1/2}}\int_{-\infty}^{N_C} e^{-x^2/2}\,dx, \tag{10.3.20}$$

$$N_C = \frac{\bar{C}g}{(\varepsilon/2)^{1/2}}(\pi/\bar{\gamma})^{1/4}\frac{1}{(1 - C)^{1/2}} \tag{10.3.21}$$

in which

$$C = (\beta/2)(1/\pi\bar{\gamma})^{1/2}\exp\{[(\beta - \bar{\lambda}_0)^2 - \bar{\lambda}_0^2]/\gamma\} \times 2\pi E^2((\beta - \bar{\lambda}_0)/\gamma^{1/2}), \tag{10.3.22}$$

where

$$\beta \equiv 1/\tau_N \tag{10.3.23}$$

and

$$E(a) \equiv (1/2\pi)^{1/2}\int_a^{\infty} e^{-x^2/2}\,dx. \tag{10.3.24}$$

Thus we have a small correction factor $(1 - C)^{-1/2}$ for coloured noise. The reason for this factor being small is, once again, that the process of selection occurs in the vicinity of $\bar{\lambda} \simeq 0$; here the time scale of evolution is very large – due to the critical slowing – and hence the difference between coloured noise and white noise is not felt by the system. A comparison of the theory and analogue electronic simulation is shown in Figure 10.3(a); as expected, the effects of coloured noise are small. (In the electronic simulation, the natural unit of time is the integrator time constant τ_i – see Chapter 9 by P. V. E. McClintock and F. Moss in Volume 3; $\tau_N < 1$ implies $\tau_N < \tau_i$.)

Coloured noise makes a significant difference in the process of selection, however, if instead of making $\bar{\lambda}$ sweep through the critical point we keep it fixed at $\bar{\lambda}_F > 0$, with $\alpha = 0$ at $\tau = 0$. In this case there is no critical slowing; the

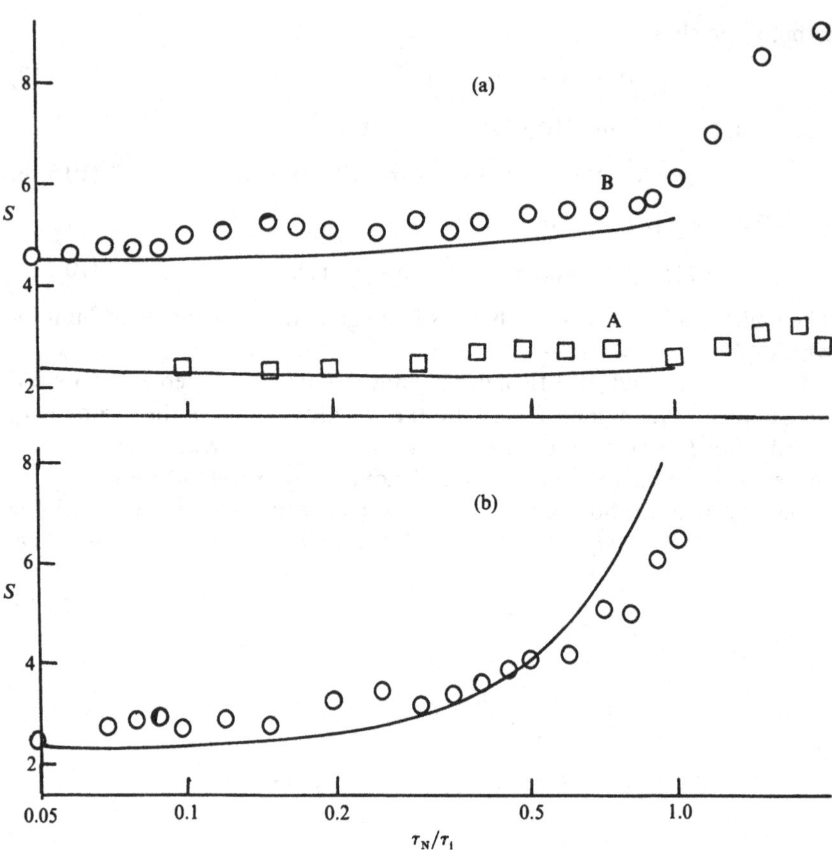

Figure 10.3. A comparison of theory and analogue simulation for the effects of coloured noise. Here 'selectivity' $S \equiv P_+/P_-$ is plotted as a function of ratio noise correlation time τ_N and the relaxation time τ_i (which is the integrator time constant of the circuit). τ_i is taken to be the unit of time. Solid curves are from theory. Simulation results are from 2000 trajectories. (a) $\bar{\lambda}$ is swept from -1 to $+1$: $\bar{\gamma} = 5.0 \times 10^{-3}$, $\varepsilon = 2.6 \times 10^{-2}$, $g = 0.016$ and $\tau_i = 100\,\mu$s for the set *A*; $\bar{\gamma} = 2.0 \times 10^{-1}$, $\varepsilon = 4.9 \times 10^{-2}$, $g = 0.1$ and $\tau_i = 1000\,\mu$s for the set *B*. Theoretical curve from (10.3.20) and (10.3.21). (b) $\bar{\lambda}$ is kept fixed at $+1$ and $\alpha = 0$ at $\tau = 0$. Here $\bar{C}g = 0.1$, $\varepsilon = 4.9 \times 10^{-2}$. Theoretical curve is from (10.3.25) and (10.3.20). (From Kondepudi, Moss and McClintock, 1986b.)

process of selection occurs on a time scale $1/\lambda_F$ (i.e. the linear relaxation time) and if $\tau_N \sim 1/\lambda_F$ we can expect strong effects. It can be shown that (Kondepudi, Moss and McClintock, 1986a, b; Mangel, 1981; Suzuki, 1981) in this case P_+ is given by (10.3.20) where

$$N_C = (\bar{C}g/\bar{\lambda}_F)(e - 1)/Q^{1/2} \tag{10.3.25}$$

in which:

$$Q = \varepsilon/(1 + \eta)^2[C_1 + C_2 + C_3] \tag{10.3.26}$$

Table 10.3. *Effects of increasing noise strength $\bar{\varepsilon}$.*

$\bar{\varepsilon}$	P_+	
	Simulation	Theory
10^{-5}	0.819	0.8134
10^{-4}	0.820	0.8134
10^{-3}	0.830	0.8134
10^{-2}	0.792	0.8134
10^{-1}	0.690	0.8134

Theoretical value begins to be a poor approximation for $\bar{\varepsilon} > 10^{-2}$. Here $\bar{\gamma} = 2 \times 10^{-3}$ and $\bar{C}g/(\bar{\varepsilon})^{1/2} = 0.1$.

$$C_1 = (e^2 - 1)/2\lambda_F \tag{10.3.27}$$

$$C_2 = (2/\beta)[\exp(-(1-\eta)/\eta) - 1]/(1-\eta) \tag{10.3.28}$$

$$C_3 = (1/2\beta)(1 - \exp(-1/\eta)), \tag{10.3.29}$$

where

$$\eta = \lambda_F \tau_N \quad \text{and} \quad \beta = 1/\tau_N. \tag{10.3.30}$$

A comparison of theory and analogue simulation is shown in Figure 10.3(b).

From these results we see that the probability, P_+, that the system will evolve to the state α_+, as given by (10.3.11) and (10.3.12) is valid in a wide range of conditions. It is also very clear that slow passage through the critical point can greatly enhance the sensitivity of the system to small biases – even if they are smaller than the r.m.s. value of the fluctuations.

We must now turn to the limitations of the theory and see for what values of the parameters $\bar{C}g$, $\bar{\varepsilon}$ and $\bar{\gamma}$ the theoretical prediction of P_+ becomes poor. Since the theory is based on the assumption that $P(\alpha, \tau)$ is a Gaussian in the vicinity of $\bar{\lambda} \simeq 0$, we must look at ranges of values of $\bar{C}g$, $\bar{\varepsilon}$ and $\bar{\gamma}$ for which we may expect this Gaussian approximation to be invalid. Table 10.3 shows a comparison of the theoretical predictions and the results of numerical simulation of (10.3.4) as $\bar{\varepsilon}$ increases and $\bar{\gamma}$ decreases. For $\bar{\gamma} \simeq 10^{-3}$, only when $\bar{\varepsilon} \simeq 10^{-2}$ do we see a significant deviation.

10.4 State selection in two- and three-mode symmetry breaking

The basic features that have emerged from the previous section are quite general and can be extended to higher-dimensional systems in which two or more modes α_k interact. In the case of two- and three-mode systems, the symmetry-breaking transition can lead to one of the four, six or eight states of broken symmetry. The general features are the following: (i) The evolution of the probability distribution in the vicinity of the transition point and in the regions where $\alpha_k \ll 1$ essentially determines the state selection probabilities. (ii) In this region, only the linear and the constant terms in the Langevin equation are of importance. (iii) Using terms up to the first order, the selection probabilities can be calculated to a very good approximation by computing the fractions of the probability distribution that are in the domain of attraction of each state as the system goes beyond the transition point. Linear terms in the Langevin equation generally lead to Gaussian probability distributions and hence we can express the selection probabilities in terms of Gaussian integrals.

Let us consider a system with two 'critical modes' α_1 and α_2. This case may arise when there is an underlying symmetry group, D_4, of a square, i.e. there are two mutually perpendicular directions that are symmetric. There will be two modes, each corresponding to one direction. In general, these modes will interact in a mutually destructive or constructive way. The equations of such a system, covariant under the group D_4 are of the form:

$$\frac{d\alpha_1}{d\tau} = \bar{\lambda}\alpha_1 - \bar{B}\alpha_2^2\alpha_1 - \bar{A}\alpha_1^3 \tag{10.4.1}$$

$$\frac{d\alpha_2}{d\tau} = \bar{\lambda}\alpha_2 - \bar{B}\alpha_1^2\alpha_2 - \bar{A}\alpha_2^3. \tag{10.4.2}$$

Here we shall assume $\bar{A}$ and $\bar{B}$ to be positive; $\bar{B} > 0$ means that the interaction between the two modes is mutually destructive. Also, we use the natural units of time τ, as defined in the previous section.

The steady states of this system fall into three classes:

(*a*) $\alpha_1 = \alpha_2 = 0$;

(*b*) $\alpha_1 = \pm(\bar{\lambda}/\bar{A})^{1/2}$, $\alpha_2 = 0$ and $\alpha_1 = 0$, $\alpha_2 = \pm(\bar{\lambda}/\bar{A})^{1/2}$;

(*c*) $\alpha_1 = \pm(\bar{\lambda}/(\bar{B}+\bar{A}))^{1/2}$, $\alpha_2 = \pm(\bar{\lambda}/(\bar{B}+\bar{A}))^{1/2}$.

For $\bar{\lambda} < 0$, only the state (*a*) is admissible. For $\bar{\lambda} > 0$, (*a*) becomes unstable and the states (*b*) and (*c*) are admissible. However, if $\bar{B} > \bar{A}$, the states (*b*) are stable but (*c*) are unstable and if $\bar{B} < \bar{A}$, the states (*c*) are stable but the states (*b*) are unstable. This tells us that if the mutual interaction term is strong, then one mode dominates, but if it is weak both mode amplitudes α_1 and α_2 attain equal values. This type of mode competition can be realized, for example, in laser systems (Tehrani and Mandel, 1978), in hydrodynamic systems (Le Gal, Pocheau and Croquette, 1985; Newell and Whitehead, 1969), and in magnetic solids that break the D_4 symmetry. With the inclusion of small biases $\bar{C}g_1$ and $\bar{C}g_2$, and fluctuations, we have the equations:

$$\frac{d\alpha_1}{d\tau} = \bar{\lambda}\alpha_1 - \bar{B}\alpha_2^2\alpha_1 - \bar{A}\alpha_1^3 + \bar{C}g_1 + (\bar{\varepsilon})^{1/2} f(\tau) \tag{10.4.3}$$

$$\frac{d\alpha_2}{d\tau} = \bar{\lambda}\alpha_2 - \bar{B}\alpha_1^2\alpha_2 - \bar{A}\alpha_2^3 + \bar{C}g_2 + (\bar{\varepsilon})^{1/2} f(\tau). \tag{10.4.4}$$

Let us denote the probability of selection of the $\alpha_1 = 0$, $\alpha_2 \neq 0$ state by P_{0+}, the state $\alpha_1 = \alpha_2 = (\bar{\lambda}/(\bar{B} + \bar{A}))^{1/2}$ by P_{++}, the state $\alpha_1 = -\alpha_2 = +(\bar{\lambda}/(\bar{B} + \bar{A}))^{1/2}$ by P_{+-}, etc. Following the ansatzes (i)–(iii) stated previously, we can arrive at the following results: for $\bar{A} > \bar{B}$ the probabilities $P_{\pm\pm}$ that are of interest are simply the products of one-mode probabilities:

$$P_{\pm\pm} = P_{\pm}^{(1)} P_{\pm}^{(2)}, \tag{10.4.5}$$

where

$$P_{+}^{(k)} = \frac{1}{(2\pi)^{1/2}} \int_{-\infty}^{N_k} e^{-x^2/2}\, dx \tag{10.4.6}$$

and

$$N_k = \frac{\bar{C}g_k}{(\bar{\varepsilon}/2)^{1/2}} (\pi/\bar{\gamma})^{1/4} \tag{10.4.7}$$

$$P_{-}^{(k)} = (1 - P_{+}^{(k)}). \tag{10.4.8}$$

In the case of strong mutual interaction for which $\bar{B} > \bar{A}$, the probabilities of interest are $P_{\pm 0}$ and $P_{0\pm}$. These probabilities can be expressed in a way similar to (10.4.6)–(10.4.8) with the following identification:

$$P_{0+} = P_{++}, \quad P_{0-} = P_{--} \tag{10.4.9}$$

$$P_{+0} = P_{+-}, \quad P_{-0} = P_{-+} \tag{10.4.10}$$

and

$$N_K = \frac{\bar{C}g'_k}{(\bar{\varepsilon}/2)^{1/2}} (\pi/\bar{\gamma})^{1/4}, \tag{10.4.11}$$

where

$$g'_1 = \frac{g_1 + g_2}{2^{1/2}} \quad \text{and} \quad g'_2 = \frac{g_1 - g_2}{2^{1/2}}. \tag{10.4.12}$$

That is, the two cases are related by a 45° rotation about the origin. Figure 10.4(a) shows a comparison between the theoretical predictions and numerical simulations of (10.4.3) and (10.4.4).

A three-mode system arises when the system has a cubic symmetry, as in the case of Fe crystal (Briss, 1964; Cracknell, 1975). The equations for such a system are:

$$\frac{d\alpha_1}{d\tau} = \bar{\lambda}\alpha_1 - \bar{B}(\alpha_2^2 + \alpha_3^2)\alpha_1 - \bar{A}\alpha_1^3 + \bar{C}g_1 + (\bar{\varepsilon})^{1/2} f(\tau) \tag{10.4.13}$$

$$\frac{d\alpha_2}{d\tau} = \bar{\lambda}\alpha_2 - \bar{B}(\alpha_3^2 + \alpha_1^2)\alpha_2 - \bar{A}\alpha_2^3 + \bar{C}g_2 + (\bar{\varepsilon})^{1/2} f(\tau) \tag{10.4.14}$$

$$\frac{d\alpha_3}{d\tau} = \bar{\lambda}\alpha_3 - \bar{B}(\alpha_1^2 + \alpha_2^2)\alpha_3 - \bar{A}\alpha_3^3 + \bar{C}g_3 + (\bar{\varepsilon})^{1/2} f(\tau). \tag{10.4.15}$$

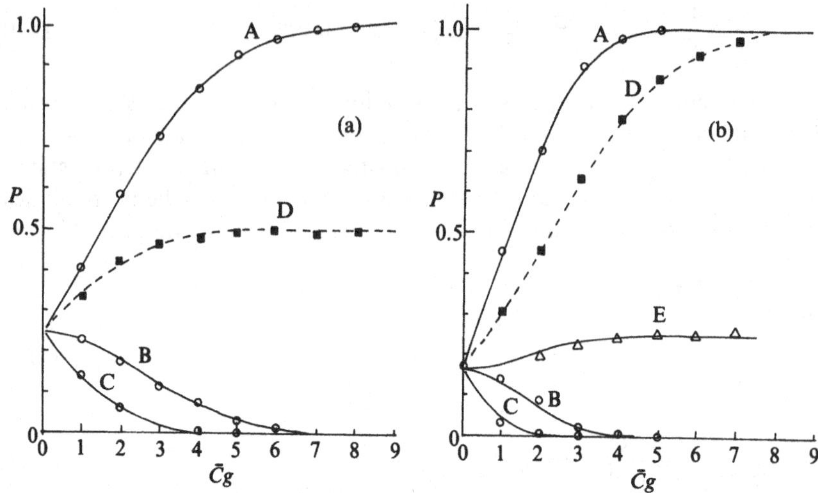

Figure 10.4. State selection probabilities as a function of the bias $\bar{C}g$ in two- and three-mode systems. (a) Two-mode probabilities: $\bar{\varepsilon}^{1/2} = 10^{-4}$, $\bar{\gamma} = 2 \times 10^{-2}$ and $\bar{C}g$ is in units of 10^{-5} and along the α_1 axis. The curves are obtained from theory. The circles and squares are obtained from numerical simulation of (10.4.3) and (10.4.4) with an ensemble of 2000 trajectories. Curves A is $P_{\pm 0}$, curve B is $P_{0\pm}$ and curve C is P_{-0} for the case $\bar{A} = 2.0$ and $\bar{B} = 4.0$. Curve D is $P_{++}(=P_{+-})$ when $\bar{A} = 2.0$ and $\bar{B} = 1.0$. (b) Three-mode probabilities: $\bar{\varepsilon}^{1/2} = 10^{-4}$ and $\bar{C}g$ is in units of 10^{-5} directed along the α_1 axis. The circles and squares are obtained from numerical simulation of (10.4.13)–(10.4.15) with an ensemble of 2000 trajectories. The curves are obtained from theoretical result (10.4.17). For curves A, B and C, the coefficient $\bar{A} = 2.0$ and $\bar{B} = 5.0$, and $\bar{\gamma} = 2 \times 10^{-3}$; curve A is P_{+00}, curve B is $P_{0\pm\pm}$ and curve C is P_{-00}. Curve D is P_{+00} for $\bar{\gamma} = 2 \times 10^{-2}$, and all other parameters are the same as for curves A, B and C. Curve E is P_{+++} when $\bar{\gamma} = 2 \times 10^{-2}$, $\bar{A} = 2.0$ and $\bar{B} = 1.0$. The triangles for curve E are the results of 1000 trajectories. (From Kondepudi and Gao, 1987.)

The macroscopic steady states of this system fall into four groups:

(a') $\alpha_1 = \alpha_2 = \alpha_3 = 0$;
(b') $\alpha_K = \pm(\bar{\lambda}/(2\bar{B} + \bar{A}))^{1/2}$, $K = 1, 2, 3$;
(c') $\alpha_1 = \alpha_2 = 0$, $\alpha_3 = \pm(\bar{\lambda}/\bar{A}))^{1/2}$, and similar states obtained through the interchange of the α's;
(d') $\alpha_1 = 0$, $\alpha_K = \pm(\bar{\lambda}/(\bar{B} + \bar{A}))^{1/2}$, $K = 2, 3$, and similar states obtained through the interchange of the α's.

The state (a') is stable for $\bar{\lambda} < 0$ and unstable for $\bar{\lambda} > 0$. States (b')–(d') are admissible only for $\bar{\lambda} > 0$. The states (d') are unstable for both $\bar{A} > \bar{B}$ and $\bar{B} > \bar{A}$. For the case $\bar{A} > \bar{B}$, states (b') are stable but (c') are unstable, and for $\bar{B} > \bar{A}$, (c') are stable but (b') are unstable.

In the case when the mutual interaction is weak, $\bar{A} > \bar{B}$, there are eight states, (b'), to which the system can evolve when $\bar{\lambda}$ becomes positive. If we denote the selection probabilities of these eight states by $P_{\pm\pm\pm}$, where P_{+++} is the probability for the state $\alpha_1 = \alpha_2 = \alpha_3 = +(\bar{\lambda}/\bar{A})^{1/2}$, etc., then, following the procedure outlined above, we can arrive at the conclusion:

$$P_{+++} = P_+^{(1)} P_+^{(2)} P_+^{(3)}$$

$$P_{+-+} = P_+^{(1)} P_-^{(2)} P_+^{(3)} \tag{10.4.16}$$

etc., where $P_{\pm}^{(K)}$ are given by (10.4.6)–(10.4.8), i.e. the system behaves as if there is no interaction as far as state selection probabilities are concerned.

The case of $\bar{B} > \bar{A}$ is mathematically more involved, though the main idea of calculating the state selection probabilities is the same. In this case there are six states (c') to which the system can evolve. To simplify the mathematics let us assume $g_2 = g_3 = 0$. We denote the probabilities for the states $\alpha_1 = \pm(\bar{\lambda}/\bar{A})^{1/2}$, $\alpha_2 = \alpha_3 = 0$ by $P_{\pm 00}$, for the states $\alpha_1 = \alpha_3 = 0, \alpha_2 = \pm(\bar{\lambda}/\bar{A})^{1/2}$ by $P_{0\pm 0}$ and so on. Then one can show (see Kondepudi and Gao, 1987) that

$$P_{+00} = P_+(N) - \frac{e^{-f}}{(\beta^2+1)^{1/2}} P_+(N/(\beta^2+1)^{1/2}) \tag{10.4.17}$$

in which:

$$f \equiv N^2\beta^2/2(1+\beta^2), \quad N \equiv \frac{\bar{C}g_1}{(\bar{\varepsilon}/2)^{1/2}}(\pi/\bar{\gamma})^{1/4}, \quad \beta = (4/\pi)^{1/2}$$

and

$$P_+(N) = \frac{1}{(2\pi)^{1/2}} \int_{-\infty}^{N} e^{-x^2/2}\,dx.$$

Probabilities for other states can similarly be obtained. A comparison of the probabilities calculated from (10.4.16) and (10.4.17) and those obtained from numerical simulation of equations (10.4.13)–(10.4.15) is shown in Figure 10.4(b). As can be seen, the agreement is rather good. Just as in the case of one mode, we may expect the effects of multiplicative noise (as was shown in Kondepudi and Gao, 1987) and coloured noise to be not very significant. And there is, of course, a wide range outside which the approximations made in deriving the state selection probabilities became poor.

10.5 Concluding remarks

The state selection process we have discussed has consequences in electronic and other switching devices. The fundamental aspects of computational switches have been discussed by Bennett (1982), Keyes and Landauer (1970) and Landauer and Woo (1971). In these discussions, the switching process often involves a transition from a single-well potential to a double-well potential as some parameter such as $\bar{\lambda}$ is changed. The process is identical to the one discussed; in our case we have the small switching signal $\bar{C}g$ and noise of r.m.s. value $(\bar{\varepsilon})^{1/2}$. In situations where the switching signal is rather weak and one has to overcome the effects of noise, we see from our study that the system can be made sensitive to the signal by passing through the transition point slowly; the sensitivity to the signal is enhanced at the expense of slowing the switching time. To get a more concrete idea of the advantages and disadvantages of the method, let us consider the case in which $\bar{C}g/(\bar{\varepsilon})^{1/2} \simeq 1$, i.e. the r.m.s. value of the noise is as big as the signal. If the switch were operated by keeping $\bar{\lambda}$ fixed at a value $+1$ with the initial state $\alpha = 0$, the switching time will equal the relaxation time of the switch, making the switching time one unit in our time units; from (10.3.20) and (10.3.25), in the white noise limit $Q \simeq \bar{\varepsilon}$, we see that $P_+/P_- \simeq 6.3$. Now if we sweep through the critical point at a rate $\bar{\gamma} \simeq 10^{-2}$, the switching time per bit is two orders of magnitude larger but $P_+/P_- \simeq 10^5$: the reliability of the switch is increased by four orders of magnitude. If $\bar{\gamma} \simeq 10^{-3}$, then we find $P_+/P_- \simeq 10^7$. Thus one can gain reliability at the expense of speed.

Another consequence of sensitivity due to slow passage through the transition point is the possible link between the particular biomolecular chirality observed on Earth, mentioned in the introduction, and the parity violation in weak interactions (Mason, 1984). One can envisage a chiral-symmetry-breaking transition in the primordial oceans (Decker, 1974; Mason, 1984). Such a transition can take place if there is suitable chiral auto-catalysis and if the system is driven far from thermodynamic equilibrium due to an inflow of reactants into the primordial oceans. An inflow is what is thought to have occurred due to synthesis of precursor molecules by various processes in the early atmosphere; not much is known about chiral auto-catalysis. In this scenario, the concentration of the precursor reactants will correspond to the parameter $\bar{\lambda}$ (Kondepudi and Nelson, 1983, 1984a), and slow passage through the transition point will correspond to slow increase of these concentrations. If there were no systematic biases, then we must expect the dominance of left- and right-handed molecules to occur with equal probability. Hence, on the surface of the Earth we must expect to have regions dominated by one kind or the other – somewhat like the domains in a ferromagnet. On the other hand, if there was a systematic bias, and if there was a slow growth of concentration of reactants, then our theory tells us that the dominance of the favored handedness can occur with high probability even if the bias is very small. Consequently only the favored handedness will dominate over the entire planet.

As a possible source of bias, parity violation in weak interaction has been of much interest, ever since its discovery (see Thieman, 1981). However, most of the previous considerations (Thieman, 1981) were of the opinion that the chemical effects of weak interactions are too small to be of significance. We have investigated the question (Kondepudi and Nelson, 1985) from the viewpoint of sensitivity due to slow passage through the transition point. The results show that if the production rates of left- and right-handed molecules differ by only one part in 10^{17}, under prebiotic conditions, a slow growth of concentration on a time scale of 10^4–10^5 years could result in the dominance of the favoured handedness with a probability larger than 98%. The choice of a reaction rate difference of one part in 10^{17} was motivated by the recent computation of Mason and Tranter (1983, 1984) that showed a difference of $\Delta E \simeq 10^{-17} kT$ in the ground state energy of mirror-image amino acid molecules. Their results also showed that the naturally dominant L-amino acids have the lower energy.

While we are still far from a satisfactory understanding of the origin of biomolecular chirality, we have now an understanding of the processes, and their time scales, that could link parity violation at a fundamental level to the observed handedness of life. In the laboratory, reaction rate differences as small as one part in 10^{13} seem to be within detectable range (Kondepudi and Nelson, 1984b).

Acknowledgements

I wish to thank the US Department of Energy, Basic Energy Sciences, for providing support for this work through the grant DE-AS05-81ER10947. I would also like to thank M-J. Gao for performing some of the numerical simulations.

References

Bennett, C. H. 1982. *Int. J. Theor. Phys.* **21**, 905–40.

Bentley, R. 1969. *Molecular Asymmetry in Biology*, vol. 1. New York: Academic Press.

Briss, R. R. 1964. *Symmetry and Magnetism.* Amsterdam: North-Holland.

Chandrasekhar, S. 1943. *Rev. Mod. Phys.* **15**, 1–89.

Cracknell, A. P. 1975. *Magnetism in Crystalline Materials.* New York: Pergamon.

Decker, P. 1974. *J. Mol. Evol.* **4**, 49–65.

Erneux, T. and Mandel, P. 1984a. *Phys. Rev. Lett.* **53**, 1818–22.

Erneux, T. and Mandel, P. 1984b. *Philos. Tran. R. Soc. London, Ser A* **313**, 285–90.

Erneux, T. and Mandel, P. 1986. *SIAM J. Appl. Math.* **46**, 1–15.

Haken, H. 1977. *Synergetics: An Introduction.* Berlin: Springer Verlag.

Joshua, M., Goldburg, W. I. and Onuki, A. 1985. *Phys. Rev. Lett.* **54**, 1175–7.

Joshua, M., Maher, J. V. and Goldburg, W. I. 1983. *Phys. Rev. Lett.* **51**, 196–8.

Kapral, R. and Mandel, P. 1985. *Phys. Rev. A* **32**, 1076–81.

Keyes, R. W. and Landauer, R. 1970. *IBM J. Res. Dev.* **14**, 152–7.

Kondepudi, D. K. and Gao, M-J. 1987. *Phys. Rev.* **35**, 340–8.

Kondepudi, D. K., Moss, F. and McClintock, P. V. E. 1986a. *Physica* **21D**, 296–306.

Kondepudi, D. K., Moss, F. and McClintock, P. V. E. 1986b. *Phys. Lett.* **114A**, 68–74.

Kondepudi, D. K. and Nelson, G. W. 1983. *Phys. Rev. Lett.* **50**, 1023–6.

Kondepudi, D. K. and Nelson, G. W. 1984a. *Physica* **125A**, 465–96.

Kondepudi, D. K. and Nelson, G. W. 1984b. *Phys. Lett.* **106A**, 203–6.
Kondepudi, D. K. and Nelson, G. W. 1985. *Nature* (*London*) **314**, 438–41.
Kondepudi, D. K. and Prigogine, I. 1981. *Physica* **107A**, 1–24.
Kondepudi, D. K., Prigogine, I. and Nelson, G. 1985. *Phys. Lett.* **111A**, 29–32.
Landauer, R. and Woo, J. W. F. 1971. *J. Appl. Phys.* **42**, 2301–8.
Le Gal, P., Pocheau, A. and Croquette, V. 1985. *Phys. Rev. Lett.* **54**, 2501–4.
Mangel, M. 1981. *Phys. Rev. A* **24**, 3226–38.
Mason, S. F. 1984. *Nature* (*London*) **311**, 19–23.
Mason, S. F. and Tranter, G. E. 1983. *JCS Chem. Commun.*, pp. 117–19.
Mason, S. F. and Tranter, G. E. 1984. *Mol. Phys.* **53**, 1091–111.
Miller, S. W. and Orgel, L. E. 1974. *The Origins of Life on Earth.* New Jersey: Prentice-Hall.
Newell, A. C. and Whitehead, J. A. 1969. *J. Fluid Mech.* **38**, 279–303.
Nicolis, G. and Prigogine, I. 1977. *Self-Organization in Nonequilibrium Systems.* New York: Wiley.
Ruelle, D. 1973. *Arch. Ration. Mech. Anal.* **51**, 136–52.
Sattinger, D. H. 1977. *SIAM J. Math. Anal.* **8**, 179–201.
Sattinger, D. H. 1979. *Group Theoretic Methods in Bifurcation Theory*, Lecture Notes in Mathematics 762. Berlin: Springer-Verlag.
Suzuki, M. 1981. In *Order and Fluctuations in Nonequilibrium Statistical Mechanics* (G. Nicolis and G. Dewel, eds.), pp. 299–365. New York: Wiley.
Tehrani, M. M. and Mandel, L. 1978. *Phys. Rev. A* **17**, 677–93.
Thieman, W., ed. 1981. *Origins of Life* **11**, 1–94. (This volume of *Origins of Life* was devoted to the origin of biomolecular chirality.)
Van Kampen, N. G. 1985. *Phys. Rep.* **124**, 69–160.

11 Noise in a ring-laser gyroscope

K. VOGEL, H. RISKEN and W. SCHLEICH

11.1 Introduction and overview

In the year 1851 Foucault demonstrated that the slow rotation of the plane of vibration of a pendulum could be used as evidence of the earth's own rotation. The first optical experiment to detect the earth's rotation was performed by Michelson and Gale (Michelson, 1925a, b) using an unusually large size for an interferometer: 0.4 miles × 0.2 miles. Nowadays high precision measurements of the earth's rotation are performed by using radio telescopes in very long baseline interferometry (Johnson *et al.*, 1979). Moreover, a recent proposal (Small and Chow, 1982) take advantage of the ultra high sensitivity of a ring-laser gyroscope (Aronowitz, 1965, 1971; Chow *et al.*, 1985; Menegozzi and Lamb, 1973; Privalov and Fridrikhov, 1969) of 10 m diameter to monitor changes in earth rate or universal time. The underlying principle of such a device is the optical analog of the Foucault pendulum, the so-called Sagnac effect (Post, 1967; Sagnac, 1913a, b). The frequencies of two counterpropagating waves in a ring interferometer are slightly different when the interferometer is rotated about an axis perpendicular to its plane. Since this frequency difference is proportional to the rotation rate it provides a direct measure of the rotation of the system.

Ring-laser gyroscopes of this size would also allow tests (Schleich and Scully, 1984; Scully, Zubairy and Haugan, 1981) of metric gravitation theories (Misner, Thorne and Wheeler, 1973). In an ultrasensitive ring-laser placed on the rotating earth, several 'effective rotations' arise as indicated in Figure 11.1. In order to detect these general relativistic corrections a ring-laser gyroscope device must be capable of measuring rotation rates as slow as 10^{-7} of the earth's rotation rate. At this point it becomes important that the linear Sagnac relationship between frequency difference and rotation rate is not true in real life.

Due to backscattering from the laser mirrors the two counterpropagating beams couple in a nonlinear way (Aronowitz, 1971). As a result the frequency difference vanishes for small rotation rates. This so-called lock-in effect limits crucially the accuracy of ring-laser gyroscopes. One method to overcome this

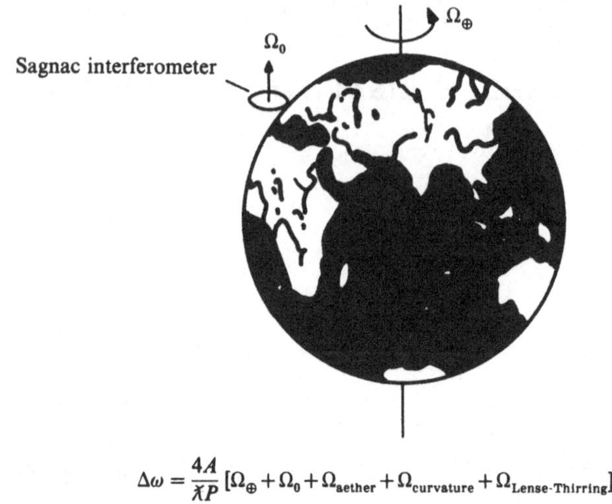

Figure 11.1. A Sagnac ring-laser interferometer fixed on the rotating earth is sensitive to its rotation rate Ω_0, the rotation of the earth $\Omega_\oplus$ and three effective rotation rates as indicated in the figure. In the equation, A denotes the area of the ring, $\bar{\lambda} = \lambda/(2\pi)$ is the reduced wavelength, and P is the perimeter of the interferometer.

problem consists of dithering (Killpatrick, 1967) the ring-laser, i.e. adding an external (controlled) rotation rate such that the gyroscope is operating most of the time outside of the dead band. Another technique to avoid locking came to light in the course of a general investigation of quantum noise, that is spontaneous emission noise from the laser atoms (Cresser *et al.*, 1982a, b). It has been shown (Cresser *et al.*, 1982a) that white noise is useful in reducing the dead band. Since the spontaneous emission noise is not easy to influence, we are therefore led to adding external noise, which necessarily has a nonvanishing correlation time and is thus colored. In the present chapter we confine ourselves to the discussion of this kind of noise and highlight our present work (K. Vogel, 1985, unpublished diploma thesis, University of Ulm; Vogel *et al.*, 1987a–c) on noise-color-induced effects in the laser gyro. Special emphasis is put in this treatment on the matrix continued fraction method (Risken, 1984)– the powerful tool allowing us to calculate in a formally exact way all properties of the gyro problem of interest and in particular the mean beat frequency. A review on quantum fluctuations in the ring-laser gyroscope can be found in Klimontovich, Kovalev and Landa (1972), Landa (1970) and Singh (1984). For the discussion of the dithering technique in the presence of noise we refer to the extensive literature on this subject (Bambini and Stenholm, 1984, 1985a, b, 1987; Hutchings and Stjern, 1978; Khoshev, 1977a, b, 1979, 1980; Kruglik, Kutsak and Kuznetsov, 1972; Kruglik, Pestov and Pokrovskii, 1975; Kruglik, Pestov, Pokrovskii and Kutsak, 1970a, b; Roland and Agrawal, 1981; Schleich, Cha and Cresser, 1984; Schleich and Dobiasch, 1984).

The present chapter is organized as follows: in Section 11.2, we briefly review the basic elements of ring-laser theory such as the Sagnac effect (Section 11.2.1), the lock-in effect (Section 11.2.2) and the influence of quantum noise due to the spontaneous emission of the laser atoms (Section 11.2.3). We then (Section 11.3) introduce the Langevin and Fokker–Planck equations for the phase difference between the two counterpropagating waves in the ring-laser gyroscope. We express the steady state distribution in terms of scalar continued fractions (Section 11.3). Moreover, the lock-in characteristic of the gyro in the presence of white noise is given in terms of scalar continued fractions and a reduction of the dead band due to the noise is shown to occur. In Section 11.4 we turn to colored noise, that is noise with a nonzero correlation time. We derive the corresponding Langevin and Fokker–Planck equations (Section 11.4.1). An approximate analytical expression for the steady state distribution derived in Section 11.4.2 shows already the 'skewing' of the steady state distribution. The two-dimensional Fokker–Planck equation is then (Section 11.4.3) expanded in an appropriate complete set of functions and cast into a three-term vector recurrence relation. In steady state we solve this equation by matrix continued fractions. Based on these expressions we, in Section 11.4.4, discuss the influence of noise color on the steady state distributions and conclude in Section 11.4.5 by presenting the mean beat frequency characteristic in the presence of colored noise. Section 11.5 is finally a summary.

11.2 Ring-laser theory: a pico-review

A ring-laser (Sargent, Scully and Lamb, 1974) consists of a ring interferometer formed by three or more mirrors and a laser medium inside the cavity as shown in Figure 11.2. In a linear, two mirror laser the amplitudes and phases of the counterpropagating waves are equal and the electric field is, therefore, a standing wave. In a ring-laser configuration, however, the two oppositely directed running waves may have different amplitudes, phases and frequencies. In particular, a clockwise rotation of the laser about an axis perpendicular to the plane of the mirrors causes a frequency difference between the two counterpropagating waves – the so-called Sagnac effect (Post, 1967; Sagnac, 1913a, b). Heterodyning the two waves and measuring the beat note as shown in Figure 11.2, therefore, provides information about the rotation of the system. This leads to the use of ring-lasers as rotation sensors, first demonstrated by Macek and Davis (1963).

In this section, we briefly review the basic elements of ring-laser theory necessary for the understanding of the remainder of this chapter. Starting from a nonrelativistic explanation (Post, 1967) of the Sagnac effect presented in Section 11.2.1, the locking of the ring-laser (Aronowitz, 1971) caused by backscattering from the laser mirrors is explained in Section 11.2.2. We conclude by discussing the influence of quantum noise due to spontaneous

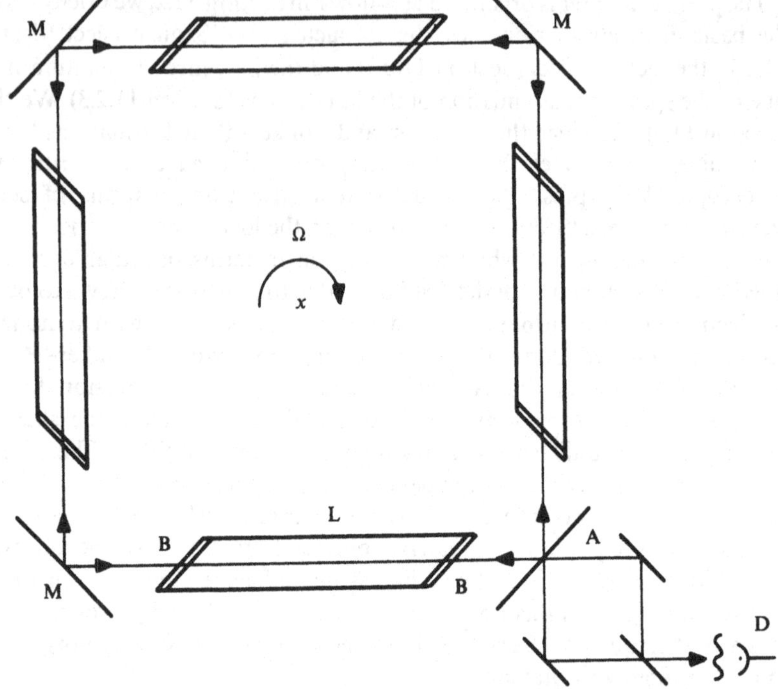

Figure 11.2. The frequencies of two counterpropagating electromagnetic waves in a ring-laser (laser medium inside the resonator) get slightly shifted when the interferometer is rotated with a rate Ω about an axis perpendicular to its surface. This frequency difference is measured by heterodyning the two beams. A = beamsplitter; B = Brewster window; D = detector; L = laser medium; M = mirror.

emission of the laser atoms on the locking equation (Cresser, 1982; Cresser *et al.*, 1982a, b). In this chapter we do not derive the complete set of semiclassical, self-consistent phase and intensity equations (Haken, 1970; Lamb, 1964) but rather refer to the literature (Aronowitz, 1965; Gauert, 1984; Sargent, Scully and Lamb, 1974).

11.2.1 Sagnac effect

In this section we give a simple explanation (Post, 1967) of the optical analog of the Foucault pendulum–the Sagnac effect.

For simplicity, we consider a circular ring interferometer of radius ρ rotating with a rate Ω around an axis perpendicular to its plane. Such an interferometer may be constructed using a glass fiber as indicated in Figure 11.3. Consider two counterpropagating beams starting from point A at time $t = 0$. Since the interferometer rotates, the two beams have to traverse different path lengths $P^{(\pm)}$ in order to reach the point A again: the beam co-directional with the

backscattering coefficient. (In the remainder of this chapter we do not distinguish between the rotation rate Ω and a and call a the rotation rate of the interferometer.)

A detector measuring the beat signal between the two counterpropagating waves (see Figure 11.2) averages over time and thus only the mean beat frequency

$$\langle\dot{\phi}\rangle_t \equiv \lim_{t\to\infty}\left\{\frac{1}{t}\int_0^t \dot{\phi}(t')\,\mathrm{d}t'\right\} \tag{11.2.6}$$

is of interest.

Equation (11.2.5) may be solved by separation of variables (Sargent, Scully and Lamb, 1974) and substituted back into (11.2.6) to yield

$$\langle\dot{\phi}\rangle_t = \begin{cases} 0 & |a| \leqslant b \\ (a^2 - b^2)^{1/2} & |a| \geqslant b. \end{cases} \tag{11.2.7}$$

We note that in the limit of large rotation rates, i.e. $|a| \gg b$, (11.2.7) reduces to

$$\langle\dot{\phi}\rangle_t \simeq a - \frac{1}{2}\frac{b^2}{a}$$

and the gyroscope responds approximately in a linear way as is shown in Figure 11.4. However, for rotation rates a smaller than the backscattering coefficient b, the beat note vanishes even for nonzero rotation rates. This implies that the frequencies of the two counterpropagating waves are equal. Therefore, the phase difference between the two waves remains constant in time. This phenomenon, called locking of the laser, is the fundamental problem in ring-laser gyroscope technology, since in this dead band (see Figure 11.4) the device is insensitive to rotation.

11.2.3 Quantum noise

In the framework of semiclassical laser theory (Haken, 1970; Sargent, Scully and Lamb, 1974) usually employed to describe ring-laser operation, the atoms are treated quantum mechanically, whereas the electric field is considered to be a classical quantity. However, the field can also be quantized giving rise to purely quantum mechanical effects such as vacuum fluctuations and spontaneous emission. The last effect is of special interest to us: an excited atom can reach its ground state by spontaneously emitting a photon. The emitted electric field having a random phase superposes with the field present in the cavity whose phase is, therefore, no longer well determined: it becomes a stochastic quantity. A ring-laser when employed as a rotation sensor measures the phase or frequency difference between the counterpropagating waves. Quantum noise does, therefore, have an influence on the accuracy of determining the rotation rate (Scully, 1985). Moreover, note that this limitation in accuracy is purely due to the quantum nature of light.

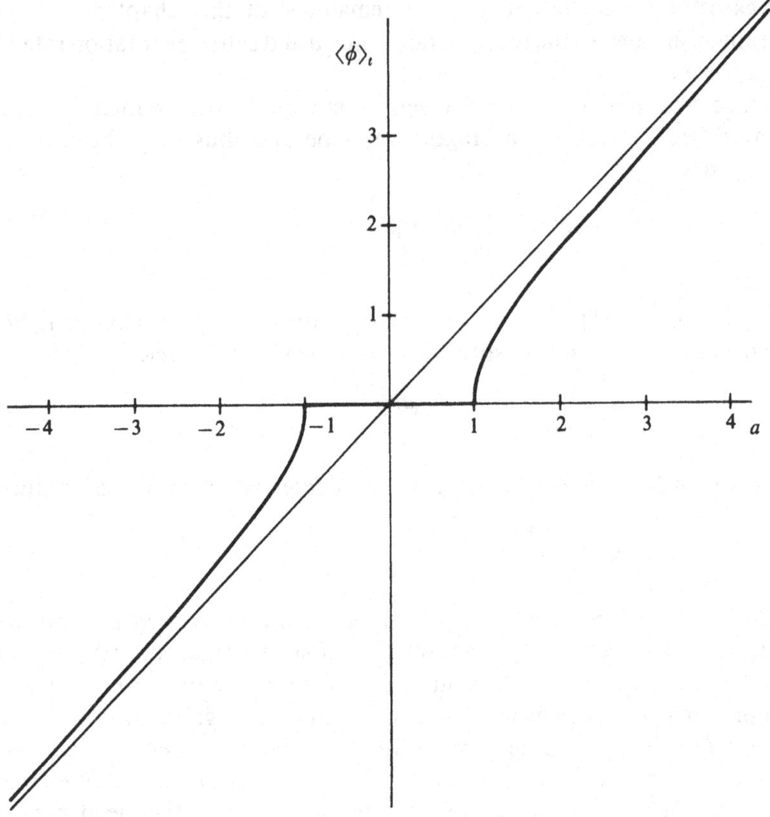

Figure 11.4. Lock-in curve of ring-laser gyroscope. Due to backscattering from the mirrors the linearity between rotation rate and mean beat frequency $\langle \dot{\phi} \rangle_t$ as predicted by Sagnac effect breaks down and a dead band around zero rotation rate emerges.

It has been shown (Cresser *et al.*, 1982b) that the fluctuations in the phase difference between the two counterpropagating waves may be taken into account by adding a fluctuating force F to (11.2.5) which thus reads

$$\dot{\phi} = a + b \sin \phi + F(t). \tag{11.2.8}$$

Here, F is assumed to be Gaussian with mean zero, i.e.

$$\langle F(t) \rangle = 0 \tag{11.2.9}$$

and the symbol $\langle \ \rangle$ denotes an average over the noise F. To fully describe this stochastic process the Markov approximation is made, i.e. the noise is assumed to be delta-correlated, i.e.

$$\langle F(t) F(s) \rangle = 2D\delta(t-s), \tag{11.2.10}$$

where $D = \nu/(2Q\bar{n})$ denotes the diffusion constant of spontaneous emission

(Scully and Lamb, 1967). (Here, ν is the frequency of light being in a resonator with quality factor Q and $\bar{n}$ is the average number of photons in the cavity.)

The phase diffusion in a ring-laser gyroscope due to spontaneous emission of the laser atoms is thus determined by the stochastic differential equation (11.2.8) together with the properties of F being Gaussian and satisfying (11.2.9) and (11.2.10). Similar stochastic differential equations also arise in a number of other physical situations, such as a laser with injected signal (Chow, Scully and Van Stryland, 1975), Josephson junctions (Bohr, Bak and Jensen, 1984; Büttiker, Harris and Landauer, 1983; Gwinn and Westervelt, 1985; Ivanchenko and Zil'berman, 1968), self-locking in a laser (Haken, Sauermann, Schmidt and Vollmer, 1967), charge density waves (Grüner and Zettl, 1985) and in radio engineering (Stratonovich, 1967).

11.3 Fokker–Planck equation for ring-laser gyro with white noise

In order to study the influence of quantum noise on the mean beat frequency, i.e. in order to perform the average of $\langle \dot{\phi} \rangle$ over the noise F, two (equivalent) approaches are possible: the Langevin and the Fokker–Planck approaches (Louisell, 1973; Risken, 1984). In the Langevin approach the equation of motion for ϕ is solved in terms of quadratures of the fluctuating force F and is substituted back into (11.2.8) to perform the average using (11.2.9) and (11.2.10) together with the assumption of F being Gaussian. However, this approach fails in most of the cases, because one is unable to solve the equation of motion which, at least in problems of interest, is nonlinear, see (11.2.8).

A more promising approach is the Fokker–Planck approach in which a linear partial differential equation for a probability distribution is derived. The average is then performed in the well known way. This technique works well because, in many cases of interest, one is able to solve the Fokker Planck equation in an exact way using infinite (matrix) continued fractions (Risken, 1984; Risken and Vollmer 1980).

The Fokker–Planck equation corresponding to the Langevin equation (11.2.8) reads (see, for instance, Risken, 1984, for a derivation)

$$\frac{\partial P}{\partial t} = -\frac{\partial}{\partial \phi}[(a + b\sin\phi)P] + D\frac{\partial^2 P}{\partial \phi^2}. \tag{11.3.1}$$

This equation and its steady state solutions in terms of quadratures have been discussed in great detail by Haken *et al.* (1967) and Stratonovich (1967). We therefore proceed directly to the continued fraction technique.

In steady state, i.e. for $\partial P/\partial t \equiv 0$ we make the ansatz

$$P_{ss}(\phi) = \frac{1}{(2\pi)^{1/2}} \sum_{m=-\infty}^{+\infty} s_m e^{im\phi} \tag{11.3.2}$$

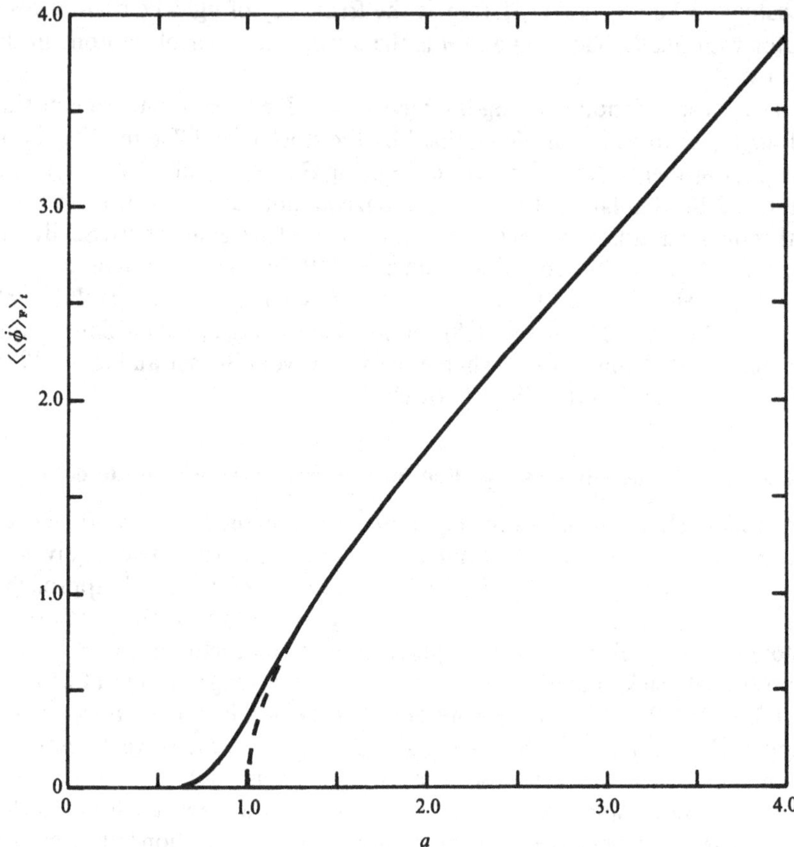

Figure 11.5. Mean beat frequency $\langle\langle\dot{\phi}\rangle_F\rangle_t$ in the presence of white noise as a function of rotation rate a for $b = 1$ and $D = 0.1$ (solid line) compared to noise-free response curve, (11.2.7) (dashed line). White noise clearly decreases the width of the dead band.

which when substituted into (11.3.1) yields the recurrence relation

$$0 = -\frac{b}{2}s_{m+1} + (\mathrm{i}a + mD)s_m + \frac{b}{2}s_{m-1}. \tag{11.3.3}$$

With the iteration

$$s_m = r_m s_{m-1} \tag{11.3.4}$$

we find for the scalar continued fraction (Risken, 1984)

$$r_m = -[2(\mathrm{i}a + mD) - br_{m+1}]^{-1}b. \tag{11.3.5}$$

From the normalization condition of P_{ss} and (11.3.2) we derive for the start value $s_0 = 1/(2\pi)^{1/2}$ and from the reality condition $s_{-m} = s_m^*$. Thus all coefficients are known.

We are now able to discuss the influence of white noise on the mean beat frequency $\langle\langle\dot{\phi}\rangle_F\rangle_t$. Making use of (11.2.9) we find from (11.2.8)

$$\langle\langle\dot{\phi}\rangle_F\rangle_t = a + b\langle\langle \sin\phi\rangle_F\rangle_t = a - (2\pi)^{1/2} b\,\mathrm{Im}(s_1), \qquad (11.3.6)$$

where in the last step we have performed the average over the noise in steady state with the help of the distribution (11.3.2) and used the fact that $s_{-1} = s_1^*$.

Figure 11.5 shows the mean beat frequency averaged over the noise, i.e. $\langle\langle\dot{\phi}\rangle_F\rangle_t$ as a function of the rotation rate based on (11.3.6) in comparison with the corresponding deterministic locking curve, (11.2.7). We note that far away from the locking region ($|a| \gg b$) the two curves coincide. However, for rotation rates $a \simeq b$ we note a new feature: due to noise, unlocking tends to occur at smaller rotation rates.

11.4 Colored noise in the ring-laser gyroscope

In the preceding section we showed that white noise can indeed reduce the width of the dead band and the tendency of the gyro to lock. We are therefore led into deliberately introducing additional noise. External noise, however, has necessarily a nonzero correlation time and is thus colored noise. Various approximate analytical schemes to treat this problem exist (Hänggi, 1978; Hänggi, Mroczkowski, Moss and McClintock, 1985; Moss, Hänggi, Mannella and McClintock, 1986; Sancho, San Miguel, Katz and Gunton, 1982; Schenzle and Graham, 1983). In the present approach we introduce in Section 11.4.1 an additional stochastic variable via an Ornstein–Uhlenbeck process and thus find a two-dimensional Fokker–Planck equation. This equation is cast into a three term vector recurrence relation in Section 11.4.3 and is solved in steady state with the help of matrix continued fractions. This technique even allows us to find a formally exact expression for the lock-in characteristic of a ring-laser gyro in the presence of colored noise. The continued fraction approach is facilitated by approximate analytical expressions for the steady state distribution discussed in Section 11.4.2.

11.4.1 Langevin and Fokker–Planck equations

Controlled external noise can be realized experimentally (Kruglik, Blazhnov, Kuznetsov and Kutsak, 1972; Kruglik, Kutsak and Kuznetsov, 1972) by mounting one of the resonator mirrors onto a piezoceramic element driven by a noise generator. The resulting mirror displacements ΔP are typically of the order of 0.1 μm with a band width (i.e. the inverse correlation time τ_c) ranging from a few hertz to kilohertz. The maximum value of b is a few-hundred hertz. Therefore external noise indeed has a correlation time $\tau_c > 0$ (colored noise). The mirror fluctuations cause the perimeter P of the gyro to fluctuate. From (11.2.4) in the form

$$a = 2\frac{P}{\lambda}\Omega$$

we realize that fluctuations in P, that is $P \to P + \Delta P$, correspond to fluctuations ε in the rotation rate

$$a \to a + 2\frac{\Delta P}{\lambda}\Omega = a + \varepsilon(t).$$

The equation of motion for the phase difference ϕ between the counterpropagating waves in a gyroscope thus reads

$$\dot{\phi} = a + b\sin\phi + \varepsilon(t). \tag{11.4.1}$$

Here we assume for $\varepsilon(t)$ Gaussian, colored noise of strength D defined by

$$\langle \varepsilon(t)\varepsilon(s)\rangle = (D/\tau_c)\,\mathrm{e}^{-|t-s|/\tau_c} \tag{11.4.2}$$

with

$$\langle \varepsilon(t)\rangle = 0. \tag{11.4.3}$$

A Fokker–Planck equation corresponding to (11.4.1), (11.4.2) and (11.4.3) can be obtained by introducing the additional stochastic variable ε via the process

$$\dot{\varepsilon} = -\frac{1}{\tau_c}\varepsilon + F(t), \tag{11.4.4}$$

where

$$\langle F(t)F(s)\rangle = (2D/\tau_c^2)\delta(t-s)$$

and

$$\langle F(t)\rangle = 0.$$

It is straightforward to show that the stochastic variable, ε, defined in this way, satisfies indeed the correlation functions (11.4.2) and (11.4.3). The two-dimensional Fokker–Planck equation corresponding to the Langevin equations (11.4.1) and (11.4.4) then reads (Risken, 1984)

$$\frac{\partial P}{\partial t} = -\frac{\partial}{\partial\phi}[(a + b\sin\phi + \varepsilon)P] + \frac{1}{\tau_c}\frac{\partial(\varepsilon P)}{\partial\varepsilon} + (D/\tau_c^2)\frac{\partial^2 P}{\partial\varepsilon^2}. \tag{11.4.5}$$

Since ϕ denotes the phase difference between two waves we impose periodic boundary conditions for ϕ and since ε is generated by an Ornstein–Uhlenbeck process, (11.4.4), natural boundary conditions for ε are adequate.

11.4.2 Approximate expressions for the steady state solution

In Sections 11.4.4 and 11.4.5 formally exact expressions for the steady state distribution $P_{ss} = P_{ss}(\phi, \varepsilon)$ and the mean beat frequency $\langle\langle\dot{\phi}\rangle_\varepsilon\rangle_t$ are obtained in terms of infinite matrix continued fractions. Whereas this method is well suited for numerical analysis, it makes it difficult to get some insight into the functional dependence of P_{ss} on the various parameters such as the correlation time τ_c of the noise. Therefore it is worthwhile to consider approximate analytical expressions which can be checked against the matrix continued fraction treatment. For this reason we consider in this section an approximation of P_{ss} valid well in the dead band for weak noise.

Within the locked region, i.e. for $|a| < b$, and in the absence of noise the phase is known (Cresser *et al.*, 1982b) to settle down at the stable steady state value

$$\phi_{ss} = \pi + \arcsin\frac{a}{b}.$$

In the presence of weak noise we expect the phase not to diffuse very far away from its equilibrium point. We thus try the ansatz

$$\phi = \phi_{ss} + \theta$$

with $|\theta| \ll 1$. Since

$$a + b\sin\phi \approx -\gamma\theta,$$

where

$$\gamma = (b^2 - a^2)^{1/2}$$

the Fokker–Planck equation (11.4.5) reduces to

$$\frac{\partial P}{\partial t} = \gamma\frac{\partial}{\partial\theta}(\theta P) - \varepsilon\frac{\partial P}{\partial\theta} + \frac{1}{\tau_c}\frac{\partial}{\partial\varepsilon}(\varepsilon P) + (D/\tau_c^2)\frac{\partial^2 P}{\partial\varepsilon^2}.$$

This equation describes an Ornstein–Uhlenbeck process and can thus be solved exactly, and after minor algebra we arrive at

$$P_{ss}(\phi, \varepsilon) = (\gamma\tau_c)^{1/2}\frac{\gamma + 1/\tau_c}{2\pi D/\tau_c}\exp\left\{-\frac{\gamma + 1/\tau_c[\varepsilon^2 - 2\gamma\varepsilon(\phi - \phi_{ss})}{2D/\tau_c^2}\right.$$
$$\left. + \gamma(\gamma + 1/\tau_c)(\phi - \phi_{ss})^2]\right\} \qquad (11.4.6)$$

where we have replaced the periodic boundary conditions for ϕ by natural ones.

Note that the rotation rate a (via γ) as well as the nonvanishing correlation time τ_c cause a 'coupling' between the two variables ε and ϕ leading to an asymmetry and a kind of 'skewing' in the probability distribution. This noise-color-induced effect is the main result and will be the subject of a detailed discussion in the later sections.

11.4.3 *Fokker–Planck equation as a differential recurrence relation*

In this section we make the first step towards the matrix continued fraction treatment and cast the Fokker–Planck equation (11.4.5) into a differential vector recurrence relation.

In order to satisfy the boundary conditions for ϕ and ε we make the ansatz

$$P(t, \phi, \varepsilon) = \frac{1}{(2\pi)^{1/2}}\mathcal{H}_0(\varepsilon)\sum_{m=0}^{\infty}\sum_{n=-\infty}^{+\infty} S_{m,n}(t)\mathcal{H}_m(\varepsilon)e^{in\phi}, \qquad (11.4.7)$$

where

$$\mathscr{H}_m(\varepsilon) = N_m \exp\left(-\frac{\varepsilon^2}{4D/\tau_c}\right) H_m(\varepsilon/(2D/\tau_c)^{1/2}). \tag{11.4.8}$$

The normalization factors N_m are chosen to be

$$N_m = (m!\,2^m)^{-1/2}(2\pi D/\tau_c)^{-1/4}$$

and the H_m are the familiar Hermite polynomials (Magnus, Oberhettinger and Soni, 1966).

Substituting this ansatz into the Fokker–Planck equation (11.4.5) and projecting onto the coefficients $S_{m,n}$ results in the differential recurrence relation

$$\dot{S}_{m,n} = -(ina + m/\tau_c)S_{m,n} + \frac{nb}{2}(S_{m,n+1} - S_{m,n-1})$$
$$- in(D/\tau_c)^{1/2}((m+1)^{1/2}S_{m+1,n} + m^{1/2}S_{m-1,n}). \tag{11.4.9}$$

From the fact that the conditional probability P is real, we deduce the symmetry relation

$$S_{m,-n} = S^*_{m,n}, \tag{11.4.10}$$

and thus (11.4.9) has to be solved only for $n \geqslant 0$. Defining the vector $\mathbf{S}_m$ via its components

$$(\mathbf{S}_m)_n \equiv S_{m,n} \tag{11.4.11}$$

equation (11.4.9) can be cast into a three term vector recurrence relation

$$\dot{\mathbf{S}}_m = \boldsymbol{B}_m\mathbf{S}_{m+1} + \boldsymbol{A}_m\mathbf{S}_m + \boldsymbol{C}_m\mathbf{S}_{m-1} \tag{11.4.12}$$

where the matrices $\boldsymbol{A}_m$, $\boldsymbol{B}_m$ and $\boldsymbol{C}_m$ are defined by

$$(\boldsymbol{A}_m)_{n,n'} := -(ina + m/\tau_c)\delta_{n,n'} + \frac{nb}{2}(\delta_{n+1,n'} - \delta_{n-1,n'}), \tag{11.4.13a}$$

$$(\boldsymbol{B}_m)_{n,n'} := -in(D/\tau_c)^{1/2}(m+1)^{1/2}\delta_{n,n'}, \tag{11.4.13b}$$

$$(\boldsymbol{C}_m)_{n,n'} := -in(D/\tau_c)^{1/2}m^{1/2}\delta_{n,n'}. \tag{11.4.13c}$$

11.4.4 Steady state distributions

We now solve the recurrence relation, (11.4.12), in steady state using matrix continued fractions. In the steady state, that is $\dot{\mathbf{S}}_m = 0$, the recurrence relation

$$0 = \boldsymbol{B}_m\mathbf{S}_{m+1} + \boldsymbol{A}_m\mathbf{S}_m + \boldsymbol{C}_m\mathbf{S}_{m-1} \tag{11.4.14}$$

can be solved by the ansatz (Risken, 1984; Risken and Vollmer, 1980)

$$\mathbf{S}_m = \boldsymbol{R}_m\mathbf{S}_{m-1} \quad \text{for } m \geqslant 1 \tag{11.4.15}$$

which reduces (11.4.14) to

$$0 = (\boldsymbol{B}_m \boldsymbol{R}_{m+1} \boldsymbol{R}_m + \boldsymbol{A}_m \boldsymbol{R}_m + \boldsymbol{C}_m)\mathbf{S}_{m-1}.$$

This equation is satisfied by the infinite matrix continued fraction

$$\boldsymbol{R}_m = -(\boldsymbol{A}_m + \boldsymbol{B}_m \boldsymbol{R}_{m+1})^{-1}\boldsymbol{C}_m. \tag{11.4.16}$$

Note that (11.4.14), (11.4.15) and (11.4.16) are the matrix analogs of the scalar expressions (11.3.3), (11.3.4) and (11.3.5). The start vector $\mathbf{S}_0$ can be obtained from (11.4.14) for $m = 0$ by noting that according to (11.4.13c) $\boldsymbol{C}_0 = 0$ and by again using (11.4.15) which results in

$$(\boldsymbol{A}_0 + \boldsymbol{B}_0 \boldsymbol{R}_1)\mathbf{S}_0 = 0 \quad \text{with} \quad S_{0,0} = 1/(2\pi)^{1/2}. \tag{11.4.17}$$

We now briefly outline the procedure for calculating the coefficients $S_{m,n}$. Substituting the matrices $\boldsymbol{A}_m$, $\boldsymbol{B}_m$ and $\boldsymbol{C}_m$ into (11.4.16) and using downward iteration (Risken, 1984) yields the matrices $\boldsymbol{R}_m$ and, in particular, $\boldsymbol{R}_1$. The start vector $\mathbf{S}_0$ is then determined from (11.4.17) by combining the results for $\boldsymbol{R}_1$ and (11.4.13a) and (11.4.13b). Substituting the thus-calculated $\mathbf{S}_0$ together with $\boldsymbol{R}_1$ into (11.4.15) we arrive at $\mathbf{S}_1$. Continuing this iteration yields $\mathbf{S}_m$.

Example solutions are shown in Figures 11.6 and 11.7. Here we have plotted $P_{ss}(\phi, \varepsilon)$ on the left half and the corresponding contours of constant probability are shown on the right half. For the sake of clarity, the phase variable ϕ is displayed over two periods from -2π to 2π. For simplicity, we have set $D = b = 1$.

The effect of the rotation rate on the steady state distribution is shown in Figure 11.6(a)–(d) for nearly white noise, that is $\tau_c = 0.1$ and in Figure 11.7(a)–(d) for $\tau_c = 1$. A noticeable effect of increasing τ_c is to markedly increase the ratio of peak height to saddle-point probability densities, as was observed (Fronzoni *et al.*, 1986; Hänggi *et al.*, 1985; Jung and Risken, 1985; Moss and McClintock, 1985) previously in the case of a bistable system. In addition, noise color has a profound influence on the shape of the contours. Whereas for small τ_c, P_{ss} is more nearly symmetric about the $\varepsilon = 0$ axis, as shown in Figure 11.6, increasing τ_c destroys this symmetry, as evident in Figure 11.7. Moreover, the contours are skewed towards an axis running from lower left to upper right for $a > 0$. (The reverse is true for $a < 0$.) Note that in Figure 11.6 the scale of ε is ± 10, so that there is small skewing, whereas in Figure 11.7 the scale of ε is ± 3.3, showing considerable skewing. This skewing of the contour lines is even more pronounced in the case of strongly colored noise $\tau_c = 10$ as shown for $a = 1.5$ in Figure 11.8. We conclude this section by noting that this 'skewing effect' is already apparent in the approximate analytical expression, (11.4.6), derived in Section 11.4.2.

11.4.5 Mean beat frequency

The noise-color-induced skewing and asymmetry of $P_{ss}(\phi, \varepsilon)$ discussed in the

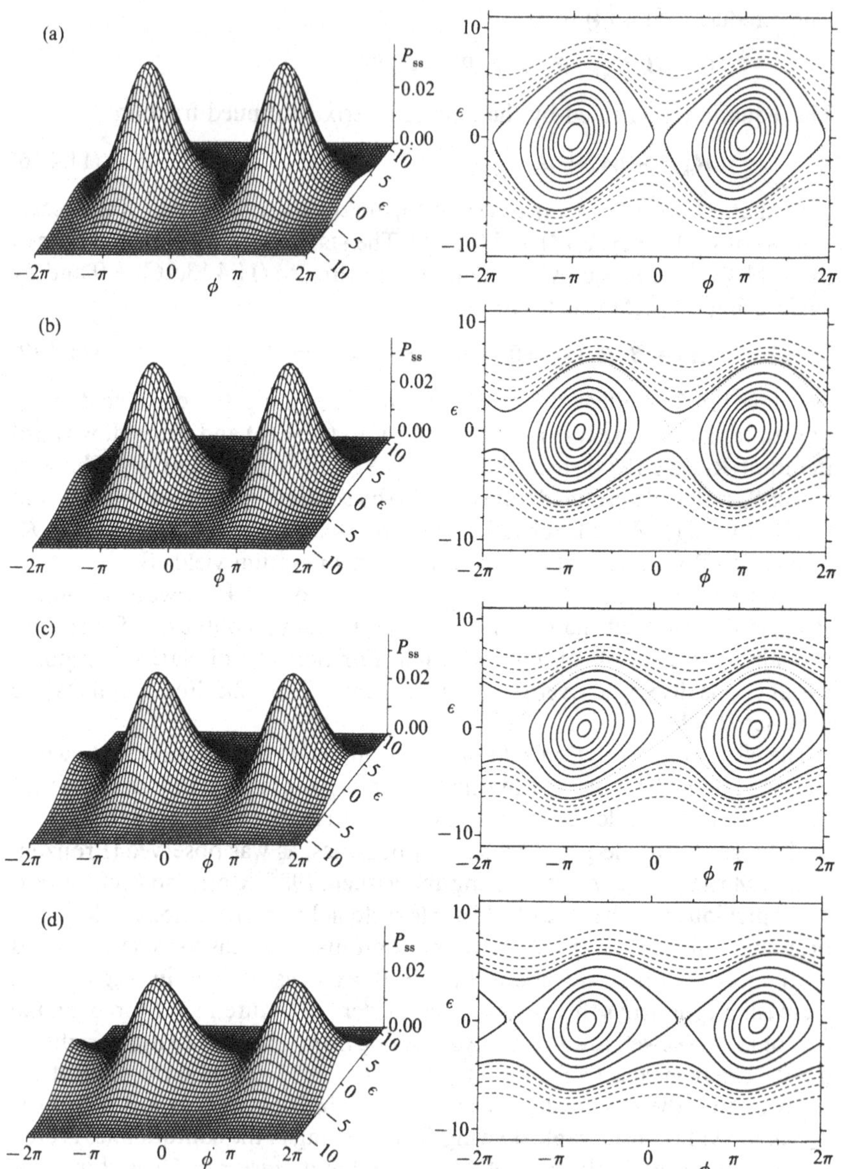

Figure 11.6. The statistical density $P_{ss}(\phi, \varepsilon)$ for $D = b = 1$ and nearly white noise, that is $\tau_c = 0.1$. Three-dimensional plots (left half) and contours of constant probability (right half) for equally spaced probabilities 0.005, 0.010... (solid lines) and 0.001, 0.002... (dashed lines). The 'separatrix' (dotted line) is for $a = 0$ at 0.00489 (a), for $a = 0.5$ at 0.00580 (b), for $a = 1$ at 0.007847 (c), and for $a = 1.5$ at 0.00995 (d).

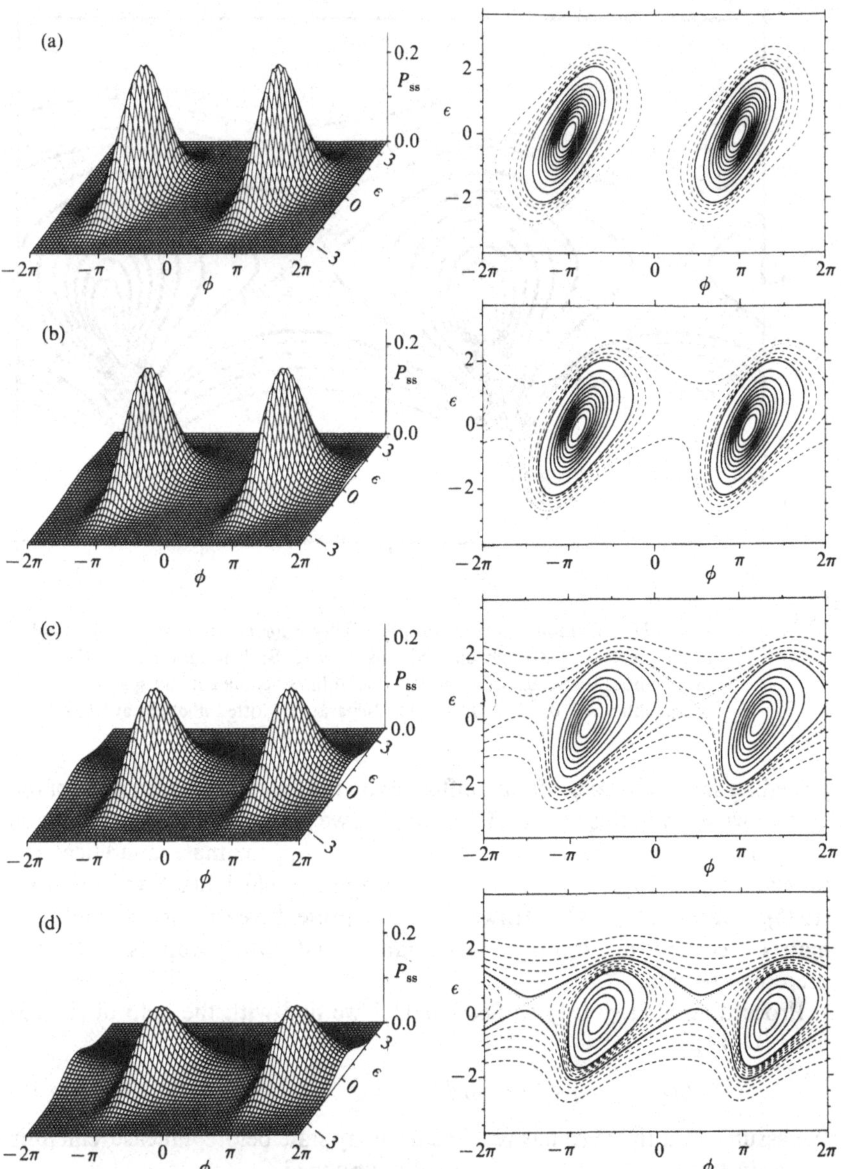

Figure 11.7. The statistical density $P_{ss}(\phi, \varepsilon)$ for $D = b = 1$ with noise color $\tau_c = 1$. Three-dimensional plots (left half) and contours of constant probability (right half) for equally spaced probabilities 0.025, 0.050... (solid lines) and 0.005, 0.010... (dashed lines). The 'separatrix' (dotted line) is for $a = 0$ at 0.00355 (a), for $a = 0.5$ at 0.00704 (b), for $a = 1$ at 0.01576 (c), and for $a = 1.5$ at 0.02564 (d).

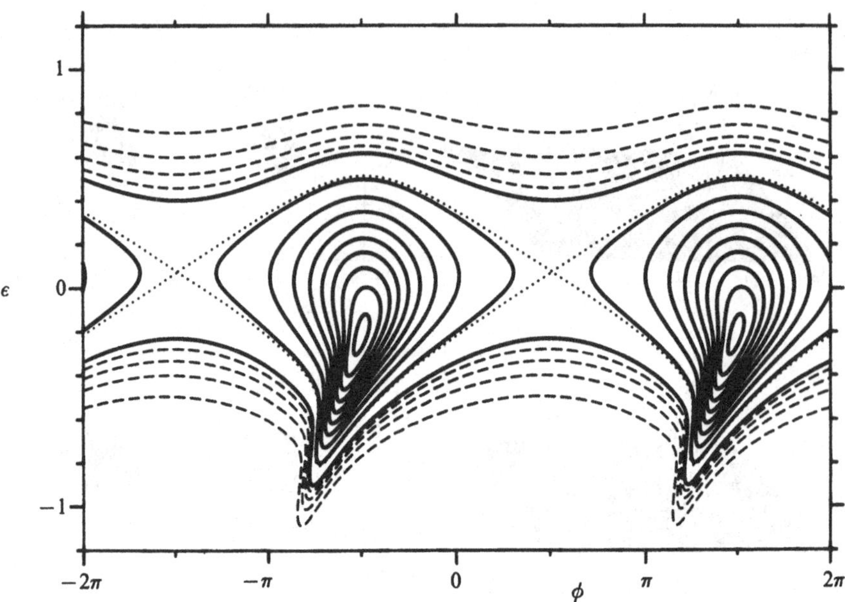

Figure 11.8. Contours of constant probability $P_{ss}(\phi, \varepsilon)$ for $D = b = 1$ and $a = 1.5$ in the presence of strongly colored noise $\tau_c = 10$. Broken lines indicate the probabilities 0.01, 0.02, 0.03 and 0.04. Solid lines denote equally spaced probabilities of 0.05, 0.10, 0.15.... The 'separatrix' (dotted line) lies at 0.09219.

preceding section has a profound influence on the locking characteristic of the gyro as we show in this section. In particular, we discuss the influence of noise color on the mean beat frequency $\langle\langle\dot{\phi}\rangle_\varepsilon\rangle_t$. Approximate analytical expressions for $\langle\langle\dot{\phi}\rangle_\varepsilon\rangle_t$ have been given by Kaplan (1966); Kutsak and Lyashko (1985); Vogel *et al.* (1987a). However, in the present treatment, we apply the matrix continued fraction technique and the results of the preceding section.

Proceeding analogously to Section 11.3 we find with the help of (11.4.3) from (11.4.1)

$$\langle\langle\dot{\phi}\rangle_\varepsilon\rangle_t = a + b\langle\langle \sin\phi\rangle_\varepsilon\rangle_t. \tag{11.4.18}$$

We assume that the gyro has reached a steady state before a measurement is made. In this case $\langle\langle \sin\phi\rangle_\varepsilon\rangle_t$ is readily obtained

$$\langle\langle \sin\phi\rangle_\varepsilon\rangle_t = \int_{-\pi}^{+\pi} \mathrm{d}\phi \int_{-\infty}^{+\infty} \mathrm{d}\varepsilon \sin\phi\, P_{ss}(\phi, \varepsilon). \tag{11.4.19}$$

Substituting (11.4.7) into (11.4.19) and performing the integrals with the help of the orthogonality of the $\mathscr{H}_m$, (11.4.8), we find from (11.4.18)

$$\langle\langle\dot{\phi}\rangle_\varepsilon\rangle_t = a - b(2\pi)^{1/2}\, \mathrm{Im}(S_{0,1}). \tag{11.4.20}$$

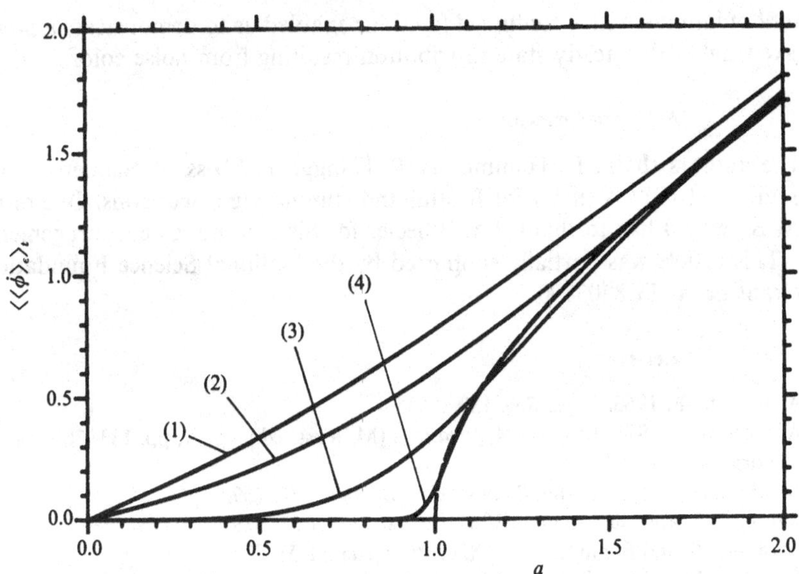

Figure 11.9. Mean beat frequency $\langle\langle\dot{\phi}\rangle_\varepsilon\rangle_t$ as a function of the rotation rate a for $D = b = 1$ and noise correlation time $\tau_c = 0$ (1), $\tau_c = 1$ (2), $\tau_c = 10$ (3), $\tau_c = 500$ (4).

Here we have also used (11.4.10). The mean beat frequency $\langle\langle\dot{\phi}\rangle_\varepsilon\rangle_t$ is thus determined by one single expansion coefficient in (11.4.7), $S_{0,1}$.

In Figure 11.9 we show the mean beat frequency $\langle\langle\dot{\phi}\rangle_\varepsilon\rangle_t$ in the presence of colored noise as a function of the rotation rate a for various correlation times τ_c. For comparison we have also depicted the noise-free lock-in curve by a broken line. We have chosen $b = D = 1$ since in that case white noise, $\tau_c = 0$, restores a merely linear relationship between $\langle\langle\dot{\phi}\rangle_\varepsilon\rangle_t$ and a. Note, however, that for increasing correlation times the tendency of the gyroscope to lock gets stronger. Moreover, we emphasize that the lock-in curve in the presence of strongly colored noise ($\tau_c = 10$) approaches the deterministic curve from below whereas the white noise curve always lies above, see Figure 11.5. We conclude by noting that Kaplan (1966) has obtained a qualitatively similar result by an heuristic argument: averaging the deterministic locking curve by a Gaussian distribution of rotation rates. Here, however, we have presented a formally exact expression for $\langle\langle\dot{\phi}\rangle_\varepsilon\rangle_t$ in terms of infinite matrix continued fractions.

11.5 Summary

In the present chapter we have reviewed the basic elements of ring-laser theory such as Sagnac effect and locking of the gyro. Special emphasis was put on the discussion of the influence of white and colored noise on the steady state distributions and on the lock-in characteristic of a ring-laser gyro. The powerful tool of infinite matrix continued fractions enabled us to demonstrate a novel 'skewing' of the steady state distribution resulting from noise color.

Acknowledgements

The authors thank D. Hammonds, P. Hänggi, F. Moss, V. Sanders, M.O. Scully and H.D. Vollmer for fruitful and stimulating discussions. One of us (W.S.) would like to thank J.A. Wheeler for his continuous encouragement.

This article was partially supported by the National Science Foundation Grant no. PHY 8503890.

References

Aronowitz, F. 1965. *Phys. Rev.* **139**, A635.

Aronowitz, F. 1971. In *Laser Applications* (M. Ross, ed.), vol. 1, pp. 133–200. New York: Academic Press.

Bambini, A. and Stenholm, S. 1984. *Opt. Commun.* **49**, 269.

Bambini, A. and Stenholm, S. 1985a. *Phys. Rev. A* **31**, 329.

Bambini, A. and Stenholm, S. 1985b. *Phys. Rev. A* **31**, 3741.

Bambini, A. and Stenholm, S. 1987. *J. Opt. Soc. Am. B* **4**, 148.

Bohr, T., Bak, P. and Jensen, M. H. 1984. *Phys. Rev. A* **30**, 1970.

Büttiker, M., Harris, E. P. and Landauer, R. 1983. *Phys. Rev. B* **28**, 1268.

Chow, W. W., Gea-Banacloche, J., Pedrotti L. M., Sanders, V. E., Schleich, W. and Scully, M. O. 1985. *Rev. Mod Phys.* **57**, 61.

Chow, W. W., Scully, M. O. and Van Stryland, E. W. 1975. *Opt. Commun.* **15**, 6.

Cresser, J. D. 1982. *Phys. Rev. A* **26**, 398.

Cresser, J. D., Hammonds, D., Louisell, W. H., Meystre, P. and Risken, H. 1982a. *Phys. Rev. A* **25**, 2226.

Cresser, J. D., Louisell, W. H., Meystre, P., Schleich, W. and Scully, M. O. 1982b. *Phys. Rev. A* **25**, 2214.

Fronzoni, L., Grigolini, P., Hänggi, P., Moss, F., Mannella, R. and McClintock, P. V. E. 1986. *Phys. Rev. A* **33**, 3320.

Gauert, R. 1984. *Optik* **67**, 21.

Grüner, G. and Zettl, A. 1985. *Phys. Rep.* **119**, 117.

Gwinn, E. G. and Westervelt, R. M. 1985. *Phys. Rev. Lett.* **54**, 1613.

Haken, H. 1970. In *Handbuch der Physik XXV/2c* (S. Flügge and L. Genzel, eds.). Berlin: Springer.

Haken, H., Sauermann, H., Schmidt, Ch. and Vollmer, H. D. 1967. *Z. Phys.* **206**, 369.

Hänggi, P. 1978. *Z. Phys. B* **31**, 407.

Hänggi, P., Mroczkowski, T. J., Moss, F. and McClintock, P. V. E. 1985. *Phys. Rev. A* **32**, 695.

Hutchings, T. J. and Stjern, D. J. 1978. In *Proceedings of the IEEE National Aerospace and Electronics Conference.* New York: IEEE.

Ivanchenko, Y. M. and Zil'berman, L. A. 1968. *Zh. Eksp. Teor. Fiz.* **55**, 2395. [*Sov. Phys. JETP* **28**, 1272 (1969).]

Johnson, K. J., Spencer, J. H., Mayer, C. H., Klepczynski, W. J., Kaplan, G.,

McCarthy, D. D. and Westerhout, G. 1979. In *Time and the Earth's Rotation* (D. D. McCarthy and J. D. H. Pilkington, eds.), pp. 221–34. Dortrecht: Reidel.
Jung, P. and Risken, H. 1985. *Z. Phys. B* **61**, 367.
Kaplan, A. A. 1966. *Radiotekhnika i Elektronika* **11**, 84. [*Radio Eng. Electron. Phys.* **11**, 1354 (1966).]
Khoshev, I. M. 1977a. *Radiotekhnika i Elektronika* **22** (1), 94. [*Radio Eng. Electron. Phys.* **22**, 135 (1977).]
Khoshev, I. M. 1977b. *Radiotekhnika i Elektronika* **22** (2), 71. [*Radio Eng. Electron. Phys.* **22**, 313 (1977).]
Khoshev, I. M. 1979. *Radiotekhnika i Elektronika* **24** (6), 108. [*Radio Eng. Electron. Phys.* **24**, 1141 (1979).]
Khoshev, I. M. 1980. *Kvantovaya Elektron.* **7**, 953. [*Sov. J. Quantum Electron.* **10**, 544 (1980).]
Killpatrick, J. 1967. *IEEE Spectrum* **4** (10), 44.
Klimontovich, Y. L., Kovalev, A. S. and Landa, P. S. 1972. *Usp. Fiz. Nauk* **106**, 279. [*Sov. Phys. USPEKHI* **15**, 95 (1972).]
Kruglik, G. S., Blazhnov, B. A., Kuznetsov, G. M. and Kutsak, A. A. 1972. *Zh. Prikl. Spektroskopii* **17**, 358. [*J. Appl. Spect.* **17**, 1100 (1972).]
Kruglik, G. S., Kutsak, A. A. and Kuznetsov, G. M. 1972. *Zh. Prikl. Skektroskopii* **16**, 58. [*J. Appl. Spect.* **16**, 44 (1972).]
Kruglik, G. S., Pestov, E. G. and Pokrovskii, V. R. 1975. *Zh. Prikl. Spektroskopii* **23** (3), 405. [*J. Appl. Spect.* **23**, 1176 (1975).]
Kruglik, G. S., Pestov, E. G., Pokrovskii, V. R. and Kutsak, A. A. 1970a.*Zh. Prikl. Spektroskopii* **12** (3), 432. [*J. Appl. Spect.* **12**, 331 (1970).]
Kruglik, G. S., Pestov, E. G., Pokrovskii, V. R. and Kutsak, A. A. 1970b. *Zh. Prikl. Spektroskopii* **13** (5), 913. [*J. Appl. Spect.* **13** (5), 1527 (1970).]
Kutsak, A. A. and Lyashko, O. M. 1985. *Zh. Prikl. Spektroskopii* **43** (2), 188. [*J. Appl. Spect.* **43**, 834 (1985).]
Lamb, W. E., Jr. 1964. *Phys. Rev.* **134**, A1429.
Landa, P. S. 1970. *Zh. Eksp. Teor. Fiz.* **58**, 1651. [*Sov. Phys. JETP* **31**, 886 (1970).]
Louisell, W. H. 1973. *Quantum Statistical Properties of Radiation.* New York: John Wiley.
Macek, W. M. and Davis, D. T. M. 1963. *Appl. Phys. Lett.* **2**, 67.
Magnus, W., Oberhettinger, F. and Soni, R. P. 1966. *Formulas and Theorems for the Special Functions of Mathematical Physics.* Berlin: Springer.
Menegozzi, L. N. and Lamb, W. E., Jr. 1973. *Phys. Rev. A* **8**, 2103.
Michelson, A. A. 1925a. *Astrophys. J.* **61**, 137.
Michelson, A. A. 1925b. *Astrophys. J.* **61**, 140.
Misner, C., Thorne, K. S. and Wheeler, J. A. 1973. *Gravitation.* San Francisco: Freeman.
Moss, F., Hänggi, P., Mannella, R. and McClintock, P. V. E. 1986. *Phys. Rev. A* **33**, 4459.
Moss, F. and McClintock, P. V. E. 1985. *Z. Phys. B* **61**, 381. See also references therein.
Post, E. J. 1967. *Rev. Mod. Phys.* **39**, 475.
Privalov, V. E. and Fridrikhov, S. A. 1969. *Usp. Fiz. Nauk* **97** (3), 377. [*Sov. Phys. USPEKHI* **12**, 153 (1969).]

Risken, H. 1984. *The Fokker–Planck Equation*, Springer Series in Synergetics, vol. 18. Berlin: Springer.

Risken, H. and Vollmer, H. D. 1980. *Z. Phys. B* **39**, 339.

Roland, J. J. and Agrawal, G. P. 1981. *Opt. Laser Technol.* **13**, 239.

Sagnac, G. 1913a. *C. R. Acad. Sci.* **157**, 708.

Sagnac, G. 1913b. *C. R. Acad. Sci.* **157**, 1410.

Sancho, J. M., San Miguel, M., Katz, S. L. and Gunton, J. D. 1982. *Phys. Rev. A* **26**, 1589.

Sargent III, M., Scully, M. O. and Lamb, W. E., Jr. 1974. *Laser Physics.* Reading, MA: Addison-Wesley.

Schenzle, A. and Graham, R. 1983. *Phys. Lett.* **98A**, 319.

Schleich, W., Cha, C.-S. and Cresser, J. D. 1984. *Phys. Rev. A* **29**, 230.

Schleich, W. and Dobiasch, P. 1984. *Opt. Commun.* **52**, 63.

Schleich, W. and Scully, M. O. 1984. In *Modern Trends in Atomic and Molecular Physics. Proceedings of the Les Houches Summer School, Session* XXXVIII (R. Stora and G. Grynberg, eds.), pp. 995–1124. Amsterdam: North-Holland.

Schleich, W., Scully, M. O. and Sanders, V. E. 1984. In *Coherence and Quantum Optics* (V. L. Mandel and E. Wolf, eds.), pp. 915–22. New York: Plenum.

Scully, M. O. 1985. *Phys. Rev. Lett.* **55**, 2802.

Scully, M. O. and Lamb, W. E., Jr. 1967. *Phys. Rev.* **159**, 208.

Scully, M. O., Zubairy, M. S. and Haugan, M. P. 1981. *Phys. Rev. A* **24**, 2009.

Singh, S. 1984. *Phys. Rep.* **108**, 217.

Small, J. G. and Chow, W. W. 1982. *Research Proposal no. 104–113.* University of New Mexico, November, 1982.

Stratonovich, R. L. 1967. *Topics in the Theory of Random Noise*, vol. 2. New York: Gordon and Breach.

Vogel, K., Leiber, Th., Risken, H., Hänggi, P. and Schleich, W. 1987a. *Phys. Rev. A* **35**, 4882.

Vogel, K., Risken, H., Schleich, W., James, M., Moss, F. and McClintock, P. V. E. 1987b. *Phys. Rev. A* **35**, 463.

Vogel, K., Risken, H., Schleich, W., James, M., Moss, F., Mannella, R. and McClintock, P. V. E. 1987c. *J. Appl. Phys.* **62**, 721.

12 Control of noise by noise and applications to optical systems

L. A. LUGIATO, G. BROGGI, M. MERRI and M. A. PERNIGO

12.1 Introduction

The consideration of multiplicative or colored noise in nonlinear dynamical systems has revealed a wide zoology of interesting phenomena (see, e.g., Graham and Schenzle, 1982; Hänggi, 1986; Horsthemke and Lefever, 1984; Sancho, San Miguel, Katz and Gunton, 1982; San Miguel and Sancho, 1981; Schenzle and Brand, 1979; Smythe, Moss and McClintock, 1983) which are well beyond the basic and classic picture that emerges from the linearized treatment (Van Kampen, 1961). In the framework of nonlinear systems, under special conditions, the presence of noise can give rise to striking qualitative deviations from the deterministic (i.e. noiseless) picture even for rather low noise levels (Broggi, Lugiato and Colombo, 1985). Furthermore, in the nonlinear domain a multiplicative noise or a colored noise can determine a behavior qualitatively different from that which arises from additive white noise; in some examples, these differences persist even in the small noise limit (Lugiato, Broggi and Colombo, 1986; Lugiato, Colombo, Broggi and Horowicz, 1986).

Modern optics provides a very appropriate framework to study not only nonlinear phenomena, but also noise effects. In fact, the theoretician can describe the behavior of optical systems by means of relatively simple models, that nonetheless are sufficiently realistic to allow a comparison with experimental data. On the other hand, in the field of optics it is possible to realize experiments on noise that allow a degree of control to some extent comparable to that attainable in analogous electronic experiments; an example is the observation of transient optical bimodality in optically pumped sodium vapor (Lange, Mitsche, Deserno and Mlynek, 1985; Mitsche, Deserno, Mlynek and Lange, 1985). The aim of this chapter is to review recent theoretical work on novel noise phenomena displayed by nonlinear optical models. Most of these phenomena are not, however, specific to the field of optics, but have quite a general nature. Our starting point is a generic stochastic differential equation for the dimensionless variable $x(t)$

$$\frac{dx}{dt} = \gamma[-f(x) + g(x)\xi(t)], \qquad (12.1.1)$$

where $f(x)$ and $g(x)$ are (generally nonlinear) functions of x and the parameter γ governs the rate of the dynamics. The random quantity $\xi(t)$ is assumed to be a Gaussian stochastic process of the Ornstein–Uhlenbeck type, (Gardiner, 1983; Risken, 1984; Stratonovich, 1967):

$$\frac{\mathrm{d}\xi}{\mathrm{d}t} = -\frac{1}{\tau}\xi + \frac{1}{\tau}\eta(t), \tag{12.1.2a}$$

where τ is the correlation time of the variable ξ and η is a Gaussian, stationary white noise such that

$$\langle \eta(t) \rangle = 0, \quad \langle \eta(t)\eta(t') \rangle = 2D\delta(t - t'). \tag{12.1.3}$$

Hence the variable ξ has the properties

$$\langle \xi(t) \rangle = 0, \quad \langle \xi(t)\xi(t') \rangle = \sigma^2 \exp\left[-\frac{|t - t'|}{\tau} \right], \tag{12.1.2b}$$

where the noise strength σ is given by

$$\sigma^2 = \frac{D}{\tau}, \tag{12.1.4}$$

and the diffusion constant D controls the noise level.

We will mainly consider parametric noise. In this case, we assume that the dynamical equation is

$$\frac{\mathrm{d}x}{\mathrm{d}t} = \gamma\phi(x, y), \tag{12.1.5}$$

where y is an adimensional parameter. If this parameter y fluctuates in time, we replace y by $y + \xi(t)$, where the quantity $\xi(t)$ represents the noise. Thus (12.1.5) becomes a stochastic equation. In the simplest situation, ϕ depends linearly on y so that

$$\phi(x, y) = -\bar{f}(x) + yg(x). \tag{12.1.6}$$

Hence if we insert (12.1.6) into (12.1.5) and replace y by $y + \xi(t)$, we obtain a stochastic equation of the form (12.1.1) with

$$f(x) = \bar{f}(x) - yg(x). \tag{12.1.7}$$

If ϕ is not linear in y, we assume that the noise level is small enough to justify the approximation

$$\phi(x, y + \xi(t)) \sim \phi(x, y) + \frac{\partial\phi(x, y)}{\partial y}\xi(t); \tag{12.1.8}$$

thus we obtain again (12.1.1) with

$$f(x) = -\phi(x, y), \quad g(x) = \frac{\partial\phi(x, y)}{\partial y}. \tag{12.1.9}$$

The features of noise in the output variable x depend obviously on the function $f(x)$ which governs the deterministic evolution, or equivalently by the corresponding potential function defined by

$$f(x) = \frac{\partial V}{\partial x} \rightarrow V(x) = \int f(x)\,\mathrm{d}x. \tag{12.1.10}$$

On the other hand, the properties of the *output* noise can be notably changed by varying the features of the noise term $g(x)\xi(t)$ which arises from the *input* variable y (i.e. its strength and correlation time, the multiplicative function $g(x)$). A purpose of this chapter is just to illustrate this 'control of noise by noise'.

It illustrates how the noise level in the output variable depends on the parameters in play and on the other features of the stochastic equation (12.1.1). According to the needs, the output noise can be reduced or enhanced. Much of this information arises from the analysis of (12.1.1) linearized around a stationary solution of the deterministic equation

$$\frac{\mathrm{d}x}{\mathrm{d}t} = -\gamma f(x). \tag{12.1.11}$$

The second part of this chapter (*nonlinear control of noise*) discusses nonlinear examples, selected from the domain of optics. It shows how the qualitative features of the output noise, as, for example, the one- or two-peaked character of the probability distribution $P(x, t)$ of the variable x, can be varied by monitoring the colored or the multiplicative character of the fluctuations, or by exploiting the features of the nonlinear dynamics.

The first part covers Sections 12.2, 12.3 and 12.4. Section 12.2 illustrates the wealth of information that we can obtain by using the adiabatic elimination principle (Haken, 1977). Special benefits are a clear 'physical' definition of white noise, and a straightforward derivation of the stationary probability distribution $P_s(x)$ in the limit of very long correlation time in the input noise. Sections 12.3 and 12.4 analyze the linearized stochastic equation and compare the scaling laws of stochastic parametric noise with those of deterministic modulation. We emphasize that the two procedures (modulation and noise) lead to scaling behaviors that display some elements in common but also some differences, a point which most often is not focused clearly.

The second part covers Sections 12.5–12.12. Most of the examples that we discuss concern systems which, in absence of noise, display bistable behavior. We begin, in Section 12.5, with the case of dispersive optical bistability with white additive noise (Lugiato and Horowicz, 1985a). We find here the expected standard bimodal (i.e. two-peaked) character of the stationary probability distribution $P_s(x)$ in the bistable regime. In Section 12.6 we treat the same problem, but with colored noise. Here, we find examples of steady-state bimodality for parameter values for which the deterministic theory does not predict bistable behavior. This phenomenon is similar to a noise-induced transition (Horsthemke and Lefever, 1984) but here it does not arise from the multiplicative but from the colored character of the noise. The same phenomenon occurs, of course, also in the case of colored multiplicative noise, as we show in the Section 12.7 where we consider the input intensity noise (Lugiato and Horowicz, 1985b).

Thus, in Sections 12.6 and 12.7 we illustrate examples in which we find a bimodal structure when we do not expect it. An opposite phenomenon is discovered in Section 12.8, in which we consider noise in the input field frequency. For this special type of multiplicative white noise, we find situations in which the deterministic theory predicts bistability, but the steady-state probability distribution $P_s(x)$ exhibits a single peak (Lugiato *et al.*, 1986).

While Sections 12.5–12.8 deal with steady-state problems, Sections 12.9–12.12 discuss transient phenomena. First, in Section 12.9, we consider absorptive optical bistability with additive white noise, and illustrate the phenomenon of transient bimodality and its universal properties (Baras, Nicolis, Malek Mansour and Turner, 1983; Broggi, Colombo, Lugiato and Mandel, 1986; Broggi and Lugiato, 1984). The same picture is shown to arise also in dispersive optical bistability and to increase the switching speed in optically bistable devices (noise switching). The same problem is discussed in Section 12.11, but for a noise source which appears quadratically in the stochastic equation, that has the structure of (12.1.1) with $\xi(t)$ replaced by $\xi^2(t)$.

Finally, in Section 12.12 we discuss the dynamical behavior of the systems with swept control parameter. Recent work (Erneux and Mandel, 1986; Mandel and Erneux, 1984) shows that, when the parameter is swept across a bifurcation point, the system 'realizes' the bifurcation with a sizable delay. On the basis of a simple laser model, we illustrate the influence of noise on the phenomenon of delayed bifurcations (Broggi *et al.*, 1986).

12.2 Two especially relevant cases

Let us now focus on (12.1.1) and (12.1.2a).

12.2.1 Fast noise limit

We consider first the case that the noise variable ξ relaxes much more quickly than the output variable x. Precisely, we assume that

$$\gamma\tau \ll 1.$$

In this limit, we can eliminate adiabatically (Haken, 1977; Lugiato, Mandel and Narducci, 1984) the noise variable by setting $d\xi/dt = 0$ in (12.1.2). Thus we obtain

$$\xi(t) = \eta(t) \tag{12.2.1}$$

or, in other words, the noise is *white*.

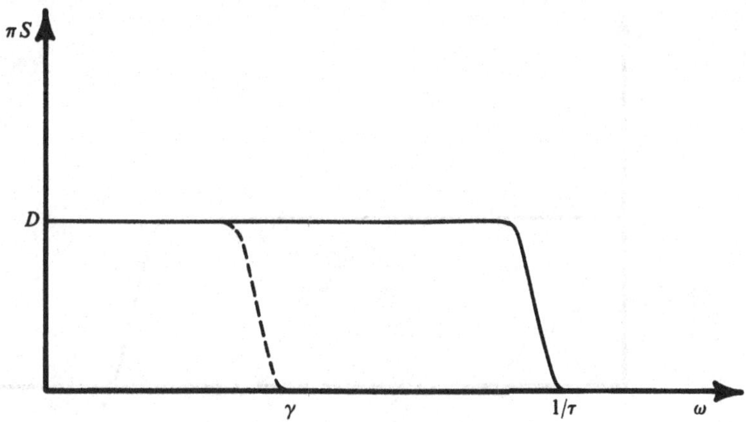

Figure 12.1. Qualitative description of the noise filtering.

The standard definition of white noise is the limit $\tau \to 0$, in which the correlation function (12.1.2b) becomes

$$\langle \xi(t)\xi(t') \rangle = D\delta(t - t').$$

In the limit $\tau \to 0$, however, the noise strength σ^2 diverges (see (12.1.4)); therefore the limit $\tau \to 0$ gives only a 'mathematical' definition of white noise. On the other hand, condition $\gamma\tau \ll 1$ identifies a 'physical' definition; i.e. a realistic noise with finite strength works as a white noise when this condition is satisfied. The reason for this fact can be immediately understood from the fact that the system (here, the variable x) filters the noise, i.e. does not react to all the frequency components of the noise spectrum. In our case, according to (12.1.3) and (12.1.4) the power spectrum of the noise is a Lorentzian (Figure 12.1):

$$\pi S(\omega) = \frac{\sigma^2/\tau}{\omega^2 + 1/\tau^2} = \frac{D}{1 + \omega^2\tau^2}. \tag{12.2.2}$$

When $\gamma\tau \ll 1$, the system reacts only to the part of noise spectrum such that $\omega < \gamma$; hence a colored noise with the spectrum (12.2.2) has the same effect of a white noise, in which $S(\omega)$ is constant and equal to D (Figure 12.1). This kind of filter action is familiar from the theory of linear electric networks; here, however, we find it in the general framework of nonlinear dynamical systems.

The previous argument indicates that, in fact, the filter action amounts to a reduction of the effective noise strength. This effect becomes evident if we introduce a noise variable $\bar{\eta}(t) = \eta(t)/D^{1/2}$, so that

$$\langle \bar{\eta}(t) \rangle = 0, \langle \bar{\eta}(t)\bar{\eta}(t') \rangle = 2\delta(t - t'). \tag{12.2.3}$$

Note that $\bar{\eta}$ is not adimensional, contrary to η. Thus, using (12.1.4), (12.1.2) becomes

$$\frac{d\xi}{dt} = -\frac{1}{\tau}\xi + \left(\frac{\sigma^2}{\tau}\right)^{1/2} \bar{\eta}(t). \tag{12.2.4}$$

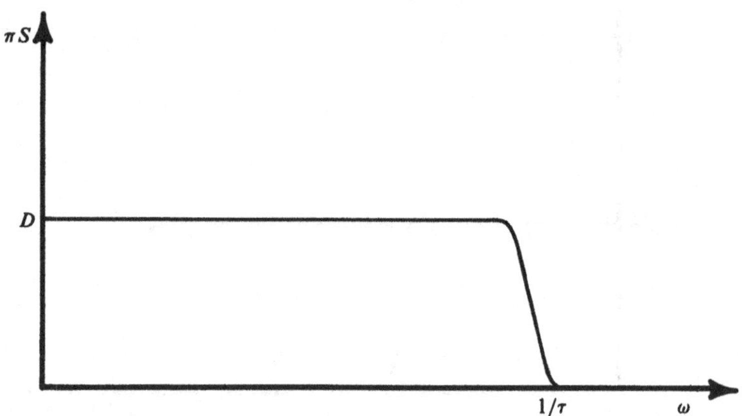

Figure 12.2. Filtering of white noise.

For $\gamma\tau \ll 1$, by setting $\mathrm{d}\xi/\mathrm{d}t = 0$, we obtain $\xi(t) = (\sigma^2\tau)^{1/2}\bar{\eta}(t)$, and (12.1.1) reduces to

$$\frac{\mathrm{d}x}{\mathrm{d}t} = -\gamma f(x) + (\sigma_x^2\gamma)^{1/2} g(x)\bar{\eta}(t), \tag{12.2.5}$$

where

$$\sigma_x^2 = \gamma\tau\sigma^2. \tag{12.2.6}$$

The parameter σ_x measures the width of the fluctuations of the variable x. In fact, using (12.2.3) and the Ito rules, the stochastic equation (12.2.5) leads to the following Fokker–Planck equation for the probability distribution $P(x, t)$ of the variable x (see, e.g., Gardiner, 1983)

$$\frac{\partial P(x,t)}{\partial t} = \gamma\left\{\frac{\partial}{\partial x} f(x)p(x,t) + \sigma_x^2 \frac{\partial^2}{\partial x^2} g(x)\right\} P(x,t), \tag{12.2.7}$$

and the stationary solution of (12.2.7) reads

$$P_s(x) = \frac{N}{g(x)} \exp\left[-\frac{1}{\sigma_x^2}\int \mathrm{d}x \frac{f(x)}{g(x)}\right], \tag{12.2.8}$$

where N is the normalization constant. Hence the width of the probability distribution x is proportional to σ_x. Relation (12.2.6) points out the noise reduction by a factor $(\gamma\tau)^{1/2} \ll 1$. Referring to Figure 12.1, (12.2.6) can be interpreted as an 'area rule', which prescribes that the square of the effective noise strength σ_x^2 is given by the area $D\gamma = \gamma\tau\sigma^2$. In this vein, (12.1.4) can also be interpreted according to this rule, as the white noise η filtered by the variable ξ (Figure 12.2.).

12.2.2 *Slow noise limit*

Let us now consider the opposite limit

$$\gamma\tau \gg 1, \tag{12.2.9}$$

in which the noise is very much colored. For the sake of definiteness, we consider the case of linear parametric noise, in which the function $f(x)$ has the form (12.1.7); the treatment can be extended immediately to the general case of (12.1.1). Let us consider the steady-state equation $f(x)=0$ of noiseless theory, which reads

$$y=\frac{\bar{f}(x)}{g(x)}, \tag{12.2.10}$$

and let us assume that, for any given value of x, this equation admits a single solution for x. This corresponds to the assumption that the function $y(x)$ defined by (12.2.10) is monotonic; let us take for definiteness the case $\mathrm{d}y/\mathrm{d}x \geqslant 0$.

In the limit (12.2.9), we can adiabatically eliminate the variable x by setting $\mathrm{d}x/\mathrm{d}t=0$ in (12.1.1). Thus, we obtain the equation

$$y+\xi(t)=\frac{\bar{f}(x)}{g(x)}, \tag{12.2.11}$$

which is identical to (12.2.10) but for the replacement $y \rightarrow y+\xi(t)$. Hence, at any given time, we can find the value of x by solving (12.2.11). This means that in the slow noise limit (12.2.9) the variable x follows adiabatically the fluctuations of the parameter y, according to the steady-state curve $x(y)$ obtained by inverting (12.2.10). This simple rule allows us now to calculate immediately the stationary solution $P_s(x)$ in the extreme colored noise limit. In fact, the stochastic equation (12.1.2a) is equivalent to the following Fokker–Planck equation for the probability distribution $P(\xi, t)$ of variable ξ (Gardiner, 1983)

$$\frac{\partial P(\xi,t)}{\partial t}=\frac{1}{\tau}\left\{\frac{\partial}{\partial \xi}\xi+\sigma^2\frac{\partial^2}{\partial \xi^2}\right\}P(\xi,t), \tag{12.2.12}$$

where we used (12.1.4). The stationary solution of (12.2.12) is given by

$$P_s(\xi)=\left(\frac{2}{\pi}\right)^{1/2}\frac{1}{\sigma}\exp\left\{-\frac{\xi^2}{2\sigma^2}\right\}. \tag{12.2.13}$$

Now, from (12.2.11) we obtain

$$\xi=\frac{\bar{f}(x)}{g(x)}-y=\frac{f(x)}{g(x)}. \tag{12.2.14}$$

The stationary distribution for the variable x is obtained by inverting (12.2.14) and multiplying the result by a factor which ensures the normalization of $P_s(x)$:

$$\begin{aligned} P_s(x) &= \exp\left\{-\frac{\xi(x)^2}{2\sigma^2}\right\}\left(\frac{1}{2\pi}\right)^{1/2}\frac{1}{\sigma}\frac{\mathrm{d}\xi}{\mathrm{d}x} \\ &= \exp\left\{-\frac{f^2(x)}{2\sigma^2 g^2(x)}\right\}\left(\frac{1}{2\pi}\right)^{1/2}\frac{1}{\sigma}\frac{f'(x)g(x)-f(x)g'(x)}{g^2(x)}. \end{aligned} \tag{12.2.15}$$

Equation (12.2.15) gives $P_s(x)$ in the limit (12.2.9) for (12.1.1), even when $f(x)$ does not have the structure (12.1.7); the only condition is that $d\xi/dx > 0$, with the function $\xi(x)$ defined as $f(x)/g(x)$.

Formula (12.2.15) was derived in Arnold *et al.* (1978), Sancho *et al.* (1982) and Stratonovich (1963).

12.3 Linearized treatment: modulation

In order to identify the scaling laws of parametric noise, let assume that the noise level is low enough to justify the linearization of the stochastic equation. Precisely, for a given value of the input parameter y let us consider a solution $\bar{x}$ of the steady-state equation (12.2.10). Let us assume that this solution is stable with respect to the deterministic equation (12.1.11), i.e. that $f'(\bar{x}) > 0$. Using (12.1.7) we obtain

$$f'(\bar{x}) = g(\bar{x})\frac{dy}{dx} \tag{12.3.1}$$

where dy/dx is obtained from (12.2.10) and is evaluated for $x = \bar{x}$. Because, for the sake of definiteness, we will assume that $g(x) > 0$, the stability condition reads equivalently

$$\frac{dy}{dx} > 0. \tag{12.3.2}$$

On setting

$$\delta x = x - \bar{x} \tag{12.3.3}$$

the linearized (12.1.1) reads

$$\begin{aligned}\frac{d}{dt}\delta x &= \gamma[-f'(\bar{x})\delta x + g(\bar{x})\xi(t)] \\ &= \gamma g(\bar{x})\left[-\frac{dy}{dx}\delta x + \xi(t)\right].\end{aligned} \tag{12.3.4}$$

A naïve argument suggests that, in this linearized situation, the steady-state variance $\langle \delta x^2 \rangle^{1/2}$ and the input variable variance $\langle \xi^2 \rangle^{1/2}$ are linked by the simple relation (see Figure 12.3)

$$\begin{aligned}\langle \delta x^2 \rangle^{1/2} &= \frac{dx}{dy}\langle \xi^2 \rangle^{1/2} \\ &= \frac{g(\bar{x})}{f'(\bar{x})}\langle \xi^2 \rangle^{1/2},\end{aligned} \tag{12.3.5}$$

where in the last passage we used (12.3.1). However, as we have seen in the previous section, the behavior of the output noise is indeed governed by the deterministic steady-state curve (12.3.1) only in the limit of strongly colored

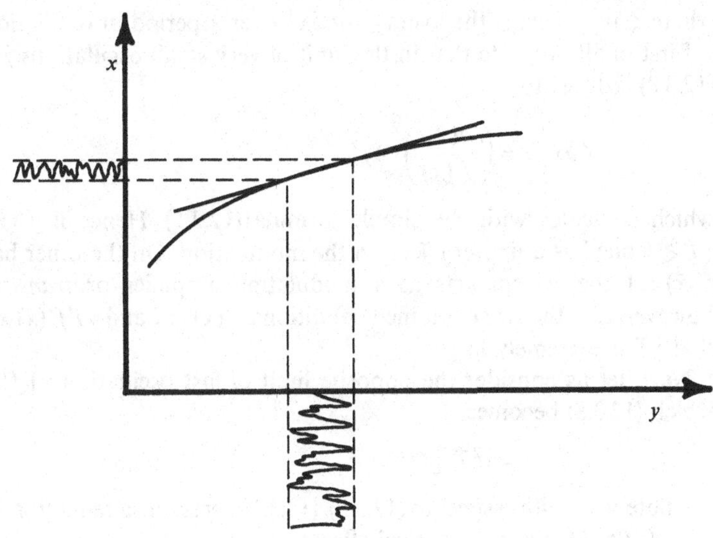

Figure 12.3. Relation between the noise in the input variable y and the noise in the output variable x according to a simple argument.

(i.e. slow) noise $\gamma\tau \gg 1$. This fact will be verified in the following section. In any case, (12.3.5) focuses on two especially interesting limit situations:

$$\frac{\mathrm{d}x}{\mathrm{d}y} \gg 1 \quad \text{and} \quad \frac{\mathrm{d}x}{\mathrm{d}y} \ll 1, \tag{12.3.6a}$$

or, equivalently and respectively, assuming that $g(\bar{x})$ has order unity,

$$f'(\bar{x}) \ll 1 \quad \text{and} \quad f'(\bar{x}) \gg 1. \tag{12.3.6b}$$

In fact, in the first case one expects that the output noise is enhanced with respect to the input noise (*noise amplifier*); on the contrary, in the second case the output noise should be reduced with respect to the input noise (*noise limiter*). Especially the second case is relevant from the practical viewpoint. We note that the situation $f'(\bar{x}) \ll 1$ arises when we approach the stability boundary, where $f'(\bar{x}) = 0$.

In this section, we will analyze the linearized equation (12.3.4) when $\xi(t)$ is not a noise variable, but a deterministic modulation given by

$$\xi(t) = \delta y(\sqrt{2})\cos\left(\frac{t}{T} + \phi\right), \tag{12.3.7}$$

where the factor $\sqrt{2}$ is introduced in order to have the time average of ξ^2 equal to δy^2. From (12.3.4) and (12.3.7), with simple calculations one obtains for times $\gamma f'(\bar{x})t \gg 1$ the simple formula

$$\langle \delta x^2 \rangle = \frac{\gamma^2 T^2 g^2(\bar{x})}{1 + (\gamma T f'(\bar{x}))^2}\delta y^2, \tag{12.3.8}$$

where $\langle \delta x^2 \rangle$ means the average of δx^2 over a period of oscillation.

First of all, we note that in the limit of very small oscillations $\gamma T f'(\bar{x}) \gg 1$ (12.3.8) reduces to

$$\langle \delta x^2 \rangle = \left(\frac{g(\bar{x})}{f'(\bar{x})} \right)^2 \delta y^2, \tag{12.3.9}$$

which coincides with the simple formula (12.3.5). Hence if $f'(\bar{x}) \gg 1$ and $\gamma T \gtrsim 1$ one has a limiter effect on the modulation. On the other hand, when $f'(\bar{x}) \ll 1$ the system acts as a modulation amplifier or *transistor*. Note, however, that the two combined conditions $f'(\bar{x}) \ll 1$ and $\gamma T f'(\bar{x}) \gg 1$ require that γT is extremely large.

Next, let us consider the opposite limit of fast oscillation $\gamma T f'(\bar{x}) \ll 1$, in which (12.3.8) becomes

$$\langle \delta x^2 \rangle = \gamma^2 T^2 g^2(\bar{x}) \delta y^2. \tag{12.3.10}$$

We note that, with respect to (12.3.9), (12.3.10) presents a reduction by a factor $(\gamma T f'(\bar{x}))^2$. This is a dynamical effect.

The condition $\gamma T f'(\bar{x}) \ll 1$ can be fulfilled by selecting $\gamma T \leqslant 1$ and $f'(\bar{x}) \ll 1$, or $\gamma T \ll 1$ and $f'(\bar{x}) \leqslant 1$. With the first choice, we do not obtain any amplification of modulation even if $f'(\bar{x}) \ll 1$. This is because for $f'(\bar{x}) \ll 1$ the dynamical evolution undergoes a slowing down, which suppresses the amplification (Mandel, 1985). With the second choice, we find a drastic reduction of modulation which has nothing to do with the limiter effect predicted by (12.3.5) for $f'(\bar{x}) \gg 1$. Here, the reduction arises from the fact that for $\gamma T \ll 1$ the system is not able to follow the fast modulation. This point is extensively discussed in Mandel and Erneux (1986).

12.4 Linearized treatment: noise

Let us now consider the set of (12.1.2) and (12.3.4). It is equivalent to the following Fokker–Planck equation for the probability distribution $P(\delta x, \xi, t)$:

$$\frac{\partial P(\delta x, \xi, t)}{\partial t} = \left\{ \gamma \frac{\partial}{\partial(\delta x)} [f'(\bar{x})\delta x - g(\bar{x})\xi] + \frac{1}{\tau} \frac{\partial}{\partial(\delta \xi)} \left[\xi + \sigma^2 \frac{\partial^2}{\partial \xi^2} \right] \right\} P(\delta x, \xi, t), \tag{12.4.1}$$

where we used (12.1.3) and (12.1.4). From (12.4.1) we obtain easily

$$\frac{\mathrm{d}}{\mathrm{d}t} \langle \delta x^2 \rangle = -2\gamma [f'(\bar{x}) \langle \delta x^2 \rangle - g(\bar{x}) \langle \delta x \xi \rangle], \tag{12.4.2}$$

$$\frac{\mathrm{d}}{\mathrm{d}t} \langle \delta x \xi \rangle = -\gamma [f'(\bar{x}) \langle \delta x \xi \rangle - g(\bar{x}) \langle \xi^2 \rangle] - \frac{1}{\tau} \langle \delta x \xi \rangle, \tag{12.4.3}$$

$$\frac{\mathrm{d}}{\mathrm{d}t}\langle \xi^2 \rangle = -\frac{2}{\tau}(\langle \xi^2 \rangle - \sigma^2). \tag{12.4.4}$$

At steady state we obtain $\langle \xi^2 \rangle = \sigma^2$ and

$$\langle \delta x^2 \rangle = \frac{g^2(\bar{x})}{(f'(\bar{x}))^2} \frac{\gamma\tau f'(\bar{x})}{1 + \gamma\tau f'(\bar{x})} \sigma^2. \tag{12.4.5a}$$

In the case $\gamma\tau f'(\bar{x}) \gg 1$ we recover again the simple formula (12.3.5), as we anticipated. Hence we can repeat here the same comments about amplification and limiter action that we gave in the previous section in the case of modulation. In fact, we must simply replace T by τ in all these considerations.

On the other hand, for $\gamma\tau f'(\bar{x}) \ll 1$, (12.4.5a) reduces to

$$\langle \delta x^2 \rangle = \frac{\gamma\tau g^2(\bar{x})}{f'(\bar{x})} \sigma^2. \tag{12.4.5b}$$

Not surprisingly after the consideration of Section 12.2.1, this expression coincides with that obtained from the linearized (12.3.4) in the white noise limit $\xi(t) = \eta(t)$. In fact, in this case (12.3.4) is equivalent to the following Fokker–Planck equation for the probability distribution $P(\delta x, t)$ (Gardiner, 1983):

$$\frac{\partial P(\delta x, t)}{\partial t} = \gamma \left\{ \frac{\partial}{\partial(\delta x)} f'(\bar{x})\delta x + D\bar{g}^2(\bar{x}) \frac{\partial^2}{\partial \xi^2} \right\} P(\delta x, t), \tag{12.4.6}$$

which at steady state and using (12.1.4), recovers (12.4.5b).

We note that, with respect to (12.3.5), (12.4.5b) presents a reduction by a factor $\gamma\tau f'(\bar{x})$. This is similar to what we found in the modulation case, and indeed (12.4.5a) presents many analogies with (12.3.8). There are, however, also differences which become important in the case $\gamma\tau f'(\bar{x}) \ll 1$ (respectively $(\gamma T f'(\bar{x}))^2 \ll 1$). In fact, (12.3.10) does not depend on $f'(\bar{x})$, whereas (12.4.6) presents a factor $(f'(\bar{x}))^{-1}$. This difference becomes important when $f'(\bar{x}) \ll 1$ or $f'(\bar{x}) \gg 1$. In the first case, for $\gamma\tau \sim 1$ this factor produces a noise amplification which is absent in the case of modulation. Therefore, noise counteracts in part the effect of the slowing down. On the other hand, for $f'(\bar{x}) \gg 1$ one obtains a noise limiter effect; we must remember, however, that the two conditions $\gamma\tau f'(\bar{x}) \ll 1$ and $f'(\bar{x}) \gg 1$ imply that $\gamma\tau$ is extremely small. Hence in this case the noise reduction arises more from the factor $\gamma\tau$ than from the factor $(f'(\bar{x}))^{-1}$.

A second difference is the following. If we let τ correspond to T, we see that (12.3.10) contains a factor $(\gamma T)^2$, whereas (12.4.6) presents a factor $\gamma\tau$. Hence the reduction in modulation obtained for $\gamma T \ll 1$ is much larger than the reduction in noise obtained for $\gamma\tau = \gamma T \ll 1$. In other words, *it is much harder to suppress noise than deterministic oscillations.*

We note that the noise bandwidth in the output variable is given by $\gamma f'(\bar{x})$. Therefore the ratio between the output and the input noise bandwidths is given again by the crucial quantity $\gamma\tau f'(\bar{x})$.

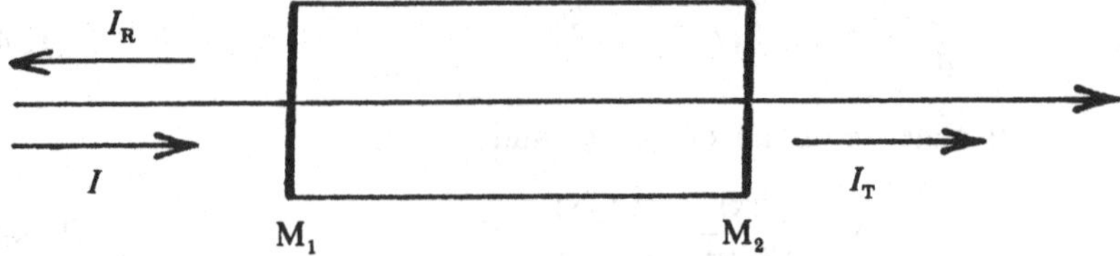

Figure 12.4. Fabry–Perot spherical cavity: M_1 and M_2 are partially transmitting mirrors; I, I_T and I_R are the incident, transmitted and reflected intensities, respectively.

We end this section with the following comments. Imagine that, starting from a situation of white noise $\gamma\tau f'(\bar{x}) \ll 1$, we increase gradually the correlation time τ, thereby increasing the degree of color of noise; we ask whether the output noise $\langle \delta x^2 \rangle$ increases or decreases. The answer depends on the procedure we follow in increasing τ. If we keep constant the strength σ^2 of the input noise, we see from (12.4.5a) that $\langle \delta x^2 \rangle$ *increases*, provided $\gamma\tau f'(\bar{x})$ remains small. If, instead, we keep the parameter $D = -\sigma^2\tau$ constant, we have from (12.4.5a) that

$$\langle \delta x^2 \rangle \sim \frac{g^2(\bar{x})}{(f'(\bar{x}))^2}\gamma D f'(\bar{x})[1 - \gamma\tau f'(\bar{x})] \tag{12.4.5c}$$

decreases when τ is increased. These opposite answers explain the apparent contradiction between the conclusion of Lugiato and Horowicz (1985a), in which D was kept constant, and of Lugiato and Horowicz (1985b), in which D was kept constant.

12.5 Dispersive optical bistability, thermal noise in the material

The last decade witnessed a florishing development of theoretical and experimental studies on the phenomenon of optical bistability (OB) (Gibbs, 1985; Lugiato, 1984; Smith, Mandel and Whenett, 1987). This arises in an optical cavity filled with a material which displays saturable absorption (absorptive OB) or has an intensity-dependent refractive index (dispersive OB) (Figure 12.4). A stationary, coherent beam near to resonance both with the cavity and with the material is injected into the cavity. The steady-state intensity of the beam transmitted by this system is a nonlinear function of the input intensity. By varying the atomic density (absorptive OB) or the frequency of the incident beam (dispersive OB), the stationary curve of transmitted intensity as a function of the input intensity becomes S-shaped (Figure 12.5). The part with negative slope is unstable, hence the system is bistable in the interval $I_{\text{down}} < I < I_{\text{up}}$, where I indicates the incident intensity. If the incident power is slowly increased from zero to beyond the bistable region, and then decreased back to zero, one obtains a hysteresis cycle. By adjusting continuously the atomic density (absorptive OB) or the input

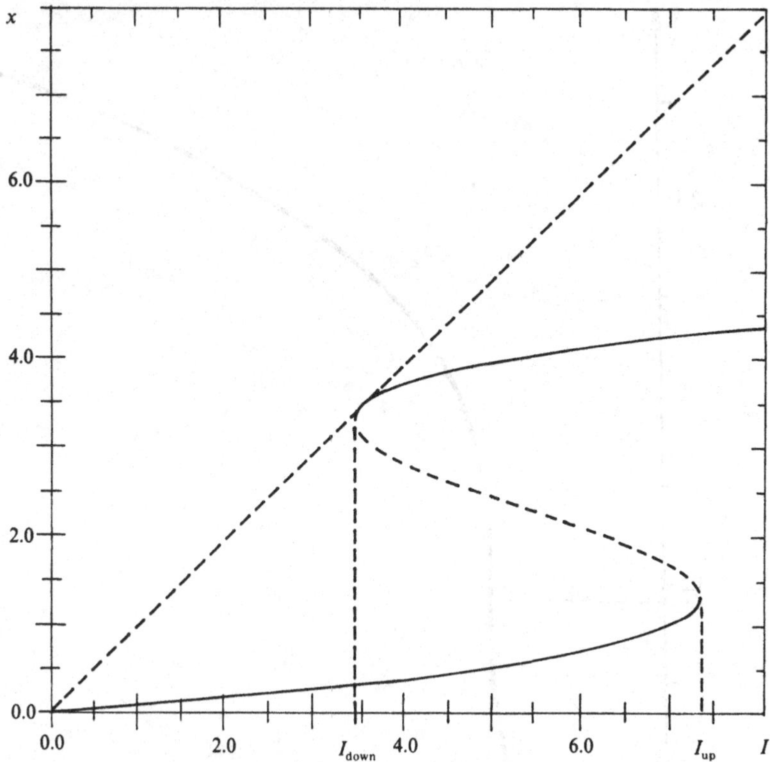

Figure 12.5. Steady-state curve (see (12.5.7)) for $\theta = 2(3^{1/2})$; the part with negative slope in unstable.

frequency (dispersive OB), the size of the hysteresis cycle can be reduced to zero and bistability disappears. Just at the boundary between bistability and monostability there is a 'critical' situation in which the steady-state curve of output versus input intensity presents an inflection point with vertical tangent (Figure 12.6). The recent investigations on the subject of optical bistability were motivated both by its intrinsic theoretical interest in the framework of nonequilibrium phase transitions and nonlinear dynamical systems (Lugiato, 1984), and by its promises in the direction of the realization of optical memories, optical transistors and eventually optical computers (Gibbs, 1985; Smith, Mandel and Whenett, 1987).

In this section we will consider the case of dispersive OB, that we describe by means of a simple model (Abraham and Firth, 1983; Lugiato and Horowicz, 1985a; Miller, 1981). In absence of noise, the dynamical equation is

$$\frac{\mathrm{d}x}{\mathrm{d}t} = -\gamma\left[x - \frac{I}{1 + F\sin^2(x - \phi_0)}\right], \tag{12.5.1}$$

where x is the normalized, intensity-dependent part of the refractive index; γ is

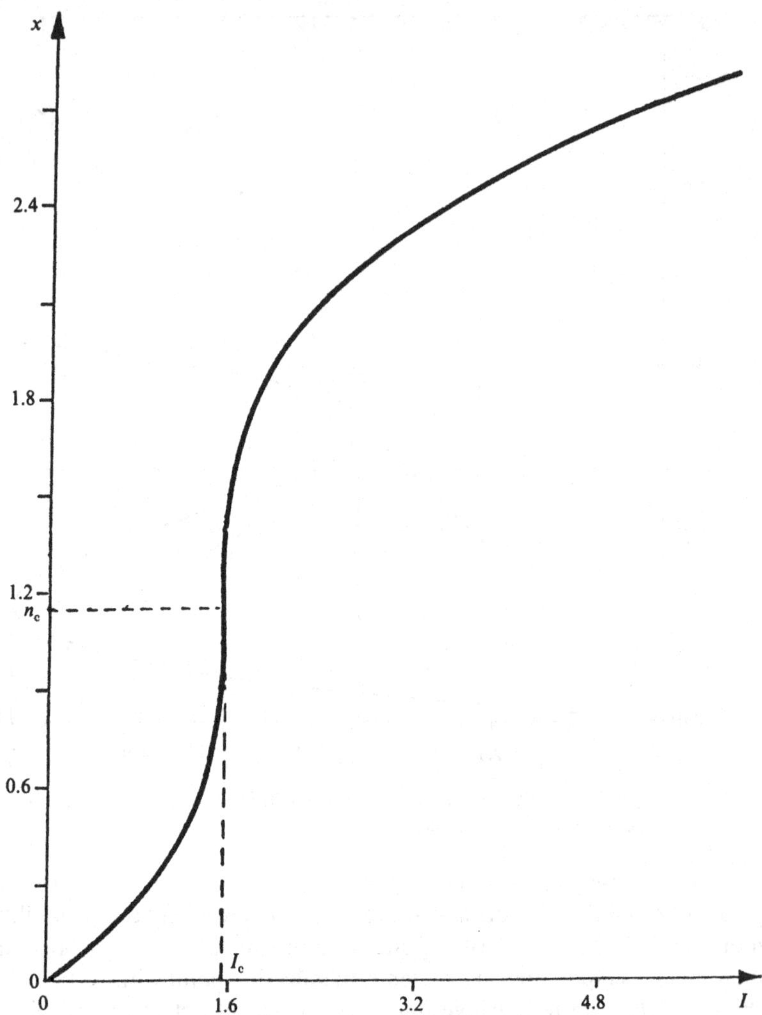

Figure 12.6. Critical steady-state curve (see (12.5.7)) for $\theta = \sqrt{3}$.

the response rate of the material; I is the normalized input intensity and ϕ_0 is the detuning parameter, which arises both from the difference between the incident field frequency and the nearest cavity frequency, and from the intensity independent part of the refractive index. The finesse factor is given by

$$F = \frac{4R}{(1-R)^2}, \tag{12.5.2}$$

where R is the reflectivity coefficient of the mirrors. The normalized transmitted intensity I_T is given by

$$I_T = \frac{I}{1 + F \sin^2(x - \phi_0)}. \tag{12.5.3}$$

In this article, we will often consider the high finesse limit $R \to 1$, i.e. $F \to \infty$, together with the limit $\phi_0 \to 0$. Because the variable x tends to zero in the limit, we can approximate $\sin(x-\phi_0) \sim x-\phi_0$ and, by setting

$$x' = F^{1/2}x, \quad t = F^{1/2}\phi_0, \quad I' = F^{1/2}I, \qquad I'_{\mathrm{T}} = F^{1/2}I_{\mathrm{T}}, \tag{12.5.4}$$

we obtain from (12.5.1) and (12.5.3), respectively,

$$\frac{\mathrm{d}x'}{\mathrm{d}t} = -\gamma\left[x' - \frac{I'}{1+(x'-\theta)^2}\right], \tag{12.5.5}$$

and

$$I'_{\mathrm{T}} = \frac{I'}{1+(x-\theta)^2}. \tag{12.5.6}$$

In the following, for the sake of simplicity, we will always omit the primes. In the case of (12.5.5), the steady-state relation is

$$I = x[1+(x-\theta)^2]; \tag{12.5.7}$$

the graph of x as a function of I, obtained from (12.5.7), exhibits bistability for $\theta > \sqrt{3}$ (Figure 12.5). The critical curve corresponds to $\theta = \sqrt{3}$ (Figure 12.6); the coordinates of the critical point are $x_c = (2\sqrt{3}/9)$ and $I_c = (8\sqrt{3}/9)$.

In the case of (12.5.1), the steady-state relation is

$$I = x[1+F\sin^2(x-\phi_0)]; \tag{12.5.8}$$

the plot of x as a function of I, obtained from (12.5.8), can exhibit in general multiple branches (Figure 12.7).

In both cases of (12.5.3) and (12.5.6) one obtains the relation

$$I_{\mathrm{T}} = x, \tag{12.5.9}$$

as one sees immediately using (12.5.8) and (12.5.7), respectively; it must be kept in mind, however, that (12.5.9) holds only at steady state and in absence of noise, whereas these limitations do not affect (12.5.3) and (12.5.6).

In this section we will consider the evolution equation (12.5.5) including also the presence of thermal fluctuations in the material, which leads to the stochastic equation (Lugiato and Horowicz, 1985a)

$$\frac{\mathrm{d}x}{\mathrm{d}t} = -\gamma\left[x - \frac{I}{1+(x-\theta)^2}\right] + \gamma\xi(t). \tag{12.5.10}$$

Clearly, this type of noise is additive, and in this section we will limit ourselves to the white noise limit $\xi(t) = \eta(t)$. Thus, we obtain the following Fokker–Planck equation for the probability distribution $P(x,t)$:

$$\frac{\partial}{\partial t}P(x,t) = \gamma\left\{\frac{\partial}{\partial x}f(x) + \sigma_x^2\frac{\partial^2}{\partial x^2}\right\}P(x,t), \tag{12.5.11}$$

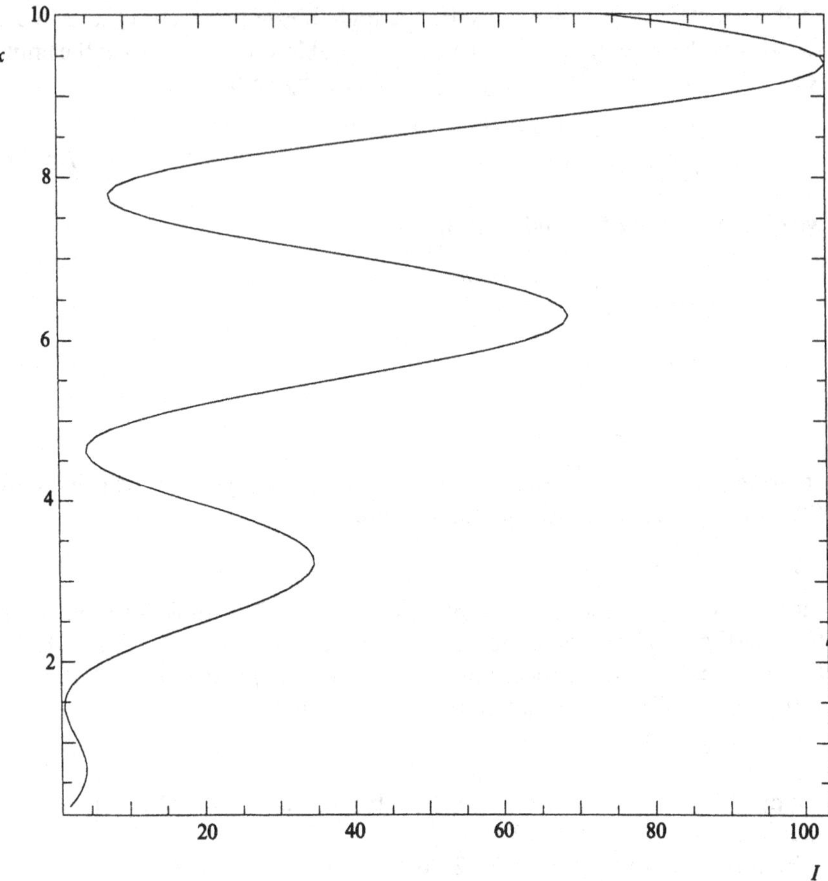

Figure 12.7. Steady-state curve (see (12.5.8)) for $F = 5$, $\phi_0 = 1.5$.

$$f(x) = x - \frac{I}{1+(x-\theta)^2}, \quad \sigma_x^2 = \gamma D. \tag{12.5.12}$$

The stationary solution of (12.5.11) has the form

$$P_s(x) = \mathcal{N} \exp(-V(x)/\sigma_x^2), \tag{12.5.13}$$

where the potential function $V(x)$, defined by (12.1.10), has explicitly the expression

$$V(x) = \tfrac{1}{2}x^2 - I tg^{-1}(x-\theta), \tag{12.5.14}$$

By definition, the local maxima and minima of the stationary distribution P_s coincide with the minima and the maxima of the potential V, respectively. Furthermore, the extrema of the function V coincide with the solutions of the steady-state equation (12.5.7). Therefore for $0 < I < I_{\text{down}}$ and $I > I_{\text{up}}$

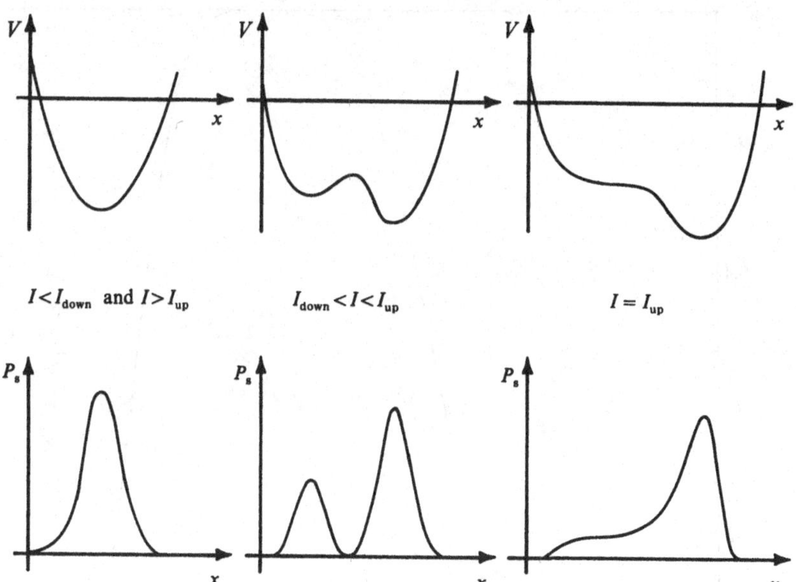

Figure 12.8. Qualitative shape of the potential V and of the stationary probability distribution P_s.

(Figures 12.5 and 12.8) the potential has a single minimum and the distribution P_s has one peak. On the other hand, for $I_{down} < I < I_{up}$ the function V presents two minima in correspondence with the two stable stationary solutions, and one local maximum in correspondence with the unstable solution. In this case the stationary distribution presents two peaks (Figure 12.8), and this phenomenon is called *bimodality*. An interesting situation occurs for $I = I_{down}$ and $I = I_{up}$, because the potential function $V(x)$ has an inflection point with horizontal tangent.

In all cases, the smaller the noise level σ_x^2, the narrower are the peaks of $P_s(x)$. Of course, the critical situation $\theta = 3^{1/2}$ is of special interest. For $I = I_c = 8(3^{1/2})/9$ the distribution P_s has a broad single peak (Figure 12.9a) because the potential V presents a flat minimum (at $x = x_c$) where the first, second and third derivatives vanish (Figure 12.9b).

12.6 Noise-induced transition from colored noise

Horsthemke and Lefever (1984) pointed out and analyzed extensively the phenomenon of noise-induced transitions. They considered the case of white noise, and showed that when this is not additive but multiplicative the number of peaks of the stationary distribution function may be different from the number of stable stationary solutions of the deterministic theory. For example, one can select values of the parameter such that the noiseless steady-state

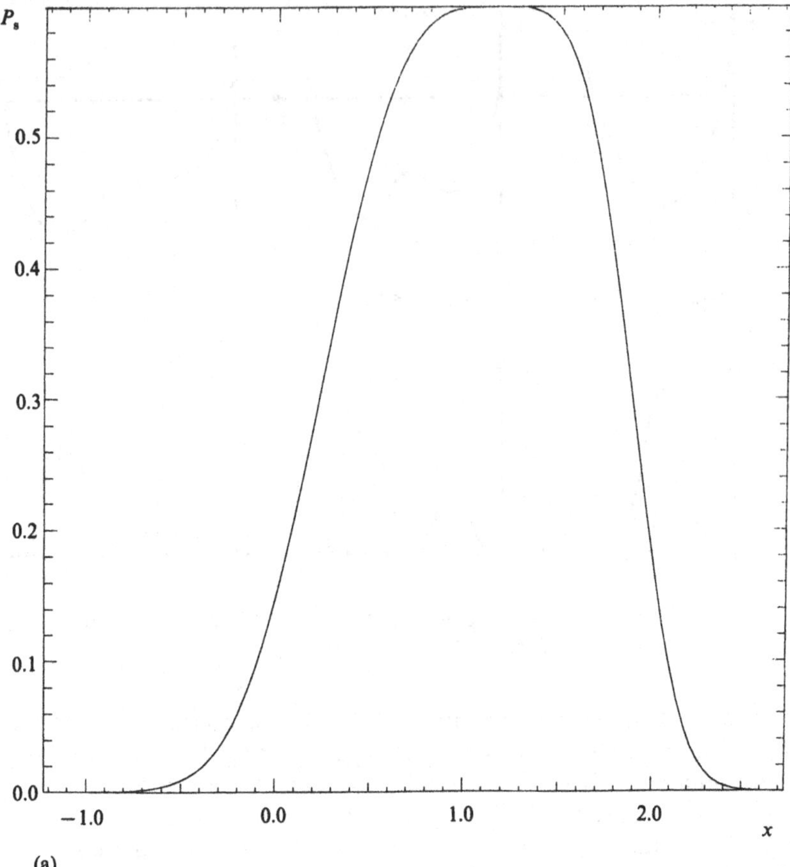

Figure 12.9. Thermal noise. We consider (12.5.13) and (12.5.14) for $\theta = \sqrt{3}$, $I = 8\sqrt{3}/9, \sigma_x^2 = 0.1$. (a) Steady-state probability distribution $P_s(x)$. (b) The function $V(x)$.

equation $f(x) = 0$ admits only one stationary solution, but in the presence of multiplicative noise the stationary distribution (12.2.8) displays two peaks instead of one.

Behavior of this kind can be found in the case of (12.5.10) if the additive noise is replaced by the multiplicative noise which arises from the fluctuations of the input intensity I (G. Broggi and R. J. Horowicz, manuscript in preparation). Here, however, we want to consider a different situation, namely we point out that similar phenomena can arise from additive noise, provided this is not white but colored.

Precisely, we consider the stochastic equation (12.5.10) out of the white noise limit, in the critical case $\theta = \sqrt{3}$, $I = I_c$. First of all, we consider the case of $\gamma\tau$ much smaller than unity, but such that some differences from the white noise limit become visible. In this situation, one can use the Fokker–Planck

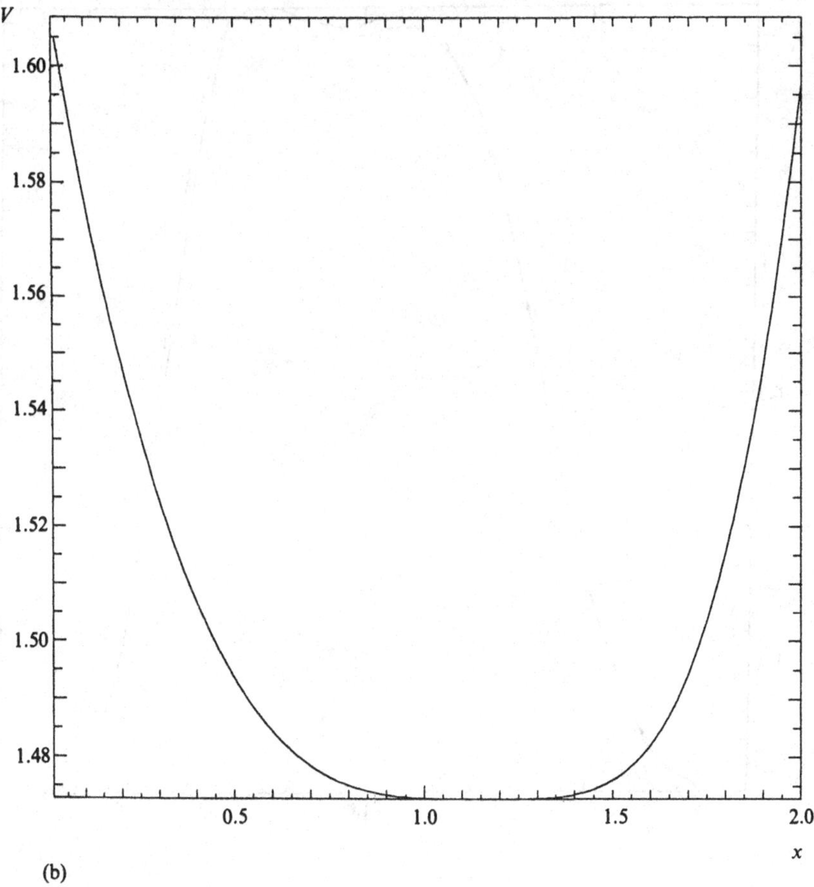

(b)

equation derived by San Miguel and Sancho (1981), which leads to the following expression for the stationary distribution P_s (Lugiato and Horowicz, 1985a):

$$\left.\begin{aligned} P_s(x) &= \frac{\mathcal{N}}{1-\gamma\tau f'(x)}\exp\left[-\frac{u(x)}{\sigma_x^2}\right], \\ U(x) &= \frac{1+\gamma\tau}{2}x^2 - Itg^{-1}(x-\theta) \\ &\quad + \gamma\tau\frac{I}{1+(x-\theta)^2}\left[\frac{I}{2[1+(x-\theta)^2]} - x\right]. \end{aligned}\right\} \quad (12.6.1)$$

As shown by Figure 12.10, for $\sigma_x^2 = 0.1$ and $\gamma\tau = 0.2$ the distribution presents two peaks instead of one. The same trend is found in the numerical solution of the set of stochastic equations (12.5.10) and (12.1.2a), as shown in Figures 12.10

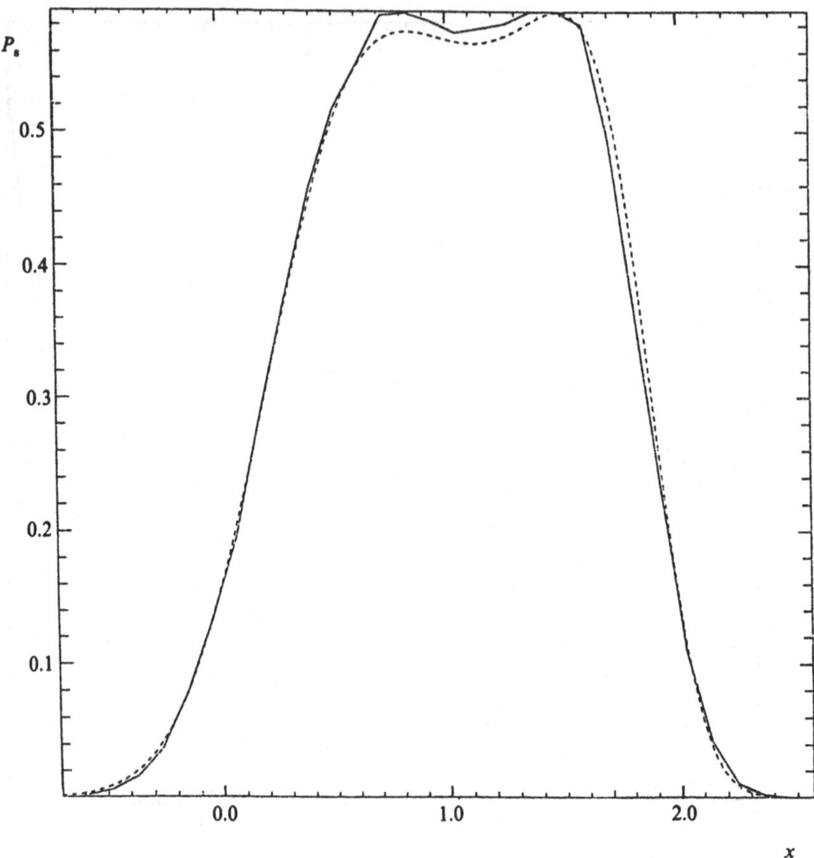

Figure 12.10. Thermal noise. The steady-state distribution for $\theta = \sqrt{3}$, $I = 8\sqrt{3}/9, \sigma_x^2 = 0.1, \gamma\tau = 0.2$ obtained numerically (full line) compared with that given by (12.6.1) (dashed line).

and 12.11. Clearly, the bimodal character becomes more pronounced when $\gamma\tau$ is increased and the results are fully confirmed by an analogical simulation of (12.5.10) (P. Grigolini *et al.*, 1988).

The behavior for $\gamma\tau \to \infty$ can be analyzed using (12.2.15) with $f(x)$ given by (12.5.12) and $g(x) = 1$. Of course, this limit must be performed keeping σ, and not σ_x^2, constant. Otherwise for $\gamma\tau \to \infty, \sigma^2 = (\sigma_x^2/\gamma\tau)$ would vanish, and this would amount to a noiseless situation. For $\theta = \sqrt{3}$ and $I = I_c$, the exponential factor in (12.2.15) presents a very flat maximum for $x = x_c$, while the prefactor vanishes for $x = x_c$ because $f(x_c) = f'(x_c) = f''(x_c) = 0$. Hence, the distribution P_s presents two peaks, and a minimum at $x = x_c$ (Figure 12.12). Roughly speaking, this phenomenon arises for the following reasons. In the case of strongly colored noise, a fluctuation from the deterministic stationary value does not come back immediately. Furthermore, in the critical situation, the

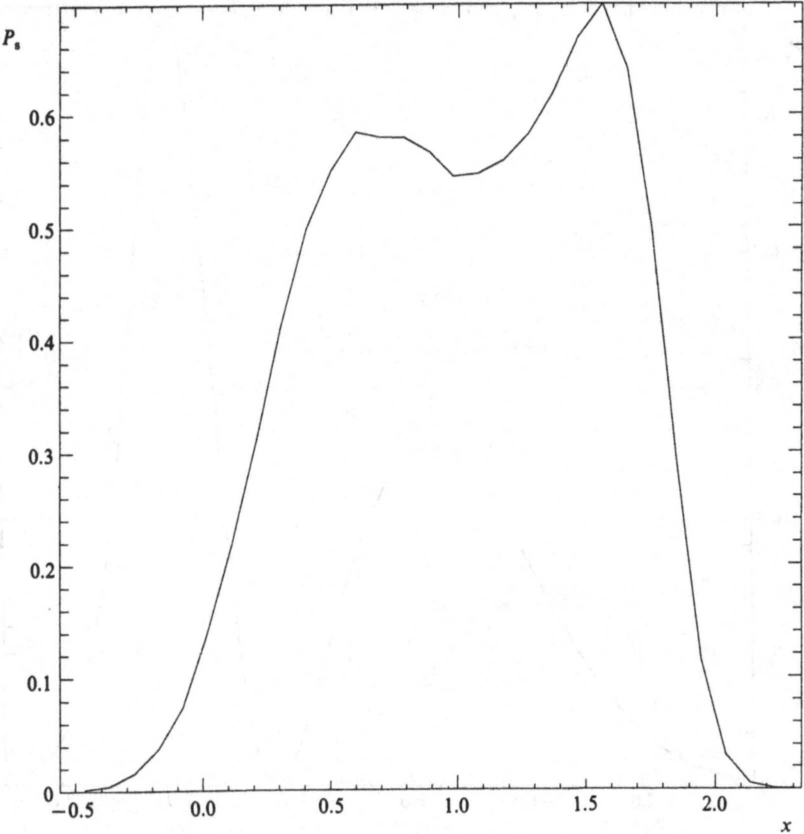

Figure 12.11. Thermal noise. Steady-state distribution for $\theta = \sqrt{3}$, $I = 8\sqrt{3}/9$, $\sigma_x^2 = 0.1$, $\gamma\tau = 1$, obtained numerically.

regression of fluctuations is especially slow. These circumstances allow the system to remain quite apart from the deterministic steady-state value for long times, thus giving rise to the two peaks of probability that surround this value. For a fixed value of $\gamma\tau$, the bimodal structure of $P_s(x)$ for $\theta = \sqrt{3}$, $I = I_c$ disappears when the noise amplitude σ is reduced enough. However, the reduction of σ necessary to obtain the disappearance becomes larger and larger as $\gamma\tau$ increases.

12.7 Dispersive bistability, input intensity noise: steady-state behavior

From the analysis of (12.2.15), it is evident that the noise-induced transition from colored noise, discovered in the previous section, is not specific of the stochastic equation considered there, but has a general nature. In fact, it arises whenever the parameter is selected in such a way that there is a critical point x_c

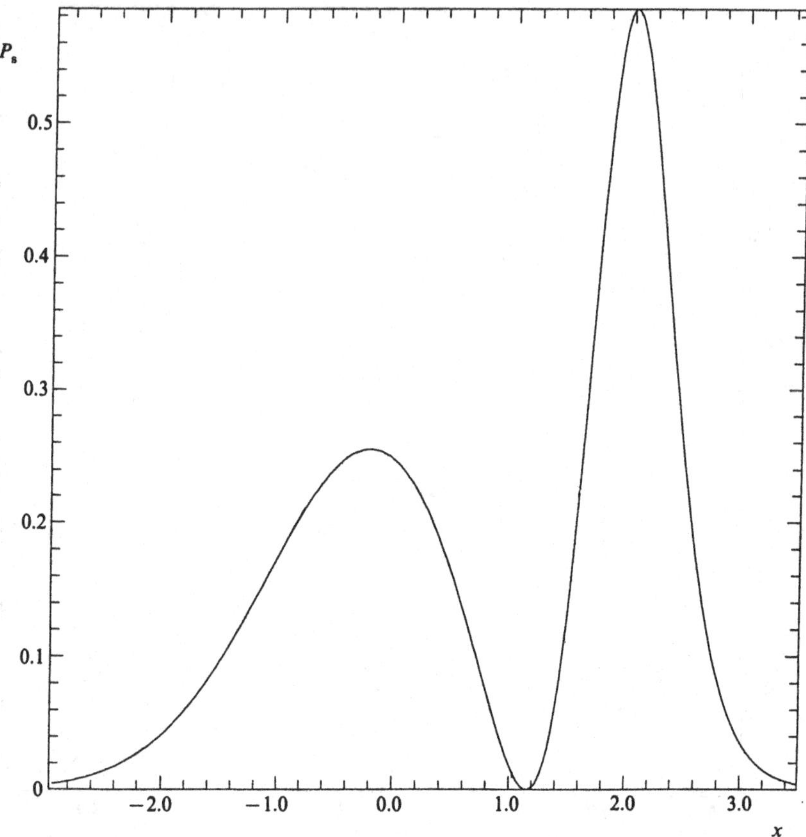

Figure 12.12. Thermal noise. Steady-state distribution for $\theta = \sqrt{3}$, $I = 8\sqrt{3}/9$, $\sigma_x^2 = 1$, obtained in the limit $\gamma\tau \to \infty$ from (12.2.15).

such that $f(x_c) = f'(x_c) = f''(x_c) = 0$. This ensures that the prefactor $d\xi/dt$ vanishes and has a minimum for $x = x_c$. Hence, in this situation, the stationary distribution P_s becomes certainly bimodal when $\gamma\tau$ is increased enough.

In particular, it is clear that this result holds not only for additive but also for multiplicative noise, because (12.2.15) covers also the case $g(x) \neq 1$. An example of this type is obtained considering a type of noise different from the thermal fluctuations selected in the previous section. In (12.5.5) we assume that the input intensity I exhibits a random jitter in time. If in (12.5.5) we replace $I \to I + \xi(t)$, we obtain the stochastic equation with multiplicative noise

$$\frac{dx}{dt} = -\gamma\left[x - \frac{I}{1 + (x - \theta)^2}\right] + \gamma\frac{1}{1 + (x - \theta)^2}\xi(t). \tag{12.7.1}$$

The stationary distribution $P_s(x)$ can be calculated analytically in the white noise limit (Lugiato and Horowicz, 1985b), and numerically in the case of

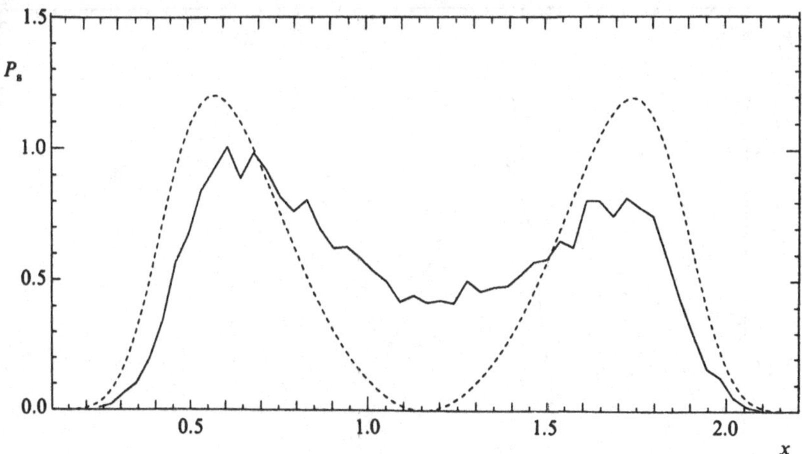

Figure 12.13. Input intensity noise, $\theta = \sqrt{3}, I = 8\sqrt{3}/9$, $\sigma_x^2 = 0.06$ (a) $\gamma\tau = 10$, numerical; (b) $\gamma\tau \to \infty$, from (12.2.15).

colored noise by solving the set of stochastic equations (12.7.1) and (12.1.2a). In Figure 12.13 we see the comparison between the distribution $P_s(x)$ for $\theta = \sqrt{3}$, $I = I_c$, $\sigma^2 = 0.06$ and $\gamma\tau = 10$, and the curve (12.2.15) ($\gamma\tau \to \infty$) for the same values of θ, I and σ^2. The two curves exhibit a qualitative similarity, but the situation $\gamma\tau = 10$ still seems removed from the asymptotic situation $\gamma\tau \to \infty$. Finally, Figure 12.14 shows the stationary probability distribution $P_s(I_T)$, obtained numerically using (12.5.6) with $I \to I + \xi(t)$.

12.8 Dispersive optical bistability, frequency noise: destruction of steady-state bimodality

Let us consider the case of multiplicative white noise. As is evident from (12.2.8), owing to the factor $[g(x)]^{-1}$ the peaks of the stationary probability distribution $P_s(x)$ are not governed only by the potential function $V(x)$, (12.1.10) and do not coincide with the stable solution of the deterministic steady-state equation $f(x) = 0$. However, when the noise parameter σ_x is small enough the positions of the maxima of $P_s(x)$ approach the stable solutions of the equation $f(x) = 0$, and the deterministic picture is recovered. This feature encompasses phenomena like noise-induced transitions, which disappear in the low noise limit. In this section we will analyze a special type of multiplicative noise, which leads to a behavior of $P_s(x)$ which deviates dramatically from the deterministic picture, and this difference persists even in the low noise limit. Precisely, we consider the model (12.5.5) assuming that the frequency of the input field fluctuates in time. This is described assuming that θ is a fluctuating parameter, i.e. we replace θ by $\theta + \eta(t)$, where the white noise $\eta(t)$ has the usual properties (12.1.3). We expand the function $[1 + (x - \theta - \eta(t))]^{-1}$ in power series of η and neglect all the terms nonlinear in η,

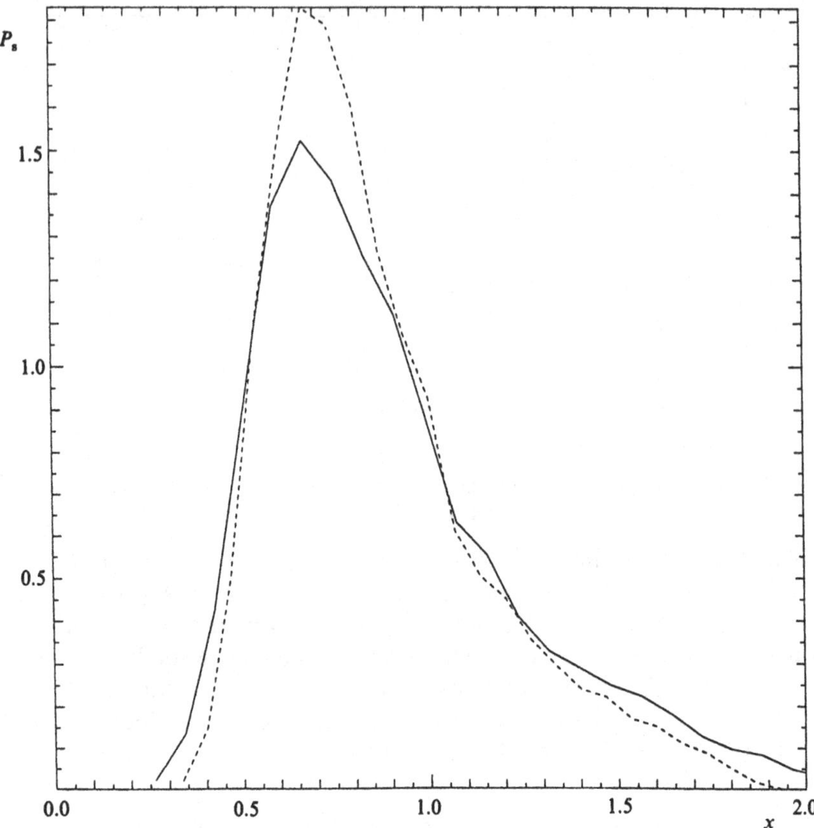

Figure 12.14. Input intensity noise, $\theta = \sqrt{3}$, $I = 8\sqrt{3}/9$, $\sigma_x^2 = 0.06$, $\gamma\tau = 1$. The stationary probability distribution for the transmitted intensity (full line) is compared with the stationary probability for the material variable n (dashed line).

because they give very small contributions. That is, we substitute in (12.5.5)

$$\frac{1}{1+(x-\theta)^2} \to \frac{1}{1+(x-\theta)^2} - \frac{2(x-\theta)}{[1+(x-\theta)^2]^2}\eta(t), \tag{12.8.1}$$

obtaining the stochastic equation with multiplicative noise

$$\frac{\mathrm{d}x}{\mathrm{d}t} = -\gamma\left(x - \frac{I}{1+(x-\theta)^2}\right) + 2\gamma I \frac{x-\theta}{[1+(x-\theta)^2]^2}\eta(t). \tag{12.8.2}$$

Finally, by using the Ito rules (Gardiner, 1983) (12.8.2) is translated into the following Fokker–Planck equation for the probability distribution $P(x, t)$ of the variable x:

$$\frac{\partial P(x,t')}{\partial t'} = \left[\frac{\partial}{\partial x}\left(x - \frac{I}{1+\zeta^2}\right) + d\frac{\partial^2}{\partial x^2}\frac{\zeta^2}{(1+\zeta^2)^4}\right]P(x,t'), \tag{12.8.3}$$

where we defined

$$t' = \gamma t, \tag{12.8.4}$$

$$\zeta = x - \theta, \tag{12.8.5}$$

$$d = 4D\gamma I^2 = 4\sigma_x^2 I^2. \tag{12.8.6}$$

The stationary solution of the Fokker–Planck equation for frequency noise is as follows (Lugiato *et al.*, 1986). Defining the function

$$\Phi(x, I) = \frac{(1+\zeta^2)^4}{\zeta^2} \exp[-U_f(x, I)], \tag{12.8.7}$$

$$U_f(x, I) = \frac{1}{d}\left\{\frac{\zeta^8}{8} + \theta\frac{\zeta^7}{7} + \frac{2}{3}\zeta^6 + \frac{4\theta - I}{5}\zeta^5 + \frac{3}{2}\zeta^4 + (2\theta - I)\zeta^3 + 2\zeta^2 + (4\theta - 3I)\zeta + \ln|z| - (\theta - I)\frac{1}{\zeta}\right\}, \tag{12.8.8}$$

we have

$$\left.\begin{aligned}
&\text{(a) for } I < \theta \quad P_s^f(x, I) = \begin{cases} \mathcal{N}'\phi(x, I) & \text{for } x < \theta, \\ 0 & \text{for } x \geqslant \theta, \end{cases} \\
&\text{(b) for } I = \theta \quad P_s^f(x, I) = \delta(x - \theta), \\
&\text{(c) for } I > \theta \quad P_s^f(x, I) = \begin{cases} 0 & \text{for } x \leqslant \theta, \\ \mathcal{N}'\phi(x, I) & \text{for } x > \theta. \end{cases}
\end{aligned}\right\} \tag{12.8.9}$$

The region for which $P_s^f(x, I)$ vanishes is shown in Figure 12.15, for all values of I. The peculiar structure of the stationary distribution (12.8.9) arises from the fact that the diffusion coefficient in the Fokker–Planck equation (12.8.3) vanishes as ζ^2 for $\zeta = 0$ (i.e. $x = \theta$), which originates a divergence in $U_f(x, I)$. This feature produces a steady-state solution that is not analytical in x, but is continuous with all its derivatives.

The special property of distribution (12.8.9) which distinguishes it from the thermal case (additive noise) of (12.5.10), as well as from all other distributions which describe a bistable system, is the following. For $d \leqslant 5$, the distribution (12.8.9) has *one peak* even in the range $\theta \leqslant I \leqslant I_{up}$, in which the deterministic theory predicts *bistability*.

Let us explain the physical reasons which give rise to the configuration (12.8.9). The key role is played by the point $\zeta = 0$, which in the deterministic steady-state theory corresponds to the situation $x = \theta = I = I_T$, as we see from (12.8.5), (12.5.7) and (12.5.6). Hence, this is the *bleaching point* from which all the incident light is transmitted. The frequency fluctuations are governed by the transmission curve $[1 + (x - \theta)^2]^{-1}$, as we see from (12.8.1). At the point $x - \theta = 0$ this curve attains its maximum (Figure 12.16), and therefore the effect of frequency fluctuations vanishes to first order (i.e. in the linear approximation with respect to the fluctuating quantity $\eta(t)$).

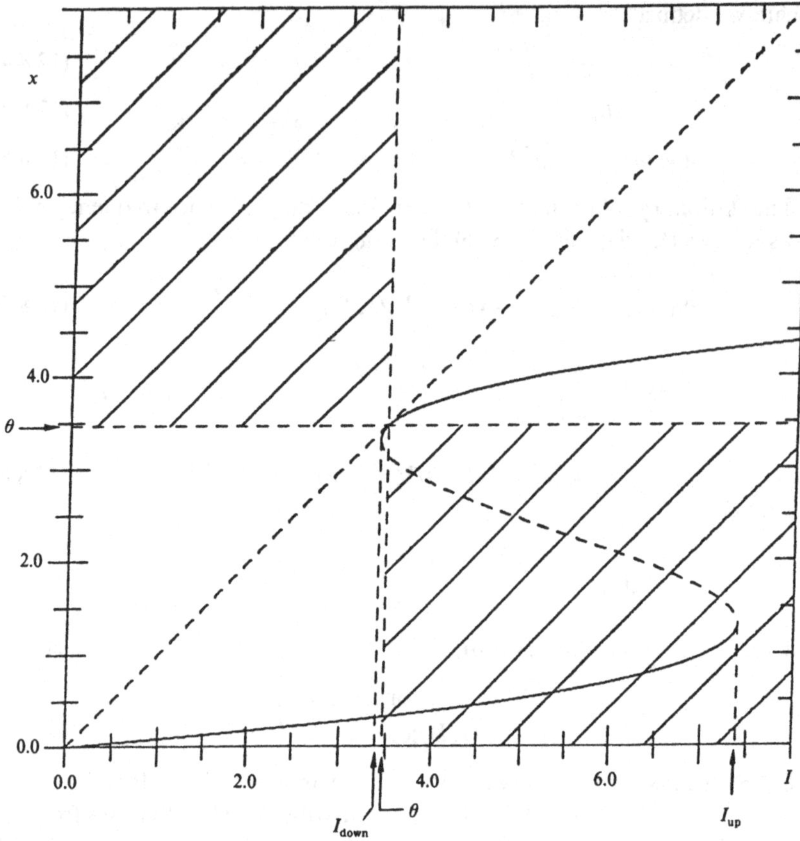

Figure 12.15. Purely dispersive bistability stationary curve for $\theta = 2\sqrt{3}$. At steady state, the normalized transmitted intensity coincides with x. The negative slope part of the curve is unstable. The hatched regions indicate the range of n in correspondence of which the stationary probability distribution $P_s^f(x)$ vanishes, for each value of the input intensity I.

From (12.8.3), we see that the point $\zeta = 0$ plays the role of a *Maxwell Devil* in this problem. In fact, let us consider the probability current in the Fokker–Planck equation (12.8.3):

$$j(x, t') = \left(\frac{I}{1+\zeta^2} - x\right)P(x, t') - d\frac{\partial}{\partial x}\frac{\zeta^2}{(1+\zeta^2)^4}P(x, t'). \qquad (12.8.10)$$

For $\zeta = 0$ $(x = \theta)$ it becomes

$$j(\theta, t') = (I - \theta)P(\theta, t'),$$

hence for $I > \theta$ the flux is always towards the right, whereas for $I < \theta$ it is towards the left. This one-sided flux across the point $I = \theta$ biases the evolution of the probability distribution and, for long times, gives rise to the configur-

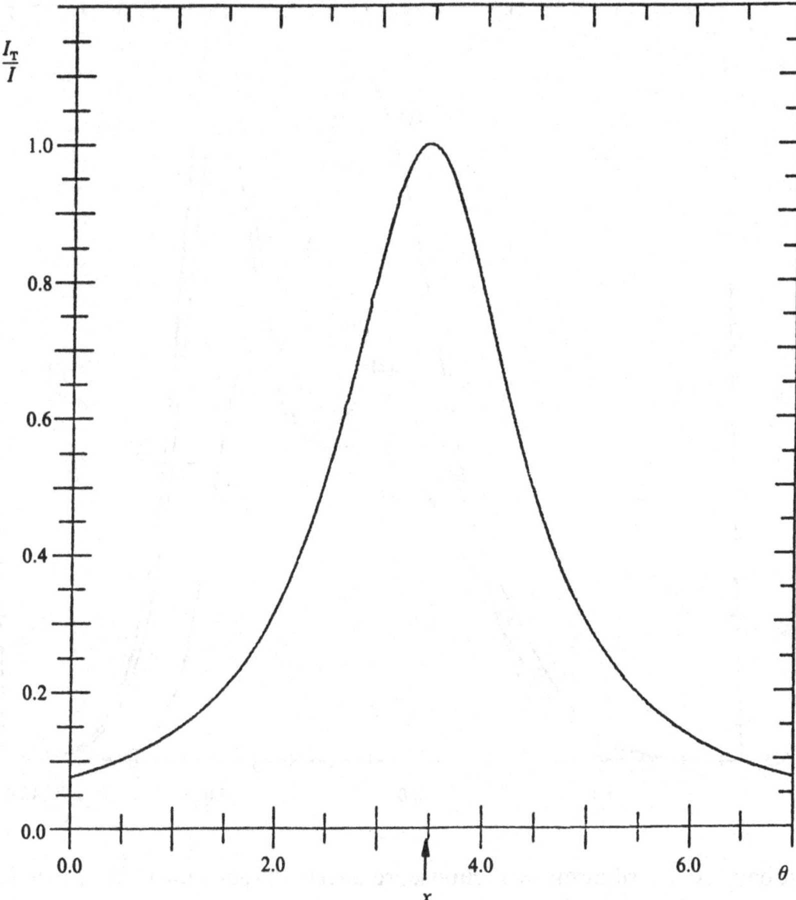

Figure 12.16. The transmission of the cavity, defined by the ratio of the intensity I_T of the transmitted beam to that of the incident beam I, is given by the function $[1+(x+\theta)^2]^{-1}$. In correspondence with the maximum $\theta = x$ the system is insensitive to the fluctuations of θ (i.e. to the frequency noise).

ation described by (12.8.9). Completely different is the situation of the thermal Fokker–Planck equation (12.5.10), for which the sign of the flux is determined by the shape of the probability distribution, for any value of x. That is, the flux is to the right or to the left, according to the configuration of $P(x, t')$.

In the case of pure frequency noise for $\theta \leqslant I < I_{up}$ we are sure that, if we wait long enough, the system performs a transition to the upper branch and will never come back, whereas in the case of thermal noise there is always, as usual, some probability of performing the opposite transition. However, it must be noted that for small noise level the transition from the lower to the upper branch also takes an extremely long time in the case of pure frequency noise. For this reason, in order to show the approach to steady state of the

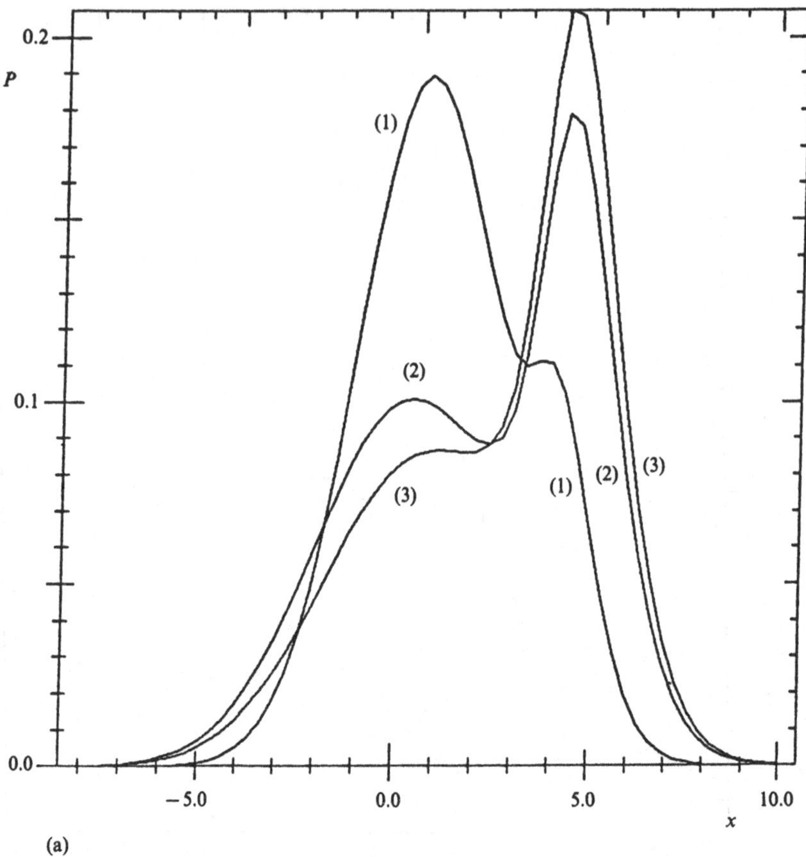

(a)

probability distribution, we choose relatively large values of the noise parameter d.

The results shown in Figure 12.17 have been obtained by solving numerically (12.8.3) and (12.5.10). Figure 12.17(a) exhibits the time evolution of $P(x, t')$ in the case of thermal noise, for a value of I which gives a sizable steady-state probability of finding the system in the lower branch. On the contrary in the case of frequency noise (Figure 12.17b and c) for the same value of I, the probability distribution gradually flows entirely to the right of the point $x = \theta$.

We end this section with three remarks. For $d \geqslant 5$, the distribution $P_s(x)$ given by (12.8.9) may present two peaks for $I > \theta$. One of them is, as usual, related to the deterministic stationary solution in the upper branch of the steady-state curve $x(I)$ obtained from the equation $f(x) = 0$ for the given value of I. The second peak is, however, completely unrelated from the deterministic stationary solution in the lower branch, and lies somewhat to the right of the point $x = \theta$. Therefore this picture does not correspond to a recovery of the standard bimodal structure in the bistable situation. Other details can be found in Lugiato *et al.* (1986).

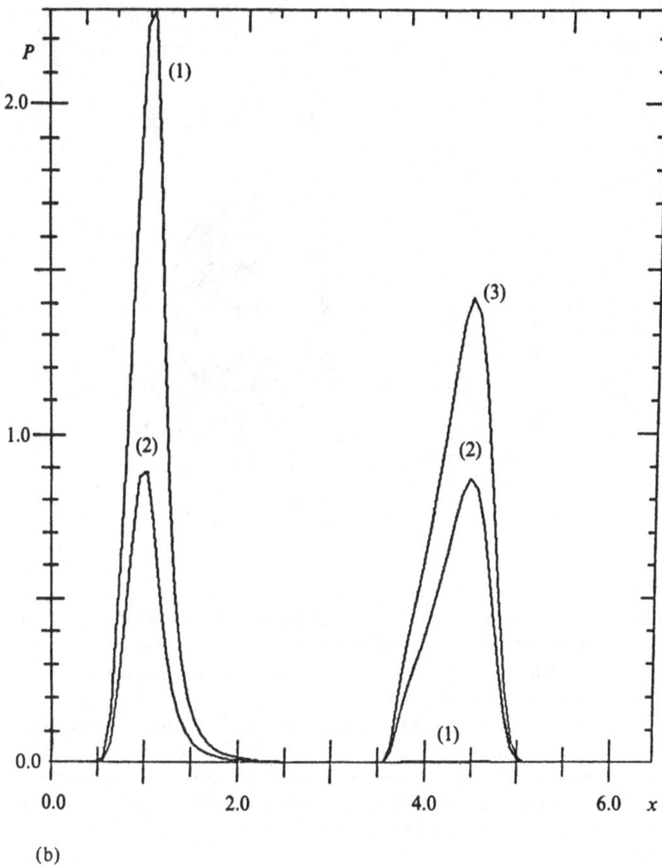

(b)

The second remark is that the time evolution described in Figure 12.17(c) has been recently observed in an analog simulation of the stochastic equation (12.8.2) (R. Mannella and P.V.E. McClintock, private communication).

The third observation is that the phenomenon of destruction of steady-state bimodality has also been discovered in other systems (Robinson, Moss and McClintock, 1985). In these cases, however, the phenomenon arises from a completely different mechanism and is an example of standard noise-induced transition, which requires a large noise level.

12.9 Absorptive optical bistability, white noise: transient bimodality

In the study of the dynamical behavior of dissipative systems, the standard picture is that the probability distribution associated with the system has a width inversely proportional to the square root of the system size, and its mean value follows the deterministic motion, i.e. the time evolution obtained by neglecting noise (Van Kampen, 1961). Hence the really interesting situations

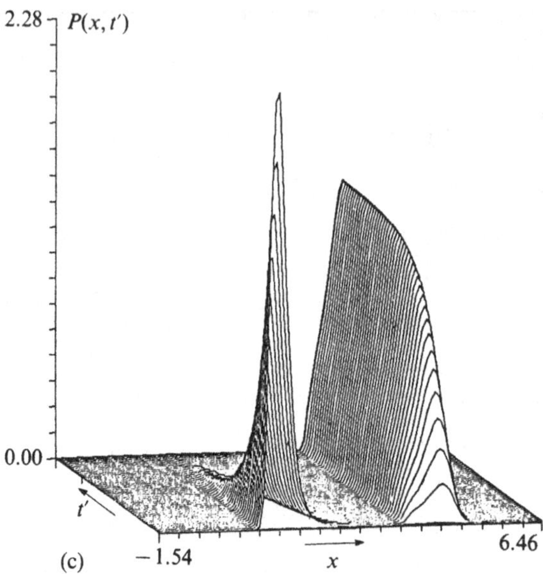

Figure 12.17. Time evolution of the probability distribution for $\theta = 2\sqrt{3}$, $y^2 = 7$ and with the initial condition $P(x, 0) = \delta(x - \bar{x})$, where $\bar{x} = 0.969$ is the lower-branch stationary value given by the deterministic theory (see Figure 12.15). (a) Solution of (12.5.11) (thermal noise) for $\sigma_x^2 = 5$ and (1) $t' = 0.464$, (2) $t' = 2.177$, (3) $t' = 15.173$. (b) Solution of (12.8.3) (frequency noise) for $d = 5$ and (1) $t' = 5.00$, (2) $t' = 154.705$, (3) $t' = 1132.562$. In both (a) and (b), curve (3) practically coincides with the stationary probability distribution.
(c). Three-dimensional diagram of the evolution of the probability distribution $P(x, t')$ for the case of frequency noise, $d = 5$.

are those in which anomalous fluctuations lead to deviations from this picture. One such situation was identified in recent work by Nicolis and collaborators (Baras *et al.*, 1983; Francowicz, Malek Mansour and Nicolis, 1984; Francowicz and Nicolis, 1983) concerning explosive chemical reactions and combustion, which predicted the phenomenon of *transient bimodality*. Broggi and Lugiato (1984) discovered the same phenomenon in the framework of optical bistability, and suggested an experimental observation which was soon achieved by Lange and his group (Lange *et al.*, 1985; Mitsche *et al.*, 1985).

In this section, we consider the case of absorptive OB described by the deterministic equation (Lugiato, 1984)

$$\frac{\mathrm{d}x}{\mathrm{d}t} = -\kappa\left(x - y - \frac{2Cx}{1 + x^2}\right), \tag{12.9.1}$$

where x and y are the normalized output and input fields, respectively (i.e. $x = I_T^{1/2}$, $y = I^{1/2}$). The parameter κ is the cavity damping constant and C is the

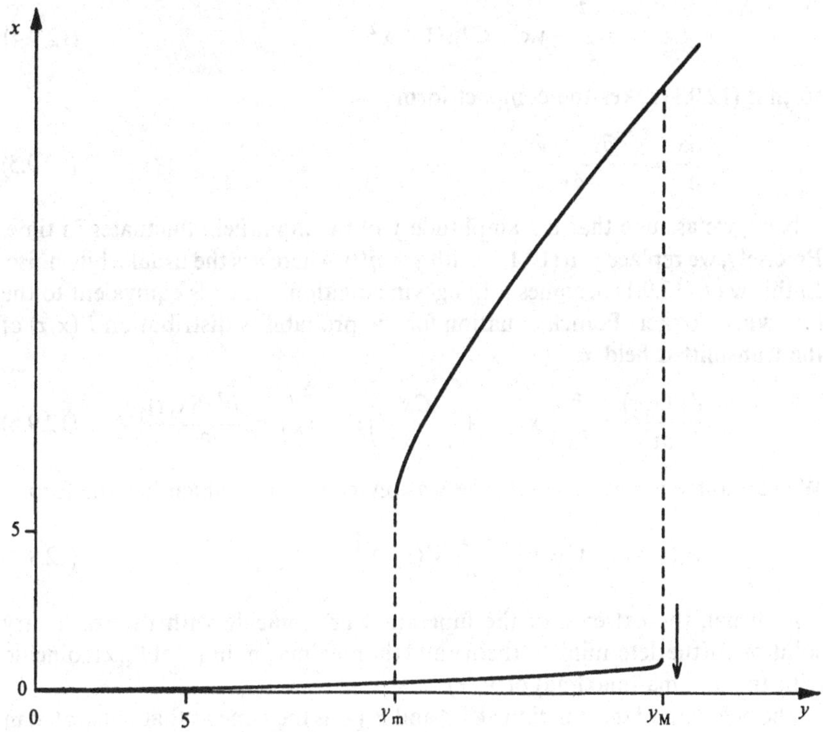

Figure 12.18. Purely absorptive bistability. Hysteresis cycle of normalized incident field y for $C = 20$. The arrow indicates a value of the incident field slightly larger than the up-switching threshold $y_M = 21.0264$.

so-called bistability parameter proportional to the atomic density. The steady-state behavior is described by the equation

$$y = x + \frac{2Cx}{1 + x^2} \tag{12.9.2}$$

which, for a given y, admits either one or three solutions. The steady-state curve of x versus y, obtained from (12.9.2), exhibits a hysteresis cycle for $C > 4$ (Figure 12.18). For $y < y_m$ and $y > y_M$, (12.1.7) admits one stationary solution $\bar{x}$, whereas for $y_m < y < y_M$ it admits three stationary solutions $x_1 < x_2 < x_3$, where x_1 and x_3 are stable and belong to the hysteresis cycle, and x_2 is unstable and therefore physically not realizable. It is convenient to introduce the normalized time

$$\tau = \kappa t \tag{12.9.3}$$

and the potential $V(x, y)$ defined as follows:

$$V(x, y) = \int \mathrm{d}x \left(x - y + \frac{2Cx}{1 + x^2} \right)$$

$$=\frac{x^2}{2}-yx-C\ln(1+x^2), \tag{12.9.4}$$

so that (12.9.1) takes the compact form

$$\frac{dx}{d\tau}=-\frac{\partial V(x,y)}{\partial x}. \tag{12.9.5}$$

Next, we assume that the amplitude y of the input field fluctuates in time. Precisely, we replace y in (12.1.6) with $y+\eta(t)$, where η is the usual white noise. In this way, (12.9.1) becomes a Langevin equation, which is equivalent to the following Fokker–Planck equation for the probability distribution $P(x,\tau)$ of the transmitted field x:

$$\frac{\partial P(x,\tau)}{\partial\tau}=\frac{\partial}{\partial x}\left(x-y+\frac{2Cx}{1+x^2}\right)P(x,\tau)+\sigma_x^2\frac{\partial^2 P(x,\tau)}{\partial x^2}. \tag{12.9.6}$$

We can immediately calculate the stationary solution which has the form

$$P_s(x,y)=\mathcal{N}\exp\left[-\frac{1}{\sigma_x^2}V(x,y)\right]. \tag{12.9.7}$$

As usual, the extrema of the function $V(x)$ coincide with the stationary solutions of the deterministic theory and the maxima (minima) of $P_s(x)$ coincide with the minima (maxima) of $V(x)$.

The behavior of the functions $V(x)$ and $P_s(x)$ is the same as that illustrated in Figure 12.8 if we simply replace I, I_{up} and I_{down} by y, y_M and y_m, respectively. The bimodal structure of the stationary probability distribution for $y_m<y<y_M$ is the very signature of optical bistability from a statistical viewpoint. However, it is tremendously difficult to observe this stationary configuration experimentally, because it is attained only for extremely long times. In fact, in order to approach this distribution, the system must have the time to perform transitions from one metastable state to the other, and the lifetime of the two metastable states is exceedingly long. The average transition time can be calculated using a classic method introduced by Kramers (1940). Assume that initially the system sits at the bottom of the lefthand well of the function V in Figure 12.8; in order to perform a transition to the righthand well it must overcome the potential barrier centered at the local maximum of the potential. Now, Kramers' method says that the average transition time is proportional to $\exp(\Delta V/\sigma_x^2)$, where ΔV is the height of the barrier, defined as the difference between the values of V at the local maximum and at the lefthand minimum. Because σ_x^2 is small, the transition time turns out to be enormous. The only exception is when the value of the input field is very near to the boundary of the bistable region, so that the potential barrier ΔV is small.

Hence, it is not surprising that the only observation of a two-peaked stationary distribution in the framework of OB was not obtained using an all-optical system like that described in Section 12.5, but a hybrid electro-optical

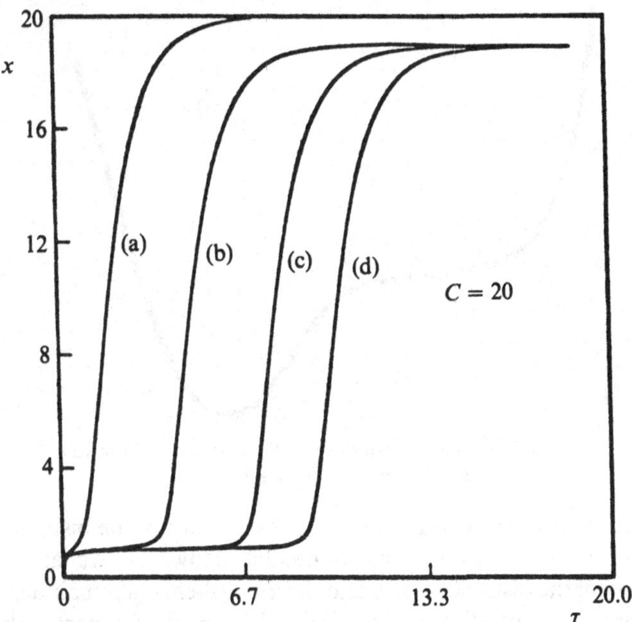

Figure 12.19. Time evolution of the transmitted field for $C = 20$ and (a) $y = 22$, (b) $y = 21.1$, (c) $y = 21.05$, (d) $y = 21.04$.

system in which fluctuations are artificially and skilfully introduced and controlled (McCall *et al.*, 1985).

Let us now show that there is another possibility of obtaining a bimodal probability distribution in OB. This time, however, the bimodality does not occur at steady state conditions, but only transiently. Nevertheless, it appears in such a way as to be accessible to the experimental observation.

Let us proceed as follows. We switch on the incident field stepwise to a value that lies slightly beyond the bistable region (Figure 12.18). The deterministic theory predicts the time evolution shown in Figure 12.19: a long lethargic stage of slow evolution is followed by a rapid switching to the stationary state. This critical slowing down behavior was first predicted by Bonifacio and Lugiato (1976, 1978) and has been the object of intense theoretical study (Benza and Lugiato, 1979; Bonifacio and Meystre, 1978; Grynberg and Cribier, 1983; Mandel and Erneux, 1982). Its experimental observation has been attained in a variety of systems: electro-optical (Garmire, Marburger, Allen and Winful, 1979), microwave (Barbarino *et al.*, 1982), sodium atomic beams (Grant and Kimble, 1983), sodium vapor (Mitsche, Deserno, Mlynek and Lange, 1983), rubidium (Cribier, Giacobino and Grynberg, 1983), semiconductors (Al Attar, Mackenzie and Firth, 1986).

It is easy to understand the origin of this phenomenon. We have seen before that in the bistable region the potential V has two minima. For a value of y as

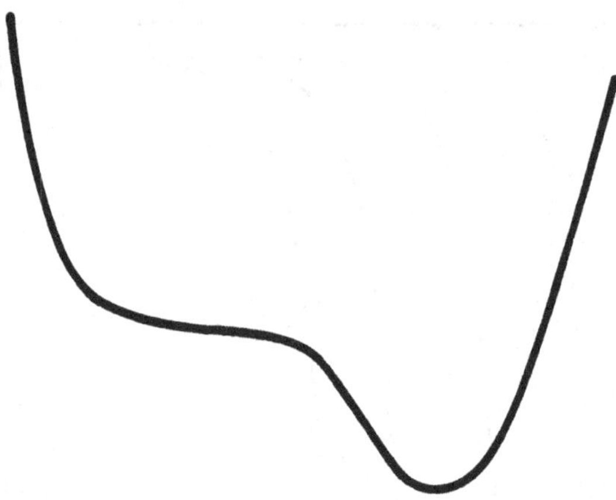

Figure 12.20. Qualitative shape of the potential V for a value of y corresponding to the arrow in Figure 12.18.

indicated by the arrow in Figure 12.18, we are out of the bistable region and therefore V presents only one minimum. However, we are near to the boundary of the bistable region, and therefore there is a remainder of the other minimum which just disappeared, and which leaves a flat part in the potential, as we see in Figure 12.20. When, during the deterministic time evolution, the system arrives at the flat part of V, the evolution slows down and becomes quick again only when the system reaches the edge of the potential well.

Let us now consider the time evolution of the system, both in terms of stochastic theory and also through numerical solutions of the Fokker–Planck equation (12.9.4) (Broggi and Lugiato, 1984; Broggi, Lugiato and Colombo, 1985). The probability distribution is initially a delta function $P(x,0) = \delta(x)$, because for $\theta < 0$ the incident field is switched off. The distribution drifts to the right and, because there is diffusion, it immediately acquires a nonzero width (curve a in Figure 12.21). Later, the distribution forms a long tail towards the right (curve b), and subsequently it develops a second peak (curve c). Finally, the peak disappears (curves c, d) and $P(x, \tau)$ approaches the stationary one-peaked configuration. In fact, in this case the steady-state distribution has one peak because we are out of the bistable region.

This type of time evolution of $P(x, \tau)$ is pictorially represented by the three-dimensional diagram shown in Figure 12.22, from which we see that the double-peaked character of the probability distribution persists for a sizable amount of time.

This phenomenon, in which the distribution becomes bimodal during the transient approach to a one-peaked distribution, has been called *transient bimodality*. Also the reason for this phenomenon becomes clear if we consider the role of the potential V shown in Figure 12.20. When the distribution arrives at the flat part of the potential due to the slowing down it sits there for a long time almost without any movement in its center of mass. However,

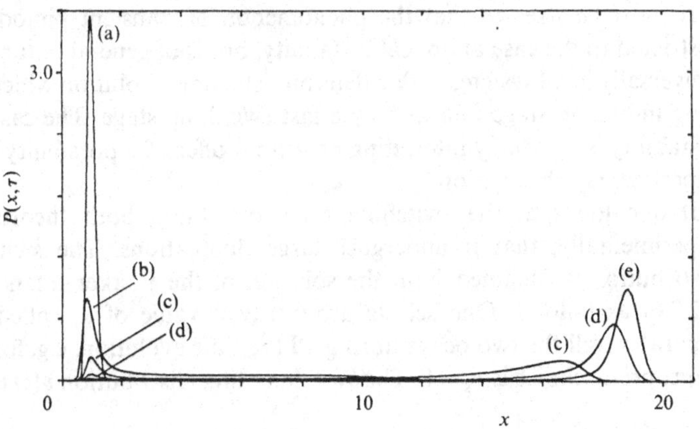

Figure 12.21 The probability distribution $P(x, \tau)$ is shown for $C = 20$, $y = 21.04$, $\sigma_x^2 = 0.1$ and (a) $\tau = 0.2$, (b) $\tau = 1.8$, (c) $\tau = 3.6$, (d) $\tau = 5.4$, and (e) $\tau = 7.2$.

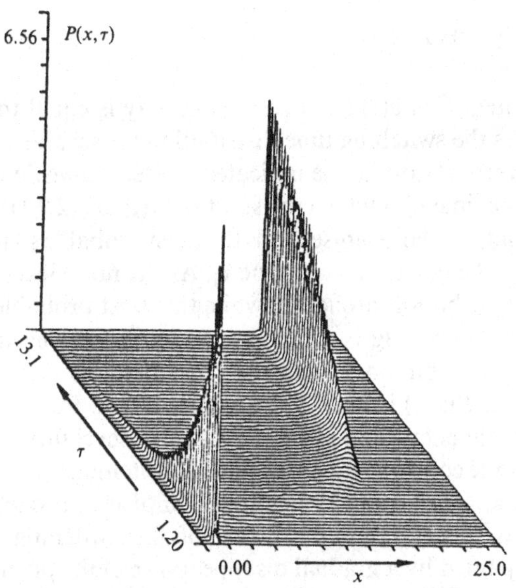

Figure 12.22. Time evolution of the probability distribution for $C = 20$, $y = 21.1$, $\sigma_x^2 = 0.005$.

because of diffusion, the distribution broadens with time and, due to the asymmetry of the potential, it develops a tail in the direction of the well. As soon as the leading edge of the tail reaches the boundary of the potential well, it is quickly transferred to the bottom, thereby giving rise to the second peak. Only later the remaining mass of the probability distribution falls into the well, thus restoring a one-peaked distribution.

It must be stressed that the phenomenon of transient bimodality is not restricted to the case of optical bistability, but is of general nature and arises universally in all systems with a deterministic time evolution which presents a long induction stage followed by a fast switching stage. The case of optical bistability is especially interesting because it offers the possibility of a precise experimental observation.

If one looks at the switching time, one finds, both theoretically and experimentally, that it undergoes large fluctuations. The switching time distribution is obtained from the solution of the Fokker–Planck equation (12.9.6) as follows. One selects arbitrarily a value of x, called x_s, which separates well the two peaks during all the time evolution; e.g. for $C = 20$ an appropriate choice is $x_s = 8$. The switching time distribution $v(\tau)$ is defined as

$$v(\tau) = \frac{dP_2(\tau)}{d\tau},$$

where $P_2(\tau)$ is the area of the righthand peak, given by

$$P_2(\tau) = \int_{x_8}^{\infty} dx\, P(x, \tau).$$

By definition, the integral of $v(\tau)$ from zero to infinity is equal to unity.

Figure 12.23 shows the switching time distribution for several values of the noise level σ_x^2. The vertical dotted line indicates the deterministic value of the switching time, i.e. the time τ_s at which the solution $x(\tau)$ of (12.9.1) for $x(0) = 0$ is equal to x_s. For relatively large noise levels the most probable switching time is much shorter than the deterministic value τ_s. As the noise level is reduced, the switching time distribution broadens, while the most probable switching time increases. This broadening is accompanied by an enhancement of the phenomenon of transient bimodality, in the sense that the time interval, in which the distribution $P(x, \tau)$ is bimodal, becomes larger. For $\sigma_x^2 = 10^{-2}$ we find an inversion of tendency: on decreasing the noise level further, the most probable switching time continues to grow, but the switching time distribution now becomes narrower and more symmetrical, until it approaches a delta function centered at the deterministic value of the switching time. This narrowing is accompanied by a gradual disappearance of the phenomenon of transient bimodality, until, when the noise level is small enough, the distribution $P(x, \tau)$ remains always a single peak which trivially follows the deterministic time evolution.

An interesting feature of this analysis is that, even if transient bimodality disappears when noise becomes too small, it persists for quite low values of the noise level. This fact was relevant with respect to ensuring the feasibility of an experimental observation.

After discussing the dependence on the noise level, another aspect that is of interest is the dependence on the difference $(y - y_M)$ between the value y of the input field and the bistability threshold, or critical value, y_M (see Figure 12.18).

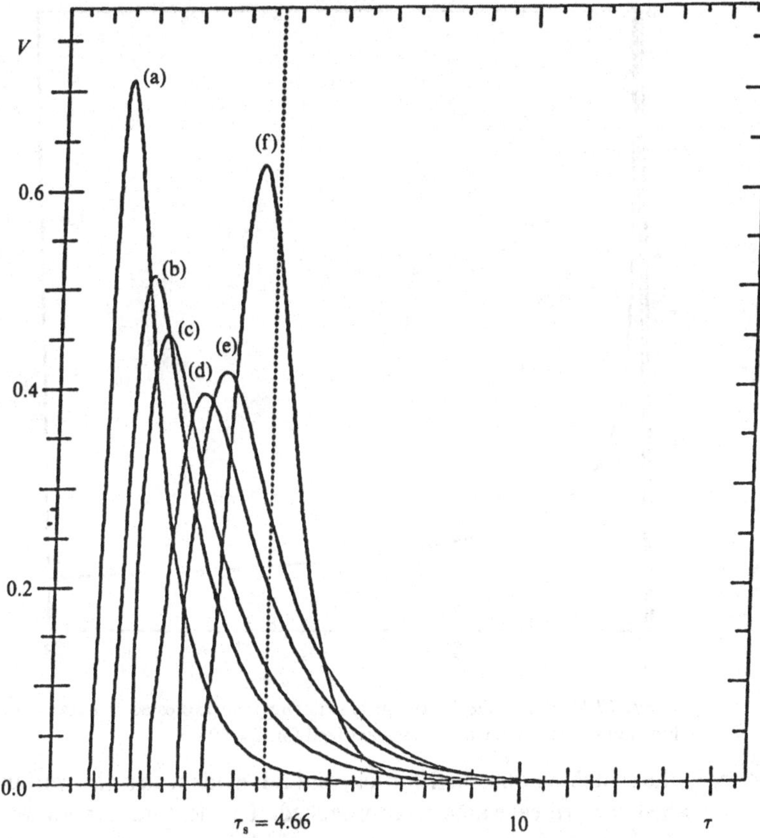

Figure 12.23. Switching time distributions for $C = 20$, $y = 12.1$ and (a) $\sigma_x^2 = 0.5$, (b) $\sigma_x^2 = 0.1$, (c) $\sigma_x^2 = 0.05$, (d) $\sigma_x^2 = 0.01$, (e) $\sigma_x^2 = 0.005$, (f) $\sigma_x^2 = 0.001$.

This is connected with scaling laws in the critical slowing down.

According to the deterministic theory the switching time diverges when the critical value is approached (see Figures 12.19 and 12.24). Grynberg and Cribier (1983) analyzed this divergence and found that it scales as $(y - y_M)^{-1/2}$. If one includes noise, the picture changes qualitatively because the divergence disappears (Figure 12.25). This is due to the fact that in the presence of noise the two branches of the hysteresis cycle are metastable. When we are near the bistability threshold y_M and the noise level is large enough, the system tends to switch to the upper branch because the lifetime of the lower transmission state is not too long. Hence the transition can be observed and the curve of the average switching time can be continued smoothly into the bistable region.

However, we must not forget that the average switching time provides an incomplete picture of the situation because of the large fluctuations in the switching time itself. For this reason we indicate also in Figure 12.26 the

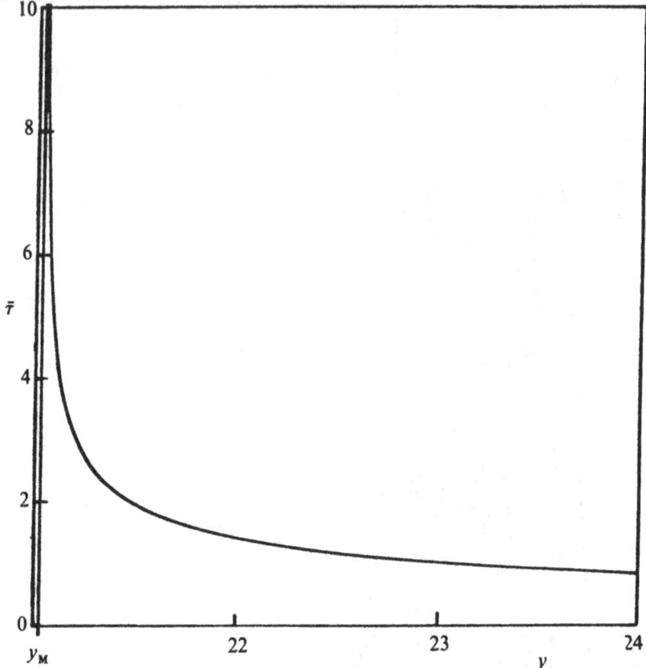

Figure 12.24. Plot of the deterministic switching time (expressed in units κ^{-1}) as a function of the value y of the incident field for $C = 20$.

variance of the switching time distribution, and it is evident that this variance increases when the critical value is approached. This feature is relevant in connection with most of the experiments on critical slowing down performed up to now, that were carried out without being aware of the fact that the large fluctuations in the switching time are intrinsic to the critical slowing down process, and are not due to imperfections in the experimental apparatus.

All the features concerning the transient bimodality, the switching distribution, the dependence on the parameters σ_x^2 and $(y - y_M)$ have been fully confirmed by experimental observations (Lange *et al.*, 1985; Mitsche *et al.*, 1985). The phenomenon of transient optical bimodality has been recently observed in a laser with a saturable absorber by Arimondo (1988), as predicted in Broggi, Lugiato and Colombo (1985).

The relations which link transient bimodality with other phenomena like the decay of unstable and metastable states are discussed in Broggi, Lugiato and Colombo (1985).

12.10 Dispersive optical bistability, colored noise: noise switching and transient bimodality

From the arguments given in the previous section, it is evident that the phenomenon of transient bimodality is not specific to the type of noise, white

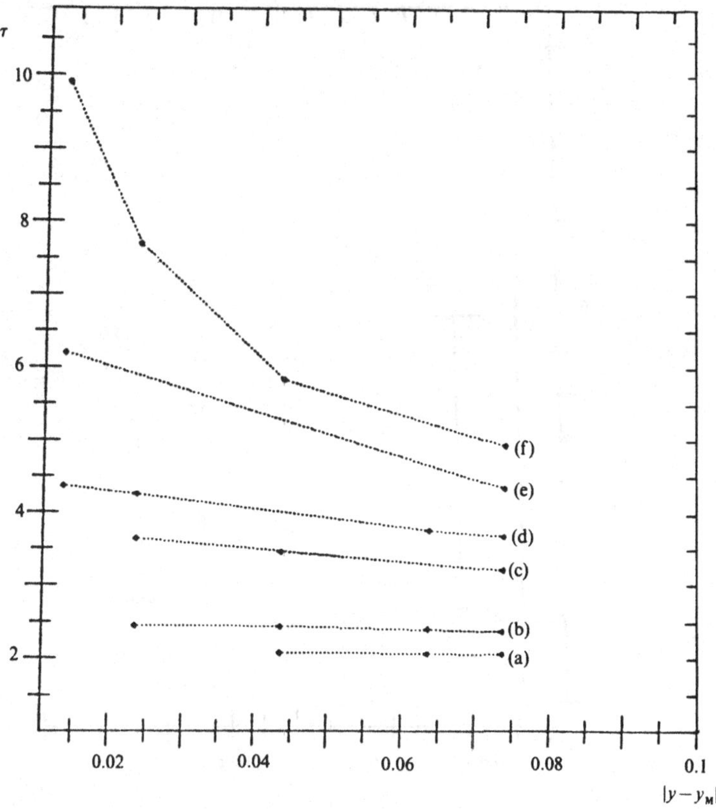

Figure 12.25. Most probable delay time as a function of $(y - y_M)$ for (a) $\sigma_x^2 = 0.1$, (b) $\sigma_x^2 = 0.05$, (c) $\sigma_x^2 = 0.01$, (d) $\sigma_x^2 = 0.005$, (e) $\sigma_x^2 = 0.001$, (f) $\sigma_x^2 = 0$ (deterministic case).

and additive, that we considered there. We will explicitly show this fact using an example of colored and multiplicative noise. We consider here the evolution equation (12.5.1), which describes the dynamics of micron-sized optically bistable devices which utilize semiconductors (Smith, Mandel and Whenett, 1987). We assume that the incident intensity I fluctuates in time, hence, we replace I by $I + \xi(t)$, where $\xi(t)$ obeys (12.1.2a). Thus, we obtain the stochastic equation with multiplicative noise

$$\frac{dx}{dt} = -\gamma\left[x - \frac{I}{1 + F\sin^2(x - \phi_0)}\right] + \gamma\frac{\xi(t)}{1 + F\sin^2(x - \phi_0)}. \tag{12.10.1}$$

On the basis of (12.10.1), we will discuss also the matter of *noise switching.* In order to illustrate this point, it is suitable to premise here that, in order to make

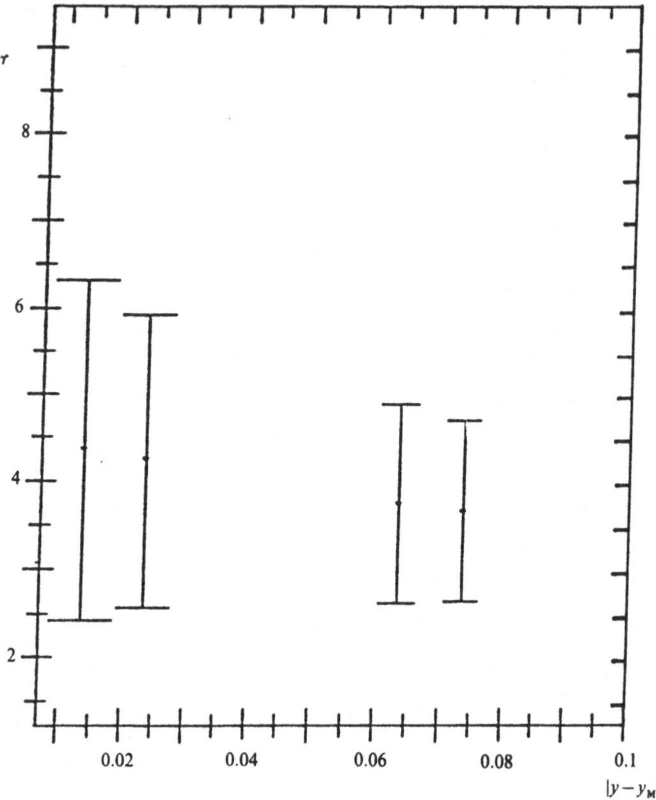

Figure 12.26. Most probable delay time as a function of $(y - y_M)$ for $\sigma_x^2 = 0.005$. The vertical bar represents the full-width half maximum of the corresponding switching time curves.

the system switch from the lower to the upper branch of the hysteresis cycle when this is wanted, one operates the device with two distinct input beams, a *holding beam* and an *address beam.* The holding beam has an intensity I_H constant in time and slightly smaller than the up-switching threshold I_{up}. The address beam is used only when we want the system to switch, and is such that the total intensity $I = I_H + I_A$ is larger than I_{up}. In order to minimize the energy requirements, one would like to keep $I_H + I_A$ as close to I_{up} as possible. However, in this way one meets the problem of critical slowing down (see the preceding section): the switching time becomes longer and longer as the difference $\varepsilon = (I_H + I_A) - I_{up}$ decreases, and diverges for $\varepsilon \to 0$. Usually, this problem is solved by overdriving the system, i.e by keeping ε large enough, but this implies an increase of the energy necessary to obtain switching. We analyzed a different possibility, namely to use an address beam which is not coherent but noisy. In fact, noise produces a random overdriving which may increase the switching speed. Precisely, we assume that the address beam intensity is given by $I_A + \xi(t)$, where I_A is the average value and ξ is the usual stochastic variable which appears in (12.10.1). In order to ensure the positivity of the intensity of the address beam $I_A + \xi(t)$, we choose the mean square fluctuation σ to be equal to $I_A/2$. Thus the dynamics of the system is governed by (12.10.1) with $I = I_H + I_A$. The total average intensity I is set slightly larger than I_{up}. Because the effects of noise increase with its amplitude $\sigma = I_A/2$, it is convenient to keep the holding intensity I_H removed from the up-switching threshold; this is suitable also in order to avoid drift problems in the holding intensity. We select $I_H = I_{down} + \frac{2}{3}(I_{up} - I_{down})$, where I_{down} is the down-switching threshold. The initial condition $x(0)$ is the stationary lower transmission value of x corresponding to the holding intensity I_H.

Clearly, in practice, the fluctuations in the holding intensity I_H must be as small as possible, otherwise they might cause a spontaneous random switching of the device even in absence of the address beam, and the system would be unreliable. In fact we assume here that the holding beam is noiseless. On the contrary, the fluctuations in the address beam can be helpful because they may speed the switching process up. We have already seen in Section 12.4 that noise can in part counteract the critical slowing down. In order to enhance this noise effect, the product $\gamma\tau$ in this case must be chosen to be of the order of unity or larger; a smaller $\gamma\tau$ would reduce the effective noise strength σ_x (see (12.2.6)). Therefore it is convenient to use an address beam with a bandwidth $1/\tau$ of intensity fluctuations of the order of the response rate γ of the system.

We simulated numerically the stochastic equations (12.10.1) and (12.1.2a) for $F = 10, \phi_0 = 1.5, I = 4.325$. In this situation we have $I_{down} = 1.483$ and $I_{up} = 4.3196613$. Figure 12.27 exhibits the time evolution of the mean value of the variable x, obtained by averaging over 1000 realizations of the stochastic equations; the time variable is normalized to γ^{-1}. Curve (a) corresponds to the deterministic situation $\xi(t) = 0$; the other curves are obtained for the same value $\sigma = 1/2[I_{in} - I_{up} + 1/3(I_{up} - I_{down})]$, but for three different values $\gamma\tau = 0.1, 1$ and 10. Clearly, noise produces a drastic reduction of the switching time. The optimal choice, for which the switching process is fastest on the average, is definitely $\gamma\tau = 1$. When the noise is more strongly colored, as in the case $\gamma\tau = 10$, the average $\langle x \rangle$ is slower in reaching its asymptotic value. This 'tail' effect is due to the unfavorable noise stories, in which the initial fluctuation goes in the wrong direction and I decreases instead of increases; when the noise is slow it takes some time to readjust the situation. Figure 12.28 plots the average switching time as a function of $\gamma\tau$; the case $\gamma\tau = 0$ corresponds to absence of noise.

It is interesting to compare the effect of noise with that of a deterministic modulation. In this case, in (12.10.1) we set $\xi(t) = \sigma \sin((2\pi/T)t + \bar{\phi})$. The quantities F, ϕ_0, I and σ have the same values as before. Figure 12.29 shows the time evolution of $\langle x \rangle$, obtained by averaging over the value of the initial phase $\bar{\phi}$ of ξ, taken by uniform sampling in the interval $0 \leqslant \bar{\phi} < 2\pi$. The case $T = 1$ shows that random noise is more efficient than deterministic modulation

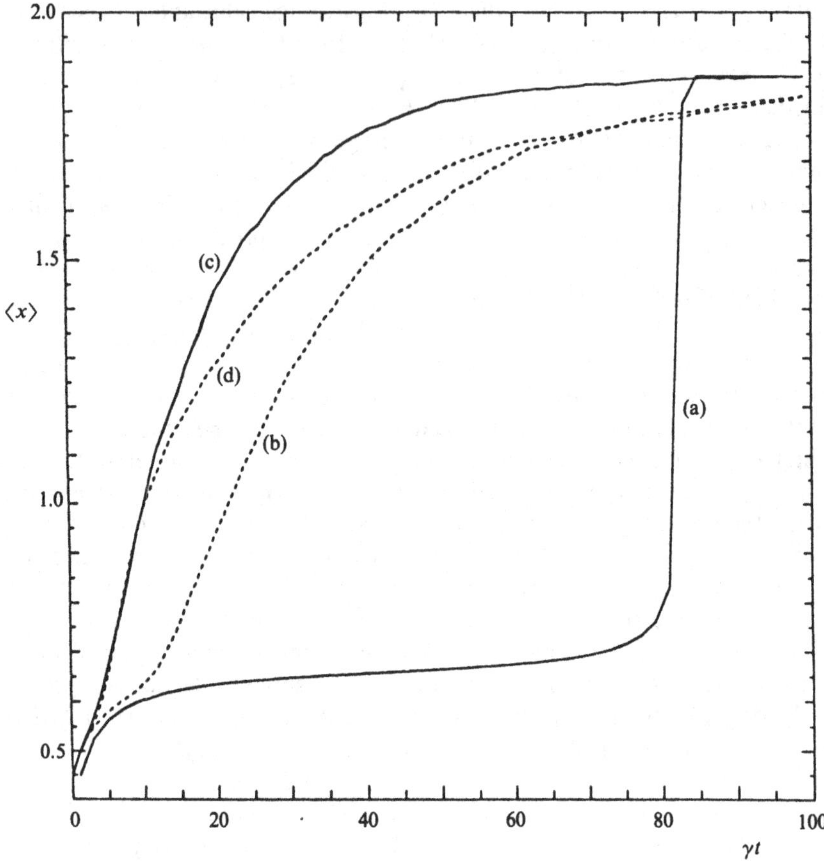

Figure 12.27. Time evolution of $\langle x \rangle$ in the absence (a) and in the presence (b, c, d) of noise. The value of $\gamma\tau$ is 0.1, 1 and 10 for (b), (c) and (d), respectively.

in speeding the switching process up. The switching time becomes drastically shorter for a slow modulation, in agreement with the analysis of Mandel and Erneux (1986); here the situation is not much different from that of a constant address beam.

Figure 12.30 exhibits the time evolution of the probability distribution of the variable x in the case of random noise, obtained for the same values of the parameters used in Figure 12.27. In all cases the phenomenon of transient bimodality is evident. We see also that in the case $\gamma\tau = 1$ the lefthand, low transmission peak decays most rapidly, thereby confirming that this is the best choice of the parameter τ to obtain a fast switching.

12.11 A case of nonlinear noise

In the preceding section, we considered an address beam which is in part coherent (I_A) and in part incoherent ($\xi(t)$). We now consider the case of a

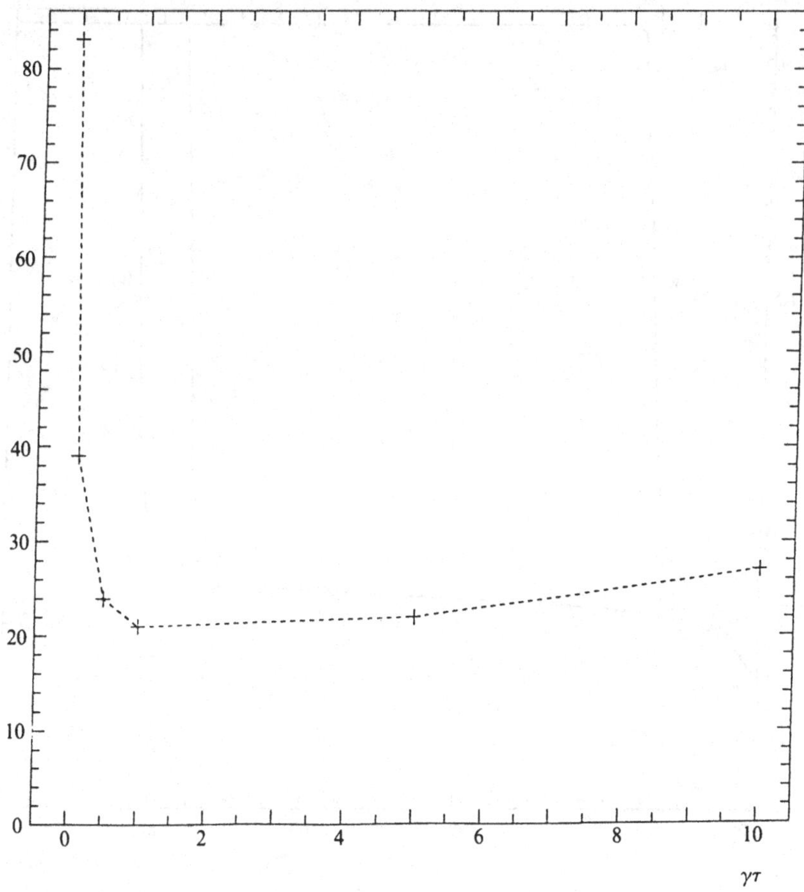

Figure 12.28. Graph of the average switching time as a function of $\gamma\tau$.

completely incoherent address beam. Precisely, we set the intensity of the address beam equal to $\xi^2(t)$; this choice ensures the positivity of the intensity. Then, instead of (12.10.1) we consider the following stochastic differential equation with nonlinear noise:

$$\frac{\mathrm{d}x}{\mathrm{d}t} = -\gamma\left[x - \frac{I_{\mathrm{H}}}{1 + F\sin^2(x - \phi_0)}\right] + \gamma\frac{\xi^2(t)}{1 + F\sin^2(x - \phi_0)}, \tag{12.11.1}$$

where the variable ξ still obeys (12.1.2a). The parameters F, ϕ_0 and I_{H} are chosen as in Figure 12.27, while the parameter σ is chosen in such a way that $\langle I \rangle = I_{\mathrm{H}} + \langle \xi^2 \rangle = I_{\mathrm{H}} + \sigma^2$ has the same value 4.325 as in Figure 12.27. Figure 12.31 shows that this type of noise shortens the average switching time notably even for $\gamma\tau = 0.1$. This reduction, in comparison with curve (b) of

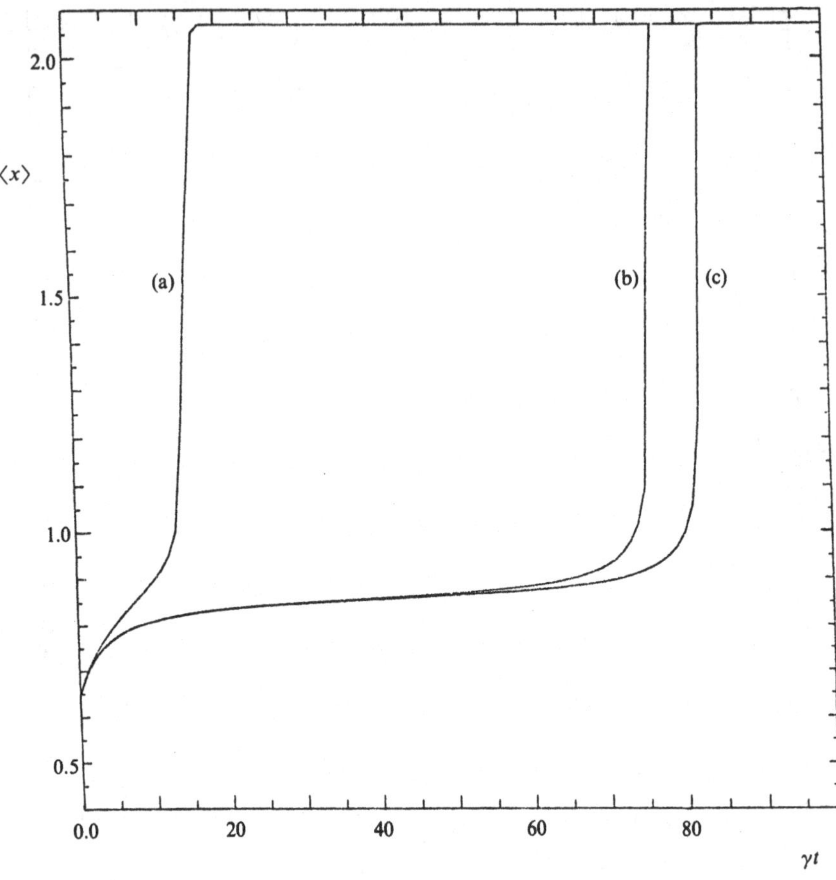

Figure 12.29. External modulation of the input intensity. The time evolution of $\langle x \rangle$ (see text) is shown for (a) $T = 10$, (b) $T = 1$, (c) $T = 0.1$.

Figure 12.27, is explained by the fact that in the case of Figure 12.27 the variance of the input intensity is $\sigma = (I - I_{\mathrm{H}})/2 \simeq 0.45$, whereas in Figure 12.31 the variance is $(\langle \xi^4 \rangle - \langle \xi^2 \rangle^2)^{1/2} = 2^{1/2}(\langle I \rangle - I_{\mathrm{H}}) \simeq 1.26$, i.e. nearly three times larger. For $\gamma\tau = 1$ the two kinds of noise give nearly identical results (Figure 12.32).

12.12 Delayed bifurcations in swept parameter systems: noise effects

In this section, we discuss the case of systems in which a control parameter is swept linearly in time. This problem has been extensively analyzed in the framework of deterministic nonlinear dynamics, especially by Mandel and collaborators (Erneux and Mandel, 1986; Kapral and Mandel, 1985; Mandel and Erneux, 1984). A particularly interesting situation arises when the swept parameter crosses a value which, in the case of constant parameter,

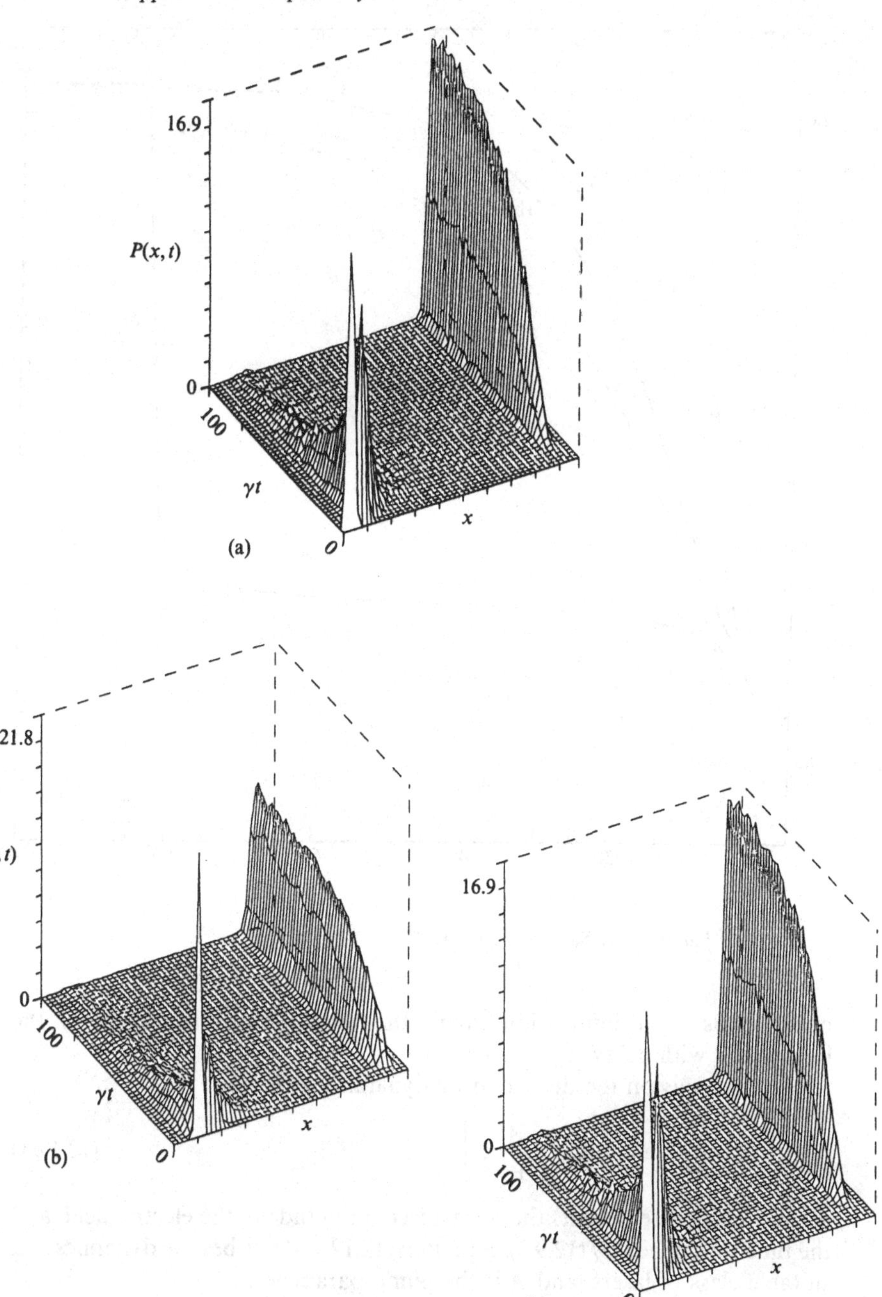

Figure 12.30. Time evolution of the probability distribution of the variable x for (a) $\gamma\tau = 1$, (b) $\gamma\tau = 0.1$, (c) $\gamma\tau = 10$. As for Figure 12.27, the other parameters are given in the text. The time variable is normalized to γ^{-1}.

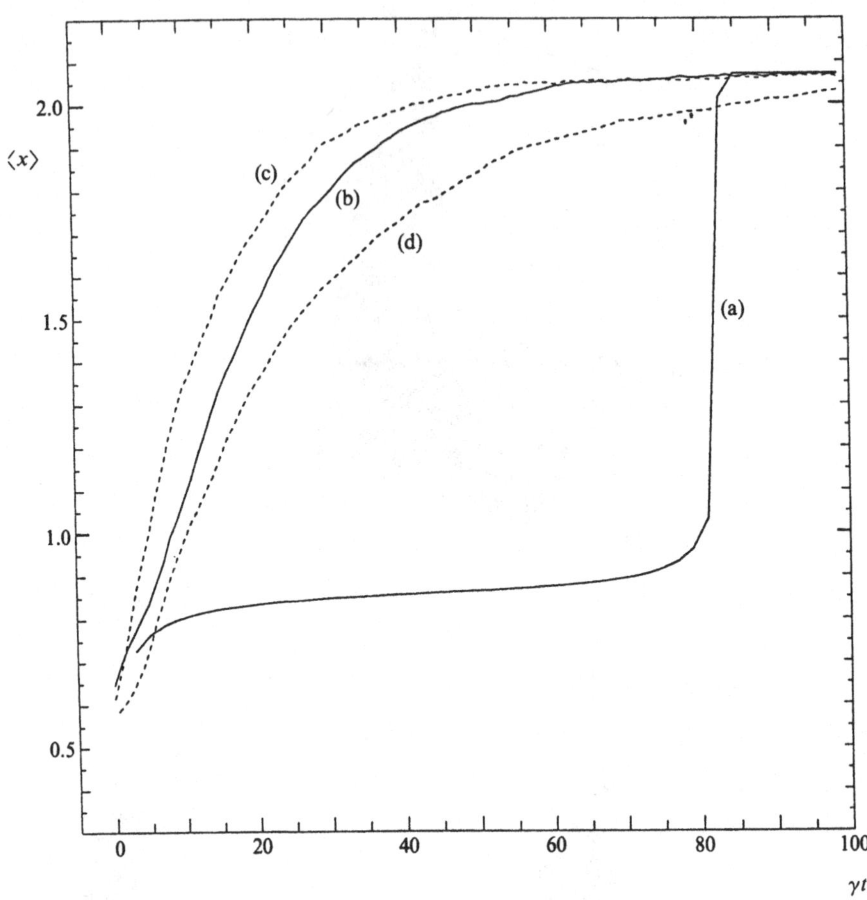

Figure 12.31. Same as Figure 12.27, but for nonlinear noise.

corresponds to a bifurcation point: the system 'feels' the effect of the bifurcation with delay.

We will focus on the deterministic dynamical equation

$$\frac{\mathrm{d}x}{\mathrm{d}\tau} = x\left[-1 + \frac{A}{1+x^2}\right]. \tag{12.12.1}$$

As in (12.9.1), x represents the normalized amplitude of the electric field, and the time τ is defined by (12.9.3). Equation (12.12.1) describes the dynamics of a suitable class of lasers, and A is the pump parameter.

When A is constant in time, (12.1.1) admits two kinds of stationary solutions:

$$x = 0 \quad \text{and} \quad x^2 = A - 1.$$

The trivial solution $x = 0$ is stable for $A < 1$, and the value $A = 1$ corresponds to

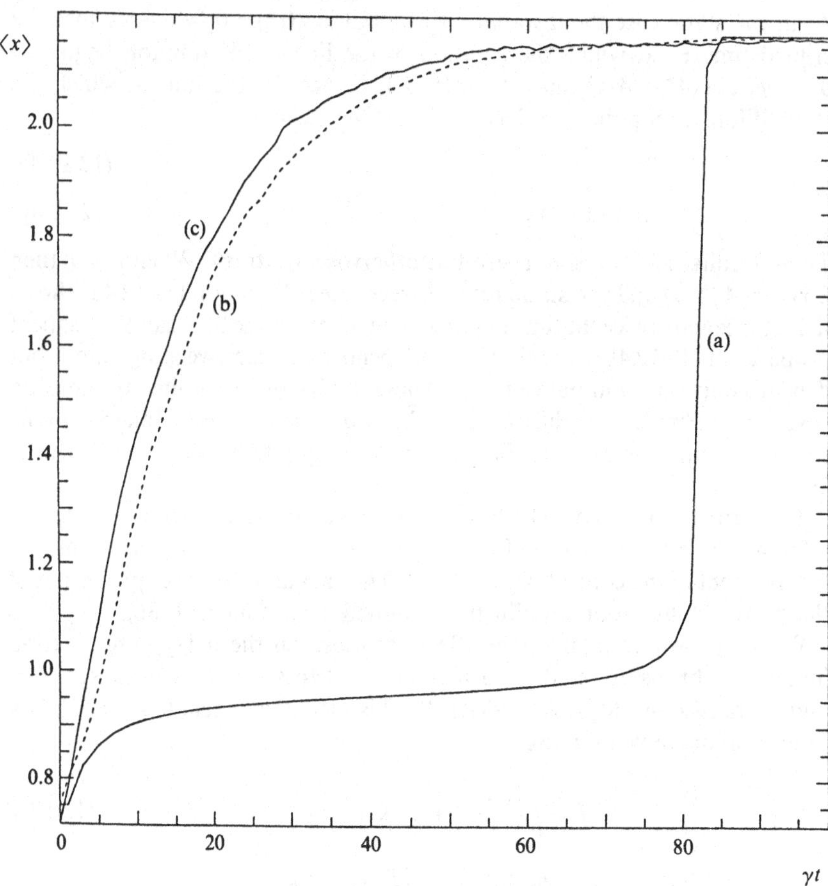

Figure 12.32. Curve (a) represents the deterministic time evolution, curves (b) and (c) show the time evolution of $\langle x \rangle$ for $\gamma\tau = 1$ in the case of linear (dashed line) and nonlinear (full line) noises.

the laser threshold. For $A = 1$, the stable solution $x^2 = A - 1$ bifurcates from the trivial solution. A common way to study experimentally a steady bifurcation point like $A = 1$ for (12.12.1) is to sweep A across $A = 1$:

$$A = A(\tau) = A_0 + v\tau, \quad A_0 < 1, \quad v > 0. \tag{12.12.2}$$

This dynamic procedure differs significantly from the static which assumes that A is time independent. To examine this difference, we analyze the (dynamic) stability of the trivial solution $E = 0$ with A given by (12.12.2). Linearizing (12.12.1) around $E = 0$ gives $\mathrm{d}E/\mathrm{d}t = E(-1 + A)$. The solution of this equation is

$$E(\tau) = E(0)\exp[-\lambda(\tau)], \tag{12.12.3}$$

$$\lambda(\tau) = \tau\left[A_0 + \frac{v\tau}{2} - 1\right].$$

As in the static case the instability threshold is $\lambda(\tau^*) = 0$, which defines the critical time τ^* at which the solution of the linearized equation begins to diverge. Let $A^* \equiv A(\tau^*)$ and $\bar{A} \equiv A(\bar{\tau}) = 1$. Hence $\bar{\tau}$ is the time at which the static bifurcation point $A = 1$ is reached. One easily finds

$$\tau^* = 2\bar{\tau} \tag{12.12.4a}$$

$$A^* - \bar{A} = \bar{A} - A_0. \tag{12.12.4b}$$

These results hold if v is nonzero but otherwise arbitrary. When $v = 0$, then $\lambda(\tau) = \tau(A_0 - 1)$ and the static result is recovered. Equation (12.12.4a) shows that τ^* is exactly twice the time required to reach the static bifurcation. A salient property of (12.12.4b) is that A^* is independent of the sweeping rate v but depends only on the initial value A_0. These results are not limited to the good cavity case. Similar conclusions have been drawn from the Haken–Lorenz equations with arbitrary atomic and cavity decay rates (Mandel and Erneux, 1984).

The critical property which is responsible for large delays in steady bifurcations is the existence of a steady solution ($E = 0$ in this case) which is independent of the control parameter A. This fact and the other properties of the delayed bifurcation are illustrated in detail in Mandel (1986).

We want now to analyze the effects of noise on the delayed bifurcation (related problems are studied in Kondepudi, Moss and McClintock, (1986) and Pieranski and Malecki (1987)). To this extent, we include in (12.12.1) a term of additive white noise.

$$\frac{\mathrm{d}x}{\mathrm{d}\tau} = x\left[-1 + \frac{A(\tau)}{1 + X^2}\right] + \eta(\tau), \tag{12.12.5}$$

$$\langle \eta(\tau) \rangle = 0, \quad \langle \eta(\tau)\eta(\tau') \rangle = 2D\delta(\tau - \tau'),$$

where $A(\tau)$ is given by (12.12.2). We integrated numerically (Broggi *et al.*, 1986) the Fokker–Planck equation, equivalent to the stochastic equation (12.12.5), for the probability distribution $P(x, \tau)$:

$$\frac{\partial P(x,\tau)}{\partial \tau} = \left\{\frac{\partial}{\partial x}\left[E\left(1 - \frac{A(\tau)}{1 + E^2}\right)\right] + \sigma_x^2 \frac{\partial^2}{\partial x^2}\right\} P(x,\tau),$$

$$\sigma_x^2 = kD. \tag{12.12.6}$$

We solved (12.12.6) numerically following the Crank–Nicolson discretization method, for several values of A_0, q, v, and with the initial condition $P(x, 0) = \delta(x)$, using a numerical code essentially identical to that utilized in Broggi and Lugiato (1984) and Broggi, Lugiato and Colombo (1985). A first result is that the dependence of the evolution of the mean electric field intensity $\langle x^2(A(\tau)) \rangle$ on the initial value $A(0) = A_0$, given in the deterministic case by (12.12.4b), disappears in the presence of a white noise. In fact, if one compares the evolution of $\langle x^2(A(\tau)) \rangle$ for the same values of the noise amplitude σ_x^2 and

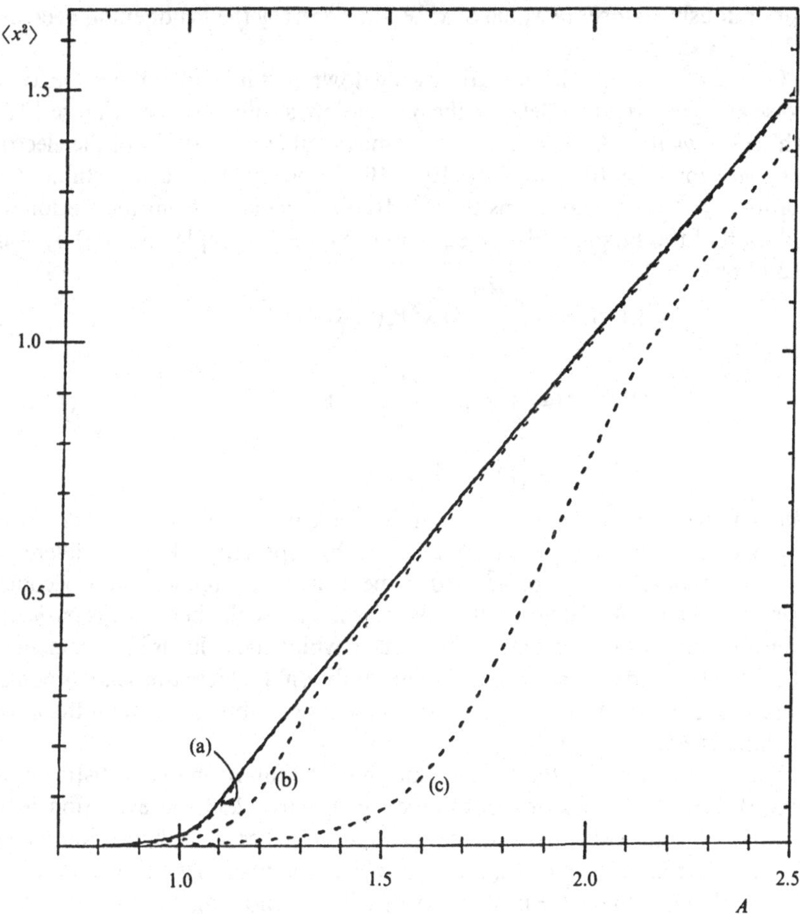

Figure 12.33. Evolution of the mean value of the electric field intensity $\langle x^2(A(t))\rangle$ for $\sigma_x^2 = 10^{-3}$ and (a) $v = 10^{-3}$, (b) $v = 10^{-2}$, (c) $v = 10^{-1}$. The full line depicts the stationary mean value $\langle x^2(A(t))\rangle_s$ of electric field for $\sigma_x^2 = 10^{-3}$.

the sweep velocity v and for two different values $A_0^{(2)} > A_0^{(1)}$, one finds that, in both cases, the evolution appears completely identical, apart from a short transient starting from $A_0^{(2)}$. For example, curve (c) in Figure 12.33 was obtained both for $A_0 = 0$ and for $A_0 = -0.5$. This difference from the deterministic result (12.12.4a, b) can be easily understood on examining the facts that give rise to the delay in the deterministic case. The first fact is the critical slowing down at $A \simeq 1$. The second fact is that, starting from an initial value $x(0)$ with $A_0 < 1$, the value of x decreases exponentially for $\tau < \bar{\tau}$ (see (12.12.3)). Hence when $A(\tau) = \bar{A} = 1$, x is smaller the larger $\bar{A} - A_0$ is, which implies that the delay increases with $\bar{A} - A_0$. The dependence on A_0 disappears in the case of white noise because the fluctuations continuously restore the value of x, i.e. they destroy the exponential decrease of x for $\tau < \bar{\tau}$.

On the other hand, the critical slowing down is still valid in the presence of noise and gives rise to a delay in the bifurcation, as illustrated by Figure 12.33 which shows the time evolution of the mean value $\langle x^2(A(\tau))\rangle$ of the electric intensity for $\sigma_x^2 = 10^{-3}$ and $v = 10^{-1}, 10^{-2}$ and 10^{-3}. It also exhibits the stationary electric field intensity $\langle x^2(A(\tau))\rangle_s$ calculated from the stationary solution of the Fokker–Planck equation obtained by replacing $A(\tau)$ by A in (12.12.6):

$$\langle x^2(A(\tau))\rangle_s = \int_{-\infty}^{+\infty} \mathrm{d}x\, x^2 P_s(x, A(\tau))$$

$$P_s(x, A) = \mathcal{N} \exp\left[\frac{-U_A(x)}{\sigma_x^2}\right], \tag{12.12.7}$$

$$U_A(x) = \tfrac{1}{2}[x^2 - A\ln(1 + x^2)].$$

As expected, the presence of white noise decreases the delay with respect to the deterministic value given by (12.12.4a, b), especially when the difference $\bar{A} - A_0$ becomes large. For $\sigma_x^2 = 10^{-3}$ the delay is negligible, and it becomes sizable when $v \gg \sigma_x^2$. Hence in the case of white noise the behavior is basically complementary to the deterministic case, in which the delay is independent of the velocity v and depends on the initial condition A_0. Here the delay becomes appreciable only when the sweep rate becomes suitably larger than the noise parameter σ_x^2.

Let us now consider the time evolution of the full probability distribution $P(x, \tau)$. For the values of v considered in Figure 12.33 the evolution is as follows. Within a short transient $P(x, \tau)$ approaches a configuration which practically coincides with $P_s(E, A(\tau))$. When A approaches unity, due to the critical slowing down, the probability distribution no longer follows the value of A adiabatically and becomes quite different from $P_s(x, A(\tau))$ (see Figure 12.34). Only after $A(\tau)$ exceeds unity by a suitable amount dependent on v does $P(x, \tau)$ return to a configuration close to $P_s(E, A(\tau))$.

Another aspect of interest in this problem is the bifurcation time distribution. In the deterministic theory, the bifurcation time is defined as the time value such that the output intensity x^2 reaches a prefixed value x_{th}^2. The choice of x_{th}^2 is arbitrary provided $x_{\mathrm{th}}^2 > x^2(0)$; we take $x_{\mathrm{th}}^2 = 0.1$. In the presence of noise the bifurcation time becomes stochastic. The distribution of bifurcation times normalized to unity, $\bar{W}(\tau)$, can be defined as

$$\bar{W}(\tau) = \frac{\mathrm{d}\bar{P}(\tau)}{\mathrm{d}\tau}, \quad \bar{P}(\tau) = 2\int_{x_{\mathrm{th}}}^{+\infty} \mathrm{d}x\, P(x, \tau). \tag{12.12.8}$$

As before, it is more convenient to use the variable A instead of time, which allows us to compare directly the distributions for different values of rate v. From (12.12.2) the bifurcation probability distribution, when expressed in

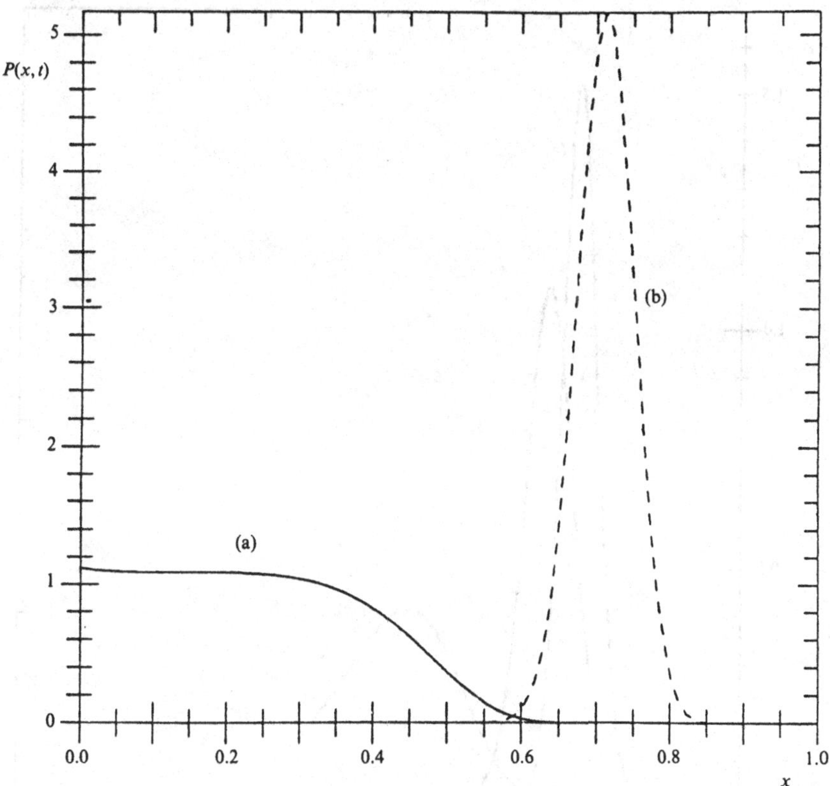

Figure 12.34. Curve (a) shows the probability distribution $P(x, t)$ for $\sigma_x^2 = 10^{-3}$, $v = 10^{-1}$, $A_0 = 0$ and $t = 15095$. Curve (b) is the distribution $P_s(x, A(t))$ for the same values of the parameters. Note that the distribution are even functions of x.

terms of A, is given by

$$W(A) = \frac{1}{v}\bar{W}\left[\frac{A - A_0}{v}\right]. \tag{12.12.9}$$

Figure 12.35 shows the distribution of $W(A)$ for $\sigma_x^2 = 10^{-3}$ and three values of v. Clearly, the width of the distribution increases with the velocity.

In conclusion, we have shown that, even if the critical slowing down is intrinsically accompanied by an enhanced sensitivity to noise, the phenomenon of delayed bifurcation persists in the presence of noise. Recent work by Mannella, Moss and McClintock (1987) provides a study of the stochastic equation (12.12.5) by analog electronic simulation. This analysis fully confirms the results of Broggi *et al.* (1986) in the case of white noise and extends them to the situation of colored noise. Up to $\gamma\tau = 10$, the data for colored noise show very limited deviations from the white noise picture.

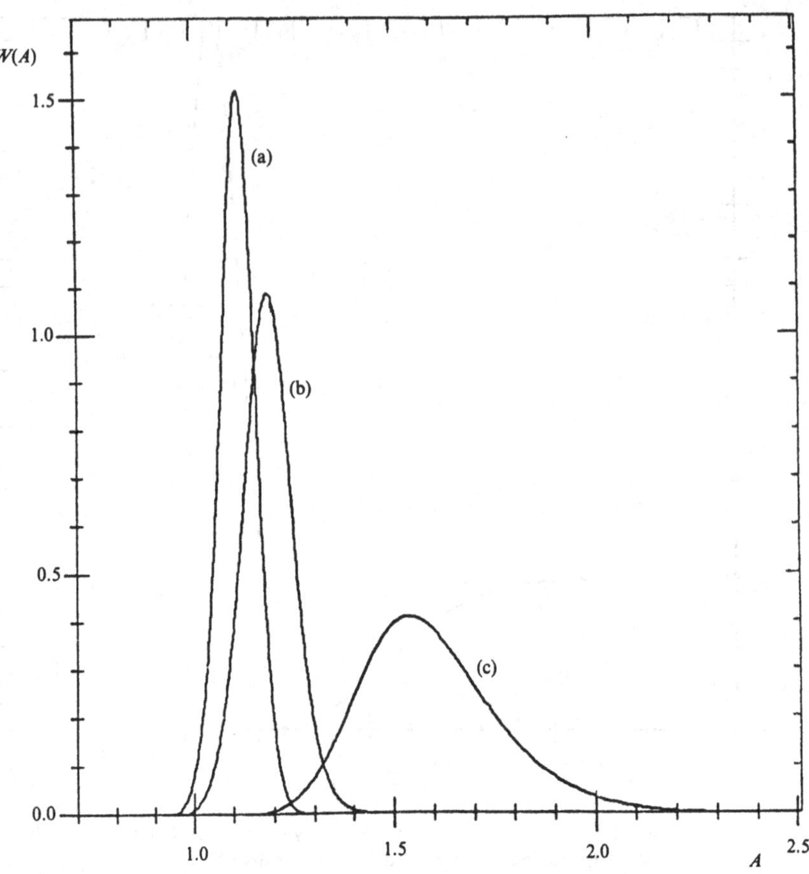

Figure 12.35. Dynamical bifurcation probability distributions $W(A)$ (see text) for $\sigma_x^2 = 10^{-3}$ and (a) $v = 10^{-3}$, (b) $v = 10^{-2}$, (c) $v = 10^{-1}$.

In the framework of optical systems, the existence of delayed bifurcations has been experimentally confirmed by Arimondo *et al.* (1986). A recent analysis (Arimondo, 1988) has also verified the qualitative behavior of the bifurcation probability distribution $W(A)$ as a function of the sweep velocity v, described in Figure 12.35.

Acknowledgements

This work was carried out in the framework of the EEC twinning project on 'Dynamics of Nonlinear Optical Systems'. One of us (GB) also acknowledges support by the Swiss National Science Foundation.

References

Abraham, E. and Firth, W. J. 1983. *Optica Acta* **30**, 1541.

Al Attar, H. A., Mackenzie, H. A. and Firth, W. J. 1986. *J. Opt. Soc. Am.* **B3**, 1157.

Arimondo, E. 1988. In *Synergetics and Dynamic Instabilities* (G. Caglioti, H. Haken and L. A. Lugiato, eds). Amsterdam: North-Holland – Società Italiana di Fisica. (In press.)

Arimondo, E., Gabbanini, C., Menchi, E., Dangoisse, D. and Glorieux, P. 1986. In *Optical Instabilities* (R. W. Boyd, M. G. Raymer and L. M. Narducci, eds.), p. 277. Cambridge University Press.

Arnold, L., Horsthemke, W. and Lefever, R. 1978. *Z. Phys. B* **29**, 367.

Baras, F., Nicolis, G., Malek Mansour, M. and Turner, J. W. 1983. *J. Stat. Phys.* **32**, 1.

Barbarino, S., Gozzini, A., Maccarrone, F., Longo, I. and Stampacchia, R. 1982. *Nuovo Cimento B* **71**, 183.

Benza, V. and Lugiato, L. A. 1979. *Lett. Nuovo Cimento* **26**, 405.

Bonifacio, R. and Lugiato, L. A. 1976. *Opt. Comm.* **19**, 172.

Bonifacio, R. and Lugiato, L. A. 1978. *Phys. Rev. A* **18**, 1129.

Bonifacio, R. and Meystre, P. 1978. *Opt. Comm.* **29**, 131.

Broggi, G., Colombo, A., Lugiato, L. A. and Mandel, P. 1986. *Phys. Rev. A* **33**, 3635.

Broggi, G. and Lugiato, L. A. 1984. *Phys. Rev. A* **29**, 2949.

Broggi, G., Lugiato, L. A. and Colombo, A. 1985. *Phys. Rev. A* **32**, 2803.

Cribier, S., Giacobino, E. and Grynberg, G. 1983. *Opt. Comm.* **47**, 170.

Erneux, T. and Mandel, P. 1986. *J. Appl. Math.* **46**, 1.

Frankowicz, M., Malek Mansour, M. and Nicolis, G. 1984. *Physica A* **125**, 237.

Frankowicz, M. and Nicolis, G. 1983. *J. Stat. Phys.* **33**, 595.

Gardiner, C. W. 1983. *Handbook of Stochastic Methods.* Berlin: Springer.

Garmire, E., Marburger, J. H., Allen, S. D. and Winful, H. G. 1979. *Appl. Phys. Lett.* **34**, 374.

Gibbs, H. M. 1985. *Optical Bistability: Controlling Light by Light.* New York: Academic Press.

Graham, R. and Schenzle, A. 1982. *Phys. Rev. A* **26**, 1676.

Grant, D. E. and Kimble, H. J. 1983. *Opt. Comm.* **44**, 415.

Grigolini, P., Lugiato, L. A., Mannella, R., McClintock, P. V. E., Merri, M. and Pernigo, M. A. 1988. *Phys. Rev. A* (in press).

Grynberg, G. and Cribier, S. 1983. *J. de Phys. Lett.* **44**, 4449.

Hänggi, P., 1986. *J. Stat. Phys.* **42**, 105.

Haken, H. 1977. *Synergetics – An Introduction.* Berlin: Springer.

Horsthemke, W. and Lefever, R. 1984. *Noise Induced Transitions.* Berlin: Springer.

Kapral, R. and Mandel, P. 1985. *Phys. Rev. A* **32**, 1076.

Kondepudi, D. K., Moss, F. and McClintock, P. V. E. 1986. *Physica D* **21**, 296.

Kramers, H. 1940. *Physica* **7**, 284.

Lange, W., Mitsche, F., Deserno, R. and Mlynek, J. 1985. *Phys. Rev. A* **32**, 1271.

Lugiato, L. A. 1984. In *Progress in Optics* (E. Wolf, ed.), vol. XXI. Amsterdam: North Holland.

Lugiato, L. A., Broggi, G. and Colombo, A. 1986. In *Frontiers in Quantum Optics* (E. R. Pike and S. Sarkar, eds.), p. 231. Bristol: Hilger.

Lugiato, L. A., Colombo, A., Broggi, G. and Horowicz, R. J. 1986. *Phys. Rev. A* **33**, 4469.

Lugiato, L. A. and Horowicz, R. J. 1985a. *J. Opt. Soc. Am. B* **2**, 971.

Lugiato, L. A. and Horowicz, R. J. 1985b. *Opt. Comm.* **54**, 184.

Lugiato, L. A., Mandel, P. and Narducci, L. M. 1984. *Phys. Rev. A* **29**, 1438.

McCall, S. L., Ovadia, S., Gibbs, H. M., Hopf, F. A. and Kaplan, D. L. 1985. *IEEE J. Quantum Electron.* **QE21**, 1441.

Mandel, P. 1985. *Opt. Comm.* **54**, 181.

Mandel, P. 1986. In *Frontiers in Quantum Optics* (E. R. Pike and S. Sarkar, eds.), pp. 430–52. Bristol: Hilger.

Mandel, P. and Erneux, T. 1982. *Opt. Comm.* **42**, 362.

Mandel, P. and Erneux, T. 1984. *Phys. Rev. Lett.* **53**, 1818.

Mandel, P. and Erneux, T. 1986. *IEEE J. Quantum Electron.* **QE21**, 1352.

Mannella, R., Moss, F. and McClintock, P. V. E., 1987. *Phys. Rev. A* **35**, 2560.

Miller, D. A. B. 1981. *IEEE J. Quantum Electron.* **QE17**, 306.

Mitsche, F., Deserno, R., Mlynek, J. and Lange, W. 1983. *Opt. Comm.* **46**, 135.

Mitsche, F., Deserno, R., Mlynek, J. and Lange, W. 1985. *IEEE J. Quantum Electron.* **QE21**, 1435.

Pieranski, P. and Malecki, M. 1987. *Nuovo Cimento* **9D**, 757.

Risken, H. 1984. *The Fokker–Planck Equation: Methods of Solution and Applications.* Berlin: Springer.

Robinson, S. D., Moss, F. M. and McClintock, P. V. E. 1985. *J. Phys. A* **18**, 289.

Sancho, J. M., San Miguel, M., Katz, S. L. and Gunton, J. D. 1982. *Phys. Rev. A* **26**, 1589.

San Miguel, M. and Sancho, J. M. 1981. In *Stochastic Nonlinear Systems* (L. Arnold and R. Lefever, eds.), p. 137. Berlin: Springer.

Schenzle, A. and Brand, H. 1979. *Phys. Rev. A* **20**, 1628.

Smith, S. D., Mandel, P. and Wherrett, B. 1987. *The European Joint Optical Bistability Project.* Amsterdam: North Holland.

Smythe, J., Moss, F. and McClintock, P. V. E. 1983. *Phys. Rev. Lett.* **51**, 1062.

Stratonovich, L. 1963. *Topics in the Theory of Random Noise*, vol. I. New York: Gordon and Breach.

Stratonovich, L. 1967. *Topics in the Theory of Random Noise*, vol. II. New York: Gordon and Breach.

Van Kampen, N. G. 1961. *Can. J. Phys.* **30**, 551.

13 Transition probabilities and spectral density of fluctuations of noise driven bistable systems

M. I. DYKMAN, M. A. KRIVOGLAZ and S. M. SOSKIN

13.1 Introduction

One of the important problems of physical kinetics is the investigation of relaxation and fluctuation phenomena in systems which have two or more stable states. Bi- and multistable systems are studied in various fields of physics, and the causes of multistability and the types of stable states are different in different cases. For systems moving in static potential fields (disregarding the interactions that give rise to relaxation and fluctuations in a system) multistability takes place if a potential has several minima. In this case the stable states are the equilibrium states. A number of systems of this type are investigated in solid state physics; in particular, diffusing atoms and impurity centers that reorient within a unit cell (see Narayanamurti and Pŏhl, 1970).

Multistability may also arise in systems driven by an external periodic field. The constrained vibrations correspond to the stable states in this case (the attractors with a more complicated structure may also appear here). In particular, nonlinear oscillators of various physical nature refer to such systems (see Landau and Lifshitz, 1976). It is well known that in a certain frequency range the dependence of the amplitude A of the constrained vibrations of a nonlinear oscillator on the resonant external field amplitude h may be S-shaped (cf. curve (c) in Figure 13.1). In the range of the non-single-valued dependence $A(h)$ the states with the largest and smallest A are stable. In the absence of noise the oscillator appears in one or another state depending on the 'history' of the field amplitude or frequency variation, i.e. the hysteresis is present when the field is varied.

Bi- and multistability of nonlinear systems in an external periodic field is studied intensively at present in nonlinear optics (Gibbs, 1985). Several stable equilibrium or vibrational states may arise also in nonequilibrium nonlinear systems driven by stationary energy sources of other types. The examples of such systems are well known in fluid dynamics, radiophysics, chemical kinetics, laser physics, etc. (see Haken, 1983).

The interaction of a multistable system with a medium (in particular, with a thermal bath) leads to relaxation of the system and to fluctuations in it.

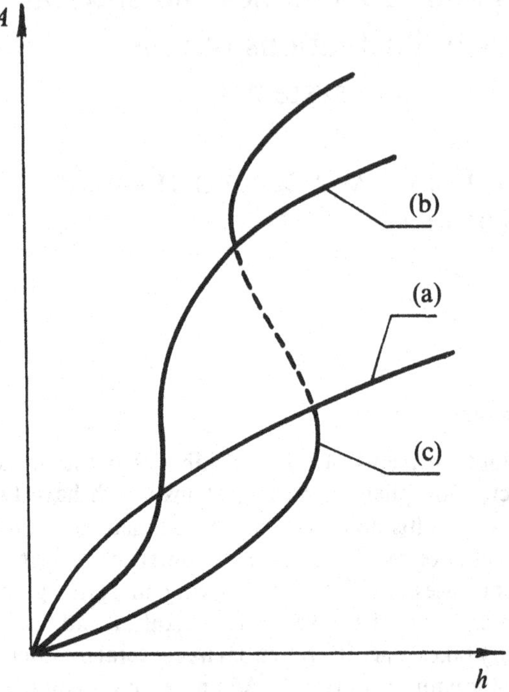

Figure 13.1. Schematic dependence of the amplitude A of the constrained vibrations of a nonlinear oscillator on the amplitude h of an external resonance force. Curves (a)–(c) correspond to different frequency detunings $\omega_h - \omega_0$. (b) corresponds to the critical value of $\omega_h - \omega_0$ starting with which the plot of A versus h becomes S-shaped. The unstable stationary states are shown as a dashed line.

Fluctuations may also arise due to other noise sources acting on a system. The presence of several stable states gives rise to a number of features of relaxation and fluctuation phenomena in a system. These features are due to a great extent to the sufficiently large fluctuations being able to result in transitions between the stable states.* As a result of fluctuational transitions the stationary distribution of a system over the states is established and the dependence of the characteristics of a system on its parameters becomes single-valued. In particular, for sufficiently slow parameter variation the hysteresis described above for a nonlinear oscillator does not arise.

Since large fluctuations are needed for the transitions between the states to occur the transition probabilities W are small for small intensity $\mathscr{B}$ of a random force acting on a system. The value of W in the case of a system performing Brownian motion in a static potential was obtained by Kramers (1940). In this case $\ln W \propto -\Delta U/\mathscr{B}$, where ΔU is the potential barrier height

* We use the term 'stable states' for those states that are stable with respect to small fluctuations.

(when relaxation and fluctuations result from coupling to a thermostat, $\mathscr{B} \propto T$ and $W \propto \exp(-\Delta U/T)$). Transition probabilities may be obtained also for Markov systems in which the potentiality conditions are fulfilled and there are no flows in the steady state.

In the general case of a nonpotential motion or motion in a variable field the calculation of W at small $\mathscr{B}$ is rather a complicated problem. To solve it when random forces are of the white-noise type or, to be more general, are Gaussian it is convenient (Dykman and Krivoglaz, 1979) to use the Feynman approach (Feynman and Hibbs, 1965) to the description of fluctuations in dynamical systems approach. This is based on the path integral method. It allows us to reduce the problem to an investigation of the extreme trajectory of a certain auxiliary dynamical system and to show the transition probability to be given by the expression

$$W = \text{const} \cdot \exp(-P/\mathscr{B}). \tag{13.1.1}$$

The quantity P in (13.1.1) is proportional to the action along the extreme trajectory, and $PT/\mathscr{B}$ plays a role of activation energy. The value of P is determined by the parameters of a dynamical system and does not depend on random force intensity. In particular, for a nonlinear oscillator in a strong resonant field, the quantities P were obtained explicitly (Dykman and Krivoglaz, 1979).

Both P and the constant pre-exponential factor in (13.1.1) may be obtained in a simple explicit form for the problem of an escape from a metastable state near bifurcation point where this state coalesces with an unstable stationary state (Dykman and Krivoglaz, 1980a). As the bifurcation point is approached the escape probability and fluctuations as a whole increase rapidly.

In a bistable system (to be precise we assume in the following that the number of stable states is equal to two) the populations w_1 and w_2 of the stable states 1 and 2 are inversely proportional to the probabilities W_{12} and W_{21} of the transitions from these states. In the general case, the values of P_1 and P_2 in (13.1.1) for the transitions $1 \to 2$ and $2 \to 1$ are different, and $|P_1 - P_2| \gg \mathscr{B}$ for small $\mathscr{B}$. Therefore w_1 and w_2 differ by many times in almost the entire range of the parameters of the system, and one of w_i is close to unity while another is close to zero. Only in the extremely narrow range of parameters where $P_1 \simeq P_2$ the values of w_1 and w_2 are of the same order of magnitude. Owing to the sharp exponential dependence (13.1.1) of W on P, the behavior of the system under its parameters passing through the range where $P_1 \simeq P_2$ is perceived as a smeared first-order phase transition. In particular, in case of a nonlinear oscillator in a strong resonant field the vibration amplitude A changes sharply from the value corresponding to one branch of $A(h)$ to that for another branch (see Figure 13.1), with a corresponding sharp change in the absorption of energy from the resonant field. In the transition region fluctuations in bistable systems acquire a number of characteristic features.

The important characteristic of relaxation and fluctuations in a system is the spectral density $Q'(\omega)$ of fluctuations of the generalized coordinates. For underdamped systems far from the transition region $Q'(\omega)$ has peaks at the eigenfrequencies of vibrations about that stable state whose population $w \simeq 1$. In the transition region besides the peaks corresponding to vibrations near both stable states $Q'(\omega)$ has an additional extremely narrow peak due to fluctuational transitions between the states. Its width is of the order of the transition probabilities $W_{12} \sim W_{21}$. It is much smaller than inverse characteristic relaxation times. For systems moving in a static bistable potential, the peak induced by fluctuational transitions occurs at zero frequency. In systems performing forced vibrations at the strong-field frequency ω_h the peak lies at the same frequency ω_h. The corresponding extremely narrow peak at ω_h is present also in the spectrum of the absorption of an additional weak field (Dykman and Krivoglaz, 1979).

As the random force intensity $\mathscr{B}$ increases, the form of $Q'(\omega)$ for underdamped systems changes substantially (Dykman and Krivoglaz, 1971, 1984; Dykman, Soskin and Krivoglaz, 1984). Even when $\mathscr{B}$ is rather small, the shape of the peaks due to small-amplitude vibrations about the stable states changes strongly because the vibration nonlinearity reveals itself. The width of the peak induced by fluctuational transitions increases exponentially with $\mathscr{B}$ (see (13.1.1)). In addition, the relatively broad peak of $Q'(\omega)$ caused by features of the motion in the vicinity of the unstable equilibrium state appears.

In what follows, two relatively simple but nontrivial bistable systems, a nonlinear oscillator driven by a sufficiently strong resonant field and an oscillator moving in a static potential with two minima, are analyzed in considerable detail. The results for these model systems describe many properties of the physical systems mentioned above.

In Section 13.2 the probabilities of transitions between stable states of a nonequilibrium system are considered using the path integral method. The spectra of fluctuations at low noise intensities, including the narrow peaks induced by fluctuational transitions, are analyzed in Section 13.3. In Section 13.4 the features of the spectral density of fluctuations in underdamped, essentially nonlinear systems with one and two stable states are investigated.

13.2 Probabilities of transitions between stable states of a nonequilibrium system

In the analysis of transition probabilities for small intensities of a random force acting on a dynamical system the most important thing is to calculate the argument of the exponential in (13.1.1), while it suffices to estimate only the order of magnitude of the pre-exponential factor. The method given in Section 13.2.1 permits the calculation of the transition probabilities to logarithmic accuracy.

We illustrate this method and its application by considering the example of an underdamped nonlinear Duffing oscillator driven by a relatively strong resonant force $h\cos\omega_h t$ and a weak random force $f(t)$ which presents white noise (Dykman and Krivoglaz, 1979). A Duffing oscillator in a resonant field is frequently used as a model; in particular in the analysis of optical bistability (Flytsanis and Tang, 1980; Goldstone and Garmire, 1984).

The equation of motion for the normal coordinate q of the oscillator is of the form

$$\left.\begin{aligned} &\ddot{q}+2\Gamma\dot{q}+\omega_0^2 q+\gamma q^3 = h\cos\omega_h t + f(t) \quad (t>0)\\ &\langle f(t)f(t')\rangle = 2\mathscr{B}\delta(t-t'). \end{aligned}\right\} \tag{13.2.1}$$

Here ω_0 is the eigenfrequency of small-amplitude vibrations in the absence of friction and external forces, Γ is the friction coefficient, and γ is the nonlinearity parameter. We suppose the oscillator to be underdamped,

$$\Gamma \ll \omega_0, \tag{13.2.2}$$

and that the periodic force is resonant and not too strong, so that $|\omega_h-\omega_0|$, $\gamma\langle q^2\rangle \ll \omega_h$. (We assume hereafter that $\gamma>0$; the results may be generalized immediately also to the case $\gamma<0$.)

When these conditions are fulfilled it is convenient to transform from the fast oscillating variables $q, \dot{q}$ to the slowly varying (over the time $\sim\omega_h^{-1}$) dimensionless envelopes u_1, u_2,

$$\left.\begin{aligned} q &= (3\gamma/8\omega_h\Gamma)^{-1/2}(u_1\cos\omega_h t - u_2\sin\omega_h t),\\ \dot{q} &= -\omega_h(3\gamma/8\omega_h\Gamma)^{-1/2}(u_1\sin\omega_h t + u_2\cos\omega_h t). \end{aligned}\right\} \tag{13.2.3}$$

Using the ideas of the averaging method from nonlinear vibration theory (see Bogolyubov and Mitropolsky, 1961) and neglecting small fast oscillating corrections to u_1, u_2 one can write the equations for u_1, u_2 in the form

$$\left.\begin{aligned} &\dot{u}_n = v_n(\mathbf{c};\mathbf{u}) + f_n(t), \quad \langle f_n(t)f_m(t')\rangle = 2\alpha\Gamma\delta_{nm}\delta(t-t'),\\ &\alpha = 3\gamma\mathscr{B}/16\omega_h^3\Gamma^2. \end{aligned}\right\} \tag{13.2.4}$$

Here $\mathbf{u}\equiv(u_1,u_2), \mathbf{c}\equiv(c_1,c_2)$. The functions v_1, v_2 are cubic polynomials in u_1, u_2 (the explicit expression for $v(u_1+\mathrm{i}u_2, u_1-\mathrm{i}u_2) = (v_1+\mathrm{i}v_2)/\Gamma$ is given by Dykman and Krivoglaz, 1979). The quantities c_1 and c_2 determine the characteristic dimensionless parameters of the dynamical system,

$$c_1 = \Gamma/(\omega_h-\omega_0), \;\; c_2 = 3\gamma h^2/32\omega_h^3|\omega_h-\omega_0|^3. \tag{13.2.5}$$

The dimensionless parameter α, characterizing noise intensity in (13.2.4), is supposed to be small.

The bistability of the nonlinear oscillator can arise (in the absence of a random force) due to the dependence of the effective vibration frequency $\omega_{\mathrm{eff}} \simeq \omega_0 + \frac{3}{8}\gamma A^2/\omega_0$ on the vibration amplitude A. As a consequence of this dependence, forced vibrations of both large amplitude (for which ω_{eff} is close to ω_h and the resonance condition is fulfilled very well) and also of small

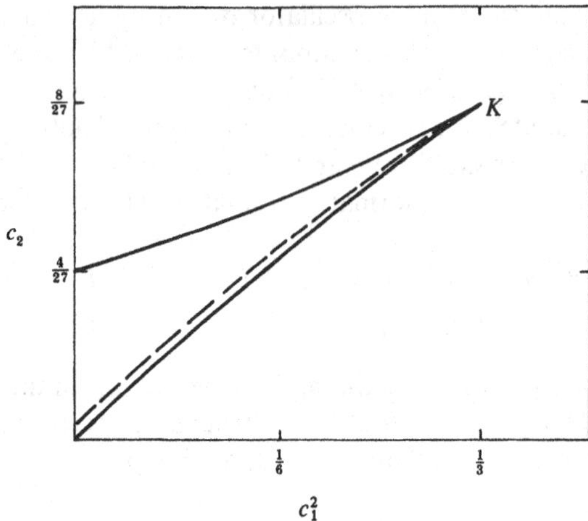

Figure 13.2. Range of coexistence of two stable states of a nonlinear oscillator (solid lines are the loci of the bifurcation points, K is the spinode). The dashed line corresponds to the parameter values at which the probabilities of the transitions $1 \to 2$ and $2 \to 1$ are equal and thus the kinetic phase transition occurs (Dykman and Krivoglaz, 1979).

amplitude (with considerably larger $|\omega_h - \omega_{\mathrm{eff}}|$) may turn out to be stable and 'self-consistent' provided that $\gamma(\omega_h - \omega_0) > 0$. The region of the parameters c_1^2, c_2 where the bistability occurs is bounded by the full curves in Figure 13.2.

Equations of the type (13.2.4) also describe the dynamics of more complicated (than an oscillator) systems. In this case the number of the variables u_n may exceed two, while the intensities of the random forces may differ for different n.

The phase portrait of an oscillator in the variables u_1 and u_2 for the parameter range where the bistability occurs is shown schematically in Figure 13.3. The dashed line in the figure is the separatrix between the attraction regions of the foci f_1 and f_2. The saddle point, s, lies on this line. Such a phase portrait is typical for a wide class of bistable systems. In the absence of a random force, the system, located at the initial instant at some general position point, will approach, over a characteristic relaxation time $t_{\mathrm{r}}(t_{\mathrm{r}} \sim \Gamma^{-1})$, that focus (or node) in whose attraction region it was located initially. (Examples of phase trajectories are shown in Figure 13.3.)

In the presence of a weak random force the system moves, with overwhelming probability, practically along the same trajectory. On approaching the focus, the system stays near it for a long time, greatly exceeding t_{r}, undergoes small fluctuations. Ultimately it will experience a sufficiently large fluctuation, as a result of which the phase trajectory will cross the separatrix. After this the system will approach another focus over a time $\sim t_{\mathrm{r}}$ and then will fluctuate near it. This just means a transition to a new stable state.

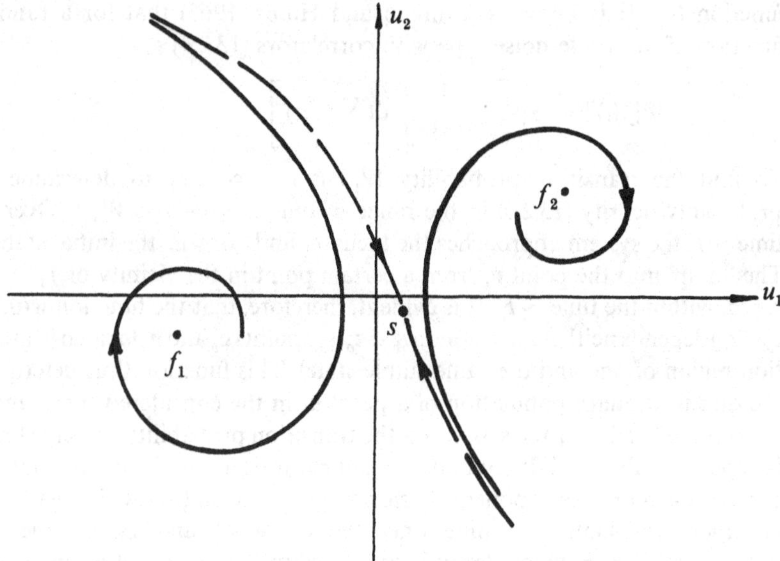

Figure 13.3. Schematic phase portrait of a nonlinear dynamical system with two stable states of the focus type. The dashed line is the separatrix, s is the saddle point, f_1 and f_2 are foci. Arrows indicate the direction of motion.

13.2.1 *Application of the path-integral method to the calculation of transition probabilities*

With the aim of calculating the probabilities W_{ij} of fluctuational transitions we shall consider the probability density $w(\mathbf{u}_b; \mathbf{u}_a; t_b - t_a)$ of the transition of the system from a certain point $\mathbf{u}_a$ (in the phase space), at which the system was located at instant t_a, to a point $\mathbf{u}_b$ at instant t_b. To logarithmic accuracy the value of, say, W_{12} is determined by the function w for $\mathbf{u}_a$ located near the focus f_1 and $\mathbf{u}_b$ located near the separatrix.

It is convenient to write $w(\mathbf{u}_b; \mathbf{u}_a; t_b - t_a)$ in the form of a path integral

$$w(\mathbf{u}_b; \mathbf{u}_a; t_b - t_a) = \int_{\mathbf{u}(t_a)=\mathbf{u}_a} \mathscr{D}\mathbf{f}(t)\mathscr{P}[\mathbf{f}(t)]\,\delta(\mathbf{u}(t_b) - \mathbf{u}_b)$$
$$\times \left\{ \int \mathscr{D}\mathbf{f}(t)\mathscr{P}[\mathbf{f}(t)] \right\}^{-1}, \quad \mathscr{D}\mathbf{f}(t) \equiv \prod_n \mathscr{D}f_n(t). \tag{13.2.6}$$

In the calculation of the path integral it is assumed that at $t = t_a$ the system is located at the point $\mathbf{u}_a$. As seen from (13.2.6), contributions to the path integral are made only by such realizations of the random force $\mathbf{f}(t)$ which transfer the system from $\mathbf{u}_a$ to $\mathbf{u}_b$ over the time $t_b - t_a$.

The functional $\mathscr{P}[\mathbf{f}]$ determines the probability distribution of the random function $\mathbf{f}(t)$. It is known (Feynman and Hibbs, 1965) that for a random function of the white-noise type with correlators (13.2.4)

$$\mathscr{P}[\mathbf{f}(t)] = \exp\left[-\frac{1}{4\alpha\Gamma}\int \mathrm{d}t \sum_n f_n^2(t)\right]. \tag{13.2.7}$$

To find the transition probability W_{12} it is necessary to determine the probability density (13.2.6) in the range of time $t_r \ll t_b - t_a \ll W_{12}^{-1}$. Over the time $\sim t_r$ the system approaches the focus f_1 and 'forgets' the initial state $\mathbf{u}_a$. The 'jump' into the point $\mathbf{u}_b$ from a certain point in the vicinity of f_1 is also rapid, within the time $\sim t_r$. It is evident, therefore, that the function $w(\mathbf{u}_b; \mathbf{u}_a; t_b - t_a)$ depends neither on $\mathbf{u}_a$ nor on $t_b - t_a$ (for points $\mathbf{u}_b$ and $\mathbf{u}_a$ located in attraction region of one and the same stable state). This function thus determines the quasistationary population of a point $\mathbf{u}_b$ in the considered time range.

For $\mathbf{u}_b$ located near the separatrix the transition probability density (13.2.6) is exponentially small. It contains a small parameter α in the denominator of the expression in the exponent. Performing the calculation with logarithmic accuracy, we shall determine only the exponent and ignore the pre-exponential factor, which depends on α weakly. To do this, it is convenient, following Feynman, to change in (13.2.6) from integration over the random force trajectories $\mathscr{D}\mathbf{f}(t)$ to integration over the trajectories of the system $\mathscr{D}\mathbf{u}(t)$,

$$\mathscr{P}[\mathbf{f}(t)]\mathscr{D}\mathbf{f}(t) = \tilde{\mathscr{P}}[\mathbf{u}(t)]\mathscr{D}\mathbf{u}(t). \tag{13.2.8}$$

The functional $\tilde{\mathscr{P}}[\mathbf{u}(t)]$ determines the probability distribution of the random function $\mathbf{u}(t)$. According to (13.2.4), (13.2.7) and (13.2.8)

$$\begin{gathered}\tilde{\mathscr{P}}[\mathbf{u}(t)] = \exp(-S/4\alpha\Gamma)J[\mathbf{u}(t)],\\ S = \int \mathrm{d}t\, \mathscr{L}(\mathbf{u}, \dot{\mathbf{u}}), \quad \mathscr{L}(\mathbf{u}, \dot{\mathbf{u}}) = \sum_n [\dot{u}_n - v_n(\mathbf{c}; \mathbf{u})]^2,\end{gathered} \tag{13.2.9}$$

where $J[\mathbf{u}]$ is the Jacobian for the transformation (13.2.8). It is obvious from (13.2.4) that $J[\mathbf{u}]$ is independent of α and influences only the pre-exponential factor in $w(\mathbf{u}_b; \mathbf{u}_a; t_b - t_a)$.

Within the adopted accuracy, it suffices to single out in the integral over $\mathscr{D}\mathbf{u}(t)$ the main exponential factor, which corresponds to the extremal path $\mathbf{u}(t)$. This path is evident from (13.2.6), (13.2.8) and (13.2.9) to be determined by the condition that the functional S be minimum. At instant t_b the extremal path passes through a point $\mathbf{u}_b$, while it starts at $\mathbf{u}_{f_1}$ (according to the physical picture described above). We note that the functional S can be regarded as the action of a certain auxiliary particle, and $\mathscr{L}$ is its Lagrangian.

Within the logarithmic accuracy the transition probability W_{12} is determined by the maximum value of $w(\mathbf{u}_b; \mathbf{u}_a; t_b - t_a)$ for the points $\mathbf{u}_b$ located on the separatrix (on approaching the separatrix, the system with the probability $\sim\frac{1}{2}$ will go to another stable state). The extremum with respect to $\mathbf{u}_b$ is reached in the saddle point, $\mathbf{u}_b = \mathbf{u}_s$, and the transition probability is given by

$$W = \text{const}\cdot\exp(-R/\alpha), \quad R = \frac{1}{4\Gamma}\min\int_0^t dt\,\mathscr{L}(\mathbf{u},\dot{\mathbf{u}}),$$

$$\mathbf{u}(0) \simeq \mathbf{u}_f, \quad \mathbf{u}(t) \simeq \mathbf{u}_s \tag{13.2.10}$$

(Dykman and Krivoglaz, 1979). The quantity R in (13.2.10) is calculated under the additional condition

$$\mathscr{E}(\mathbf{u},\dot{\mathbf{u}}) = 0, \quad \mathscr{E} = -4\Gamma\frac{\partial R}{\partial t} = \sum_n [\dot{u}_n^2 - v_n^2(\mathbf{c};\mathbf{u})]. \tag{13.2.11}$$

The equality $\mathscr{E} = 0$ is obviously the condition of R be extremal with respect to the duration of the motion along the path $\mathbf{u}(t)$. The criterion of the applicability of (13.2.10) is the inequality

$$R \gg \alpha. \tag{13.2.12}$$

In the case of thermal fluctuations in a system (which are due to coupling to a thermostat), $\alpha \propto T$ and the quantity RT/α is independent of T and equals the activation energy of the transition.

It is obvious from (13.2.9)–(13.2.11) that the suggested method reduces the calculation of the probabilities of transitions between stable states for a nonlinear oscillator driven by a resonant field, or for a system of a more general type, to the solution of the variational problem of finding a minimum of the action S (13.2.9) of an auxiliary particle which moves from the point $\mathbf{u}_f$ to $\mathbf{u}_s$. The function $\mathscr{E}(\mathbf{u},\dot{\mathbf{u}})$ (13.2.11), is the energy of this particle (Landau and Lifshitz, 1976). The auxiliary particle has twice as many degrees of freedom as the initial dynamical system (its coordinates are u_n and the momenta are $\dot{u}_n - v_n$).

13.2.2 Fluctuational transitions between stable states in concrete systems

The explicit expression for R may be obtained easily in the case when the functions v_n in (13.2.4) satisfy the 'potentiality condition' (see Haken, 1983), $v_n = -\partial U/\partial u_n$. In this case the equations of motion for an auxiliary particle are solved by the substitution $\dot{u}_n = -v_n$, and as a result $R = \Gamma^{-1}[U(\mathbf{u}_s) - U(\mathbf{u}_f)]$.

Equations (13.2.10) and (13.2.11) permit us to find the transition probabilities in the nontrivial case of a system whose motion is nonpotential. An example of such a system is a Duffing oscillator in a resonant field. The values of the effective 'activation energy' R may be obtained here in the explicit form in a number of limiting cases (Dykman and Krivoglaz, 1979, 1980a). The dependences of R_1 and R_2 on the characteristic field intensity $c_2 \propto h^2$ at relatively large frequency detuning, $\omega_h - \omega_0 \gg \Gamma(c_1 \ll 1)$, are shown in Figure 13.4 (the states 1 and 2 correspond to the smaller and the larger

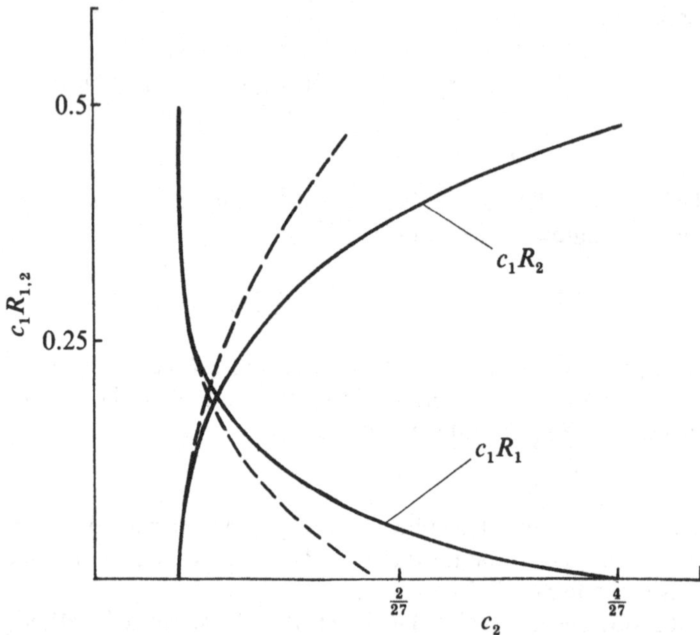

Figure 13.4. Dependences of the characteristic 'activation energies' R_1 and R_2 on c_2 for the transitions $1 \rightarrow 2$ and $2 \rightarrow 1$ between the stable states of a nonlinear oscillator in a strong resonant field at relatively large frequency detuning, $|c_1| \ll 1$ (Dykman and Krivoglaz, 1979).

amplitudes of the constrained oscillations). For small c_2 (but $c_2 \gg c_1^2$) these dependences are steep,

$$c_1 R_1 = \tfrac{1}{2} - \zeta' c_2^{1/4}, \quad c_1 R_2 = 2c_2^{1/2}, \quad c_1^2 \ll c_2 \ll 1; \zeta' \simeq 0.98. \quad (13.2.13)$$

The dashed curves in Figure 13.4 are plots of the expressions (13.2.13). Similar results were obtained recently for $c_1 \ll 1$ (Dmitriev and Dyakonov, 1986; Maslova, 1986) by another approach.

The arguments $R_{1,2}/\alpha$ of the exponentials in (13.2.10) for the transition probabilities at $c_1 \ll 1$ are seen from (13.2.5), (13.2.13) and from Figure 13.4 to depend on the external field amplitude h only through the parameter $c_2 \propto h^2|\omega_h - \omega_0|^{-3}$, and R_1 decreases monotonically with increasing h, while R_2 increases. The frequency detuning $\omega_h - \omega_0$ enters in both c_1 and c_2, and $R \propto |\omega_h - \omega_0|$ for fixed c_2. In addition, $R/\alpha \propto (\alpha\Gamma)^{-1}$. If friction and noise result from the coupling of an oscillator to a thermostat, then $\alpha \propto T\Gamma^{-1}$ and thus $R/\alpha \propto T^{-1}$, while Γ drops out from R/α.

The approach to the calculation of transition probabilities stated above may be generalized directly to the case when random forces acting on a dynamical system depend on dynamical variables (in a nonsingular way) or are Gaussian, but not δ-correlated (in the latter case the expressions of the type (13.2.9) and

(13.2.10) for the action S and R include retardation). The approach may be also generalized easily to systems described by higher-order equations.

We shall illustrate the latter by taking as an example a system performing Brownian motion in a static potential $U(q)$,

$$\ddot{q}+2\Gamma\dot{q}+\frac{\mathrm{d}U}{\mathrm{d}q}=f(t), \quad \langle f(t)f(t')\rangle = 2\mathscr{B}\delta(t-t'). \tag{13.2.14}$$

Writing the transition probability density $w(q_b, \dot{q}_b; q_a, \dot{q}_a; t_b - t_a)$ in the form (13.2.6) and changing from integration over $\mathscr{D}f(t)$ to integration over $\mathscr{D}q(t)$ similarly to (13.2.8), we obtain, that within the logarithmic accuracy the probability of the transition from an equilibrium position $q = q_f$ to a saddle point $q = q_s$ is

$$W = \text{const}\cdot\exp(-P/\mathscr{B}), \quad P = \tfrac{1}{4}\min\int_0^t \left(\ddot{q}+2\Gamma\dot{q}+\frac{\mathrm{d}U}{\mathrm{d}q}\right)^2 \mathrm{d}t,$$

$$q(0) = q_f, \quad q(t) = q_s, \quad \dot{q}(0) = \dot{q}(t) = 0. \tag{13.2.15}$$

The extremal trajectory of the functional P, which satisfies the boundary conditions and the condition $\partial P/\partial t = 0$, is described by the equation

$$\ddot{q}-2\Gamma\dot{q}+\frac{\mathrm{d}U}{\mathrm{d}q}=0 \tag{13.2.16}$$

(i.e. it corresponds to a motion with negative friction). Substituting (13.2.16) into (13.2.15) we obtain the well-known result

$$W = \text{const}\cdot\exp\left\{-\frac{2\Gamma}{\mathscr{B}}[U(q_s)-U(q_f)]\right\}. \tag{13.2.17}$$

13.2.3 Transition probabilities near bifurcation points

The additive-noise induced fluctuations in dynamical systems acquire peculiar features if the system is near a bifurcation point, i.e. if its parameters are close to values at which, e.g., new equilibrium positions or limit cycles appear or coalesce. These features are quite universal, they depend not on details of a system but on a type of bifurcation. The problem of fluctuations near bifurcation points was first considered for the bifurcation points corresponding to soft excitation of a limit cycle within a model of the Van der Pol oscillator (see, e.g., Lax, 1967, 1968; Risken, 1970; Rytov, 1955).

Besides the bifurcation points where the roots $\Lambda_{1,2}$ of the characteristic equation for a dynamical system pass through the imaginary axis and a limit cycle is excited, as in the case of the Van der Pol oscillator, there are also quite general bifurcation points of the marginal type, where $\Lambda_1 = 0$ ($\Lambda_i \neq 0$ at $i \neq 1$) and two singular points (e.g. a node and a saddle) mutually annihilate (or arise) in phase space. The value of the set of the system parameters $\mathbf{c} = (c_1, c_2, \ldots)$ corresponding to a marginal point will be denoted by $\mathbf{c}_M$.

In the range of $\mathbf{c}$ close to $\mathbf{c}_M$ the inequality $|\Lambda_1| \ll |\mathrm{Re}\,\Lambda_2|, |\mathrm{Re}\,\Lambda_3|, \ldots$ holds, i.e. one of the motions in the system (say that described by the variable u_1) is slow (u_1 is a 'soft mode'). This results in the strong increase of fluctuations, if the system is located in the corresponding region of phase space. The smallness of Λ_1 makes it possible to use an adiabatic approximation for the description of the fluctuations and to reduce the multidimensional problem, generally speaking, to a one-dimensional one if the random forces are small (Dykman and Krivoglaz, 1980a).

In the region of phase space adjacent to the singular points emerging at $\mathbf{c} = \mathbf{c}_M$ the 'fast' dynamical variables of the system $u_2, u_3, \ldots$ within a time $\sim t_c (t_c = \max(|\mathrm{Re}\,\Lambda_2|^{-1}, |\mathrm{Re}\,\Lambda_3|^{-1}, \ldots), |\Lambda_1| t_c \ll 1$ relax to their equilibrium values (for a given u_1) and then fluctuate about these values, following adiabatically the slow variation of u_1. The equation of motion for $u \equiv u_1$ near the marginal point is of the form

$$\dot{u} = -\frac{\mathrm{d}U_M}{\mathrm{d}u} + f(t), \quad U_M(u) = u(\tfrac{1}{3}bu^2 - \varepsilon),$$

$$\langle f(t) f(t') \rangle = 2\mathscr{B}\delta(t - t') \quad (u \equiv u_1, f(t) \equiv f_1(t)). \tag{13.2.18}$$

The parameter ε in (13.2.18) characterizes a distance to the bifurcation point in space of the parameters $\mathbf{c}$ (ε, b, $\mathscr{B}$ may be easily expressed in terms of the parameters of the initially multidimensional system (cf. Dykman and Krivoglaz, 1980a)). Only the main terms of the expansion of $U_M(u)$ in u are kept in (13.2.18). The adiabatic approximation used in the derivation of (13.2.18) is accurate to corrections $\sim (\mathscr{B}b^2)^{1/3} t_c \propto \mathscr{B}^{1/3}$ in the most favorable case, $\varepsilon = 0$.

At $\varepsilon b > 0$ in the region of small $|u|$ the system has a stable equilibrium point $u^{(0)}$ and a saddle point $u^{(s)}$,

$$u^{(0)} = (\varepsilon/b)^{1/2}\,\mathrm{sign}\,b, \quad u^{(s)} = -(\varepsilon/b)^{1/2}\,\mathrm{sign}\,b, \tag{13.2.19}$$

which correspond to a local minimum and maximum of the potential $U_M(u)$ (see Figure 13.5). At $\varepsilon = 0$ the points $u^{(0)}$ and $u^{(s)}$ merge, and at $\varepsilon b < 0$ the system has no stable states with small $|u|$. Obviously, the point $u^{(0)}$ corresponds to a metastable state at sufficiently small ε.

The one-dimensional Markov process $u(t)$ can be investigated by standard methods. In particular, when $(\varepsilon/b)^{1/2} |\varepsilon| \gg \mathscr{B}$ it is of interest to determine the probability W_M of the escape of the system from the metastable state (in this case W_M is small compared with the reciprocal time $\Lambda_1 \sim |b|(\varepsilon/b)^{1/2}$ characterizing the motion of the system near the equilibrium position). With allowance for physical picture of a motion, the problem of calculation of the escape probability W_M for the one-dimensional process $u(t)$ reduces in a natural way to the well-known (Kolmogorov and Leontovich, 1933) first passage time problem. This makes it possible to determine explicitly not only the exponent, but also the pre-exponential factor in the expression for W_M (Dykman and

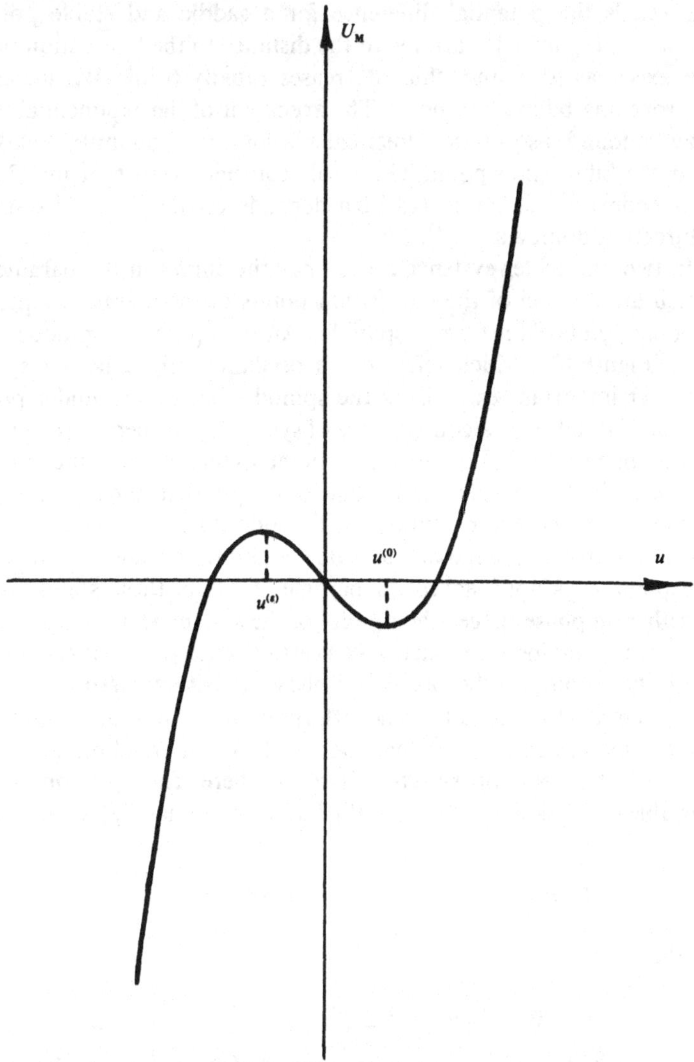

Figure 13.5. The potential for the 'slow' variable of a system near the marginal point.

Krivoglaz, 1980a):

$$W_M = \frac{1}{\pi}(\varepsilon b)^{1/2} \exp(-P_M/\mathscr{B}),$$

$$P_M = \tfrac{4}{3}(\varepsilon/b)^{1/2}|\varepsilon| \qquad (P_M \gg \mathscr{B}). \tag{13.2.20}$$

According to (13.2.20), the escape probability W_M near a marginal point depends on the noise intensity $\mathscr{B}$ in an activation way. The 'activation energy'

P_M equals the potential difference for a saddle and stable points, $P_M = U_M(u^{(s)}) - U_M(u^{(0)})$. It varies with the distance to the bifurcation point along the axis ε as $|\varepsilon|^{3/2}$ and, thus, decreases rapidly (while W_M increases) with approaching bifurcation point. The argument of the exponential in (13.2.20) may be found also by the direct calculation of the quantity $R/4\alpha\Gamma$ (13.2.10) near the bifurcation point. The result coincides with that in (13.2.20). The pre-exponential factor in (13.2.20) depends on the small 'distance' to the bifurcation point ε as $|\varepsilon|^{1/2}$.

In two-parameter systems, $\mathbf{c} = (c_1, c_2)$, the curves in the parameter plane, which are the loci of the bifurcation points of the marginal type, can have singular points of the form of spinodes. An example of the spinode is the point K in Figure 13.2 which refers to an oscillator driven by a resonant field. Another important example is the spinode that occurs under polarization optical bistability as a consequence of symmetry properties (Dykman, 1986).

The shape of the bifurcation curve in the vicinity of the spinode K ($c_1 = c_{1K}$, $c_2 = c_{2K}$) in the general case coincides with that shown in Figure 13.2. In the parameter region bounded by the solid curves in Figure 13.2 the system has two stable states and one unstable equilibrium state – a saddle point. As the point K is approached (in parameter space) these states come closer together (in phase space), the rigidity of the system decreases, and as a result the fluctuations increase sharply. In a certain sense, the point K is analogous to the critical point on the gas–liquid phase-transition curve.

To describe the fluctuations near the point K one may also use the adiabatic approximation and thus reduce the problem to a one-dimensional one. The error due to this approximation is $\propto \mathscr{B}^{1/4}$ here. The equation for the 'slow' variable $u \equiv u_1$ is similar to (13.2.18), but the potential $U_K(u)$ is the polynomial of fourth degree,

$$U_K(u) = u(\tfrac{1}{4}du^3 - \tfrac{1}{2}\varepsilon_1 u - \varepsilon_2) \quad (d > 0). \tag{13.2.21}$$

In the range

$$\varepsilon_1 > 0, \quad 3|\varepsilon_2|(3d)^{1/2}/2\varepsilon_1^{3/2} \leqslant 1 \tag{13.2.22}$$

the potential $U_K(u)$ has two minima separated by a region in which $U_K(u)$ has a local maximum. The extrema of $U_K(u)$ correspond to stable states and a saddle point. They coalesce at $\varepsilon_1 = \varepsilon_2 = 0$.

Not too close to the point K the probabilities W_{ij} of transitions between the stable states are small compared with the reciprocal relaxation times. To logarithmic accuracy they are given by

$$W_{ij} = \text{const}\cdot\exp\{-\mathscr{B}^{-1}[U_K(u^{(s)}) - U_K(u^{(0i)})]\}, \quad i, j = 1, 2, \tag{13.2.23}$$

where $u^{(0i)}$ is the position of the ith minimum of $U_K(u)$.

As the point K is approached the quantities W_{ij} increase sharply. In the immediate vicinity of the point K the concept of the transition probability becomes meaningless, since all relaxation and fluctuation processes have the same time scale (see Section 13.3.3).

The results for the transition probabilities near the marginal and spinode points (the bifurcation points of codimension 1 and 2, cf. Arnold, 1978) were applied to the problem of a nonlinear oscillator in a resonant field (Dykman and Krivoglaz, 1980a). The dependences of these probabilities (in particular, of the activation energies) on the oscillator parameters in the vicinity of the bifurcation curves shown in Figure 13.2 were obtained in the explicit form.

It should be noted that the probability of reaching the phase-space point far from points corresponding to stable states was investigated by another method in a mathematical paper by Ventsel and Freidlin (1970) for a certain type of Markov system. The approach presented above in Section 13.2.2 makes it possible to obtain simply the results of this investigation and to generalize them to include, in particular, the case of Markov processes of more general type, as well as the case of non-Markov processes. It allowed also to obtain the expression for the probabilities of transitions between stable states in the closed form and to calculate these probabilities for a concrete physical system, a Duffing oscillator in a strong resonant field.

To calculate transition probabilities and to find a stationary distribution of a system whose motion presents Markov process the Einstein–Fokker–Planck equation may be used as well. This method is particularly suited to the case of potential motion. It was shown recently that both in this case and in the case of nonpotential motion, the determination of a stationary distribution of a bistable system and the calculation of transition probabilities to logarithmic accuracy may be reduced for small noise intensities to the solution of a nonlinear partial differential equation (Ben-Jacob, Bergman, Matkowsky and Schuss, 1982; Graham and Shenzle, 1981). This equation is the Hamilton–Jacobi equation for the action S, (13.2.9), of an auxiliary particle (see Section 13.2.2). A similar equation arises also when transition probabilities are calculated by solving the first passage time problem for Markov process (Matkowsky and Schuss, 1983; Schuss, 1980; Shenoy and Agarwal, 1984; Talkner and Hänggi, 1984). The latter method permits to find also the pre-exponential factor in the expression for W.

13.3 Spectral density of fluctuations in bistable systems at low noise intensities

The time correlation functions $\langle L(t)M(t')\rangle$ ($\langle\ldots\rangle$ denotes the ensemble averaging) of the dynamical variables L, M and their spectral distributions are analyzed for bistable systems hereafter. These quantities not only characterize quite completely the relaxation and fluctuations, but describe also the generalized susceptibilities (cf. Landau and Lifshitz, 1980) and, thus, may be determined directly by experiment. We note however that when dissipation and a random force acting on a system are not connected (or not only connected) with coupling of a system to a thermostat or when a system is driven by a periodic field, the fluctuation–dissipation relations become more complicated and lose the universality. For Markov systems these relations are considered in detail by Risken (1984).

In the absence of external periodic fields the correlators $\langle L(t)M(t')\rangle$ under stationary conditions depend on $t-t'$ only. Supposing a system to be ergodic we can write the time correlation function $Q_{LM}(t)$ of variables L, M in the form

$$Q_{LM}(t) = \lim_{T\to\infty} (2T)^{-1} \int_{-T}^{T} \mathrm{d}\tau [L(t+\tau) - \langle L(t+\tau)\rangle] \times [M(\tau) - \langle M(\tau)\rangle]. \tag{13.3.1}$$

For systems subjected to a periodic force $h\cos\omega_h t$ the distribution function under stationary conditions depends periodically on time (such systems are, generally speaking, nonergodic) and therefore the correlators $\langle L(t+\tau)M(\tau)\rangle$ depend periodically on τ. However, for a number of applications it is of interest (cf. Dykman, 1978; Dykman and Krivoglaz, 1984) to find the addend in $\langle L(t+\tau)M(\tau)\rangle$ that does not depend on τ. This addend is given by (13.3.1) as well.

Besides $Q_{LM}(t)$ we shall consider also the spectral distributions

$$Q_{LM}(\omega) = \frac{1}{\pi}\int_0^\infty \mathrm{d}t\, \mathrm{e}^{\mathrm{i}\omega t - \varepsilon t} Q_{LM}(t) \quad (\varepsilon \to +0). \tag{13.3.2}$$

The quantity Re $Q_{LL}(\omega)$ is the spectral density of fluctuations of a dynamical variable L.

In Sections 13.3.1 and 13.3.2 we suppose that the intensity of noise acting on a dynamical system is small, so that the probabilities of transitions between stable states are much smaller than all reciprocal relaxation times (in the absence of noise). In this case within an overwhelming part of time a system fluctuates about equilibrium positions, and it is convenient to write $Q_{LM}(\omega)$ in the form

$$Q_{LM}(\omega) = \sum_i w_i Q_{LM}^{(i)}(\omega) + Q_{LM}^{(\mathrm{tr})}(\omega). \tag{13.3.3}$$

Here w_i is the stationary population of the ith stable state, and $Q_{LM}^{(i)}(\omega)$ is the partial spectrum formed by small fluctuations about the ith state. The term $Q_{LM}^{(\mathrm{tr})}(\omega)$ at small noise intensities is formed by transitions between the states (see Section 13.3.2).

The populations w_i are determined by the balance equation, and in the case when transition probabilities are given by (13.2.10), $W_{ij} \propto \exp(-R_i/\alpha)$, we obtain for a bistable system

$$\frac{w_1}{w_2} = \frac{W_{21}}{W_{12}} = \mathrm{const}\cdot\exp\left(\frac{R_1 - R_2}{\alpha}\right), \quad \alpha \ll R_{1,2}. \tag{13.3.4}$$

It is evident from (13.3.4) that for almost all values of the dynamical system parameters, excluding the narrow range where $R_1 \simeq R_2$, the population ratio w_1/w_2 is either exponentially small or large, and only one addend contributes to $\sum_i w_i Q_{LM}^{(i)}(\omega)$ in (13.3.3) (for this addend $w_i \simeq 1$).

13.3.1 Contribution from small fluctuations about stable states

To calculate $Q^{(j)}(\omega)$ in the limit of very small noise intensities (see below) one may linearize the equations of motion of a system in the vicinity of the jth stable equilibrium state. The respective linearized equations may be easily solved. For systems described by (13.2.4) we obtain with account taken of (13.3.2)

$$Q_{u_n u_m}^{(j)}(\omega) = \alpha \sum_p (\Gamma/\mathrm{Re}\,\Lambda_p^{(j)})(\Lambda_p^{(j)} + \mathrm{i}\omega)^{-1}(S_{pn}^{(j)})^* S_{pm}^{(j)}. \tag{13.3.5}$$

Here $\Lambda_p^{(j)}$ are the eigenvalues of the matrix $\|\partial v_n(\mathbf{c};\mathbf{u})/\partial u_m\|$ calculated for the jth stable state, $\hat{S}^{(j)}$ is the unitary transformation that diagonalizes this matrix.

The form of the function $Q^{(j)}(\omega)$ is seen from (13.3.5) to depend on the relation between the real and imaginary parts of the roots $\Lambda_p^{(j)}$ of the characteristic equation (note that for stable states $\mathrm{Re}\,\Lambda_p^{(j)} < 0$). At $|\mathrm{Re}\,\Lambda_p^{(j)}| \gg |\mathrm{Im}\,\Lambda_p^{(j)}|$ the distribution (13.3.5) is smooth, and the spectral densities of fluctuations $\mathrm{Re}\,Q_{u_n u_n}^{(j)}(\omega)$ are of the form of the Lorentzian peak with the maximum at low frequency (as compared with $|\mathrm{Re}\,\Lambda_p^{(j)}|$) and the halfwidth $|\mathrm{Re}\,\Lambda_p^{(j)}|$ (or of the superposition of such peaks, if the number of the dynamical variables u_n exceeds two).

If for some p the opposite condition is fulfilled,

$$|\mathrm{Re}\,\Lambda_p^{(j)}| \ll |\mathrm{Im}\,\Lambda_p^{(j)}|, \tag{13.3.6}$$

the spectral densities of fluctuations $\mathrm{Re}\,Q_{u_n u_n}^{(j)}(\omega)$ have distinct peaks at frequencies $\mathrm{Im}\,\Lambda_p^{(j)}$. The shape of these peaks is also Lorentzian, and their halfwidth is $|\mathrm{Re}\,\Lambda_p^{(j)}|$.

For a nonlinear oscillator performing constrained vibrations in a resonant field the values of $\Lambda_{1,2}^{(j)}$ may be obtained explicitly with account taken of (13.2.2)–(13.2.4), $\Lambda_{1,2}^{(j)} = -\Gamma[1 \pm \mathrm{i}(\lambda_j^2 - 1)^{1/2}]$, where λ_j are given in eqn. (25) of the paper by Dykman and Krivoglaz (1979). The 'weak damping' condition (13.3.6) is fulfilled for a relatively large detuning of the field frequency relative to the eigenfrequency of the oscillator, $|\omega_h - \omega_0| \gg \Gamma$ (then $\lambda_j^2 \gg 1$).

The spectral densities of fluctuations of the oscillator coordinates and momenta in the range of resonance, $\omega_h \simeq \omega$, are expressed according to (13.2.3) in terms of the functions $Q_{u_n u_m}(\omega - \omega_h)$. In terms of these functions one can express also the coefficient $\mu(\omega)$ of the absorption (amplification) of an additional weak resonant field at frequency ω by the oscillator. At small noise intensities the quantity $\mu(\omega)$ may be put into a form similar to that of (13.3.3). When condition (13.3.6) is satisfied, the partial contributions to $\mu(\omega)$ have distinct Lorentzian peaks with a halfwidth Γ (see Dykman and Krivoglaz,

1979, 1984 for details). The emergence of peaks of this type was shown by Bonifacio and Lugiato (1978) (see also Lugiato, 1984) when analyzing light transmission spectra for bistable optical systems.

If the weak damping condition (13.3.6) is satisfied, the range of noise intensities, where the expression (13.3.5) holds, appears to be much narrower than that determined by the inequality $\alpha \ll R$, (13.3.4). Indeed, we have obtained (13.3.5) allowing only for the terms linear in the displacements from the equilibrium positions. In this approximation the frequencies $|\mathrm{Im}\,\Lambda_p^{(j)}|$ of vibrations about equilibrium positions are independent of the vibration amplitudes. This does not, of course, hold true once the nonlinearity of the motion becomes significant. The dependence of the vibration frequency on the amplitude results in the frequency being modulated by fluctuations of the amplitude. This gives rise to the specific modulational broadening of the peaks of the spectral densities of fluctuations (Dykman and Krivoglaz, 1971).

It is evident that the approximation (13.3.5) describes the spectrum at the frequency $\simeq \mathrm{Im}\,\Lambda_p^{(j)}$ correctly when the characteristic frequency straggling due to fluctuations is small compared with the frequency uncertainty due to damping $|\mathrm{Re}\,\Lambda_p^{(j)}|$. In most cases for underdamped systems this condition reduces to the inequality

$$\frac{\alpha}{R_j} \ll \left|\frac{\mathrm{Re}\,\Lambda_p^{(j)}}{\mathrm{Im}\,\Lambda_p^{(j)}}\right| \ll 1 \qquad (13.3.7)$$

(see Section 13.4.1 for details). The restriction (13.3.7) is substantially stronger than (13.3.4). The effects arising when the left inequality in (13.3.7) does not hold are discussed below.

13.3.2 *Narrow spectral peaks caused by noise-induced transitions between stable states*

In contrast with the addends $Q_{LM}^{(i)}(\omega)$ in (13.3.3), which are formed by small fluctuations, the addend $Q_{LM}^{(\mathrm{tr})}(\omega)$ is formed by large fluctuations causing transitions between the states of a system. At small noise intensities, when $\alpha \ll R$, the probability of such fluctuations is small. The peaks of $Q_{LM}^{(i)}(\omega)$ and $Q_{LM}^{(\mathrm{tr})}(\omega)$ are formed within substantially different time intervals, i.e. the different ranges of integration over t in (13.3.2) contribute to $Q_{LM}^{(i)}(\omega)$ and $Q_{LM}^{(\mathrm{tr})}(\omega)$. The peaks of $Q_{LM}^{(i)}(\omega)$ are formed within a characteristic relaxation time $|\mathrm{Re}\,\Lambda_p^{(i)}|^{-1}$ (cf. (13.3.5)), while those of $Q_{LM}^{(\mathrm{tr})}(\omega)$ relate, in effect, to the times corresponding to the transition probabilities, $t \sim W_{12}^{-1} + W_{21}^{-1} \gg |\mathrm{Re}\,\Lambda_p^{(i)}|^{-1}$.

The values of the time correlation function $Q_{LM}(t)$ in the region $t \gg |\mathrm{Re}\,\Lambda_p^{(i)}|^{-1}$ and, consequently, the values of $Q_{LM}^{(\mathrm{tr})}(\omega)$ differ qualitatively for the values of the system parameters far from the range of the kinetic phase transitions and for those close to this range. It is easy to see that in the first case, when $|R_1 - R_2| \gg \alpha$, the function $Q_{LM}(t)$ is exponentially small for corresponding t. Indeed, in this case with a probability close to unity the system fluctuates about the certain stable states, while an average time passed in the vicinity of another stable state is exponentially small. The correlation of such fluctuations decays within a time $\sim |\mathrm{Re}\,\Lambda_p^{(i)}|^{-1}$ (with the proper i).

For the parameter region where $|R_1 - R_2| \lesssim \alpha$ and the populations w_1, w_2 are thus of the same order of magnitude the form of $Q_{LM}(t)$ is quite different. In this region a system located initially, e.g., in the state 1, within a time $t \sim W_{12}^{-1} \sim W_{21}^{-1}$ changes to state 2 with an appreciable probability. The transition results in the finite change of the values of dynamical variables L, M (from those corresponding to the state 1, $L \simeq L_1$, $M \simeq M_1$, to those corresponding to state 2, $L \simeq L_2$, $M \simeq M_2$). The probability of the opposite transition is of the same order of magnitude. Taking the transitions into account, we obtain

$$Q_{LM}(t) = w_1 w_2 (L_1 - L_2)(M_1 - M_2)\exp[-(W_{12} + W_{21})t],$$
$$t \gg |\mathrm{Re}\,\Lambda_p^{(1,2)}|^{-1}$$

and respectively,

$$Q_{LM}^{(\mathrm{tr})}(\omega) = \frac{1}{\pi} w_1 w_2 (W_{12} + W_{21} - \mathrm{i}\omega)^{-1} \times (L_1 - L_2)(M_1 - M_2). \tag{13.3.8}$$

In the more general case, when a system is subjected to a field $h \cos \omega_h t$ and the values of $L(t)$, $M(t)$ in the stable states depend periodically on time,

$$L_j(t) = \sum_n L_{jn} \exp(\mathrm{i}n\omega_h t),$$
$$M_j(t) = \sum_n M_{jn} \exp(\mathrm{i}n\omega_h t), \quad j = 1, 2$$

the function $Q_{LM}^{(\mathrm{tr})}(\omega)$ takes the form

$$Q_{LM}^{(\mathrm{tr})}(\omega) = \frac{1}{\pi} w_1 w_2 \sum_n [W_{12} + W_{21} - \mathrm{i}(\omega - n\omega_h)]^{-1} \times (L_{1n}^* - L_{2n}^*)(M_{1n} - M_{2n}). \tag{13.3.9}$$

It is obvious from (13.3.8) and (13.3.9) that in bistable systems the spectral density of fluctuations $\mathrm{Re}\, Q_{LL}^{(\mathrm{tr})}(\omega)$ has the extremely narrow Lorentzian peak at zero frequency and similar peaks at the external field frequency ω_h and multiple frequencies. The halfwidth of the peaks $W_{12} + W_{21}$ equals the sum of the transition probabilities. It is much smaller than the reciprocal relaxation time of a dynamical system. The intensity of the peak is proportional to

$$w_1 w_2 = W_{12} W_{21} (W_{12} + W_{21})^{-2} \sim \exp[-|R_1 - R_2|/\alpha].$$

It is exponentially small everywhere excluding the range of the smeared phase transition.

In the case of a nonlinear oscillator in a strong resonant field the most intensive peak of the spectral density of fluctuations of the coordinates (momenta) is located at the field frequency ω_h. The corresponding extremely narrow peak exists as well in the spectrum $\mu(\omega)$ of the absorption (amplification) of an additional weak field by an oscillator (Dykman and Krivoglaz, 1979, 1984). The intensity of this peak is also proportional to $\exp[-|R_1 - R_2|/\alpha]$. The emergence of such a specific peak in the vicinity of the first-order kinetic phase transition in a system far from thermal equilibrium may underlie a way of detecting the transition itself. The peak of $\mu(\omega)$ makes it possible also to compare with high accuracy the frequencies of the strong and weak fields. The frequency error is $\sim W$ here, and thus can be smaller by many orders of magnitude than not only the strong field frequency ω_h itself, but also the small friction coefficient Γ of the nonlinear oscillator.

The relation between the parameters of a nonlinear oscillator and strong resonant field that corresponds to the kinetic phase transition may be obtained by solving the variational problem (13.2.10), (13.2.11) and (13.2.4) for R_1, R_2. The results of Sections 13.2.2 and 13.2.3 (see also Dykman and Krivoglaz, 1979) give this relation in the explicit form in the region of a comparatively large frequency detuning, $|\omega_h - \omega_0| \gg \Gamma$, and also in the parameter region near the spinode K in Figure 13.2. For the dimensionless parameters c_1 and c_2, (13.2.5), it takes the form: $c_2 \simeq 0.013$ at $c_1 \to 0$, and $c_2 = c_{2K}[1 - \frac{3}{2}\sqrt{3}(c_{1K} - c_1)]$ at $c_1 \simeq c_{1K} = \sqrt{3}$, $c_2 \simeq c_{2K} = 8/27$. Allowing for these results and for the monotonic character of the 'phase-transition curve' $c_2(c_1^2)$, which is obvious from qualitative arguments, we have interpolated this curve by that shown dashed in Figure 13.2.

13.3.3 *Fluctuations in close vicinity to the spinode point on the bifurcation curve*

The spectral density of fluctuations has the characteristic narrow peak in the case when the parameters of the system lie in the immediate vicinity of the spinode point K on the bifurcation curve (see Figure 13.2). Since the two stable states and the saddle point are very close to one to another in this parameter range (see Section 13.2.3), the system is 'soft' and the relaxation times are large (the so-called 'critical slowing down' occurs). At point K itself the damping is substantially nonexponential in the absence of fluctuations (the dynamics is obvious from (13.2.18) and (13.2.21) to be described by the equation $\dot{u} = -du^3$). Respectively, for the parameter values close enough to the point K even a weak random force causes such large fluctuations, that the probabilities of transitions between the stable states appear to be of the same order of magnitude as the reciprocal relaxation times, and thus the concept of a transition probability becomes meaningless. The fluctuations may be characterized here with the aid of the time correlation function $Q_{uu}(t)$ of the slow variable u and its spectral distribution $Q_{uu}(\omega)$.

The evolution of the correlator $Q_{uu}(t)$ is determined by the set of the damping decrements $\lambda_n \mathscr{B}^{1/2}$,

$$Q_{uu}(t) = \sum_n a_n \exp(-\lambda_n \mathscr{B}^{1/2} t) \quad (t > 0, \lambda_n > 0). \tag{13.3.10}$$

The values of λ_n are independent of the random force intensity $\mathscr{B}$, and thus the 'rate' of damping is proportional to the small quantity $\mathscr{B}^{1/2}$. The behavior of $Q_{uu}(t)$ at large times is determined by the lowest nonzero decrement $\lambda_1 \mathscr{B}^{1/2}$,

$$Q_{uu}(t) \simeq a_1 \exp(-\lambda_1 \mathscr{B}^{1/2} t), \quad \lambda_1 \mathscr{B}^{1/2} t \gg 1. \tag{13.3.11}$$

The quantities λ_n may be obtained by reducing the Einstein–Fokker–Planck equation for the random process $u(t)$ in (13.2.18) and (13.2.21) to the eigenvalue problem (see, e.g., Lax, 1967; Tomita, Ito and Kidachi, 1976; Van Kampen, 1965). The dependence of λ_1 on the parameters of the system obtained in this manner is shown in Figure 13.6 (Dykman and Krivoglaz, 1980a, 1984). The parameters g_1 and g_2 in Figure 13.6 are proportional to the parameters ε_1 and ε_2 in (13.2.21), which characterize the distance to the spinode point,

$$g_1 = \mathscr{B}^{-1/2} \varepsilon_1 d^{-1/2}, \quad g_2 = \mathscr{B}^{-1/4} \varepsilon_2 d^{-1/4}. \tag{13.3.12}$$

In particular, at the spinode point itself ($\varepsilon_1 = \varepsilon_2 = 0$) we have $\lambda_1 \simeq 1.37 d^{1/2}$. It is seen from Figure 13.6 that the dependence of λ_1 on ε_1 is, generally speaking, nonmonotonous and has rather sharp maximum for sufficiently large $|\varepsilon_2|$. The asymptotic expressions for λ_1 may be obtained in the explicit form (Dykman and Krivoglaz, 1984).

Some other questions concerning fluctuations near bifurcation points were considered by Mangel (1979). The fluctuation phenomena arising in the case of a multiplicative noise were analyzed in detail by Horsthemke and Lefever (1984).

13.4 Spectral density of fluctuations in underdamped systems

It follows from the results given above that the form of the spectral density of fluctuations in a bistable system changes extremely strongly with increasing intensity of a random force acting on a system. The features of the spectra caused by the bistability are manifested most distinctly when the characteristic frequencies of a system (in particular, the frequencies of vibrations about stable states) exceed substantially the characteristic reciprocal relaxation times, i.e. when dissipation is small. In this case the system motion is quasiconservative, there is a certain quantity of the type of energy (or action) that varies in time relatively slowly (the adiabatic invariant). Just this case is investigated in the following.

Since the considered features of the spectral density of fluctuations are quite general we shall analyze them within the simple model of a one-dimensional

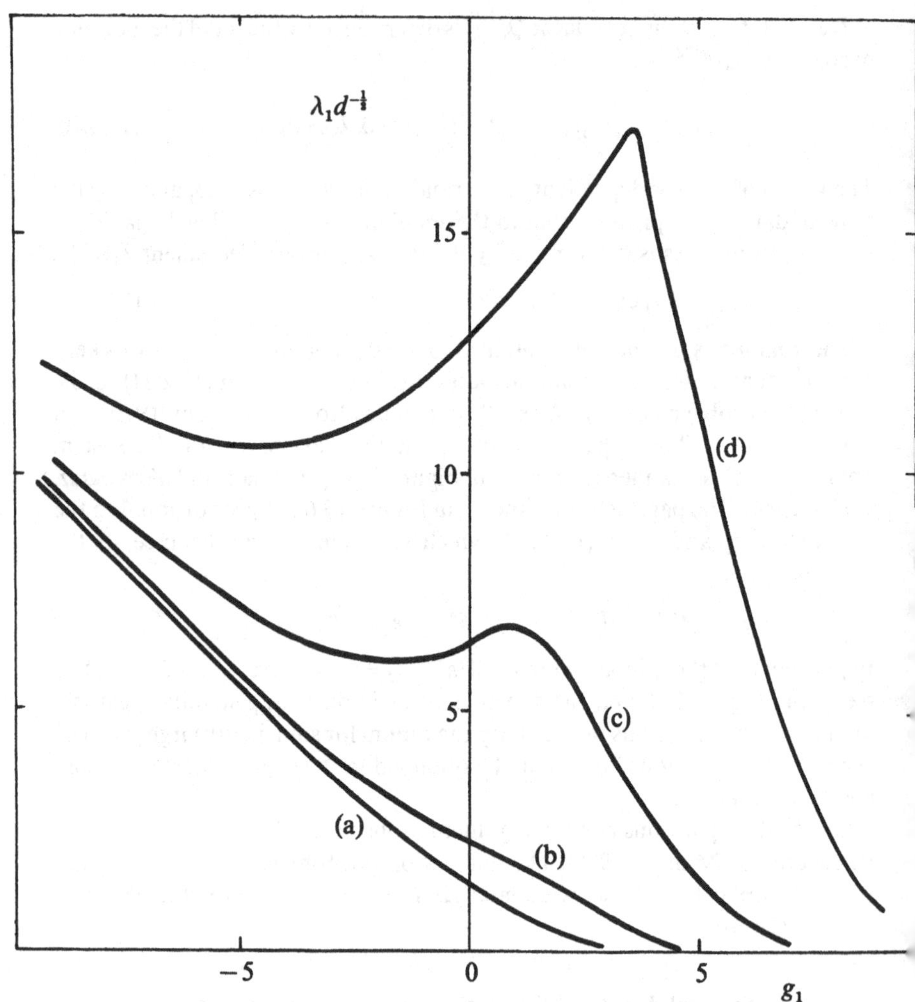

Figure 13.6. Dependence of the parameter λ_1, which determines the lowest nonzero decrement $\lambda_1 \mathscr{B}^{1/2}$ in the immediate vicinity of the spinode point K, on g_1. Curves (a)–(d) correspond to the values $g_2 = 0$, 2, 5, and 10 (Dykman and Krivoglaz, 1980a, 1984).

oscillator performing Brownian motion in a static potential $U(q)$ which has two minima and a local maximum between them (see Figure 13.7). The known example of such a potential is the potential of the double-well Duffing oscillator,

$$U_D(q) = -\tfrac{1}{2}\kappa^2 q^2 + \tfrac{1}{4}\gamma q^4 \quad (\gamma > 0). \tag{13.4.1}$$

This potential is symmetric, its wells have equal depths and curvatures.

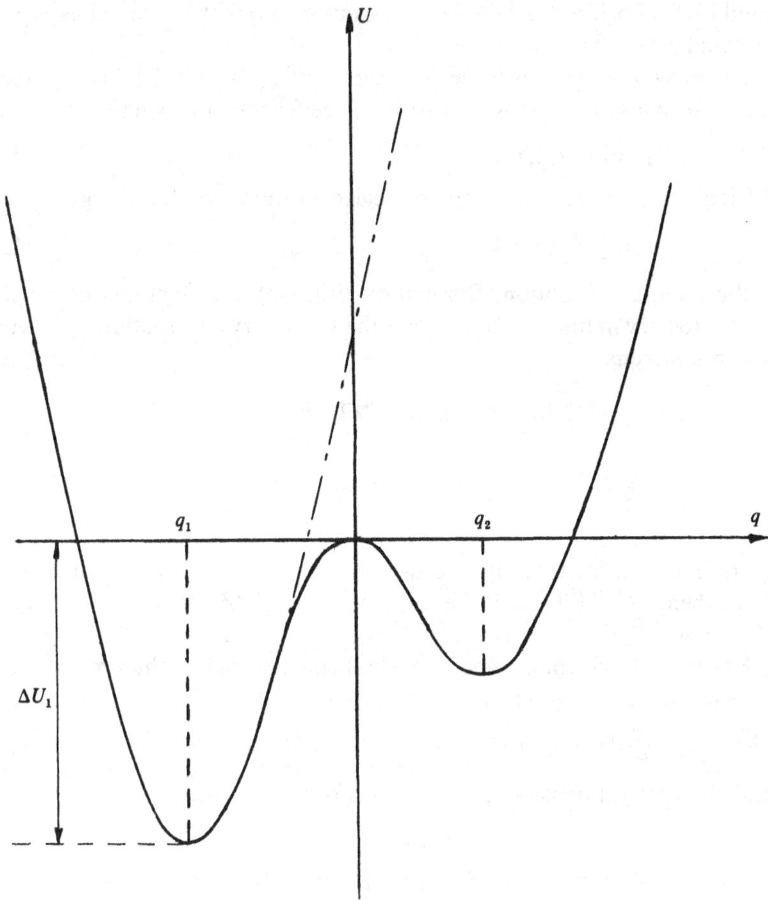

Figure 13.7. The double-well potential. For an underdamped bistable oscillator at low noise intensities the peak in the spectrum of fluctuations due to small-amplitude vibrations about q_1 coincides in shape with that for an oscillator moving in the single-well potential, whose part to the right of q_1 is shown as a dot–dashed line.

In the general case the potential $U(q)$ is asymmetric and parameters of the wells are different. It is essential that, as a rule, near minima, $q = q_i$ ($i = 1, 2$), and near the local maximum, $q = 0$, the potential $U(q)$ is parabolic,

$$U(q) \simeq U_i + \tfrac{1}{2}\omega_i^2(q - q_i)^2, \quad q \simeq q_i, \quad U_i \equiv U(q_i),$$
$$U(q) \simeq -\tfrac{1}{2}\kappa^2 q^2, \quad |q| \ll |q_i|. \tag{13.4.2}$$

Here ω_i are the eigenfrequencies of small-amplitude vibrations about the equilibrium positions. At the local maximum, $q = 0$, the potential is supposed to equal zero.

The Brownian motion of the oscillator with a potential $U(q)$ is described by (13.2.14). We shall suppose the friction coefficient to be small,

$$\Gamma \ll \omega_1, \omega_2, \kappa. \tag{13.4.3}$$

If friction and a random force are neglected the oscillator energy

$$E = \tfrac{1}{2}p^2 + U(q) \tag{13.4.4}$$

is the integral of motion. Dissipation (friction) and fluctuations cause the energy to vary in time, and as a result the stationary distribution w_{st} is worked out in a system,

$$w_{st} = w_{st}(E) = Z^{-1} \exp(-2E\Gamma/\mathscr{B}), \qquad Z = \iint dq\, dp \exp(-2E\Gamma/\mathscr{B}). \tag{13.4.5}$$

If friction and fluctuations are due to coupling of a system to a thermostat, then $\mathscr{B} = 2\Gamma T$ and the distribution (13.4.5) is Gibbsian, $w_{st}(E) = Z^{-1}\exp(-E/T)$.

For the sake of concreteness we shall consider below the time correlation function of the coordinates

$$Q(t) \equiv Q_{qq}(t) = \langle [q(t) - \langle q \rangle][q(0) - \langle q \rangle] \rangle \tag{13.4.6}$$

and the spectral density of fluctuations of the coordinates

$$Q'(\omega) \equiv \mathrm{Re}\, Q_{qq}(\omega) = \frac{1}{\pi} \mathrm{Re} \int_0^\infty dt \exp(i\omega t) Q(t). \tag{13.4.7}$$

The general structure of the function $Q'(\omega)$ in the limit of low noise intensity $\mathscr{B}$ was considered in Section 13.3. It follows from the results of Section 13.3 that for the potential shown in Figure 13.7 at

$$F \gg 1, \quad F = 2\Delta U_1 \Gamma/\mathscr{B}, \quad \Delta U_1 = U(0) - U(q_1) = -U(q_1) \tag{13.4.8}$$

$Q'(\omega)$ has the distinct peak caused by vibrations about the equilibrium position q_1 which corresponds to the lowest minimum of $U(q)$. This peak is located at frequency $\simeq \omega_1$.

The shape of the peak in question within the range $|\omega - \omega_1| \ll \omega_1$ at $\exp(-F) \ll 1$ is determined by relatively small displacements $q - q_1$, $|q - q_1| \ll |q_1|$, and therefore is determined by the form of the potential $U(q)$ near the equilibrium position q_1. In fact, this shape coincides with that of the peak of the spectral density of fluctuations $\tilde{Q}(\Omega_1)$ ($\Omega_1 = \omega - \omega_1$) of the displacements $q - q_1$ of an oscillator moving in a single-well potential coinciding with $U(q)$ near the minimum (but differing from $U(q)$ at large

$|q - q_1|$, cf. the dot–dashed line in Figure 13.7),

$$Q'(\omega) \simeq w_1 \tilde{Q}(\Omega_1), \quad \Omega_1 = \omega - \omega_1, \quad |\Omega_1| \ll \omega_1,$$

$$w_1 = \frac{\pi \mathscr{B}}{\omega_1 \Gamma} w_{st}(U_1), \tag{13.4.9}$$

where

$$\tilde{Q}(\Omega_1) = \frac{1}{\pi} \operatorname{Re} \int_0^\infty dt \exp(i\omega t) \langle [q(t) - q_1][q(0) - q_1] \rangle_1,$$

$$\Omega_1 = \omega - \omega_1. \tag{13.4.10}$$

($\langle \cdots \rangle_1$ denotes statistical averaging for a single-well oscillator; in the 'degenerate' case of a double-well potential with symmetric wells the r.h.s. of (13.4.9) should be doubled.)

The spectrum of fluctuations of a substantially nonlinear oscillator with a single-well potential is of interest not only in connection with the problem of a bistable oscillator, but is of great interest also in itself.

13.4.1 *Features of the spectrum of fluctuations of a nonlinear oscillator with a single-well potential*

To analyze the spectral density of fluctuations $\tilde{Q}(\Omega_1)$ of an underdamped oscillator at not too low noise intensities the nonlinear in $q - q_1$ terms in the expansion of $U(q)$ should be taken into account,

$$U(q) \simeq U_1 + \tfrac{1}{2}\omega_1^2 (q - q_1)^2 + \tfrac{1}{3}\zeta_1 (q - q_1)^3 + \tfrac{1}{4}\gamma_1 (q - q_1)^4 + \dots. \tag{13.4.11}$$

These terms result, in particular, in the dependence of the effective frequency ω_{eff} of eigenvibrations on the vibration amplitude (or energy):

$$\omega_{eff} \simeq \omega_1 + \frac{3}{4\omega_1} \tilde{\gamma}_1 \overline{(q - q_1)^2} \simeq \omega_1 + \tfrac{3}{4}\tilde{\gamma}_1 \omega_1^{-3} (E - U_1)$$

$$(\tilde{\gamma}_1 = \gamma_1 - \tfrac{10}{9}\zeta_1 \omega_1^{-2}). \tag{13.4.12}$$

Here the bar denotes averaging over the period of vibrations.

Random-force-induced fluctuations of the oscillator energy are seen from (13.4.12) to result in fluctuations of the vibration frequency. The characteristic frequency spread $\delta\omega$ is determined by the width of the distribution in energy, $\delta\omega = |3\tilde{\gamma}_1 / 4\omega_1^3| \mathscr{B}/\Gamma$ according to (13.4.5).

The shape of $\tilde{Q}(\Omega_1)$ is determined by competition between the two spectrum-broadening mechanisms, the broadening due to friction and that due to frequency modulation by a random force. Thus $\tilde{Q}(\Omega_1)$ depends on the ratio of the modulational broadening $\delta\omega$ to the friction coefficient Γ, i.e. on the parameter

$$\rho = \tfrac{3}{16}\tilde{\gamma}_1 \mathscr{B}/\Gamma^2\omega_1^3. \tag{13.4.13}$$

The quantity ρ is of the order of the ratio of two small parameters, $F^{-1} = \mathscr{B}/2\Gamma\,\Delta U_1$ and Γ/ω_1. Therefore even when (13.4.8) is fulfilled $|\rho|$ may be more or less than unity.

At $|\rho| \ll 1$ the spectrum broadening is due mainly to friction and the function $\tilde{Q}(\Omega_1)$ is described by the Lorentzian distribution with the halfwidth Γ. The corrections caused by nonlinearity can be treated here by perturbation theory (Ivanov, Kvashnina and Krivoglaz, 1965; Krivoglaz and Pinkevich, 1970). The perturbation theory in oscillator nonlinearity applicable for $|\rho| \ll 1$ was developed by Sture, Nordholm and Zwanzig (1974) and by Rodriguez and Van Kampen (1976).

The analysis of the oscillator relaxation and of the spectrum of fluctuations is most complicated in the region $|\rho| \sim 1$, where both broadening mechanisms (damping and frequency modulation) make contributions of the same order of magnitude. The spectrum $\tilde{Q}(\Omega_1)$ in this region is substantially non-Lorentzian, and the time correlation function of the displacements from the equilibrium position $q - q_1$ decays substantially nonexponentially.

The displacement correlator and the spectrum of fluctuations were calculated (Dykman and Krivoglaz, 1971) by the special method based on the averaging method and some properties of the Gaussian random processes (we note that the process q(t) itself at $|\rho| \gtrsim 1$ is essentially non-Gaussian; see also Dykman and Krivoglaz, 1984). The time correlation functions were also obtained by another method based on the solution of the Einstein–Fokker–Planck equation for a nonlinear oscillator (Dykman and Krivoglaz, 1980b). The same solution in a somewhat different form was obtained recently by Renz (1985).

The resulting expression for $\tilde{Q}(\Omega_1)$ takes on the form of an integral of an elementary function,

$$\left.\begin{aligned}
&\tilde{Q}(\Omega_1) = \frac{1}{\pi} G_1 \operatorname{Re} \int_0^\infty dt \exp(i\Omega_1 t)\tilde{Q}^*(t), \quad G_1 = \mathscr{B}/4\Gamma\omega_1^2,\\
&\tilde{Q}(t) = \exp(\Gamma t)\psi^{-2}(t), \quad \psi(t) = \cosh at + \frac{\Gamma}{a}(1 - 2i\rho)\sinh at,\\
&a = \Gamma(1 - 4i\rho)^{1/2} \quad (\operatorname{Re} a > 0).
\end{aligned}\right\} \tag{13.4.14}$$

It is obvious from (13.4.14) that the shape of the spectrum $G_1^{-1}\tilde{Q}(\Omega_1)$ as a function of Ω_1/Γ is determined by the single parameter ρ. At $\rho \to 0$ (13.4.14) goes over into the Lorentzian distribution $(G_1/\pi)\Gamma/(\Gamma^2 + \Omega_1^2)$. At small $|\rho|$ the maximum of $\tilde{Q}(\Omega_1)$ shifts by $\simeq 4\rho\Gamma$, while the asymmetric part of $\tilde{Q}(\Omega_1)$ is of the order of ρ^3. As $|\rho|$ increases the deviation of $\tilde{Q}(\Omega_1)$ from a Lorentzian distribution becomes more and more pronounced. At $|\rho| \gg 1$ the shape of the peak of $\tilde{Q}(\Omega_1)$ is substantially asymmetric,

$$\left.\begin{aligned}
&\tilde{Q}(\Omega_1) \simeq \tilde{Q}_0(\Omega_1),\\
&\tilde{Q}_0(\Omega_1) = G_1 \frac{|\Omega_1|}{\bar{\Omega}_1^2} \exp\left(-\frac{\Omega_1}{\bar{\Omega}_1}\right)\theta\left(\frac{\Omega_1}{\bar{\Omega}_1}\right),\\
&\rho\Omega_1 \gg \Gamma|\rho|^{3/2};\\
&\tilde{Q}(\Omega_1) \sim \left|\frac{\omega_1}{\bar{\gamma}_1}\right|^{3/2} G_1^{-1/2}\Gamma^{1/2}, \quad |\Omega_1| \lesssim \Gamma|\rho|^{1/2};\\
&\bar{\Omega}_1 = 2\rho\Gamma, \quad |\rho| \gg 1.
\end{aligned}\right\} \qquad (13.4.15)$$

($\theta(x)$ is the step function.)

According to (13.4.15) $\tilde{Q}(\Omega_1)$ near the maximum is determined at $|\rho| \gg 1$ by only the mechanism of frequency modulation (it is evident in particular that $\bar{\Omega}_1 = \langle E - U_1 \rangle_1 \partial\omega_{\text{eff}}/\partial E$). The terms due to dissipation are, however, essential in the wings of $\tilde{Q}(\Omega_1)$.

More detailed analytic expressions for the spectral density of fluctuations of a single-well oscillator in the limiting cases, as well as the detailed numerical results, are given by Dykman and Krivoglaz (1971); cf. also the computations by Renz and Marchesoni (1985). The respective change of shape and shift of the maximum of the peak in the spectrum with varying ρ were observed for an underdamped single-well oscillator by Fronzoni, Grigolini, Mannella and Zambon (1985, 1986) in analog experiments and were ascribed to the modulational broadening mechanism.

13.4.2 *Spectral distribution of a bistable oscillator in the absence of dissipation*

If the damping Γ is much smaller than the characteristic modulational frequency straggling $\delta\omega$, the spectral density $Q'(\omega)$ near the maximum (or the maxima, but excluding the maximum at zero frequency, see below) is determined just by the modulational mechanism. In this case within a time $\sim(\delta\omega)^{-1}$ needed to form a peak of $Q'(\omega)$ the dissipation effects do not succeed is manifesting themselves and may be neglected. The corresponding 'dissipationless' approximation allows us to investigate the spectrum of both the single- and double-well oscillators over a wide range of frequencies ω and noise intensities $\mathscr{B}$ (note that this approximation neglects the terms $\sim \Gamma/\delta\omega$, while $\mathscr{B}/\Gamma$ is regarded as finite).

In the absence of dissipation and noise a bistable oscillator performs periodic (but, generally speaking, anharmonic) vibrations with a given energy E either within one of the wells ($j = 1, 2$) or over the barrier ($j = 0$):

$$q(t) = \sum_{n=-\infty}^{\infty} q_{nj}(E) \exp[in(\omega_j(E)t + \varphi_j)], \quad j = 0, 1, 2. \qquad (13.4.16)$$

Here $\omega_j(E)$ is the energy-dependent vibration frequency (in (13.4.2) $\omega_j \equiv \omega_j(U_j)$), $2|q_{nj}|$ is the amplitude of vibrations at the nth harmonic.

The determination of the coordinate correlator (13.4.6), neglecting dissipation, reduces to the averaging of the product $[q(t)-\langle q\rangle]\cdot[q(0)-\langle q\rangle]$, calculated by (13.4.16), over the phase and energy with the weight $\omega_j^{-1}(E)w_{st}(E)$ and to the subsequent summation over j. Performing then the Fourier-transform (13.4.7) we get the following expession for $Q'(\omega)=Q'_0(\omega)$ in this approximation:

$$Q'_0(\omega)=\sum_{n,j} Q^{(nj)}(\omega)+\mathscr{M}\delta(\omega),$$

$$Q^{(nj)}(\omega)=2\pi\omega^{-1}w_{st}(E_{nj})|q_{nj}(E_{nj})|^2|d\omega_j(E_{nj})/dE_{nj}|^{-1}, \qquad (13.4.17)$$

where the energies E_{nj} are determined by the equation

$$n\omega_j(E_{nj})=\omega, \qquad (13.4.18)$$

while

$$\mathscr{M}=2\pi\sum_j\int dE\, w_{st}(E)\omega_j^{-1}(E)q_{0j}^2(E)$$

$$-\left[2\pi\sum_j\int dE\, w_{st}(E)\omega_j^{-1}(E)q_{0j}(E)\right]^2. \qquad (13.4.19)$$

Equations (13.4.17)–(13.4.19) may be easily understood: in the absence of dissipation the contribution to the spectral density of fluctuations at a frequency ω is made only by those vibrations whose eigenfrequency $\omega_j(E)$ or its overtones are equal to ω. This contribution is proportional to the squared amplitude of the corresponding vibrations $|q_{nj}(E_{nj})|^2$, to the occupation factor $w_{st}(E_{nj})$ and to the spectral density of vibrations at ω/n, $|d\omega_j(E_{nj})/dE_{nj}|^{-1}$ (if it diverges, additional peaks of $Q'(\omega)$ appear: Soskin, 1987). The singular term $\mathscr{M}\delta(\omega)$ in (13.4.17) is connected with the terms q_{0j} in (13.4.16) that do not vary in time but depend on E and j. The term $\mathscr{M}\delta(\omega)$ is smeared when relaxation is taken into account (see below; it should be noted that if (13.4.18) has several solutions for given j, n, the summation over these solutions ought to be carried out in (13.4.17)).

At $F=2\Gamma\Delta U_1/\mathscr{B}\gg 1$ the oscillator performs mainly small-amplitude vibrations about the equilibrium position q_1, and the main contribution to (13.4.17) in the range $\omega\simeq\omega_1$ is made by the term $Q^{(11)}(\omega)$ which is due to the fundamental tone of these vibrations (if $|U_1-U_2|\lesssim\mathscr{B}/\Gamma$, and ω_1 and ω_2 are closely spaced, the term $Q^{(12)}(\omega)$ is also essential). For small $\omega-\bar{\omega}_1(U_1)$ in solving (13.4.18) $\omega_1(E_{11})=\omega$ it suffices to retain only the linear in $E-U_1$ term in the expansion of $\omega_1(E)$. To the lowest order in F^{-1}, the expression for $Q^{(11)}(\omega)$ goes over into that for $w_1\tilde{Q}_0(\Omega_1)$ (see (13.4.9), (13.4.15); apparently, the result of the dissipationless approximation (13.4.17) is valid near the maximum of $Q'(\omega)$ provided $|\rho|\gg 1$). However, in contrast to (13.4.14) and (13.4.15), the expressions (13.4.17)–(13.4.19) are not limited to the range of

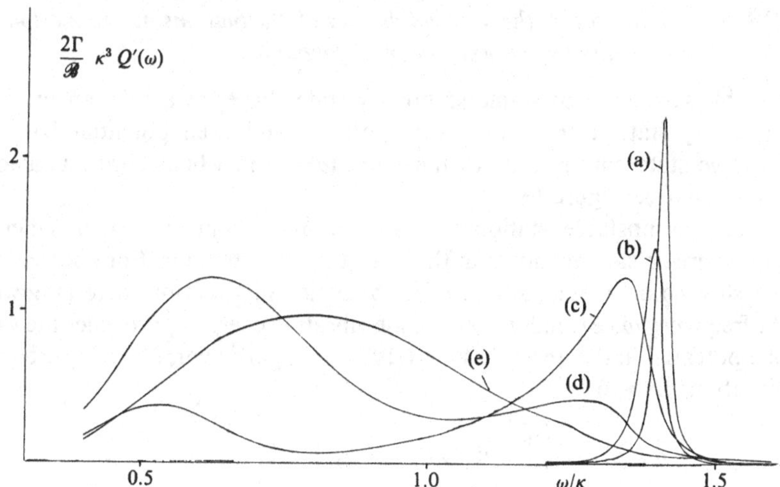

Figure 13.8. Spectral density of fluctuations of an underdamped Duffing oscillator performing Brownian motion in the double-well potential (13.4.1). The low-frequency parts of the spectra, which contain the zero-frequency peak, are not plotted. Curves (a)–(e) correspond to the values $F^{-1} = 2\gamma\mathscr{B}/\Gamma\kappa^4 = 0.02$, 0.06, 0.24, 0.6, 1.6; $\Gamma/\kappa = 0.01$. The ordinates of curves (a) and (b) are decreased by a factor of three.

small $F^{-1}, |\omega - \omega_1|/\omega_1$, and thus give the essential corrections to (13.4.14) and (13.4.15) when these parameters are not too small.

For many actual double-well potentials $\omega_1 \equiv \omega_1(U_1)$ is the highest eigenfrequency of the vibrations in well 1. Respectively, $Q^{(11)}(\omega)$ vanishes for $\omega \geqslant \omega_1$. The important contribution to $Q'(\omega)$ in this frequency range is due to the effects of dissipation. In particular, at $|\omega - \omega_1| \lesssim \Gamma|\rho|^{1/2}$, according to (13.4.15), $Q'(\omega)$ depends nonanalytically on the noise intensity $\mathscr{B}$ (and on the friction coefficient Γ for fixed $\mathscr{B}/\Gamma$), $Q'(\omega) \propto (\mathscr{B}/\Gamma)^{-1/2}\Gamma^{1/2}$. On the whole the fluctuation spectrum $Q'(\omega)$ near the frequency ω_1 at $F \gg 1$ may be shown to be described accurately up to terms of order Γ/ω_1 by the expression

$$Q'(\omega) = Q'_0(\omega) + w_1[\tilde{Q}(\Omega_1) - \tilde{Q}_0(\Omega_1)]\left(1 + \frac{\omega - \omega_1}{2\omega_1}\right)^{-2},$$

$$|\omega - \omega_1| \ll \omega_1. \tag{13.4.20}$$

In the case of the double-well Duffing oscillator the second addend in the r.h.s. of (13.4.20) should be doubled. For this case the evolution of the considered peak of $Q'(\omega)$ (it is located at $\omega \simeq \omega_1 = \kappa\sqrt{2}$) with increasing $\mathscr{B}/\Gamma\,\Delta U_1$ is seen from Figure 13.8 (cf. in particular curves (a)–(c)). Curve (a) is close to the Lorentzian ($\rho \simeq -0.25$ here). Curve (b) ($\rho \simeq -0.75$) has a considerably higher width and is slightly asymmetric. Curve (c) ($\rho \simeq -3$) is strongly asymmetric in the range of the considered peak.

13.4.3 *Features of the spectral density of fluctuations due to motion near the local maximum of a potential*

Bistable systems with stable stationary states have, as a rule, an unstable stationary state. For an oscillator with a double-well potential $U(q)$ the unstable state corresponds to the point $(q=0)$ in which $U(q)$ has a local maximum (see Figure 13.7).

Near an unstable stationary state an underdamped system suffers a characteristic slowing down in the absence of fluctuations. For an oscillator this slowing down manifests itself in the period of vibrations diverging while the frequency $\omega_j(E)$ tends to zero as the vibration energy approaches the value of a potential in the unstable state $U(0)=0$. In particular, for the overbarrier vibrations $(E>0)$

$$\omega_0(E)=\pi\left\{\int_{q_t^{(1)}}^{q_t^{(2)}} dq[2E-2U(q)]^{-1/2}\right\}^{-1}$$
$$\simeq \pi\kappa \ln^{-1}(C_0/E), \quad E\to 0, \qquad (13.4.21)$$

where q_t are the turning points, $U(q_t^{(1,2)})=E$, $C_0\sim\kappa^2 q_{1,2}^2$.

It is evident from (13.4.17), (13.4.18) and (13.4.21) that in the dissipationless approximation at $e^{-F}\ll 1$ the features of the oscillator motion near the unstable stationary state give rise to the specific quite universal form of $Q_0'(\omega)$ in the range $\omega\ll\pi\kappa,\ \omega_1$:

$$Q_0'(\omega)=w_{st}(0)C_0'\exp\left[-\frac{\pi\kappa}{\omega}-\frac{2\Gamma C_0}{\mathscr{B}}\exp\left(-\frac{\pi\kappa}{\omega}\right)\right],$$
$$\exp\left(-\frac{\pi\kappa}{\omega}\right)\ll 1, \qquad (13.4.22)$$
$$C_0'\equiv C_0'(\omega)\sim\frac{C_0^2}{\omega\kappa^3},\quad w_{st}(0)\propto\exp\left(-\frac{2\Gamma\,\Delta U_1}{\mathscr{B}}\right)\ll 1$$

(the detailed derivation of (13.4.22) and the explicit form of C_0, C_0' for the Duffing oscillator are given by Dykman, Soskin and Krivoglaz, 1984).

The intensity of the spectrum (13.4.22) is proportional to the population of the oscillator states near the top of the barrier and thus $\propto\exp(-F)$. The shape of the spectrum is determined by the competition between two factors, the reciprocal density of vibration frequencies $|d\omega_0(E)/dE|^{-1}$ and the state occupation factor $w_{st}(E)$ (see (13.4.17)). With rising frequency $\omega_0(E)=\omega$ the former increases exponentially (see (13.4.22)), while the latter decreases extremely sharply (cf. (13.4.5); $E\propto\exp[-\pi\kappa/\omega_0(E)]$).

As a result of this competition a quite narrow peak of $Q_0'(\omega)$ is formed. The position of its maximum is given by

$$\omega_m=\pi\kappa[\ln(2\Gamma C_0/\mathscr{B})]^{-1},\quad \ln(2\Gamma C_0/\mathscr{B})\gg 1. \qquad (13.4.23)$$

Such a peak for the double-well Duffing oscillator is seen in Figure 13.8(c) (the dissipation was neglected when calculating the spectral curves (c)–(e) in the range $\omega \leqslant \kappa$).

As the noise intensity $\mathscr{B}$ increases the intensity of the peak of $Q'(\omega)$ due to the overbarrier motion increases exponentially (see (13.4.22)). The width of the peak increases rapidly also. For sufficiently high $\mathscr{B}/\Gamma\Delta U_1$ this peak practically overlaps the peaks of $Q'(\omega)$ due to vibrations about the minima of $U(q)$ and a common peak is formed in $Q'(\omega)$ (see Figure 13.8).

The spectral density in the dissipationless approximation $Q'_0(\omega)$ for the Duffing oscillator was calculated by Onodera (1970). In Onodera's paper the term $\mathscr{M}\delta(\omega)$ in (13.4.17), which describes the peak of $Q'_0(\omega)$ at zero frequency was omitted; see Dykman, Soskin and Krivoglaz (1984). In the latter paper it was shown also, in particular, that the slowing down of the motion near the local maximum of a potential means that the combined influence of a weak random force, together with friction, is able to substantially modify the character of motion; the corresponding additions to $Q'(\omega)$ in the low-frequency range are proportional to $\Gamma^{1/2}(\mathscr{B}/\Gamma)^{1/2}w_{st}(0)$. The numerical calculation of $Q'(\omega)$ for the Duffing oscillator for several values of $\mathscr{B}/\Gamma\Delta U_1$ and Γ/κ was carried out by Voigtlaender and Risken (1985).

13.4.4 *The range of very small frequencies*

As mentioned in Section 13.3.2, one of the features of bistable systems driven by low-intensity noise is the extremely narrow peak of $Q'(\omega)$ at zero frequency. This peak is due to fluctuational transitions between stable states. For an oscillator moving in a double-well potential it has noticeable intensity when $|U(q_1)-U(q_2)|\Gamma/\mathscr{B} \lesssim 1$ and its width $\sim \Gamma\exp(-F)$ (see (13.3.8) and (13.4.8)).

The shape of the zero-frequency peak of the spectral density of fluctuations is determined, however, not only by fluctuational transitions, i.e. by large fluctuations, but also by fluctuations with other scales. In particular in the wing of the peak, $\omega \sim \Gamma \gg \Gamma\exp(-F)$, where the contribution to $Q'(\omega)$ made by fluctuational transitions is exponentially small, the spectrum is formed by the small fluctuations about the equilibrium positions, namely by the comparatively slow (with a characteristic time scale $\sim \Gamma^{-1} \gg \omega_j^{-1}$) fluctuations of the oscillator coordinate averaged over the vibration period, $q_{0j}(E)$ (see (13.4.16)). Such fluctuations are connected with the energy fluctuations. This mechanism yields the following term in $Q'(\omega)$:

$$\left.\begin{aligned} &\delta Q'(\omega) = \frac{1}{\pi} D \frac{2\Gamma}{\omega^2 + 4\Gamma^2}, \\ &D = \left(\frac{\mathscr{B}}{2\Gamma}\right)^2 \sum_{j=1,2} w_j[(dq_{0j}/dE)_{E=U_j}]^2, \\ &\omega_{1,2} \gg \omega \gg \Gamma\exp(-F), \quad \exp(-F) \ll 1,\ F = 2\Gamma\,\Delta U_1/\mathscr{B}. \end{aligned}\right\} \qquad (13.4.24)$$

The spectral density (13.4.24) is of the form of the Lorentzian peak with a halfwidth 2Γ. This peak is much less intensive than that at $\omega \simeq \omega_1$, the ratio of their intensities $D/G \sim F^{-1} \ll 1$ (see (13.4.24), (13.4.14)). In essence the emergence of the peak of the type (13.4.24) was shown by Krivoglaz and Pinkevich (1966) in considering the single-well quantum oscillator with an asymmetric potential. It is just the asymmetry of the potential $U(q)$ near the minima $q = q_j (j = 1, 2)$ that gives rise to the dependence of q_{0j} on E.

At $|U(q_1) - U(q_2)| \lesssim \mathscr{B}/\Gamma$ the spectrum $Q'(\omega)$ in the intermediate frequency range $\Gamma \exp(-F) < \omega < \Gamma$ goes over from the extremely narrow peak to the function (13.4.24). With rising noise intensity $\mathscr{B}$ the width of the peak increases sharply and the shape of the peak changes as a whole. The integral intensity of the peak is determined by the parameter $\mathscr{M}$, see (13.4.19).

Thus, depending on the noise intensity $\mathscr{B}$ and the relation between the parameters of the potential wells, the spectral density of fluctuations can have different number of peaks. At small $\mathscr{B}$ and different well depths ($U(q_2) - U(q_1) \gg \mathscr{B}/\Gamma$) $Q'(\omega)$ has a peak near the frequency ω_1 of the vibrations about the lowest minimum of $U(q)$ (and, generally speaking, weak peaks with relative intensities $\propto \mathscr{B}$ at frequencies $\simeq n\omega_1$; $n = 0, 2, 3, \ldots$). At close values of the well depths $Q'(\omega)$ has an intensive extremely narrow peak at zero frequency. At somewhat larger values of $\mathscr{B}$ the function $Q'(\omega)$ has also a distinct peak at the frequency (13.4.23) which is caused by the overbarrier vibrations. With increasing $\mathscr{B}$ the intensity and width of this peak increase, and it shifts and coalesces with the peaks due to the intrawell vibrations. Experimentally such transformation of the spectrum with rising $\mathscr{B}$, including the emergence of three distinct peaks of $Q'(\omega)$ in a certain interval of $\Gamma \Delta U_1/\mathscr{B}$, was confirmed recently by Mannella, McClintock and Moss (1987; see also Chapter 9 of Volume 3) using the analog electronic circuit which simulated an underdamped double-well Duffing oscillator.

Finally, we note that in terms of quantum theory an anharmonicity of a vibration subsystem results in nonequidistance of the energy levels and therefore in a difference in the Bohr frequencies. If this difference exceeds the damping the spectral density of fluctuations has a fine structure (Ivanov, Kvashnina and Krivoglaz, 1965). For an oscillator coupled to a bath, $Q'(\omega)$ may be calculated in the explicit form with allowance for a fine structure (Dykman and Krivoglaz, 1973). In bistable systems quantum effects lead also to a change of the form of $Q'(\omega)$ at small ω. In particular, the narrow peak of $Q'(\omega)$ turns out to lie at a finite frequency which depends on a tunneling probability. The tunneling in a subsystem also causes transitions between stable states. The probabilities of such transitions in the case of a nonlinear oscillator in a strong, resonant field were obtained by Dmitriev and Dyakonov (1986) neglecting relaxation. The quantum theory of transitions between the stable states of a nonequilibrium nonlinear oscillator coupled to a medium is developed by Dykman and Smelyanskii (1988).

References

Arnold, V. I. 1978. *Additional Chapters of the Theory of Ordinary Differential Equations.* Moscow: Nauka. (In Russian.)

Ben-Jacob, E., Bergman, D. J., Matkowsky, B. J. and Schuss, Z. 1982. *Phys. Rev. A* **26**, 2805–16.

Bogolyubov, N. N. and Mitropolsky, Y. A. 1961. *Asymptotic Methods in the Theory of Nonlinear Oscillators.* New York: Gordon and Breach.

Bonifacio, R. and Lugiato, L. A. 1978. *Phys. Rev. Lett.* **40**, 1023–7.

Dmitriev, A. P. and Dyakonov, M. I. 1986. *Zh. Eksp. Teor. Fiz.* **90**, 1430–40.

Dykman, M. I. 1978. *Fiz. Tverd. Tela* (*Leningrad*) **20**, 2264–72. [*Sov. Phys.-Solid State* **20**, 306 (1978).]

Dykman, M. I. 1986. *Zh. Eksp. Teor. Fiz.* **91**, 1518–30.

Dykman, M. I. and Krivoglaz, M. A. 1971. *Phys. Stat. Sol.* (*b*) **48**, 497–512.

Dykman, M. I. and Krivoglaz, M. A. 1973. *Zh. Eksp. Teor. Fiz.* **64**, 993–1005. [*Sov. Phys.-JETP* **37**, 506 (1973).]

Dykman, M. I. and Krivoglaz, M. A. 1979. *Zh. Eksp. Teor. Fiz.* **77**, 60–73. [*Sov. Phys.-JETP* **50**, 30–7 (1979).]

Dykman, M. I. and Krivoglaz, M. A. 1980a. *Physica* **104A**, 480–94.

Dykman, M. I. and Krivoglaz, M. A. 1980b. *Physica* **104A**, 495–508.

Dykman, M. I. and Krivoglaz, M. A. 1984. In *Soviet Physics Reviews* (I. M. Khalatnikov, ed.), vol. 5, pp. 265–441. New York: Harwood.

Dykman, M. I. and Smelyanskii, V. N. 1988. *Zh. Eksp. Theor. Fiz.* **94** (9).

Dykman, M. I., Soskin, S. M. and Krivoglaz, M. A. 1984. Preprint: Institute of Metal Physics 4.84 (in Russian). [*Physica* **133A**, 53–73 (1985).]

Feynman, R. P. and Hibbs, A. R. 1965. *Quantum Mechanics and Path Integrals.* New York: McGraw-Hill.

Flytsanis, C. and Tang, C. L. 1980. *Phys. Rev. Lett.* **45**, 441–5.

Fronzoni, L., Grigolini, P., Mannella, R. and Zambon, B. 1985. *J. Stat. Phys.* **41**, 553–79.

Fronzoni, L., Grigolini, P., Mannella, R. and Zambon, B. 1986. *Phys. Rev. A* **34**, 3293–3303.

Gibbs, H. 1985. *Bistability: Controlling Light with Light.* New York: Academic Press.

Goldstone, J. A. and Garmire, E. 1984. *Phys. Rev. Lett.* **53**, 910–13.

Graham, R. and Schenzle, A. 1981. *Phys. Rev. A* **23**, 1302–21.

Haken, H. 1983. *Synergetics*, 3rd edn. Berlin: Springer.

Horsthemke, W. and Lefever, R. 1984. *Noise-Induced Transitions.* Berlin: Springer.

Ivanov, M. A., Kvashnina, L. B. and Krivoglaz, M. A. 1965. *Fiz. Tverd. Tela* (*Leningrad*) **7**, 2047–57. (*Sov. Phys.-Solid State* **7**, 1652 (1965).]

Kolmogorov, A. and Leontovich, M. 1933. *Phys. Z. Sowjetunion* **4**, 1.

Kramers, H. A. 1940. *Physica* **7**, 284–304.

Krivoglaz, M. A. and Pinkevich, I. P. 1966. *Zh. Eksp. Teor. Fiz.* **51**, 1151–62.

Krivoglaz, M. A. and Pinkevich, I. P. 1970. *Ukr. Fiz. Zhurn.* **15**, 2039–49.

Landau, L. D. and Lifshitz, E. M. 1976. *Mechanics.* London: Pergamon.

Landau, L. D. and Lifshitz, E. M. 1980. *Statistical Physics*, 3rd edn., part I (revised

by E. M. Lifshitz and L. P. Pitaevskii). New York: Pergamon.
Lax, M. 1967. *Phys. Rev.* **160**, 290–307.
Lax, M. 1968. In *Statistical Physics, Phase Transitions and Superfluidity* (M. Chrétien, E. P. Gros and S. Deser, eds.). New York: Gordon and Breach.
Lugiato, L. A. 1984. *Progr. Optics* **21**, 69–216.
Mangel, M. 1979. *Physica* **97A**, 597–615, 616–31.
Mannella, R., McClintock, P. V. E. and Moss, F. 1987. *Europhys. Lett.* **4**, 511.
Maslova, N. S. 1986. *Zh. Eksp. Teor. Fiz.* **91**, 715–27.
Matkowsky, B. J. and Schuss, Z. 1983. *Phys. Lett.* **95A**, 213–15.
Narayanamurti, V. and Pohl, R. O. 1970. *Rev. Mod. Phys.* **42**, 201–36.
Onodera, Y. 1970. *Prog. Theor. Phys.* **44**, 1477.
Renz, W. 1985. *Z. Phys. B* **59**, 91–102.
Renz, W. and Marchesoni, F. 1985. *Phys. Lett.* **112A**, 124–8.
Risken, H. 1970. *Progr. Optics* **8**, 239–94.
Risken, H. 1984. *The Fokker–Planck Equation.* Berlin: Springer.
Rodriguez, R. F. and Van Kampen, N. G. 1976. *Physica* **85A**, 347–62.
Rytov, S. M. 1955. *Zh. Eksp. Teor. Fiz.* **29**, 304–14, 315–28.
Schuss, Z. 1960. *SIAM Rev.* **22**, 119–55.
Shenoy, S. R. and Agarwal, G. S. 1984. *Phys. Rev. A* **29**, 1315–25.
Soskin, S. M. 1987. Preprint N7 IPAS Ukr.SSR, Kiev.
Sture, K., Nordholm, J. and Zwanzig, R. 1974. *J. Stat. Phys.* **11**, 143–58.
Talkner, P. and Hänggi, P. 1984. *Phys. Rev. A* **29**, 768–73.
Tomita, H., Ito, A. and Kidachi, H. 1976. *Prog. Theor. Phys.* **56**, 786–800.
Van Kampen, N. G. 1965. In *Fluctuation Phenomenon in Solids* (R. E. Butgess, ed.), pp. 139–77. New York: Academic Press.
Ventsel, A. D. and Freidlin, M. I. 1970. *Usp. Mat. Nauk.* **25**, 3–55. [*Russ. Math. Surv.* **25**, 1–50 (1970).]
Voigtlaender, L. and Risken, H. 1985. *J. Stat. Phys.* **40**, 397–429.

Index